W9-ADT-923

MATHEMATICAL SURVEYS AND MONOGRAPHS SERIES LIST

Volume

1 **The problem of moments,** J. A. Shohat and J. D. Tamarkin

2 **The theory of rings,** N. Jacobson

3 **Geometry of polynomials,** M. Marden

4 **The theory of valuations,** O. F. G. Schilling

5 **The kernel function and conformal mapping,** S. Bergman

6 **Introduction to the theory of algebraic functions of one variable,** C. C. Chevalley

7.1 **The algebraic theory of semigroups, Volume I,** A. H. Clifford and G. B. Preston

7.2 **The algebraic theory of semigroups, Volume II,** A. H. Clifford and G. B. Preston

8 **Discontinuous groups and automorphic functions,** J. Lehner

9 **Linear approximation,** Arthur Sard

10 **An introduction to the analytic theory of numbers,** R. Ayoub

11 **Fixed points and topological degree in nonlinear analysis,** J. Cronin

12 **Uniform spaces,** J. R. Isbell

13 **Topics in operator theory,** A. Brown, R. G. Douglas, C. Pearcy, D. Sarason, A. L. Shields; C. Pearcy, Editor

14 **Geometric asymptotics,** V. Guillemin and S. Sternberg

15 **Vector measures,** J. Diestel and J. J. Uhl, Jr.

16 **Symplectic groups,** O. Timothy O'Meara

17 **Approximation by polynomials with integral coefficients,** Le Baron O. Ferguson

18 **Essentials of Brownian motion and diffusion,** Frank B. Knight

19 **Contributions to the theory of transcendental numbers,** Gregory V. Chudnovsky

20 **Partially ordered abelian groups with interpolation,** Kenneth R. Goodearl

21 **The Bieberbach conjecture: Proceedings of the symposium on the occasion of the proof,** Albert Baernstein, David Drasin, Peter Duren, and Albert Marden, Editors

22 **Noncommutative harmonic analysis,** Michael E. Taylor

23 **Introduction to various aspects of degree theory in Banach spaces,** E. H. Rothe

24 **Noetherian rings and their applications,** Lance W. Small, Editor

25 **Asymptotic behavior of dissipative systems,** Jack K. Hale

OPERATOR THEORY AND ARITHMETIC IN H^∞

MATHEMATICAL SURVEYS
AND MONOGRAPHS

NUMBER 26

OPERATOR THEORY AND ARITHMETIC IN H^∞

HARI BERCOVICI

American Mathematical Society
Providence, Rhode Island

1980 *Mathematics Subject Classification* (1985 *Revision*). Primary 47A45, 47A53, 47D25; Secondary 46E20, 46E25, 47A20, 47A60.

Library of Congress Cataloging-in-Publication Data

Bercovici, Hari, 1953-
Operator theory and arithmetic in *H* [infinity]/Hari Bercovici.
p. cm. -- (Mathematical surveys and monographs, ISSN 0076-5376; no. 26)
On t.p. "[infinity]" appears as the infinity symbol.
Bibliography: p.
Includes index.
ISBN 0-8218-1528-8 (alk. paper)
1. Contraction operators. 2. Fredholm operators. 3. Hilbert space. I. Title. II. Series.
QA329.2.B47 1988
515.7′24--dc19 88-10344

To the memory of Irina Gorun

Contents

Introduction

The deep relationship between linear algebra and the arithmetical properties of polynomial rings is well understood, and a highlight is naturally Jordan's classification theorem for linear transformations on a finite-dimensional vector space. The methods and results of finite-dimensional linear algebra seldom extend to, or have analogues in, infinite-dimensional operator theory. Thus it is remarkable to have a class of operators whose properties are closely related with the arithmetic of the ring H^∞ of bounded analytic functions in the unit disc and for which a classification theorem is available, analogous to Jordan's classical result. Such a class is the class C_0, discovered by B. Sz.-Nagy and C. Foiaş in their work on canonical models for contraction operators on Hilbert space. A contraction operator belongs to this class if and only if the associated functional calculus on H^∞ has a nontrivial kernel. The class C_0 is the central object of study of this monograph, but we have included other related topics where it seemed appropriate. In an effort to make the book as self-contained as possible we give an introduction to the theory of dilations and functional models for contraction operators (see Chapters 1 and 5). While this introduction is adequate for our purposes, the reader familiar with the basic book [**6**] by Sz.-Nagy and Foiaş will be able to put the subject matter of this monograph in a greater perspective. Prerequisites for this book are a course in functional analysis (Rudin [**2**], for instance, will cover most of what we need) and an acquaintance with the theory of Hardy spaces in the unit disc (either Hoffman [**1**] or Duren [**1**] covers the required material). In addition, knowledge of the trace class of operators is needed in Chapter 6 (see, for example, Gohberg and Krein [**1**]).

Quite possibly, the class C_0 is the best understood class of nonnormal operators on a Hilbert space, even though there are still unsolved problems and unexplored avenues. Besides its intrinsic interest and direct applications, operators of class C_0 are very helpful as a source of inspiration, and in constructing examples and counterexamples in other branches of operator theory. Interestingly, the class C_0 also surfaces in certain problems of control and realization theory. It is hoped that this book will be interesting for operator theorists (present or to be), as well as those theoretical engineers who are interested in the applications of operator theory.

I tried to make this book more useful by including a number of exercises for each section. The numbering of theorems, propositions, etc. is conceived such as to make cross-references easy. For instance, Theorem 8.1.8 is in §1 of Chapter 8, and it is followed by relation (8.1.9) and Lemma 8.1.10. The first numeral is omitted for references within the same chapter. Each chapter begins with a description of the material to be covered. References to the literature and historical comments are kept to a minimum in the text. There is an appendix dedicated to these questions.

My teachers, colleagues, and friends Ciprian Foiaş, Carl Pearcy, Béla Sz.-Nagy, and Dan Voiculescu encouraged me at various times to write this book. Part of the book or earlier versions of some chapters were written while I was at the University of Michigan, the Massachusetts Institute of Technology, the Mathematical Sciences Research Institute, and Indiana University. Much of the material was presented in a seminar at the University of Michigan. I am grateful to all of these institutions for their hospitality and to some of them for help in typing the manuscript.

My wife Irina, with her exceptional talent and warmth, has been an inspiration for me during most of my mathematical life. Irina helped me get through difficult times and gave me determination and ambition when I lacked them. This book is dedicated to her memory.

Hari Bercovici

CHAPTER 1

An Introduction to Dilation Theory

Any contraction, i.e., operator of norm ≤ 1, on a Hilbert space has a unitary dilation. This is Sz.-Nagy's theorem, and it was the starting point of an important branch in operator theory. In this chapter we give the basic elements of dilation theory, which will help us enter the subject proper of the book in Chapter 2. In Section 1 we present Sz.-Nagy's dilation theorem mentioned above. As a consequence we deduce the decomposition of any contraction into a direct sum of unitary and completely nonunitary parts. We also give a proof of the commutant lifting theorem, which relates the commutant of a contraction with the commutants of its isometric and unitary dilations. Section 2 contains more detailed information about the minimal isometric dilation of an operator. It is shown that the completely nonunitary summand of an isometry is a unilateral shift, and conditions are given on an operator which ensure that its minimal isometric dilation is a unilateral shift. An important result concerns the absolute continuity (with respect to Lebesgue arclength measure on the unit circle) of the minimal unitary dilation. In Section 3 we discuss the notions of cyclic multiplicity, quasisimilarity, and quasiaffine transforms. The latter two notions are weak forms of similarity. The most important result (Theorem 3.7) relates an operator T, with small cyclic multiplicity, to a simpler operator. This result is the starting point of the classification theory of operators of class C_0.

1. Unitary dilations of contractions. Let T be a contraction on the Hilbert space $\mathscr{H}$. We will use the following notation:

$$\begin{aligned} D_T &= (I - T^*T)^{1/2}, \qquad & D_{T^*} &= (I - TT^*)^{1/2}, \\ \mathscr{D}_T &= (\operatorname{ran} D_T)^-, \qquad & \mathscr{D}_{T^*} &= (\operatorname{ran} D_{T^*})^-. \end{aligned} \tag{1.1}$$

The operator D_T is called the *defect operator* of T and $\mathscr{D}_T$ the *defect space.* Using the functional calculus for selfadjoint operators, it is easy to see that the obvious relation $T(I - T^*T) = (I - TT^*)T$ implies

$$TD_T = D_{T^*}T. \tag{1.2}$$

In particular, we have $T\mathscr{D}_T \subset \mathscr{D}_{T^*}$.

Easier to understand among contractions are the isometric and unitary operators. Arbitrary contractions can be related to isometries using dilations. We recall that if $\mathscr{K}$ is a Hilbert space, $\mathscr{H} \subset \mathscr{K}$ is a subspace, $S \in \mathscr{L}(\mathscr{K})$, and $T \in \mathscr{L}(\mathscr{H})$, then S is a dilation of T (and T is a power-compression of S) provided that

$$T^n = P_{\mathscr{H}} S^n \mid \mathscr{H}, \quad n = 0, 1, 2, \ldots.$$

If, in addition, S is an isometry (unitary operator) then S will be called an isometric (unitary) dilation of T. An isometric (unitary) dilation S of T is said to be minimal if no restriction of S to an invariant subspace is an isometric (unitary) dilation of T. The following result is left as an exercise.

1.3. LEMMA. *Let S be an isometric (unitary) dilation of T. Then S is a minimal isometric (unitary) dilation of T if and only if $\bigvee_{n=0}^{\infty} S^n \mathscr{H} = \mathscr{K}$ $(\bigvee_{n=-\infty}^{\infty} S^n \mathscr{H} = \mathscr{K})$.*

The proof of the next result is motivated by the following calculation:

$$||x||^2 - ||Tx||^2 = (x, x) - (T^*Tx, x) = ||D_T x||^2, \quad x \in \mathscr{H},$$

which shows that the operator $X : \mathscr{H} \to \mathscr{H} \oplus \mathscr{H}$ defined by $Xx = Tx \oplus D_T x$ is isometric. Of course, X is not a dilation of T because it acts between two different Hilbert spaces.

1.4. THEOREM. *Every contraction $T \in \mathscr{L}(\mathscr{H})$ has a minimal isometric dilation. This dilation is unique in the following sense: if $S \in \mathscr{L}(\mathscr{K})$ and $S' \in \mathscr{L}(\mathscr{K}')$ are two minimal isometric dilations for T, then there exists an isometry U of $\mathscr{K}$ onto $\mathscr{K}'$ such that $Ux = x$, $x \in \mathscr{H}$, and $S'U = US$.*

PROOF. We first prove the uniqueness part. Thus, let $S \in \mathscr{L}(\mathscr{K})$ and $S' \in \mathscr{L}(\mathscr{K}')$ be two minimal isometric dilations of T and note that, by Lemma 1.3, we have

$$\mathscr{K} = \bigvee_{n=0}^{\infty} S^n \mathscr{H}, \qquad \mathscr{K}' = \bigvee_{n=0}^{\infty} S'^n \mathscr{H}.$$

If $\{x_j\}_{j=0}^{\infty}$ is a finitely nonzero family of vectors in $\mathscr{H}$, we have

$$\left|\left| \sum_{j=0}^{\infty} S^j x_j \right|\right|^2 = \sum_{j,k=0}^{\infty} (S^j x_j, S^k x_k).$$

Since S is an isometry, we have $(S^j x_j, S^k x_k) = (S^{j'} x_j, S^{k'} x_k)$ if $k - j = k' - j'$ and therefore

$$\begin{aligned}
\left|\left| \sum_{j=0}^{\infty} S^j x_j \right|\right|^2 &= \sum_{j \geq k} (S^{j-k} x_j, x_k) + \sum_{j<k} (x_j, S^{k-j} x_k) \\
&= \sum_{j \geq k} (S^{j-k} x_j, P_{\mathscr{H}} x_k) + \sum_{j<k} (P_{\mathscr{H}} x_j, S^{k-j} x_k) \\
&= \sum_{j \geq k} (P_{\mathscr{H}} S^{j-k} x_j, x_k) + \sum_{j<k} (x_j, P_{\mathscr{H}} S^{k-j} x_k) \\
&= \sum_{j \geq k} (T^{j-k} x_j, x_k) + \sum_{j<k} (x_j, T^{k-j} x_k),
\end{aligned}$$

where we used the fact that S is a power-dilation of T. A similar computation for S' shows that $||\sum_{j=0}^{\infty} S^j x_j|| = ||\sum_{j=0}^{\infty} S'^j x_j||$. This easily implies the existence of an isometry U of $\mathscr{K}$ onto $\mathscr{K}'$ satisfying

$$U\left(\sum_{j=0}^{\infty} S^j x_j\right) = \sum_{j=0}^{\infty} S'^j x_j$$

for every finitely nonzero sequence $\{x_j\}_{j=0}^{\infty}$ in $\mathscr{H}$. Clearly then $Ux = x$, $x \in \mathscr{H}$, and $S'U = US$, so that uniqueness is proved.

For the existence part, we define the space $\mathscr{K}_+$ by

$$\mathscr{K}_+ = \mathscr{H} \oplus \left(\bigoplus_{n=0}^{\infty} \mathscr{D}_n\right), \qquad \mathscr{D}_n = \mathscr{D}_T,\ n = 0, 1, 2, \ldots,$$

and the operator $U_+ \in \mathscr{L}(\mathscr{K}_+)$ by

$$U_+\left(x \oplus \left(\bigoplus_{n=0}^{\infty} d_n\right)\right) = Tx \oplus \left(\bigoplus_{n=0}^{\infty} e_n\right)$$

where $e_0 = D_T x$ and $e_n = d_{n-1}$, $n \geq 1$. Since $||Tx||^2 + ||D_T x||^2 = ||x||^2$, it is obvious that U_+ is an isometry. It is also clear that U_+ is an isometric dilation of T, if we identify the vector $x \in \mathscr{H}$ with the vector $x \oplus (\bigoplus_{n=0}^{\infty} 0) \in \mathscr{K}$; in fact $\mathscr{H}$ is invariant under U_+^* and $T^* = U_+^* \mid \mathscr{H}$. It remains to be shown that U_+ is minimal. It is clear that $\mathscr{H} \vee U_+\mathscr{H}$ contains all elements of the form $0 \oplus D_T x \oplus 0 \oplus \cdots$, $x \in \mathscr{H}$, so that

$$\mathscr{H} \vee U_+\mathscr{H} = \mathscr{H} \oplus \mathscr{D}_T \oplus \{0\} \oplus \cdots .$$

It now follows from the definition of U_+ that

$$\bigvee_{j=0}^{n} U_+^j \mathscr{H} = \mathscr{H} \oplus \underbrace{\mathscr{D}_T \oplus \mathscr{D}_T \oplus \cdots \oplus \mathscr{D}_T}_{n \text{ times}} \oplus \{0\} \oplus \cdots$$

and the minimality of U_+ follows from Lemma 1.3.

The following result is the counterpart of Theorem 1.4 for unitary dilations.

1.5. THEOREM. *Every contraction $T \in \mathscr{L}(\mathscr{H})$ has a minimal unitary dilation, unique in the sense specified in Theorem* 1.4.

PROOF. The uniqueness is proved using the same calculation as in the proof of Theorem 1.4, except that one would consider sums of the form $\sum_{j=-\infty}^{\infty} S^j x_j$, $x_j \in \mathscr{H}$. In order to prove the existence of a minimal unitary dilation we consider the space $\mathscr{K}$ defined as

$$\mathscr{K} = \left(\bigoplus_{j=-\infty}^{0} \mathscr{E}_j\right) \oplus \mathscr{H} \oplus \left(\bigoplus_{j=0}^{\infty} \mathscr{D}_j\right),$$

where $\mathscr{E}_{-j} = \mathscr{D}_{T^*}$ and $\mathscr{D}_j = \mathscr{D}_T$, $j = 0, 1, 2, \ldots$, and the operator $U \in \mathscr{L}(\mathscr{K})$ defined by

$$U\left(\left(\bigoplus_{j=-\infty}^{0} e_j\right) \oplus x \oplus \left(\bigoplus_{j=0}^{\infty} d_j\right)\right) = \left(\bigoplus_{j=-\infty}^{0} e_j'\right) \oplus x' \oplus \left(\bigoplus_{j=0}^{\infty} d_j'\right),$$

where $x' = Tx + D_{T^*}e_0$, $d_0' = -T^*e_0 + D_T x$, $d_j' = d_{j-1}$, $j \geq 1$, and $e_j' = e_{j-1}$, $j \leq 0$. The space $\mathscr{K}_+$, constructed in the previous proof, can be identified with $\{0\} \oplus \mathscr{K}_+ \subset \mathscr{K}$, and clearly $U_+ = U \mid \mathscr{K}_+$. It follows at once that U becomes a dilation of T upon the identification of $\mathscr{H}$ with $\{0\} \oplus \mathscr{H} \oplus \{0\} \subset \mathscr{K}$. In order to show that U is unitary it suffices to show that U and U^* are isometries. The fact that U is an isometry is equivalent to the identity

$$||Tx + D_{T^*}e_0||^2 + ||-T^*e_0 + D_T x||^2 = ||e_0||^2 + ||x||^2, \quad e_0 \in \mathscr{D}_{T^*},\ x \in \mathscr{H}.$$

The left-hand side of this identity can be rewritten as follows:

$$\begin{aligned} &||Tx||^2 + ||D_{T^*}e_0||^2 + 2\ \mathrm{Re}(Tx, D_{T^*}e_0) + ||T^*e_0||^2 \\ &\quad + ||D_T x||^2 - 2\ \mathrm{Re}(D_T x, T^*e_0) \\ &\quad = ||x||^2 + ||e_0||^2 + 2\ \mathrm{Re}[(x, T^*D_{T^*}e_0) - (x, D_T T^* e_0)], \end{aligned}$$

and the required identity follows from (1.2) applied to T^*. The minimality of U and the fact that U^* is also an isometry are left as exercises.

As noted above, the space $\mathscr{K}_+$ constructed in the proof of Theorem 1.4 can (and shall) be considered as a subspace of $\mathscr{K}$, invariant under U:

$$U_+ = U \mid \mathscr{K}_+.$$

Thus U is also a minimal unitary dilation of U_+. In fact U^* is the minimal isometric dilation of U_+^* and therefore a different proof of Theorem 1.5 would consist in showing that the minimal isometric dilation of an operator of the form U_+^* is always unitary. We chose the above proof because it is more difficult to identify the defect space of U_+^* in terms of the original operator T.

1.6. DEFINITION. A contraction $T \in \mathscr{L}(\mathscr{H})$ is said to be *completely nonunitary* if there is no invariant subspace $\mathscr{M}$ for T such that $T \mid \mathscr{M}$ is a unitary operator.

An important consequence of Theorem 1.5 is the following.

1.7. PROPOSITION. *For every contraction $T \in \mathscr{L}(\mathscr{H})$ there exist reducing subspaces $\mathscr{H}_0$, $\mathscr{H}_1$ for T such that*

(i) *$\mathscr{H} = \mathscr{H}_0 \oplus \mathscr{H}_1$;*

(ii) *$T \mid \mathscr{H}_1$ is completely nonunitary; and*

(iii) *$T \mid \mathscr{H}_0$ is a unitary operator.*

The spaces $\mathscr{H}_0$ and $\mathscr{H}_1$ are uniquely determined by conditions (i)–(iii).

PROOF. Let $U \in \mathscr{L}(\mathscr{K})$ be a minimal unitary dilation of T. Denote by $\mathscr{K}_0$ the reducing subspace for U generated by $\mathscr{K} \ominus \mathscr{H}$ and set $\mathscr{H}_0 = \mathscr{K} \ominus \mathscr{K}_0$. Obviously $\mathscr{H}_0 \subset \mathscr{H}$ and $Tx = Ux$, $x \in \mathscr{H}_0$, because $\mathscr{H}_0$ reduces U. Thus $\mathscr{H}_0$ is reducing for T and $T \mid \mathscr{H}_0 = U \mid \mathscr{H}_0$ is unitary. We now set $\mathscr{H}_1 = \mathscr{H} \ominus \mathscr{H}_0$ and prove that $T \mid \mathscr{H}_1$ is completely nonunitary. If $\mathscr{M} \subset \mathscr{H}$ is invariant for T and $T \mid \mathscr{M}$ is unitary then the equalities

$$||h|| = ||Th|| = ||P_{\mathscr{H}} Uh||, \quad h \in \mathscr{M},$$

imply that $Th = Uh$ for $h \in \mathscr{M}$. Thus $\mathscr{M}$ is invariant for U, $U \mid \mathscr{M}$ is unitary, and hence $\mathscr{M}$ is reducing for U. Now, $\mathscr{M}$ is orthogonal onto $\mathscr{K} \ominus \mathscr{H}$ and therefore onto $\mathscr{K}_0$; we deduce that $\mathscr{M} \subset \mathscr{H}_0$. This argument shows at once that $T \mid \mathscr{H}_1$ is completely nonunitary and that the decomposition $\mathscr{H} = \mathscr{H}_0 \oplus \mathscr{H}_1$ is unique with the properties (i)–(iii).

The preceding result shows that the study of general contractions can be reduced in many cases to the study of the completely nonunitary ones.

Before proving an important property of isometric and unitary dilations we study in further detail the space of the minimal isometric dilation of a contraction $T \in \mathscr{L}(\mathscr{H})$. Let us denote by $\mathscr{H}_n$, $n = 0, 1, 2, \ldots$, the subspace of $\mathscr{K}_+$ defined as

$$\mathscr{H}_n = \mathscr{H} \oplus \underbrace{\mathscr{D}_T \oplus \mathscr{D}_T \oplus \cdots \oplus \mathscr{D}_T}_{n \text{ times}} \oplus \{0\} \oplus \cdots .$$

Thus $\mathscr{H}_0 = \mathscr{H}$ and each $\mathscr{H}_n$ is invariant for U_+^*. If we set $T_n = P_{\mathscr{H}_n} U_+ \mid \mathscr{H}_n$, then T_{n+1} is a dilation of T_n for every n. The contractions T_n can be viewed differently. For an arbitrary contraction $S \in \mathscr{L}(\mathscr{H})$ we can construct a dilation $S_\sim$ of S on $\mathscr{H} \oplus \mathscr{D}_S$ defined by

$$S_\sim(x \oplus y) = Sx \oplus D_S x. \tag{1.8}$$

Clearly then $S_\sim$ is a partial isometry and

$$\mathscr{D}_{S_\sim} = \ker S_\sim = \{0\} \oplus \mathscr{D}_S.$$

Thus if we repeat this procedure, we can construct a partial isometry $S_{\sim\sim} = (S_\sim)_\sim$ which dilates $S_\sim$, acts on $\mathscr{H} \oplus \mathscr{D}_S \oplus \mathscr{D}_S$, and is defined by

$$S_{\sim\sim}(x \oplus y \oplus z) = Sx \oplus D_S x \oplus y.$$

It is clear now that the contractions T_n considered above satisfy the relations

$$T_{n+1} = (T_n)_\sim, \quad n = 0, 1, 2, \ldots,$$

up to natural unitary equivalences.

1.9. PROPOSITION. *Let $T \in \mathscr{L}(\mathscr{H})$ and $T' \in \mathscr{L}(\mathscr{H}')$ be two contractions, and let $X \in \mathscr{L}(\mathscr{H}, \mathscr{H}')$ satisfy the intertwining relation $T'X = XT$. Then there exists an operator $Y \in \mathscr{L}(\mathscr{H} \oplus \mathscr{D}_T, \mathscr{H}' \oplus \mathscr{D}_{T'})$ such that*

(i) $Y(\{0\} \oplus \mathscr{D}_T) \subset \{0\} \oplus \mathscr{D}_{T'}$;
(ii) $P_{\mathscr{H}'} Y \mid \mathscr{H} = X$;
(iii) $||Y|| = ||X||$; *and*
(iv) $T'_\sim Y = YT_\sim$, *where $T_\sim$ and $T'_\sim$ are the dilations of T and T' described by* (1.8).

PROOF. We may assume without loss of generality that $||X|| = 1$. Indeed, if $X = 0$ we take $Y = 0$ and if $X \neq 0$ we can replace X by $X/||X||$. In order to satisfy (i) and (ii), Y must have the form

$$Y(x \oplus y) = Xx \oplus (Z(x \oplus y)), \qquad x \oplus y \in \mathscr{H} \oplus \mathscr{D}_T,$$

where $Z \in \mathscr{L}(\mathscr{H} \oplus \mathscr{D}_T, \mathscr{D}_{T'})$. The condition that $||Y|| \leq 1$ is easily seen to be equivalent to

$$||Z(x \oplus y)||^2 \leq ||x||^2 - ||Xx||^2 + ||y||^2 = ||D_X x \oplus y||^2$$

and therefore there must exist $C \in \mathscr{L}(\mathscr{H} \oplus \mathscr{D}_T, \mathscr{D}_{T'})$, $||C|| \leq 1$, such that

$$Z = C(D_X \oplus I).$$

Finally, condition (iv) is easily translated into $Z(Tx \oplus D_T X) = D_{T'} Xx$, $x \in \mathscr{H}$. Thus, in order to finish the proof, it suffices to prove the existence of a contraction $C \in \mathscr{L}(\mathscr{H} \oplus \mathscr{D}_T, \mathscr{D}_{T'})$ such that

$$C(D_X Tx \oplus D_T x) = D_{T'} Xx, \quad x \in \mathscr{H}.$$

Since C can be defined to be zero on the orthocomplement of the linear manifold $\{D_X Tx \oplus D_T x : x \in \mathscr{H}\}$, we only have to prove that

$$||D_X Tx \oplus D_T x|| \geq ||D_{T'} Xx||, \quad x \in \mathscr{H},$$

or, equivalently,

$$||Tx||^2 - ||XTx||^2 + ||x||^2 - ||Tx||^2 \geq ||Xx||^2 - ||T'Xx||^2,$$

and this inequality follows from the commutation relation $T'X = XT$ and the fact that $||X|| = 1$. The proposition follows.

We can now prove the following general lifting theorem.

1.10. THEOREM. *Let $T \in \mathscr{L}(\mathscr{H})$, $T' \in \mathscr{L}(\mathscr{H}')$ be two contractions and $U \in \mathscr{L}(\mathscr{K})$, $U' \in \mathscr{L}(\mathscr{K}')$ be unitary or minimal isometric dilations of T and T', respectively. Then for every $X \in \mathscr{L}(\mathscr{H}, \mathscr{H}')$ satisfying $T'X = XT$, there exists $Y \in \mathscr{L}(\mathscr{K}, \mathscr{K}')$ such that $U'Y = YU$, $||Y|| = ||X||$, $X = P_{\mathscr{H}'} Y \mid \mathscr{H}$, $Y\mathscr{K}_+ \subset \mathscr{K}'_+$, and*

$$Y(\mathscr{K}_+ \ominus \mathscr{H}) \subset \mathscr{K}'_+ \ominus \mathscr{H}',$$

where

$$\mathscr{K}_+ = \bigvee_{n=0}^{\infty} U^n \mathscr{H}, \qquad \mathscr{K}'_+ = \bigvee_{n=0}^{\infty} U'^n \mathscr{H}'.$$

PROOF. We consider first the case when $U = U_+ \in \mathscr{L}(\mathscr{K}_+)$ [resp. $U' = U'_+ \in \mathscr{L}(\mathscr{K}'_+)$] is the minimal isometric dilation of T [resp. T']. Denote by T_n [resp. T'_n], $n \geq 0$, the compression of U_+ [resp. T'_n] to $\mathscr{H}_n = \bigvee_{j=0}^n U_+^j \mathscr{H}$ [resp. $\mathscr{H}'_n = \bigvee_{j=0}^n U'^j_+ \mathscr{H}'$], and observe that by Proposition 1.9 (and the remarks preceding it) we can find bounded operators $Y_n \in \mathscr{L}(\mathscr{H}_n, \mathscr{H}'_n)$ such that $Y_0 = X$, $P_{\mathscr{H}'_n} Y_{n+1} \mid \mathscr{H}_n = Y_n$, $Y_{n+1}(\mathscr{H}_{n+1} \ominus \mathscr{H}_n) \subset \mathscr{H}'_{n+1} \ominus \mathscr{H}'_n$, $||Y_n|| = X$, and $T'_n Y_n = Y_n T_n$ for $n \geq 0$. It obviously follows that there exists an operator

$Y_+ \in \mathcal{L}(\mathcal{K}_+, \mathcal{K}'_+)$ such that $\|Y_+\| = \|X\|$, $Y_+(\mathcal{K}_+ \ominus \mathcal{H}) \subset \mathcal{K}'_+ \ominus \mathcal{H}'$, and $P_{\mathcal{H}'_n} Y_+ \mathcal{H}_n = Y_n$, $n \geq 0$ (set, e.g., $Y_+ x = \lim_{n\to\infty} Y_n P_{\mathcal{H}_n} x$, $x \in \mathcal{K}_+$). Since we also have $U_+ x = \lim_{n\to\infty} T_n P_{\mathcal{H}_n} x$, $x \in \mathcal{K}_+$, it follows that $U'_+ Y_+ = Y_+ U_+$ and the theorem follows in this case.

Assume now that U [resp. U'] is a minimal unitary dilation of T [resp. T']. Then $U_+ = U \mid \mathcal{K}_+$ [resp. $U'_+ = U' \mid \mathcal{K}'_+$] is a minimal isometric dilation of T [resp. T'] so, by what has just been proved, there exists $Y_+ \in \mathcal{L}(\mathcal{K}_+, \mathcal{K}'_+)$ such that $U'_+ Y_+ = Y_+ U_+$, $\|Y_+\| = \|X\|$, $P_{\mathcal{H}'} Y_+ \mid \mathcal{H} = X$, and $Y_+(\mathcal{K}_+ \ominus \mathcal{H}) \subset \mathcal{K}'_+ \ominus \mathcal{H}'$. Then U^* [resp. U'^*] is a minimal isometric dilation of U_+^* [resp. U'^*_+] and $U_+^* Y_+^* = Y_+^* U'^*_+$. By the first part of the proof (applied to Y_+^*) there exists an operator $Y \in \mathcal{L}(\mathcal{K}, \mathcal{K}')$ such that $\|Y\| = \|Y_+\|$, $Y^*(\mathcal{K}' \ominus \mathcal{K}'_+) \subset \mathcal{K} \ominus \mathcal{K}_+$, and $P_{\mathcal{K}_+} Y^* \mid \mathcal{K}'_+ = Y_+^*$. It follows then that $Y_+ = Y \mid \mathcal{K}_+$ and Y satisfies all the conditions of the theorem.

Finally, if U and U' are arbitrary unitary dilations, then we can obviously write $U = U_0 \oplus U_1$ [resp. $U' = U'_0 \oplus U'_1$] where U_0 [resp. U'_0] is a minimal unitary dilation of T [resp. T'] acting on $\mathcal{K}_0$ [resp. $\mathcal{K}'_0$]. If $Y_0 \in \mathcal{L}(\mathcal{K}_0, \mathcal{K}'_0)$ is such that $\|Y_0\| = \|X\|$, $U'_0 Y_0 = Y_0 U_0$, $P_{\mathcal{H}'} Y_0 \mid \mathcal{H} = x$, and $Y_0(\mathcal{K}_+ \ominus \mathcal{H}) \subset \mathcal{K}'_+ \ominus \mathcal{H}'$, then $Y = Y_0 \oplus 0_{\mathcal{K} \ominus \mathcal{K}_0}$ will satisfy the conditions of the theorem. The proof is now complete.

A consequence of the preceding result is the following commutant lifting theorem, whose proof is left as an exercise.

1.11. THEOREM. *Let $T \in \mathcal{L}(\mathcal{H})$ be a contraction and $U \in \mathcal{L}(\mathcal{K})$ a unitary or isometric dilation of T. For every $X \in \{T\}'$ there exists $Y \in \{U\}'$ such that*

$$Y\mathcal{K}_+ \subset \mathcal{K}_+, \qquad Y(\mathcal{K}_+ \ominus \mathcal{H}) \subset \mathcal{K}_+ \ominus \mathcal{H},$$

and $P_{\mathcal{H}} Y \mid \mathcal{H} = X$, where $\mathcal{K}_+ = \bigvee_{n=0}^{\infty} U^n \mathcal{H}$.

This commutant lifting theorem gives a very useful description of $\{T\}'$, especially so when $\{U\}'$ is easy to describe.

Exercises

1. Let S be a multiplicative semigroup of operators on $\mathcal{K}$ and assume that $\mathcal{H} \subset \mathcal{K}$ is a subspace with the property that the mapping $A \mapsto P_{\mathcal{H}} A \mid \mathcal{H}$ is multiplicative on S. Show that $\mathcal{H}$ is *semi-invariant* for S, i.e., there exist subspaces $\mathcal{M}$ and $\mathcal{N}$, invariant under every operator in S, such that $\mathcal{M} \supset \mathcal{N}$ and $\mathcal{H} = \mathcal{M} \ominus \mathcal{N}$.

2. Complete the proof of Lemma 1.3.

3. Let $U \in \mathcal{L}(\mathcal{K})$ be a minimal unitary dilation of $T \in \mathcal{L}(\mathcal{H})$ and set $U_+ = U \mid \mathcal{K}_+$, where $\mathcal{K}_+ = \bigvee_{n=0}^{\infty} U^n \mathcal{H}$. Show that U is a minimal isometric dilation of U_+^*.

4. Let $V \in \mathcal{L}(\mathcal{H})$ be an isometry. Show directly that the minimal isometric dilation of V^* is a unitary operator.

5. Let $T \in \mathscr{L}(\mathscr{H})$ be a contraction and let $\mathscr{M}$ be a subspace of $\mathscr{H}$ such that $P_{\mathscr{M}}T \mid \mathscr{M}$ is a unitary operator. Prove that $\mathscr{M}$ is a reducing subspace for T.

6. Let $T \in \mathscr{L}(\mathscr{H})$ be a contraction. Show that the matrix $U = \left[\begin{smallmatrix} T & D_T \\ -D_{T^*} & T^* \end{smallmatrix}\right]$ defines a unitary operator on $\mathscr{H} \oplus \mathscr{H}$. Is U is a unitary dilation of T?

7. Under what conditions is the operator Y in Proposition 1.9 uniquely determined?

8. Describe all operators Y satisfying conditions (i), (ii), and (iv) of Proposition 1.9.

9. Prove that if X is unitary in Proposition 1.9, then Y is uniquely determined and unitary.

10. Prove that if X is an isometry in Proposition 1.9, then Y can be chosen to be an isometry. Is Y necessarily unique?

11. Is Theorem 1.10 true if U and U' are only assumed to be isometric dilations of T and T', respectively?

12. Prove Theorem 1.11.

13. Let T, $T' \in \mathscr{L}(\mathscr{H})$ be two commuting contractions ($TT' = T'T$). Prove that there exists a Hilbert space $\mathscr{K} \supset \mathscr{H}$ and commuting unitary operators U, $U' \in \mathscr{L}(\mathscr{K})$ such that U and U' are unitary dilations of T and T', respectively.

14. If X in Theorem 1.11 belongs to the double commutant $\{T\}''$, can Y always be chosen in $\{U\}''$?

2. Isometries and unitary operators. The results of the preceding paragraph show the importance of understanding the structure and relative position of the invariant subspaces of an isometry. Let us first recall that an isometry $V \in \mathscr{L}(\mathscr{H})$ is a *unilateral shift* if there is a closed subspace $\mathscr{F} \subset \mathscr{H}$ (called a *wandering space*) such that the spaces $\{V^n\mathscr{F}\}_{n=0}^{\infty}$ are mutually orthogonal and

$$\mathscr{H} = \bigoplus_{n=0}^{\infty} V^n \mathscr{F}.$$

The dimension of the Hilbert space $\mathscr{F}$ is called the *multiplicity* of V. Clearly a unilateral shift is a completely nonunitary operator; indeed $\bigcap_{n=0}^{\infty} V^n\mathscr{H} = \{0\}$. The following result of Wold and von Neumann shows that the converse is also true.

2.1. THEOREM. *Let V be an isometry on the Hilbert space $\mathscr{H}$. Then there exists a unique reducing subspace $\mathscr{H}_0$ for V such that*

(i) *$V \mid \mathscr{H}_0$ is a unilateral shift; and*
(ii) *$V \mid \mathscr{H} \ominus \mathscr{H}_0$ is a unitary operator.*

PROOF. The sequence of subspaces $\{V^n\mathscr{H}\}_{n=0}^{\infty}$ is obviously decreasing so that we have

$$\mathscr{H} = \left(\bigoplus_{n=0}^{\infty}(V^n\mathscr{H} \ominus V^{n+1}\mathscr{H})\right) \oplus \left(\bigcap_{n=0}^{\infty} V^n\mathscr{H}\right); \tag{2.2}$$

we set $\mathscr{H}_0 = \bigoplus_{n=0}^{\infty}(V^n\mathscr{H} \ominus V^{n+1}\mathscr{H}) = \bigoplus_{n=0}^{\infty} V^n\mathscr{F}$, $\mathscr{F} = \mathscr{H} \ominus V\mathscr{H}$. Thus $\mathscr{F}$ is a wandering subspace and $V \mid \mathscr{H}_0$ is a shift. Relation (2.2) shows that $\mathscr{H}_0$ is reducing and

$$\mathscr{H} \ominus \mathscr{H}_0 = \bigcap_{n=0}^{\infty} V^n\mathscr{H}.$$

Thus $V(\mathscr{H} \ominus \mathscr{H}_0) = \bigcap_{n=1}^{\infty} V^n\mathscr{H} = \bigcap_{n=0}^{\infty} V^n\mathscr{H} = \mathscr{H} \ominus \mathscr{H}_0$ and $\mathscr{H}_0$ has properties (i) and (ii) of the statement. Now $V \mid \mathscr{H}_0$ is completely nonunitary so that the uniqueness of $\mathscr{H}_0$ follows from Proposition 1.7.

2.3. REMARK. The projection onto $V^n\mathscr{H}$ is given by $V^nV^{*^n}$ and therefore the projection $P_{\mathscr{H}\ominus\mathscr{H}_0}$ equals the strong limit of the decreasing sequence $\{V^nV^{*^n} : n \geq 0\}$.

2.4. COROLLARY. *An isometry $V \in \mathscr{L}(\mathscr{H})$ is a unilateral shift if and only if $\lim_{n\to\infty} \|V^{*^n}x\| = 0$ for all $x \in \mathscr{H}$.*

PROOF. Assume V is a shift so that

$$\mathscr{H} = \bigoplus_{n=0}^{\infty} V^n\mathscr{F}, \quad \mathscr{F} \subset \mathscr{H}.$$

Then $V^{*^{n+1}}x = 0$ for $x \in V^n\mathscr{F}$. Since the sequence $\{V^{*^n}\}_{n=1}^{\infty}$ is bounded in norm and the spaces $\{V^n\mathscr{F}\}_{n=0}^{\infty}$ span $\mathscr{H}$, it follows that $\lim_{n\to\infty} \|V^{*^n}x\| = 0$ for every $x \in \mathscr{H}$. Conversely, if V is not a shift and $\mathscr{H}_0$ is as in Theorem 2.1, then $\|V^{*^n}x\| = \|x\|$ for every $x \in \mathscr{H} \ominus \mathscr{H}_0$. The corollary follows.

2.5. DEFINITION. Let $T \in \mathscr{L}(\mathscr{H})$ be a contraction and $U_+ \in \mathscr{L}(\mathscr{K}_+)$ be the minimal isometric dilation of T. Set $\mathscr{L}_* = \mathscr{K}_+ \ominus U_+\mathscr{K}_+$,

$$\mathscr{M} = \bigoplus_{n=0}^{\infty} U_+^n\mathscr{L}_*, \qquad \mathscr{R} = \mathscr{K}_+ \ominus \mathscr{M}, \qquad R = U_+ \mid \mathscr{R}.$$

The space $\mathscr{R}$ is called the *residual part* of $\mathscr{K}_+$ and the unitary operator R is called the *residual part* of U_+.

There is one important particular case in which the residual part of $\mathscr{K}_+$ is absent; the part $\mathscr{M}$ is absent only when T is a unitary operator, as we shall see shortly.

2.6. DEFINITION. A contraction $T \in \mathscr{L}(\mathscr{H})$ is said to be of class $C_{\cdot 0}$ if $\lim_{n\to\infty} \|T^{*^n}x\| = 0$ for all $x \in \mathscr{H}$; T is of class $C_{0\cdot}$ if T^* is of class $C_{\cdot 0}$. Finally, T is of class C_{00} if it is both of class $C_{\cdot 0}$ and of class $C_{0\cdot}$.

2.7. PROPOSITION. *Let $T \in \mathscr{L}(\mathscr{H})$ be a contraction with minimal isometric dilation $U_+ \in \mathscr{L}(\mathscr{K}_+)$. Then the residual part of $\mathscr{K}_+$ is $\{0\}$ if and only if T is of class $C_{\cdot 0}$.*

PROOF. Assume first that T is of class $C_{\cdot 0}$, $x \in \mathscr{H}$, and k is a natural number. If $n > k$ we have

$$U_+^{*^n} U_+^k x = U_+^{*^{n-k}} x = T^{*^{n-k}} x$$

and consequently

$$\lim_{n\to\infty} \|U_+^{*^n} U_+^k x\| = 0.$$

Since the family of vectors $\{U_+^k x : x \in \mathscr{H},\ k \geq 0\}$ spans $\mathscr{K}_+$, we conclude that U_+^* is of class $C_{\cdot 0}$ and hence it is a unilateral shift by Corollary 2.4. Thus $\mathscr{R} = \{0\}$ by the von Neumann–Wold decomposition theorem.

Conversely, we note that for $x \in \mathscr{H}$ we have by Remark 2.3

$$P_{\mathscr{R}} x = \lim_{n\to\infty} U_+^n U_+^{*^n} x = \lim_{n\to\infty} U_+^n T^{*^n} x \tag{2.8}$$

so that $\|P_{\mathscr{R}} x\| = \lim_{n\to\infty} \|T^{*^n} x\|$. If $\mathscr{R} = \{0\}$ it obviously follows that T is of class $C_{\cdot 0}$.

2.9. PROPOSITION. *With the notation of Definition 2.5, the spaces $\mathscr{L}_*$ and $\mathscr{D}_{T^*}$ have the same dimension. If U_+ is the dilation constructed in the proof of Theorem 1.4, then an isometry ϕ_* of $\mathscr{D}_{T^*}$ onto $\mathscr{L}_*$ is given by*

$$\phi_*(x) = D_{T^*} x \oplus (-T^* x) \oplus 0 \oplus 0 \oplus \cdots.$$

PROOF. We have $\mathscr{L}_* = \ker U_+^*$. Assume U_+ is constructed as in Theorem 1.4. Then every element u in $\mathscr{K}_+$ can be written as

$$u = y \oplus \left(\bigoplus_{n=0}^{\infty} d_n \right)$$

with $y \in \mathscr{H}$, $d_n \in \mathscr{D}_T$, and $\sum_{n=0}^{\infty} \|d_n\|^2 < \infty$. Then we have

$$U_+^* u = (T^* y + D_T d_0) \oplus \left(\bigoplus_{n=0}^{\infty} d_{n+1} \right)$$

so that $U_+^* u = 0$ if and only if

$$T^* y + D_T d_0 = 0 \tag{2.10}$$

and $d_n = 0$ for $n \geq 1$. Upon multiplication by T, (2.10) becomes

$$TT^* y + TD_T d_0 = 0$$

or

$$y - D_{T^*}^2 y + D_{T^*} T d_0 = 0,$$

where we have used (1.1) and (1.2). Thus we have $y = D_{T^*} x$ with $x = D_{T^*} y - T d_0$. Note that $x \in \mathscr{D}_{T^*}$ since $T\mathscr{D}_T \subset \mathscr{D}_{T^*}$. Then multiplication of (2.10) by D_T easily yields

$$T^* D_{T^*} y + d_0 - T^* T d_0 = 0$$

or, equivalently, $d_0 = -T^* x$. These calculations show that the map ϕ_* is onto $\mathscr{L}_*$. That ϕ_* is an isometry is easy to verify, thus concluding the proof.

2.11. COROLLARY. *The minimal isometric dilation of an operator T of class $C_{\cdot 0}$ is a unilateral shift of multiplicity* $\dim(\mathscr{D}_{T^*})$.

PROOF. This obviously follows from Propositions 2.7 and 2.9.

2.12. REMARK. It is interesting to note that the minimal isometric dilation of an operator $T \in \mathscr{L}(\mathscr{H})$ of class $C_{\cdot 0}$ can be obtained using a different construction. Set $\mathscr{K}_+ = \bigoplus_{n=0}^{\infty} \mathscr{D}_n$, $\mathscr{D}_n = \mathscr{D}_{T^*}$, $n = 0, 1, 2, \ldots$, and construct a linear mapping $V: \mathscr{H} \to \mathscr{K}_+$ using the formula

$$Vx = \bigoplus_{n=0}^{\infty} (D_{T^*} T^{*^n} x), \quad x \in \mathscr{H}.$$

Then it is easy to check that V is an isometry and $U_+^* V = VT^*$, where U_+ is the unilateral shift on $\mathscr{K}_+$ defined by $U_+(\bigoplus_{n=0}^{\infty} d_n) = 0 \oplus (\bigoplus_{n=1}^{\infty} d_{n-1})$. Thus, following identification of $\mathscr{H}$ to $V\mathscr{H}$, U_+ becomes an isometric dilation of T. We leave it to the reader to prove that U_+ is minimal.

We conclude this section with an important property of unitary dilations.

2.13. THEOREM. *The minimal unitary dilation of a completely nonunitary contraction is absolutely continuous* (*with respect to Lebesgue measure*).

PROOF. Let $T \in \mathscr{L}(\mathscr{H})$ be a completely nonunitary contraction, and let $U \in \mathscr{L}(\mathscr{K})$ be the minimal unitary dilation constructed in the proof of Theorem 1.5. It follows from the proof of Proposition 1.7 that $\mathscr{K}$ is the smallest reducing subspace of U containing $\mathscr{K} \ominus \mathscr{H}$. Let us denote by $\mathscr{L}$ [resp. $\mathscr{L}_*$] the subspace of all vectors $(\bigoplus_{j=-\infty}^{\infty} e_j) \oplus x \oplus (\bigoplus_{j=0}^{\infty} d_j) \in \mathscr{K}$ such that $e_j = 0$, $x = 0$, and $d_{j+1} = 0$ for all $j \geq 0$ [resp. $e_{-j-1} = 0$, $x = 0$, $d_j = 0$ for all $j \geq 0$]. Then we have

$$\mathscr{K} \ominus \mathscr{H} = \left(\bigoplus_{j=0}^{\infty} U^j \mathscr{L} \right) \oplus \left(\bigoplus_{j=0}^{\infty} U^{*^j} \mathscr{L}_* \right)$$

and hence

$$\mathscr{K} = \left(\bigoplus_{j=-\infty}^{\infty} U^j \mathscr{L} \right) \vee \left(\bigoplus_{j=-\infty}^{\infty} U^j \mathscr{L}_* \right).$$

It suffices to prove that $U \mid \bigoplus_{j=-\infty}^{\infty} U^j \mathscr{L}$ and $U \mid \bigoplus_{j=-\infty}^{\infty} U^j \mathscr{L}_*$ have absolutely continuous spectral measure, and this is obvious because these two operators are bilateral shifts.

Exercises

1. Check the formula for U_+^* used in the proof of Proposition 2.9.

2. Give an alternate proof of Proposition 2.9 using the construction of minimal unitary dilations.

3. Use Theorem 2.1 to show that every isometry has a minimal unitary dilation (which is actually a minimal unitary extension).

4. Use Proposition 2.9 and the preceding exercise to deduce the form of the minimal unitary dilation of an arbitrary contraction T.

5. Provide the details left out in Remark 2.12, that is, show that V is an isometry, $U_+^*V = VT^*$, and $\bigvee_{n=0}^{\infty} U_+^n V\mathscr{H} = \mathscr{K}_+$.

6. Show that every restriction of a unilateral shift to an invariant subspace is again a unilateral shift. What can be said about the multiplicity of a restricted shift?

7. What happens with the construction in Remark 2.12 if T is not of class $C_{\cdot 0}$? Can the construction be amended in order to yield an isometric dilation?

3. Cyclic multiplicity and quasiaffine transforms.

3.1. DEFINITION. The *cyclic multiplicity* μ_T of an operator $T \in \mathscr{L}(\mathscr{H})$ is the smallest cardinal of a subset $\mathscr{M} \subset \mathscr{H}$ with the property that $\bigvee_{n=0}^{\infty} T^n\mathscr{M} = \mathscr{H}$. The operator T is *multiplicity-free* if $\mu_T = 1$.

Thus μ_T is the number of cyclic subspaces for T that are needed to generate $\mathscr{H}$, and T is multiplicity-free if and only if it has a cyclic vector. We note that this notion of multiplicity does not coincide with that used in the reduction theory of C^*-algebras. The cyclic multiplicity extends the notion of multiplicity for shifts, as shown by the following result.

3.2. PROPOSITION. *Let $V \in \mathscr{L}(\mathscr{H})$ be a unilateral shift with wandering space $\mathscr{F}$. Then we have $\mu_V = \dim \mathscr{F}$.*

PROOF. If $\mathscr{M}$ is an orthonormal basis in $\mathscr{F}$ then $\bigvee_{n=0}^{\infty} V^n\mathscr{M} = \mathscr{H}$ so that we have $\mu_V \leq \dim(\mathscr{F})$. Conversely, let $\mathscr{M}$ be an arbitrary set such that $\bigvee_{n=0}^{\infty} V^n\mathscr{M} = \mathscr{H}$. Then $\mathscr{F} = \mathscr{H} \ominus V\mathscr{H} = (\bigvee_{n=0}^{\infty} V^n\mathscr{M}) \ominus (\bigvee_{n=1}^{\infty} V^n\mathscr{M})$ and it follows that $\mathscr{F}$ is spanned as a closed space by the set $P_{\mathscr{F}}\mathscr{M}$. Consequently $\dim(\mathscr{F}) \leq \operatorname{card}(P_{\mathscr{F}}\mathscr{M}) \leq \operatorname{card}(\mathscr{M})$ and the inequality $\dim(\mathscr{F}) \leq \mu_V$ follows at once.

3.3. LEMMA. *Let $T \in \mathscr{L}(\mathscr{H})$, $T' \in \mathscr{L}(\mathscr{H}')$, and $X \in \mathscr{L}(\mathscr{H}, \mathscr{H}')$ be such that $T'X = XT$ and $(X\mathscr{H})^- = \mathscr{H}'$. Then we have $\mu_{T'} \leq \mu_T$.*

PROOF. If $\mathscr{M} \subset \mathscr{H}$, $\operatorname{card}(\mathscr{M}) = \mu_T$, and $\bigvee_{n=0}^{\infty} T^n\mathscr{M} = \mathscr{H}$, then $X\mathscr{M} \subset \mathscr{H}'$, $\operatorname{card}(X\mathscr{M}) \leq \mu_T$, and $\bigvee_{n=0}^{\infty} T'^n X\mathscr{M} = \bigvee_{n=0}^{\infty} XT^n\mathscr{M} = \overline{X\mathscr{H}} = \mathscr{H}'$. Therefore $\mu_{T'} \leq \operatorname{card}(X\mathscr{M}) \leq \mu_T$.

3.4. COROLLARY. *If $T \in \mathscr{L}(\mathscr{H})$ is a contraction of class $C_{\cdot 0}$ then $\mu_T \leq \dim(\mathscr{D}_{T^*})$.*

PROOF. By Proposition 2.9, the minimal isometric dilation $U_+ \in \mathscr{L}(\mathscr{K}_+)$ of T is a unilateral shift of multiplicity $\dim(\mathscr{D}_{T^*})$. If P denotes the projection of $\mathscr{K}_+$ onto $\mathscr{H}$, we have $TP = PU_+$ so that $\mu_T \leq \mu_{U_+} = \dim(\mathscr{D}_{T^*})$ by the preceding results.

Let us note that contractions T of class $C_{\cdot 0}$ for which $\dim(\mathscr{D}_{T^*})$ is small are relatively easy to understand. (Indeed, they are compressions of a shift of multiplicity $\dim(\mathscr{D}_{T^*})$.) This is especially true when $\dim(\mathscr{D}_{T^*}) = 1$. It is therefore natural to try to reduce problems related with a contraction of class $C_{\cdot 0}$ to operators T with smaller $\dim(\mathscr{D}_{T^*})$.

3.5. LEMMA. *For every operator $T \in \mathscr{L}(\mathscr{H})$ of class $C_{\cdot 0}$ there exists a unilateral shift $U \in \mathscr{L}(\mathscr{H}_1)$ and an operator $X \in \mathscr{L}(\mathscr{H}_1, \mathscr{H})$ with dense range such that*

(i) $TX = XU$; *and*
(ii) $\mu_U = \mu_T$.

PROOF. Let $U_+ \in \mathscr{L}(\mathscr{K}_+)$ be the minimal isometric dilation of T, and choose a set $\mathscr{M} \subset \mathscr{H}$ such that $\operatorname{card}(\mathscr{M}) = \mu_T$ and $\bigvee_{n=0}^{\infty} T^n \mathscr{M} = \mathscr{H}$. We can now define $\mathscr{H}_1 = \bigvee_{n=0}^{\infty} U_+^n \mathscr{M}$, $U = U_+ \mid \mathscr{H}_1$, and $X = P_{\mathscr{H}} \mid \mathscr{H}_1$. The relation $TX = XU$ follows because $TP_{\mathscr{H}} = P_{\mathscr{H}} U_+$. Then $(X\mathscr{H}_1)^- = \bigvee_{n=0}^{\infty} XU_+^n \mathscr{M} = \bigvee_{n=0}^{\infty} T^n X \mathscr{M} = \bigvee_{n=0}^{\infty} T^n \mathscr{M} = \mathscr{H}$ so that X has dense range. Thus U is a unilateral shift (as the restriction of a unilateral shift), and $\mu_U \leq \operatorname{card}(\mathscr{M}) = \mu_T$. The opposite inequality, $\mu_T \leq \mu_U$, now follows from Lemma 3.3.

3.6. DEFINITION. An operator $T \in \mathscr{L}(\mathscr{H})$ is a *quasiaffine transform* of $T' \in \mathscr{L}(\mathscr{H}')$ if there exists an injective operator $X \in \mathscr{L}(\mathscr{H}, \mathscr{H}')$ with dense range in $\mathscr{H}'$ such that $T'X = XT$. We write $T \prec T'$ if T is a quasiaffine transform of T. The operators T and T' are *quasisimilar* if $T \prec T'$ and $T' \prec T$. We write $T \sim T'$ if T and T' are quasisimilar.

It is easy to check that "$\prec$" is reflexive and transitive, and therefore "$\sim$" is an equivalence relation.

We are now ready to prove the main result of this paragraph.

3.7. THEOREM. *For every contraction $T \in \mathscr{L}(\mathscr{H})$ of class $C_{\cdot 0}$ there exists a contraction $T' \in \mathscr{L}(\mathscr{H}')$ of class $C_{\cdot 0}$ such that $T' \prec T$ and*

$$\mu_{T'} = \dim(\mathscr{D}_{T'^*}) = \mu_T.$$

PROOF. Let $U \in \mathscr{L}(\mathscr{H}_1)$ and $X \in \mathscr{L}(\mathscr{H}_1, \mathscr{H})$ be provided by Lemma 3.5. We define $\mathscr{H}' = \mathscr{H}_1 \ominus \ker X$, $Y = X \mid \mathscr{H}'$, and $T' = P_{\mathscr{H}'} U \mid \mathscr{H}'$. Because $TX = XU$, $\ker X$ is invariant for U and so T' is a $C_{\cdot 0}$ operator; indeed, $T'^* = U^* \mid \mathscr{H}'$. We have $Y\mathscr{H}' = X\mathscr{H}_1$ so that Y has dense range. Moreover, Y is obviously one-to-one. Also, if $x' \in \mathscr{H}'$, we have $TYx' = TXx' = XUx' = X(Ux' - P_{\ker X} Ux') = XP_{\mathscr{H}'} Ux' = XT'x' = YT'x'$ so that $TY = YT'$. The inequalities $\mu_T \leq \mu_{T'} \leq \dim(\mathscr{D}_{T'^*})$ are obvious from Lemma 3.3 and Corollary 3.4. Finally, the wandering space $\mathscr{F}$ of U has dimension μ_T by Lemma 3.5, and

$$\begin{aligned} I_{\mathscr{H}'} - T'T'^* = I_{\mathscr{H}'} - T'U^* \mid \mathscr{H}' &= P_{\mathscr{H}'}(I - UU^*) \mid \mathscr{H}' \\ &= P_{\mathscr{H}'} P_{\mathscr{F}} \mid \mathscr{H}'. \end{aligned}$$

We conclude that

$$\dim(\mathscr{D}_{T'^*}) = \operatorname{rank}(I - T'T'^*) \leq \operatorname{rank}(P_{\mathscr{F}}) = \dim(\mathscr{F}) = \mu_T$$

and the theorem is proved.

Exercises

1. Is Corollary 3.4 true for arbitrary contractions T?

2. Let T be a contraction on an infinite-dimensional Hilbert space. Show that, if $\mu_T = 1$, we have $2U \prec T$, where U is a unilateral shift of multiplicity one.

3. Let S denote a unilateral shift of multiplicity one. Show that $S \oplus S \prec \alpha S$, $\alpha \in \mathbf{C}$, if and only if $0 < |\alpha| < 1$. (The representation of S as a multiplication operator on the Hardy space H^2 is useful here.)

4. Let $T \in \mathscr{L}(\mathscr{H} \oplus \mathscr{H})$ be defined by $T(x \oplus y) = 0 \oplus x$ and let 0 denote the zero operator on $\mathscr{H}$. If $\mathscr{H}$ has infinite dimension show that $T \oplus 0 \sim T$.

CHAPTER 2

The Class C_0

In this chapter we introduce the class C_0 and study its elementary properties. Since the class C_0 and the algebra H^∞ (of bounded analytic functions in the unit disc) are closely related, we also give some details about this algebra. Section 1 describes first the functional calculus for general completely nonunitary contractions T. This is a homomorphism $h \mapsto h(T)$ defined for h in H^∞. T is an operator of class C_0 if this homomorphism has nontrivial kernel. It is shown that operators of class C_0 have a minimal inner function, which is analogous with the minimal polynomial of a finite matrix. Moreover, it is shown that every inner function occurs as the minimal function of some operator. (This is the first appearance of the Jordan blocks, to be studied in more detail in Chapter 3.) Section 2 collects several facts about the algebra H^∞, some of which do not appear in standard texts on the subject. The greatest common inner divisor and least common inner multiple of a family of functions are defined, and the "local" structure of inner functions is studied. An important fact is the countability of any family of pairwise relatively prime inner functions with a common multiple. The main result of Section 3 is that operators of class C_0 admit a "local" characterization. More precisely, T is of class C_0 if and only if its restriction to each of its cyclic spaces is of class C_0. As a by-product of the proof we obtain the existence of maximal cyclic spaces for T (the corresponding cyclic vectors are called maximal vectors). Finally, Section 4 contains properties of operators of class C_0 related to adjoints, invariant and hyperinvariant subspaces, and functional calculus. It is shown that operators of class C_0 are decomposable.

1. Functional calculus and the class C_0. Let $T \in \mathscr{L}(\mathscr{H})$ be a completely nonunitary contraction with minimal unitary dilation $U \in \mathscr{L}(\mathscr{K})$. For every polynomial $p(\lambda) = \sum_{j=0}^{n} a_j \lambda^j$ we have then

$$p(T) = P_{\mathscr{H}} p(U) \,|\, \mathscr{H}, \tag{1.1}$$

and this formula suggests that the functional calculus $p \mapsto p(T)$ might be extended to more general functions p. More precisely, we know that the spectral measure of U is absolutely continuous with respect to Lebesgue measure on $\mathbf{T} = \{\lambda \in \mathbf{C}\colon |\lambda| = 1\}$ and hence the expression $f(U)$ makes sense for every

$f \in L^\infty = L^\infty(\mathbf{T})$. We generalize accordingly formula (1.1) and define $f(T)$ by

$$f(T) = P_{\mathscr{H}} f(U) \,|\, \mathscr{H}, \quad f \in L^\infty. \tag{1.2}$$

While the mapping $f \mapsto f(T)$ is obviously linear, it is not generally multiplicative, and it is convenient to find a subalgebra in L^∞ on which the functional calculus is multiplicative. It turns out that there is a unique maximal algebra that will do the job for all operators T; this algebra is H^∞. We recall that H^∞ is defined as the set of all bounded analytic functions on $\mathbf{D} = \{\lambda\colon |\lambda| < 1\}$. Every function $u \in H^\infty$ can be extended a.e. on $\mathbf{T}$ via taking radial limits:

$$u(e^{it}) = \lim_{r \to 1} u(re^{it}) \tag{1.3}$$

(this is a usual and harmless abuse of notation). These radial limits uniquely determine u and so H^∞ can be regarded as a subalgebra of L^∞; H^∞ is actually closed in the weak* topology of L^∞ given by its duality with L^1.

We note that the functional calculus can be defined in terms independent of the minimal unitary dilation. Indeed, if $u(\lambda) = \sum_{n=0}^\infty \lambda^n a_n$ is in H^∞ then

$$u(T) = \lim_{r \to 1} u(rT) = \lim_{r \to 1} \sum_{n=0}^\infty a_n r^n T^n, \tag{1.4}$$

where the limit exists in the strong operator topology (and even in the $*$-ultrastrong topology). However, the existence of the limit in (4.4) is normally proved in the context of unitary dilations.

We mention here an important property of the functional calculus. For every $u \in H^\infty$ we denote by $u^\sim \in H^\infty$ the function defined by $u^\sim(\lambda) = \overline{u(\overline{\lambda})}$. Then we have

$$u^\sim(T) = (u(T^*))^*, \quad u \in H^\infty. \tag{1.5}$$

This equation has an analogue for functions $u \in L^\infty$, where $u^\sim$ is defined by $u^\sim(e^{it}) = \overline{u(e^{-it})}$.

1.6. LEMMA. *Let $T \in \mathscr{L}(\mathscr{H})$ be a completely nonunitary contraction. Then the mapping $f \mapsto f(T)$ is continuous when L^∞ and $\mathscr{L}(\mathscr{H})$ are given their weak* topologies.*

PROOF. Let E_t be the spectral measure of the minimal unitary dilation U of T; thus

$$U = \int_0^{2\pi} e^{it}\, dE_t.$$

If $h, g \in \mathscr{H}$ then the function $t \mapsto (E_t h, g)$ is absolutely continuous; we denote by $\phi_{h,g} \in L^1$ its derivative. We have for $f \in L^\infty$

$$(f(T)h, g) = (f(U)h, g) = \int_0^{2\pi} f(e^{it}) \phi_{h,g}(e^{it})\, dt$$

and this clearly shows that $f \mapsto f(T)$ is continuous when L^∞ is given the weak* topology and $\mathscr{L}(\mathscr{H})$ the weak topology. To conclude the proof we note that the minimal unitary dilation of $\tilde{T} = T \oplus T \oplus T \oplus \cdots$ is $\tilde{U} = U \oplus U \oplus U \oplus \cdots$ and $f(\tilde{T}) = f(T) \oplus f(T) \oplus f(T) \oplus \cdots$. This allows us to replace the weak topology on $\mathscr{L}(\mathscr{H})$ by the weak* topology.

Sequences in H^∞ that converge in the weak* topology are described as follows.

1.7. LEMMA. *Let $\{u_n : n \geq 1\}$ be a sequence of functions in H^∞. Then $u_n \to u \in H^\infty$ in the weak* topology if and only if*

(i) $\sup \|u_n\| < \infty$; *and*

(ii) $\lim u_n(\lambda) = u(\lambda)$ *for* $\lambda \in \mathbf{D}$.

PROOF. The necessity of (i) follows from the uniform boundedness principle, and the necessity of (ii) follows from the Cauchy formula for H^∞:

$$u(\lambda) = \frac{1}{2\pi} \int_0^{2\pi} f(e^{it})(1 - \lambda e^{-it})^{-1}\, dt, \quad \lambda \in \mathbf{D},$$

which shows that $\lambda \mapsto u(\lambda)$ is a weak* continuous functional on H^∞. Conversely, if (i) holds, the weak* compactness of the unit ball of H^∞ shows that $\{u_n\}$ must have convergent subsequences. By (ii) any limit of a subsequence of $\{u_n\}$ must coincide with u and consequently $u = \lim u_n$ in the weak* topology.

1.8. REMARK. If $\{u_n : n \geq 1\}$ is a bounded sequence in H^∞ then $\{u_n\}$ must have subsequences that converge uniformly on the compact subsets of $\mathbf{D}$; this follows from the Vitali–Montel theorem. Thus, under the conditions of Lemma 4.6, u_n converges to u uniformly on compact subsets of $\mathbf{D}$.

We now come to our central object of study: the class C_0.

1.9. DEFINITION. A completely nonunitary contraction $T \in \mathscr{L}(\mathscr{H})$ is said to be of class C_0 if there exists $u \in H^\infty$, $u \not\equiv 0$, such that $u(T) = 0$; T is said to be locally of class C_0 if for every $h \in \mathscr{H}$ there exists $u_h \in H^\infty \setminus \{0\}$ such that $u_h(T)h = 0$.

Clearly an operator of class C_0 is locally of class C_0; quite interestingly, the converse of this statement also holds, as we will see in the following paragraphs.

Recall that a function $u \in H^\infty$ is inner if $|u(e^{it})| = 1$ almost everywhere on $\mathbf{T}$.

Let T be an operator of class C_0. Then the set $J = \{u \in H^\infty : u(T) = 0\}$ is an ideal in H^∞. Moreover, since $u(T) = 0$ if and only if $(u(T)h, g) = 0$, $h, g \in \mathscr{H}$, it follows that J is weak* closed, and hence it has the form $J = vH^\infty$ for some inner function v (cf. Gamelin [**1**]).

1.10. DEFINITION. The inner function v such that $vH^\infty = \{u \in H^\infty : u(T) = 0\}$ is called the *minimal function* of T and is denoted by m_T. Analogously, if T is locally of class C_0 and $h \in \mathscr{H}$, we denote by m_h the inner function defined by $m_h H^\infty = \{u \in H^\infty : u(T)h = 0\}$.

Because of the notation m_T and m_h, we will abstain in the sequel from using the letter m for generic objects. Care should also be exercised when using the symbol m_h, because m_h really depends on T as well as on h. Let us note that the functions m_T and m_h are determined only up to a constant scalar multiple of absolute value one. Therefore it will be convenient not to distinguish between two functions that differ only by such a scalar multiple, so that we can talk about "the" minimal function. Alternatively, one may impose on m_T an additional condition that determines it completely; for example, we may require that the first nonzero Fourier coefficient of m_T be positive.

In the remainder of this paragraph we describe a method for proving the existence of the minimal functions m_f that does not depend on the description of weak* closed ideals in H^∞. It will be convenient to recall here Beurling's description of the invariant subspaces of a unilateral shift of multiplicity one.

1.11. THEOREM. *Let $U \in \mathscr{L}(\mathscr{H})$ be a unilateral shift of multiplicity one. For every nonzero invariant subspace $\mathscr{M} \subset \mathscr{H}$ of T there exists an inner function $\theta \in H^\infty$ such that*

$$\mathscr{M} = \theta(U)\mathscr{H}.$$

The function θ is uniquely determined by $\mathscr{M}$ up to a constant scalar multiple of absolute value one.

The reader will realize without difficulty that the theorem above is equivalent to Beurling's original result in which

$$\mathscr{H} = H^2 = \left\{ u(\lambda) = \sum_{n=0}^{\infty} a_n \lambda^n,\ \lambda \in \mathbf{D}\colon \sum_{n=0}^{\infty} |a_n|^2 < \infty \right\}$$

and $U = S$ is the shift defined by

$$(Su)(\lambda) = \lambda u(\lambda), \quad \lambda \in \mathbf{D}, u \in H^2.$$

In this situation the operator $\theta(S)$ coincides with the Toeplitz operator T_θ defined by

$$(T_\theta u)(\lambda) = \theta(\lambda)u(\lambda), \quad \lambda \in D, u \in H^2$$

and therefore

$$\theta(S)H^2 = \theta H^2 = \{\theta u\colon u \in H^2\}.$$

1.12. LEMMA. *If $T \in \mathscr{L}(\mathscr{H})$ is of class C_0 then T is of class $C_{\cdot 0}$.*

PROOF. We have to prove that $\lim_{n\to\infty} \|T^{*^n} h\| = 0$ for each $h \in \mathscr{H}$. Assume that $u(T) = 0$ for some $u \in H^\infty \backslash \{0\}$, let $U_+ \in \mathscr{L}(\mathscr{K}_+)$ be the minimal isometric dilation of T, let $\mathscr{R}$ be the residual part of $\mathscr{K}_+$, and $R = U_+ | \mathscr{R}$. We have $TP_{\mathscr{H}} = P_{\mathscr{H}} U_+^*$ because $T^* = U_+^* | \mathscr{H}$, and this implies that

$$T(P_{\mathscr{H}} | \mathscr{R}) = (P_{\mathscr{H}} | \mathscr{R})R. \tag{1.13}$$

Consequently, we have

$$(P_{\mathscr{H}} | \mathscr{R})u(R) = u(T)(P_{\mathscr{H}} | \mathscr{R}) = 0. \tag{1.14}$$

(Note that $u(R)$ makes sense since R is absolutely continuous; cf. Theorem 1.2.13.) By the F. and M. Riesz theorem, the function $u(\xi)$ is different from zero for almost every $\xi \in \mathbf{T}$, and the spectral theorem implies that $u(R)$ has dense range. Then (1.14) is seen to imply that $P_{\mathscr{H}} \mid \mathscr{R} = 0$. Therefore we have $P_{\mathscr{R}} \mid \mathscr{H} = 0$ and, in particular,

$$0 = \|P_{\mathscr{R}} h\| = \left\| \lim_{n\to\infty} U_+^n T^{*^n} h \right\| = \lim_{n\to\infty} \|T^{*^n} h\|.$$

The lemma follows.

1.15. PROPOSITION. *Let $T \in \mathscr{L}(\mathscr{H})$ be locally of class C_0. For every vector $h \in \mathscr{H}$ there exists an inner function m_h such that*

$$\{u \in H^\infty : u(T)h = 0\} = m_h H^\infty.$$

PROOF. Fix $h \in \mathscr{H}$, $h \neq 0$. We may replace T by its restriction to the cyclic space generated by h, and hence assume that T is of class C_0. Let $U_+ \in \mathscr{L}(\mathscr{K}_+)$ denote the minimal isometric dilation of T; by Lemma 1.12 U_+ is a unilateral shift. The subspace $\mathscr{K}_0 = \bigvee_{n=0}^\infty U_+^n h$ is invariant for U_+ and the restriction $U_0 = U_+ \mid \mathscr{K}_0$ is a unilateral shift of multiplicity one. Moreover, $\mathscr{K}_1 = \mathscr{K}_0 \cap (\mathscr{K}_+ \ominus \mathscr{H})$ is invariant for U_0. Note that

$$u(T)h = P_{\mathscr{H}} u(U_+)h = P_{\mathscr{H}} u(U_0)h, \quad u \in H^\infty,$$

and therefore $u(T)h = 0$ if and only if $u(U_0)h \in \mathscr{K}_1$. The operator $u(U_0)$ is one-to-one if $u \neq 0$ and, since $u(T)h = 0$ for some such functions u, the space $\mathscr{K}_1$ is not zero. By Theorem 1.11 we can find an inner function $\theta \in H^\infty$ such that

$$\mathscr{K}_1 = \theta(U_0)\mathscr{K}_0.$$

It is trivial now to verify that

$$\{u \in H^\infty : u(T)h = 0\} = \theta H^\infty.$$

Indeed, $u(T)h = 0$ if and only if $u(U_0)h \in \mathscr{K}_1$. Since h is a cyclic vector for U_0, this happens if and only if

$$u(U_0)\mathscr{K}_0 \subset \theta(U_0)\mathscr{K}_0$$

or, equivalently, if and only if

$$uH^2 \subset \theta H^2.$$

In particular, $u = u \cdot 1 = \theta g$ for some $g \in H^2$ and a moment's thought shows that we necessarily have $g \in H^\infty$. The proposition is proved.

It seems appropriate here to give some example of operators of class C_0. Let S be the unilateral shift on H^2, and let $\theta \in H^\infty$ be an inner function. Then we consider the Hilbert space $\mathscr{H}(\theta)$ given by

$$\mathscr{H}(\theta) = H^2 \ominus \theta H^2 \tag{1.16}$$

and the operator $S(\theta) \in \mathscr{L}(\mathscr{H}(\theta))$ defined by

$$S(\theta) = P_{\mathscr{H}(\theta)} S \mid \mathscr{H}(\theta) \tag{1.17}$$

or, equivalently, $S(\theta)^* = S^* \mid \mathscr{H}(\theta)$.

1.18. PROPOSITION. *For every inner function θ, the operator $S(\theta)$ is of class C_0 and its minimal function is θ.*

PROOF. We have $\theta(S)H^2 \subset \theta H^2 = H^2 \ominus \mathscr{H}(\theta)$, so that $\theta(S(\theta)) = P_{\mathscr{H}(\theta)}\theta(S)\,|\,\mathscr{H}(\theta) = 0$. If $u(S(\theta)) = 0$ it follows that $u(S)\mathscr{H}(\theta) = u\mathscr{H}(\theta) \subset \theta H^2$. Consequently $uH^2 = u\mathscr{H}(\theta) + u\theta H^2 \subset \theta H^2$ so that u can be written as $u = \theta g$, $g \in H^2$. Clearly then we must have $g \in H^\infty$ so that $u \in \theta H^\infty$. The proposition follows at once.

Operators of the form $S(\theta)$ will be called *Jordan blocks.*

Exercises

1. Show that the functional calculus with H^∞ functions makes sense for operators T whose unitary part is an absolutely continuous unitary operator.

2. Let T be a completely nonunitary contraction and $\mathscr{A}$ an algebra, $H^\infty \subset \mathscr{A} \subset L^\infty$, with the property that the functional calculus $u \mapsto u(T)$ is multiplicative for $u \in \mathscr{A}$. Prove that $\mathscr{A} = H^\infty$.

3. Let T be a contraction and $\theta \in H^\infty$ an inner function; assume that $\lim_{r\to 1}\theta(rT) = 0$ in the strong operator topology. Is T necessarily an operator of class C_0?

4. An operator T is *algebraic* if $p(T) = 0$ for some polynomial p. Clearly an algebraic operator T such that $\|T\| < 1$ is an operator of class C_0. Prove that every operator T of class C_0 such that $\|T\| < 1$ is necessarily an algebraic operator.

5. Give an example of a nonalgebraic operator T of class C_0.

2. The arithmetic of inner functions. In this section we discuss certain properties of H^∞ that will be essential in the sequel, particularly in the proof of the fact that locally C_0 operators are in fact of class C_0.

Let θ and θ' be two functions in H^∞. We say that θ *divides* θ' (or $\theta\,|\,\theta'$) if θ' can be written as $\theta' = \theta \cdot \phi$ for some $\phi \in H^\infty$. It is clear that, if θ and θ' are inner, any such ϕ must be an inner function.

2.1. LEMMA. *For any two inner functions θ, $\theta' \in H^\infty$, the following assertions are equivalent:*

(i) *θ divides θ';*
(ii) *$\theta' H^\infty \subset \theta H^\infty$;*
(iii) *$\theta' H^2 \subset \theta H^2$;*
(iv) *$|\theta'(\lambda)| \le |\theta(\lambda)|$ for all $\lambda \in \mathbf{D}$.*

PROOF. The equivalence of (i) and (ii) is obvious, and so are the implications (i) $\Rightarrow$ (iii) and (i) $\Rightarrow$ (iv). If $\theta' H^2 \subset \theta H^2$ then we can write $\theta' = \theta\phi$ for some $\phi \in H^2$. The boundary values of ϕ must then have absolute value one almost everywhere. Consequently, $\phi \in H^\infty$ and θ divides θ' and the implication (iii) $\Rightarrow$ (i) follows. Finally, assume that (iv) holds. Then the function $\phi(\lambda) = \theta'(\lambda)/\theta(\lambda)$ only has removable singularities in $\mathbf{D}$, and $|\phi(\lambda)| \leq 1$. Thus ϕ has an analytic extension to $\mathbf{D}$, the extension belongs to H^∞, and the relation $\theta' = \theta\phi$ shows that θ divides θ'. The lemma follows.

2.2. DEFINITION. Let F be a family of functions in H^∞. An inner function θ is called the *greatest common inner divisor* of F if θ divides every element in F and if θ is a multiple of any other common inner divisor of F. The greatest common inner divisor is denoted by $\bigwedge F$ (or $\bigwedge_{i\in I} f_i$ if $F = \{f_i \colon i \in I\}$, or $f_1 \wedge f_2$ if $F = \{f_1, f_2\}$). Let G be a family of inner functions in H^∞. An inner function θ is called the *least common inner multiple* of the family G if every element of G divides θ, and θ divides any other common inner multiple of G. The least common inner divisor is denoted by $\bigvee G$ (or $\bigvee_{i\in I} g_i$ if $G = \{g_i \colon i \in I\}$, or $g_1 \vee g_2$ if $G = \{g_1, g_2\}$).

We note that the functions $\bigwedge F$ and $\bigvee G$ (if they exist) are determined only up to constant scalar multiples of absolute value one. Clearly $\bigwedge \varnothing$ and $\bigwedge\{0\}$ do not exist. The first part of the following result shows that we can talk about the greatest common inner divisor of a family $F \subset H^2$ if $\varnothing \neq F \neq \{0\}$.

2.3. PROPOSITION. (i) *Let F be a family of functions in H^∞ such that $F \neq \varnothing$ and $F \neq \{0\}$. Then $\bigwedge F$ exists.*

(ii) *Let G be a family of inner functions. If G has a nonzero common multiple in H^∞ then $\bigvee G$ exists.*

PROOF. (i) The subspace $\mathscr{M} = \bigvee_{f\in F} fH^2$ is invariant under the unilateral shift S ($(Su)(\lambda) = \lambda u(\lambda)$, $\lambda \in \mathbf{D}$, $u \in H^2$) and $\mathscr{M} \neq \{0\}$. By Beurling's theorem, we can find an inner function θ such that

$$\mathscr{M} = \theta H^2.$$

Since $\theta H^2 \supset fH^2$, $f \in F$, it follows that f can be written as $f = \theta \cdot g$ for some $g \in H^2$. A comparison of boundary values shows that in fact $g \in H^\infty$ and hence θ is a common divisor of F. If θ' is another common divisor of F, then $\theta' H^2 \supset fH^2$, $f \in F$, and consequently $\theta' H^2 \supset \mathscr{M} = \theta H^2$. Thus θ' divides θ by Lemma 2.1 and we conclude that $\theta = \bigwedge F$.

(ii) If $h \in H^\infty$ is a common multiple of the family G then

$$h \in \mathscr{N} = \bigcap_{g\in G} fH^2.$$

Thus $\mathscr{N} \neq \{0\}$ and hence $\mathscr{N} = \phi H^2$ for some inner function ϕ by Beurling's theorem. An argument quite similar to that in (i) shows that $\phi = \bigvee G$.

We remark that every finite family of inner functions has a common multiple and therefore a least common inner multiple. Thus, for example, $\bigvee \varnothing = 1$. The use of greatest common inner divisors is dictated by the fact that H^∞ is not a principal ideal domain; however, the weak*-closed ideals are principal and they are generated by inner functions. We chose not to use this fact here because it is harder to prove than Beurling's theorem.

We note a few useful properties of common inner divisors and multiples. If $u, v \in H^\infty$ and $u \mid v$ then we will denote, somewhat informally, by v/u or $\frac{v}{u}$ the unique function $w \in H^\infty$ determined by the relation $v = uw$. We will use the notation $\theta \equiv \theta'$ if θ and θ' are two inner functions that differ only by a constant scalar factor of absolute value one. Thus the relations $\theta \mid \theta'$ and $\theta' \mid \theta$ imply that $\theta \equiv \theta'$.

2.4. PROPOSITION. *If $\{\theta_i : i \in I\}$ is a family of inner divisors of the inner function $\theta \in H^\infty$ then*

$$\theta / \left(\bigvee_{i \in I} \theta_i \right) \equiv \bigwedge_{i \in I} (\theta / \theta_i).$$

PROOF. Since $\theta_j \mid \bigvee_{i \in I} \theta_i$ we conclude that $\theta / \left(\bigvee_{i \in I} \theta_i \right)$ divides θ / θ_j for each $j \in I$. If ϕ is an arbitrary common inner divisor of $\{\theta/\theta_i : i \in I\}$ then $\theta_i | \theta / \phi$ for every $i \in I$ and consequently $(\bigvee_{i \in I} \theta_i) | \theta / \phi$. This is equivalent to $\phi | \theta / (\bigvee_{i \in I} \theta_i)$ and so $\theta / (\bigvee_{i \in I} \theta_i)$ is the greatest common inner divisor of $\{\theta / \theta_i : i \in I\}$. The lemma follows.

2.5. COROLLARY. *If θ and θ' are inner functions in H^∞ then $\theta\theta' \equiv (\theta \wedge \theta')(\theta \vee \theta')$.*

PROOF. By Proposition 2.4 we have

$$\theta\theta' / (\theta \vee \theta') \equiv (\theta\theta' / \theta) \wedge (\theta\theta' / \theta') = \theta' \wedge \theta.$$

Common inner divisors and multiples are easier to understand in the context given by the canonical factorization of an inner function. For each $\alpha \in \mathbf{D}$ we introduce the *Blaschke factor* b_α defined by

$$b_\alpha(\lambda) = \frac{\overline{\alpha}}{|\alpha|} \frac{\alpha - \lambda}{1 - \lambda \overline{\alpha}}, \qquad \lambda \in \mathbf{D},$$

if $\alpha \neq 0$, and $b_0(\lambda) = \lambda$, $\lambda \in \mathbf{D}$. Then we recall that a *Blaschke product* is a function of the form

$$b(\lambda) = \prod_{0 \le j < n} b_{\alpha_j}(\lambda), \qquad \lambda \in \mathbf{D}, \tag{2.6}$$

where $n \le \infty$ and the sequence $\{\alpha_j : 0 \le j < n\} \subset \mathbf{D}$ has the property that

$$\sum_{0 \le j < n} (1 - |\alpha_j|) < \infty. \tag{2.7}$$

The convergence condition (2.7) implies that the product in (2.6) converges (if infinite) uniformly on compact subsets of $\mathbf{D}$; a Blaschke product is an inner function.

A singular inner function is determined by a positive finite measure ν on $\mathbf{T}$, singular with respect to Lebesgue measure, via the formula

$$s_\nu(\lambda) = \exp\left(- \int_{\mathbf{T}} \frac{\varsigma + \lambda}{\varsigma - \lambda} \, d\nu(\varsigma) \right), \qquad \lambda \in \mathbf{D}. \tag{2.8}$$

Now, if θ is an arbitrary inner function, there exist a Blaschke product b, a singular inner function s, and a constant γ, $|\gamma| = 1$, such that

$$\theta = \gamma b s.$$

Clearly b, s, and γ are uniquely determined by the function θ, and if θ' is another inner function, written as

$$\theta' = \gamma' b' s',$$

we have $\theta\theta' = (\gamma\gamma')(bb')(ss')$.

2.9. LEMMA. *Let ν and ν' be positive singular measures on $\mathbf{T}$, and let s_ν be defined by (2.8). Then we have $s_\nu s_{\nu'} = s_{\nu+\nu'}$.*

PROOF. This follows from the familiar properties of exp.

In order to produce an analogue of Lemma 2.9 for Blaschke products we must use the function describing the multiplicity of the zeros of an inner function. We will call a function $\mu\colon \mathbf{D} \to \mathbf{N} = \{0, 1, 2, \dots\}$ a *Blaschke function* if

$$\sum_{\alpha \in \mathbf{D}} \mu(\alpha)(1 - |\alpha|) < \infty.$$

If μ is a Blaschke function we can now introduce the Blaschke product

$$b_\mu(\lambda) = \prod_{\alpha \in \mathbf{D}} (b_\alpha(\lambda))^{\mu(\alpha)}. \tag{2.10}$$

Clearly $\mu(\alpha)$ represents the multiplicity of α as a zero of b_μ.

2.11. LEMMA. *If μ and μ' are Blaschke functions then $b_\mu b_{\mu'} = b_{\mu+\mu'}$.*

PROOF. Obviously follows from (2.10).

Let us recall now that the set of positive finite measures on $\mathbf{T}$ has a lattice structure with respect to the relation: $\nu \le \nu' \Leftrightarrow \nu(A) \le \nu'(A)$ for every Borel subset A of $\mathbf{T}$. We denote by $\nu \vee \nu'$ and $\nu \wedge \nu'$ the least upper bound and greatest lower bound of ν and ν', respectively. The measures ν and ν' are mutually singular (in symbols, $\nu \perp \nu'$) if $\nu \wedge \nu' = 0$. The set of Blaschke functions can also be organized as a lattice with respect to the relation: $\mu \le \mu' \Leftrightarrow \mu(\alpha) \le \mu'(\alpha)$ for every $\alpha \in \mathbf{D}$. Clearly then $(\mu \vee \mu')(\alpha) = \max\{\mu(\alpha), \mu'(\alpha)\}$, $(\mu \wedge \mu')(\alpha) = \min\{\mu(\alpha), \mu'(\alpha)\}$, $\alpha \in \mathbf{D}$.

2.12. PROPOSITION. *Let μ and μ' be Blaschke functions, ν and ν' singular measures on $\mathbf{T}$, γ and γ' complex numbers of absolute value one, and set $\theta = \gamma b_\mu s_\nu$, $\theta' = \gamma' b_{\mu'} s_{\nu'}$.*

(i) *$\theta \equiv \theta'$ if and only if $\mu = \mu'$ and $\nu = \nu'$.*

(ii) *$\theta | \theta'$ if and only if $\mu \le \mu'$ and $\nu \le \nu'$.*

(iii) *$\theta \vee \theta' \equiv b_{\mu \vee \mu'} s_{\nu \vee \nu'}$; $\theta \wedge \theta' = b_{\mu \wedge \mu'} s_{\nu \wedge \nu'}$.*

(iv) *If θ'' is an inner function, $\theta'' \wedge \theta \equiv \theta'' \wedge \theta' \equiv 1$, then $\theta'' \wedge \theta\theta' \equiv 1$.*

PROOF. (i) obviously follows from the uniqueness of the canonical decomposition of inner functions. If $\theta | \theta'$ and $\theta'/\theta = \gamma'' b_{\mu''} s_{\mu''}$ then we have $\mu' = \mu + \mu'' \ge \mu$

and $\nu' = \nu + \nu'' \geq \nu$ by Lemmas 2.9 and 2.11. The converse in (ii) is even more obvious. (iii) clearly follows from (ii). Finally, assume that $\theta'' \wedge \theta \equiv \theta'' \wedge \theta' \equiv 1$ and $\theta'' = \gamma'' b_{\mu''} s_{\nu''}$. Then by (ii) we have $\mu'' \wedge \mu = \mu'' \wedge \mu' = 0$ and $\nu'' \wedge \nu = \nu'' \wedge \nu' = 0$. These relations imply $\mu'' \wedge (\mu + \mu') = 0$ (if $\mu(\alpha) + \mu'(\alpha) \neq 0$ then either $\mu(\alpha) \neq 0$ or $\mu'(\alpha) \neq 0$, and either of these relations implies $\mu''(\alpha) = 0$) and $\nu'' \wedge (\nu + \nu') = 0$ (if ν is supported by A, ν' by A', and $\nu''(A) = \nu''(A') = 0$, then $\nu + \nu'$ is supported by $A \cup A'$ and $\nu''(A \cup A') \leq \nu''(A) + \nu''(A') = 0$). Again by (ii) we infer $\theta'' \wedge \theta\theta' \equiv 1$. The proposition is proved.

2.13. COROLLARY. *If $\{\theta_1, \theta_2, \ldots, \theta_n\}$ is a finite collection of inner functions in H^∞ and $\theta_i \wedge \theta_j \equiv 1$ for $i \neq j$, $1 \leq i, j \leq n$, then $\bigvee_{j=1}^n \theta_j = \theta_1 \theta_2 \cdots \theta_n$.*

PROOF. If $n = 2$ this follows from Corollary 2.5. Then the present corollary follows easily by induction. Indeed, by Proposition 2.12(iv) we have $\theta_n \wedge (\theta_1 \theta_2 \cdots \theta_{n-1}) = 1$ and, a fortiori, $\theta_n \wedge (\bigvee_{i=1}^{n-1} \theta_i) = 1$. An application of Corollary 2.5 then yields $\bigvee_{i=1}^n \theta_i = (\bigvee_{i=1}^{n-1} \theta_i) \vee \theta_n = (\bigvee_{i=1}^{n-1} \theta_i) \cdot \theta_n$ and this completes the proof.

The following result is particularly useful.

2.14. PROPOSITION. *Let $\{\theta_i : i \in I\}$ be a family of nonconstant inner divisors of the inner function $\theta \in H^\infty$. If $\theta_i \wedge \theta_j \equiv 1$ for $i, j \in I$, $i \neq j$, then I is at most countable.*

PROOF. Choose $\lambda \in \mathbf{D}$ such that $\theta(\lambda) \neq 0$. If $i_1, i_2, \ldots, i_n \in I$ then an application of Corollary 2.13 shows that $\theta_{i_1} \theta_{i_2} \cdots \theta_{i_n} | \theta$ and therefore

$$|\theta(\lambda)| \leq |\theta_{i_1}(\lambda)||\theta_{i_2}(\lambda)| \cdots |\theta_{i_n}(\lambda)|$$

by Lemma 2.1. We conclude that the infinite product $\prod_{i \in I} |\theta_i(\lambda)|$ converges to a number $\geq |\theta(\lambda)| > 0$ and hence $\sum_{i \in I}(1 - |\theta_i(\lambda)|) < \infty$. We conclude that the set $I_1 = \{i \in I : |\theta_i(\lambda)| \neq 1\}$ is at most countable. Now, if ϕ is inner and $|\phi(\lambda)| = 1$, it follows from the maximum modulus principle that ϕ must be constant. Since there are no constant functions among the θ_i, we have $I = I_1$ and the proposition follows.

We conclude this section with a few remarks on the analytic properties of inner functions. Let us note that the product in (2.6) converges for all values λ that do not belong to the closure of $\{1/\overline{\alpha}_j : 0 \leq j < n\}$ and the product is not zero except at the points $\{\alpha_j : 0 \leq j < n\}$. Thus $b(\lambda)$ can be continued analytically across an arc γ of $\mathbf{T}$ if and only if γ is disjoint from the closure of $\{\alpha_j : 0 \leq j < n\}$. For singular inner functions we note that the right-hand side in (2.8) makes sense for every $\lambda \notin \operatorname{supp} \nu$. The function s_ν can be continued analytically across an arc γ of $\mathbf{T}$ if and only if $\operatorname{supp} \nu \cap \gamma = \varnothing$.

2.15. DEFINITION. The *support* $\operatorname{supp}(\theta)$ of the inner function θ consists of those points $\lambda \in \overline{\mathbf{D}}$ such that either (i) $|\lambda| < 1$ and $\theta(\lambda) = 0$; or (ii) $|\lambda| = 1$ and θ cannot be continued analytically across λ.

Thus, if b is given by (2.6) we have $\operatorname{supp}(b) = \{\lambda_j : 0 \leq j < n\}^-$ and if s_ν is given by (2.8) we have $\operatorname{supp}(s_\nu) = \operatorname{supp} \nu$. It is also clear that $\operatorname{supp}(bs_\nu) =$

$\operatorname{supp}(b) \cup \operatorname{supp}(s_\nu)$. An important property of the support is that it can be localized in the sense of the following definition.

2.16. DEFINITION. Let θ be an inner function and A a subset of $\mathbf{C}$. A divisor θ' of θ is a *localization of* θ *to* A if $\operatorname{supp}(\theta') \subset \overline{A}$, $\operatorname{supp}(\theta/\theta') \subset (\mathbf{C} \setminus A)^-$, and $\theta' \wedge (\theta/\theta') \equiv 1$.

Clearly, if θ' is a localization of θ to A, the function θ/θ'' is a localization of θ to $\mathbf{C} \setminus A$.

2.17. PROPOSITION. *For every inner function θ and every subset A of $\mathbf{C}$ there exists a localization of θ to A.*

PROOF. It follows from the discussion above that we have only to indicate how to construct localizations of Blaschke products and singular inner functions. If b is the Blaschke product determined by the Blaschke multiplicity function μ, then a localization of b to A is the Blaschke product determined by the multiplicity function μ' defined by

$$\begin{aligned} \mu'(\lambda) &= 0 \qquad \text{if } \lambda \notin A, \\ &= \mu(\lambda) \ \text{ if } \lambda \in A. \end{aligned}$$

Analogously, if s_ν is defined by (2.8), a localization of s_ν to A is the singular inner function $s_{\nu'}$, where $\nu' = \chi_{\overline{A}}\nu$ (i.e., $\nu'(B) = \nu(B \cap \overline{A})$).

2.18. REMARK. Let θ' be a localization of θ to A. If $\operatorname{int}(A) \cap \operatorname{supp}(\theta) \neq \varnothing$ then we clearly have $\theta' \not\equiv 1$. Indeed, if $\theta' \equiv 1$, then $\operatorname{supp}(\theta) = \operatorname{supp}(\theta/\theta') \subset \operatorname{supp}(\theta) \setminus \operatorname{int}(A) \neq \operatorname{supp}(\theta)$, a contradiction.

2.19. REMARK. If $\{\theta_i : i \in I\}$ is a family of divisors of θ and each θ_i is a localization of θ to A, then $\bigvee_{i \in I} \theta_i$ is also a localization of θ to A. It follows that there exists a largest (in divisibility) localization of θ to A. We will call this localization the maximal localization of θ to A. The following property of the maximal localization θ_0 of θ to A is easy to verify. If $\phi \mid \theta$ and $\operatorname{supp}(\phi) \subset \overline{A}$, then $\phi \mid \theta_0$.

Exercises

1. Give an example of a pair $\{\theta, \theta'\}$ of inner functions such that $\theta \wedge \theta' \equiv 1$ but $\theta H^\infty + \theta' H^\infty \neq H^\infty$.

2. Assume that the inner function θ can be extended continuously to $\mathbf{D} \cup \mathbf{T}$. Prove that θ is a finite Blaschke product (i.e., a Blaschke product with finitely many factors).

3. Give examples of uncountable families $\{\theta_i : i \in I\}$ consisting of Blaschke products [resp. singular inner functions] with the property that $\theta_i \neq 1$, $i \in I$, and $\theta_i \wedge \theta_j \equiv 1$, $i \neq j$, $i, j \in I$.

4. Let $\{\theta_n : n \in \mathbf{N}\}$ be a sequence of nonconstant inner functions. Does there always exist an inner function θ such that $\theta \wedge \theta_n \not\equiv 1$ for all $n \in \mathbf{N}$? Does such a function θ exist if each θ_n is a Blaschke product [resp. a singular inner function]?

5. Let $\{\theta_1, \theta_2, \ldots, \theta_n\}$ be a finite family of inner functions. Prove that there exists a family $\{\theta'_1, \theta'_2, \ldots, \theta'_n\}$ of inner functions such that $\theta'_i | \theta_i$, $1 \le i \le n$, $\theta'_i \wedge \theta'_j \equiv 1$, $1 \le i \ne j \le n$, and

$$\bigvee_{i=1}^{n} \theta_i = \theta'_1 \theta'_2 \cdots \theta'_n.$$

Is this result true for infinite families of inner functions?

6. Let θ and θ' be two inner functions. Show that $\theta H^\infty + \theta' H^\infty = H^\infty$ whenever $\operatorname{supp}(\theta) \cap \operatorname{supp}(\theta') = \varnothing$.

3. Minimal functions and maximal vectors. The principal aim of this section is a proof of the fact that a locally C_0 operator is really of class C_0. We also prove the existence of maximal vectors as defined below.

3.1. DEFINITION. Let $T \in \mathscr{L}(\mathscr{H})$ be locally of class C_0. A vector $h \in \mathscr{H}$ is called *T-maximal* (or simply *maximal* when no confusion may arise) if $m_g \mid m_h$ for every $g \in \mathscr{H}$.

Clearly $m_h(T) = 0$ if h is a maximal vector, and hence T is necessarily of class C_0 and $m_h \equiv m_T$. We start our proof of the existence of maximal vectors with a remark about "relatively" maximal vectors. Assume that $T \in \mathscr{L}(\mathscr{H})$ is locally of class C_0 and $\mathscr{K} \subset \mathscr{H}$ is a subspace of dimension two. Then there exist functions $u \in H^\infty$ such that $u(T)\mathscr{K} = \{0\}$. Indeed, if $\{h_1, h_2\}$ is a basis of $\mathscr{K}$ then the function

$$u = m_{\mathscr{K}} = m_{h_1} \vee m_{h_2} \tag{3.2}$$

will satisfy this condition. It is easy to check that the inner function $m_{\mathscr{K}}$ defined by (3.2) does not depend on the particular basis; indeed, it can be alternatively characterized by the equality

$$m_{\mathscr{K}} H^\infty = \{u \in H^\infty : u(T)\mathscr{K} = \{0\}\}.$$

3.3. LEMMA. *Let $T \in \mathscr{L}(\mathscr{H})$ be locally of class C_0 and let $\mathscr{K} \subset \mathscr{H}$ be a subspace of dimension two. Then the set $A = \{h \in \mathscr{K} : m_h \not\equiv m_{\mathscr{K}}\}$ is the union of an at most countable family of one-dimensional subspaces of $\mathscr{K}$.*

PROOF. Clearly $0 \in A$ and $m_{\lambda h} \equiv m_h$ whenever λ is a nonzero scalar. We conclude that A is the union of a family of one-dimensional subspaces; say $A = \bigcup_{i \in I} \mathbf{C} h_i$, where h_i and h_j are independent whenever $i \ne j$, $i, j \in I$. Define $\theta_i = m_{\mathscr{K}}/m_{h_i}$, $i \in I$; we have $\theta_i \not\equiv 1$ because $h_i \in A$. If $i \ne j$, $i, j \in I$, the vectors h_i and h_j form a basis of $\mathscr{K}$. We conclude by Proposition 2.4 that

$$\theta_i \wedge \theta_j \equiv m_{\mathscr{K}}/(m_{h_i} \vee m_{h_j}) \equiv m_{\mathscr{K}}/m_{\mathscr{K}} = 1.$$

The fact that I is at most countable follows immediately now from Proposition 2.14.

The next result clears the way for an application of the Baire category theorem.

3.4. LEMMA. *Let $T \in \mathscr{L}(\mathscr{H})$ be locally of class C_0. For each $\lambda_0 \in \mathbf{D}$ and every $a > 0$ the set*

$$\sigma = \{h \in \mathscr{H} : |m_h(\lambda_0)| \geq a\}$$

is closed in $\mathscr{H}$.

PROOF. Let $\{h_n\} \subset \sigma$ be a convergent sequence, $h = \lim h_n$. An application of the Vitali–Montel theorem allows us to assume, after replacing $\{h_n\}$ by a subsequence, that the sequence $\{m_{h_n}\}$ converges in the weak* topology of H^∞ to a function $u \in H^\infty$. We certainly have $|u(\lambda)| \leq 1$, $\lambda \in \mathbf{D}$, and $|u(\lambda_0)| \geq a$. By the continuity of the functional calculus, $m_{h_n}(T)$ converges weakly to $u(T)$ and therefore, for $k \in \mathscr{H}$,

$$\begin{aligned}|(u(T)h, k)| &\leq |((u(T) - m_{h_n}(T))h, k)| + |(m_{h_n}(T)(h - h_n), k)| \\ &\leq |((u(T) - m_{h_n}(T))h, k)| + \|k\|\|h - h_n\| \to 0\end{aligned}$$

as $n \to \infty$, where we made use of the relation $m_{h_n}(T)h_n = 0$. Since k is arbitrary we conclude that $u(T)h = 0$ and hence $m_h \mid u$. We can thus write $u = m_h\phi$, $\phi \in H^\infty$, and a moment's thought shows that $|\phi(\lambda)| \leq 1$ for $\lambda \in \mathbf{D}$. Therefore $a \leq |u(\lambda_0)| \leq |m_h(\lambda_0)|$ and we conclude that $h \in \sigma$. The lemma is proved.

The following application of Baire's category argument is the key to the principal result in this section.

3.5. LEMMA. *Let $T \in \mathscr{L}(\mathscr{H})$ be locally of class C_0. Then for every $\lambda_0 \in \mathbf{D}$ the set $\{k \in \mathscr{H} : |m_k(\lambda_0)| = \inf_{h \in \mathscr{H}} |m_h(\lambda_0)|\}$ is a dense G_δ in $\mathscr{H}$.*

PROOF. If we set $a = \inf_{h \in \mathscr{H}} |m_h(\lambda_0)|$ then the complement of the set $\{k \colon |m_k(\lambda_0)| = a\}$ is the union $\bigcup_{j=1}^\infty \sigma_j$, where $\sigma_j = \{h \in \mathscr{H} \colon |m_h(\lambda_0)| \geq a + 1/j\}$. The preceding lemma implies that each σ_j is a closed set in $\mathscr{H}$, and to finish the proof it suffices to show that each σ_j has empty interior. Suppose to the contrary that σ_j contains the open ball $B = \{h \colon \|h - h_0\| < \varepsilon\}$. The definition of a implies the existence of $k \in \mathscr{H} \setminus \sigma_j$. Choose such a vector k and denote by $\mathscr{K}$ the space generated by h_0 and k. Lemma 3.3 implies the existence of $f \in \mathscr{K} \cap B$ such that $m_f = m_{\mathscr{K}}$; in particular $m_k \mid m_f$, from which we infer that $|m_f(\lambda_0)| \leq |m_k(\lambda_0)| < a + 1/j$ ($k \notin \sigma_j$). On the other hand $f \in B \subset \sigma_j$, a contradiction. The lemma follows.

3.6. THEOREM. *Let $T \in \mathscr{L}(\mathscr{H})$ be locally of class C_0. Then there exist T-maximal vectors and the set of T-maximal vectors is a dense G_δ in $\mathscr{H}$. In particular, T is an operator of class C_0 and $m_T \equiv m_h$ for each T-maximal vector h.*

PROOF. The intersection of countably many dense G_δ sets is still a dense G_δ set and therefore the set

$$M = \{h \in \mathscr{H} : |m_h(\lambda_n)| = \inf_{k \in \mathscr{H}} |m_k(\lambda_n)|, n \in \mathbf{N}\}$$

is a dense G_δ for any choice of the sequence $\{\lambda_n\} \subset \mathbf{D}$. Let us assume that the sequence $\{\lambda_n\}$ is dense in $\mathbf{D}$. If $h \in M$ and $k \in \mathscr{H}$ we then have $|m_h(\lambda_n)| \leq$

$|m_k(\lambda_n)|$, $n \in \mathbf{N}$, and this relation extends by continuity to $|m_h(\lambda)| \leq |m_k(\lambda)|$, $\lambda \in \mathbf{D}$. We conclude that $m_k \mid m_h$ and thus every element of M is a T-maximal vector. The remaining assertions of the theorem are obvious.

Let us note that this theorem allows us to give a new definition of the minimal function m_T for operators of class C_0. This new definition does not depend on the structure of weak*-closed ideals of H^∞.

The following result is a useful variation of Theorem 3.6 on the existence of maximal vectors.

3.7. THEOREM. *Let $T \in \mathscr{L}(\mathscr{H})$ be an operator of class C_0, $\mathscr{K}$ a Banach space, and $X \colon \mathscr{K} \to \mathscr{H}$ a continuous linear map such that $\bigvee_{n\geq 0} T^n X\mathscr{K} = \mathscr{H}$. Then the set $\{k \in \mathscr{K} : m_{Xk} \equiv m_T\}$ is a dense G_δ in $\mathscr{K}$.*

PROOF. The proof closely imitates that of Theorem 3.6. We provide the relevant details. For a fixed $\lambda_0 \in \mathbf{D}$ denote $a = \inf_{k\in\mathscr{K}} |m_{Xk}(\lambda_0)|$. Then the set

$$\sigma_j = \{k \in \mathscr{K} : |m_{Xk}(\lambda_0)| \geq a + \frac{1}{j}\} = X^{-1}\{h \in \mathscr{H} : |m_h(\lambda_0)| \geq a + \frac{1}{j}\}$$

is closed for $j \geq 1$ by Lemma 3.4. We then proceed as in the proof of Lemma 3.5 to show that each σ_j has empty interior. It follows that the set

$$\{k \in \mathscr{K} : |m_{Xk}(\lambda_0)| = a\}$$

is a dense G_δ in $\mathscr{K}$. Then the argument of Theorem 3.6 shows that the set

$$M = \{k \in \mathscr{K} : |m_{Xk}(\lambda)| = \inf_{h\in\mathscr{K}} |m_{Xh}(\lambda)|, \lambda \in \mathbf{D}\}$$

is a dense G_δ in $\mathscr{K}$. If $k \in M$ it follows that m_{Xk} is a multiple of every m_{Xh}, $h \in \mathscr{K}$, and hence $[m_{Xk}(T)](X\mathscr{K}) = \{0\}$. This last relation clearly implies

$$[m_{Xk}(T)]\left(\bigvee_{n\geq 0} T^n X\mathscr{K}\right) = \{0\}$$

and hence $m_{Xk}(T) = 0$, from which we deduce $m_{Xk} = m_T$. The theorem follows.

Exercises

1. Show that the space $\mathscr{K}$ introduced in the proof of Lemma 3.5 cannot be one-dimensional.

2. Let T be an algebraic operator. One can then define maximal vectors for T. Indeed, the minimal polynomial p_T is the generator of the ideal $\{u \in \mathrm{C}[X] \colon u(T) = 0\}$ of the polynomial ring $\mathbf{C}[X]$, and we can define p_h as the generator of $\{u \in \mathbf{C}[X] \colon u(T)h\} = 0$. The vector h is maximal if $p_h \equiv p_T$ (where $p_h \equiv p_T$ means that p_T/p_h is a nonzero constant). Show that the set of T-maximal vectors is open.

3. Assume that T is an operator of class C_0 and the set of T-maximal vectors is open. Is T necessarily an algebraic operator? (Cf. also the exercises to §4.)

4. With the notation of Theorem 3.7, assume that the set $\{k \in \mathscr{H} : m_{Xk} = m_T\}$ is an open set in $\mathscr{H}$. Is T necessarily an algebraic operator?

5. Let $\mathscr{F}$ be a family of inner functions in H^∞ with the property that $\theta \vee \theta' \in \mathscr{F}$ whenever θ, $\theta' \in \mathscr{F}$. Prove that $\mathscr{F}$ has a common multiple in H^∞ if and only if $\inf\{|\theta(\lambda_0)| : \theta \in \mathscr{F}\} > 0$ for some $\lambda_0 \in \mathbf{D}$. If $u = \bigvee \mathscr{F}$, show that $|u(\lambda)| = \inf\{|\theta(\lambda)| : \theta \in \mathscr{F}\}$ for all $\lambda \in \mathbf{D}$.

6. Let $T \in \mathscr{L}(\mathscr{H})$ be an operator of class C_0, h_1, $h_2 \in \mathscr{H}$. An application of Exercise 2.5 to the inner functions $\theta_1 = m_{h_1}$, $\theta_2 = m_{h_2}$ provides decompositions $\theta_1 = \theta_1'\theta_1''$ and $\theta_2 = \theta_2'\theta_2''$ such that $\theta_1' \wedge \theta_2' \equiv 1$ and $\theta_1 \vee \theta_2 \equiv \theta_1'\theta_2'$. Define $k = \theta_1''(T)h_1 + \theta_2''(T)h_2$ and show that $m_k = \theta_1 \vee \theta_2$.

7. Use the preceding exercise to provide a new proof of the existence of maximal vectors for operators of class C_0 with finite multiplicity.

4. General properties of operators of class C_0.

4.1. PROPOSITION. *An operator $T \in \mathscr{L}(\mathscr{H})$ is of class C_0 if and only if T^* is of class C_0. If T is of class C_0 we have $m_{T^*} \equiv m_T^\sim$.*

PROOF. Assume that T is of class C_0. Then we have $m_T^\sim(T^*) = (m_T(T))^* = 0$ and consequently T^* is of class C_0 and m_{T^*} divides $m_T^\sim$. Analogously, if T^* is of class C_0 then T is of class C_0 and m_T divides $m_{T^*}^\sim$. Since the mapping $u \mapsto u^\sim$ is a homomorphism of the ring H^∞ into itself and u is inner if and only if $u^\sim$ is inner, we conclude that $m_{T^*} \equiv m_T^\sim$ if T is of class C_0.

4.2. COROLLARY. *Every operator of class C_0 is also of class C_{00}.*

PROOF. Let T be an operator of class C_0. We already know from Lemma 1.12 that T is of class $C_{\cdot 0}$. The corollary follows from that lemma applied to T^*.

We continue with a few properties related to the invariant subspaces of operators of class C_0.

4.3. PROPOSITION. *Let $T \in \mathscr{L}(\mathscr{H})$ be a completely nonunitary contraction, $\mathscr{H}'$ an invariant subspace for T, and $\mathscr{H}'' = \mathscr{H} \ominus \mathscr{H}'$. Let $T = \begin{bmatrix} T' & X \\ 0 & T'' \end{bmatrix}$ be the matrix of T with respect to the decomposition $\mathscr{H} = \mathscr{H}' \oplus \mathscr{H}''$. Then T is of class C_0 if and only if T' and T'' are operators of class C_0. If T is of class C_0 then $m_{T'} \mid m_T$, $m_T'' \mid m_T$, and $m_T \mid m_{T'}m_{T''}$.*

PROOF. We have $u(T) = \begin{bmatrix} u(T') & * \\ 0 & u(T'') \end{bmatrix}$ for every $u \in H^\infty$. If $u(T) = 0$ we conclude that $u(T') = 0$ and $u(T'') = 0$ so that T', T'' are of class C_0 and $m_{T'} \mid m_T$, $m_{T''} \mid m_T$. Conversely, assume that T' and T'' are operators of class C_0, $\theta' = m_{T'}$ and $\theta'' = m_{T''}$. If $h'' \in \mathscr{H}''$ we have $0 = \theta''(T'')h'' = P_{\mathscr{H}''}\theta''(T)h''$ and therefore $\theta''(T)h'' \in \mathscr{H}'$. Consequently $(\theta'\theta'')(T)h'' = \theta'(T')\theta''(T)h'' = 0$. Since $(\theta'\theta'')(T) \mid \mathscr{H}' = \theta''(T')\theta'(T') = 0$ we conclude that $\ker(\theta'\theta'')(T) \supset \mathscr{H}' \cup \mathscr{H}''$ and this clearly implies that $(\theta'\theta'')(T) = 0$. We conclude that T is an operator of class C_0 and $m_T \mid \theta'\theta''$, thus concluding the proof.

Simple examples (e.g., $T = \begin{bmatrix} 0 & 0 \\ 0 & 0 \end{bmatrix}$) show that we do not usually have $m_T \equiv m_{T'} m_{T''}$ in Proposition 4.3. We will prove later that the relation $m_T \equiv m_{T'} m_{T''}$ holds whenever $\mathscr{H}'$ is hyperinvariant for T. We recall that a subspace $\mathscr{H}' \subset \mathscr{H}$ is hyperinvariant for T if it is invariant for every operator in the commutant $\{T\}'$ of T. The following result shows that the equality $m_T \equiv m_{T'} m_{T''}$ is indeed true for certain hyperinvariant subspaces.

If $T \in \mathscr{L}(\mathscr{H})$ is a completely nonunitary contraction and $u \in H^\infty$ then $u(T)$ belongs to the weakly closed algebra generated by T and I. Therefore the subspace $\ker u(T)$ is a hyperinvariant subspace for T.

4.4. PROPOSITION. *Let $T \in \mathscr{L}(\mathscr{H})$ be an operator of class C_0 and let θ be an inner divisor of m_T. If $T = \begin{bmatrix} T' & X \\ 0 & T'' \end{bmatrix}$ is the matrix of $\mathscr{H}$ with respect to the decomposition $\mathscr{H} = \mathscr{H}' \oplus \mathscr{H}''$, with $\mathscr{H}' = \ker \theta(T)$, then $m_{T'} \equiv \theta$ and $m_{T''} \equiv m_T/\theta$.*

PROOF. We have $\theta(T') = \theta(T) \mid \ker \theta(T) = 0$ so that $m_{T'} \mid \theta$. It is also clear that $\{0\} = m_T(T)\mathscr{H} = \theta(T)(m_T/\theta)(T)\mathscr{H}$ so that

$$(m_T/\theta)(T)\mathscr{H}'' \subset (m_T/\theta)(T)\mathscr{H} \subset \ker \theta(T) = \mathscr{H}''$$

and consequently $(m_T/\theta)(T'') = P_{\mathscr{H}''}(m_T/\theta)(T) \mid \mathscr{H}'' = 0$. We have $m_{T'} \mid \theta$, $m_{T''} \mid (m_T/\theta)$, and by Proposition 4.3, $\theta(m_T/\theta) = m_T \mid m_{T'} m_{T''}$. A moment's thought shows that these relations imply $m_{T'} \equiv \theta$ and $m_{T''} \equiv m_T/\theta$.

4.5. COROLLARY. *Let $T \in \mathscr{L}(\mathscr{H})$ be an operator of class C_0. If T is not a scalar then it has nontrivial hyperinvariant subspaces.*

PROOF. By Proposition 4.4, an inner divisor θ of m_T is uniquely determined (up to a constant coefficient) by the hyperinvariant subspace $\ker \theta(T)$. Thus, if $\theta \not\equiv 1$ and $\theta \not\equiv m_T$, $\ker \theta(T)$ is a nontrivial hyperinvariant subspace for T. Assume that m_T has no nontrivial inner divisors. Then m_T must be a Blaschke factor: $m_T(\lambda) = (a-\lambda)/(1-\bar{a}\lambda)$, $\lambda \in \mathbf{D}$, and in this case we clearly have $T = aI$.

4.6. THEOREM. *Let $T \in \mathscr{L}(\mathscr{H})$ be an operator of class C_0 and let $\{\theta_i : i \in I\}$ be a family of inner divisors of m_T. If $\phi \equiv \bigwedge_{i\in I} \theta_i$ and $\psi \equiv \bigvee_{i \in I} \theta_i$, we have $\bigcap_{i\in I} \ker \theta_i(T) = \ker \phi(T)$ and $\bigvee_{i\in I} \ker \theta_i(T) = \ker \psi(T)$.*

PROOF. The function ϕ divides θ_i and hence $\ker \phi(T) \subset \ker \theta_i(T)$ for each $i \in I$; thus $\ker \phi(T) \subset \bigcap_{i\in I} \ker \theta_i(T)$. If $h \in \bigcap_{i \in I} \ker \theta_i(T)$ then the minimal function m_h divides θ_i for each i. Consequently $m_h \mid \phi$ and $h \in \ker \phi(T)$. The equality $\ker \phi(T) = \bigcap_{i\in I} \ker \theta_i(T)$ follows at once.

For the second equality we note that there is no loss of generality in assuming that $\psi = m_T$ and hence $\ker \psi(T) = \mathscr{H}$. Indeed, if we replace T by $T \mid \ker \psi(T)$, the spaces $\ker \theta_i(T)$ and $\ker \psi(T)$ do not change. With this additional assumption we must prove that $\mathscr{H} \ominus (\bigvee_{i\in I} \ker \theta_i(T)) = \{0\}$. Now, if $h \in \mathscr{H} \ominus \ker \theta_i(T)$, it follows from Propositions 4.4 and 4.1 that $(m_T/\theta_i)^\sim(T^*)h = (\psi/\theta_i)^\sim(T^*)h = 0$.

If this happens for every $i \in I$ we have therefore

$$h \in \bigcap_{i\in I} \ker(\psi/\theta_i)^{\sim}(T^*) = \ker\Big(\bigwedge_{i\in I}(\psi/\theta_i)^{\sim}\Big)(T^*),$$

where we have used the equality proved in the first part of the proof. Since

$$\bigwedge_{i\in I}(\psi/\theta_i) \equiv \psi/\left(\bigvee_{i\in I}\theta_i\right) \equiv \psi/\psi = 1$$

it follows that the last space in (4.7) is $\{0\}$ and hence $h = 0$. The proof is now complete.

4.8. COROLLARY. *With the notation of Theorem* 4.6 *we have*

$$\bigcap_{i\in I}(\operatorname{ran}\theta_i(T))^- = (\operatorname{ran}\psi(T))^-$$

and

$$\bigvee_{i\in I}(\operatorname{ran}\theta_i(T))^- = (\operatorname{ran}\phi(T))^-.$$

PROOF. These equalities follow from the equalities $\mathscr{H} \ominus (\operatorname{ran}\theta_i(T))^- = \ker(\theta_i(T))^* = \ker\theta_i^{\sim}(T^*)$ combined with Theorem 4.6. For example

$$\begin{aligned}
\mathscr{H} \ominus \left(\bigcap_{i\in I}(\operatorname{ran}\theta_i(T))^-\right) &= \bigvee_{i\in I}(\mathscr{H} \ominus (\operatorname{ran}\theta_i(T))^-)\\
&= \bigvee_{i\in I}\ker\theta_i^{\sim}(T^*)\\
&= \ker\left(\bigvee_{i\in I}\theta_i\right)^{\sim}(T^*)\\
&= \ker\psi^{\sim}(T^*) = \mathscr{H} \ominus (\operatorname{ran}\psi(T))^-.
\end{aligned}$$

Let us recall that for any completely nonunitary contraction T the class K_T^∞ consists of those functions $u \in H^\infty$ for which $u(T)$ is a *quasiaffinity* (i.e., $\ker u(T) = \ker(u(T))^* = \{0\}$). The class K_T^∞ is important in introducing the functional calculus with rational functions:

$$(v/u)(T) = u(T)^{-1}v(T), \quad u \in K_T^\infty, \quad v \in H^\infty.$$

In general $(v/u)(T)$ is a discontinuous closed and densely defined operator.

4.9. PROPOSITION. *For every operator T of class C_0 we have $K_T^\infty = \{u \in H^\infty\colon u \wedge m_T \equiv 1\}$. Moreover, for $u \in H^\infty$, we have* $\ker u(T) = \{0\}$ *if and only if* $\ker u(T)^* = \{0\}$.

PROOF. Assume that $u \wedge m_T \equiv 1$. If $h \in \ker u(T)$ then $m_h \mid u$ and $m_h \mid m_T$ so that $m_h \equiv 1$. This clearly implies $h = 0$ and so $\ker u(T) = \{0\}$. Analogously, $u \wedge m_T \equiv 1$ implies that $u^{\sim} \wedge m_{T^*} = 1$ so that $\ker u(T)^* = \ker u^{\sim}(T^*) = \{0\}$. Conversely, if $u \wedge m_T = \theta \neq 1$ then $\ker u(T) \supset \ker\theta(T)$ and $\ker\theta(T) \neq \{0\}$ by Proposition 4.4. The proposition follows.

4.10. REMARK. The preceding proof actually shows that $\ker u(T) = \ker(u \wedge m_T)(T)$ for every $u \in H^\infty$.

The hyperinvariant subspaces considered above will help us determine the spectrum $\sigma(T)$ of an operator T of class C_0 in terms of the minimal function m_T. We will denote by $\sigma_p(T)$ the point spectrum of T, that is, the set of all eigenvalues of T.

4.11. THEOREM. *For every operator T of class C_0 we have $\sigma(T) = \operatorname{supp}(m_T)$ and $\sigma_p(T) = \operatorname{supp}(m_T) \cap \mathbf{D}$.*

PROOF. Assume first that $\lambda_0 \notin \operatorname{supp}(m_T)$. Then m_T can be analytically continued (if $|\lambda_0| = 1$) across λ_0 and $m_T(\lambda_0) \neq 0$. It follows that we have a factorization of the form

$$m_T(\lambda) - m_T(\lambda_0) = (\lambda - \lambda_0)g(\lambda), \quad \lambda \in \mathbf{D},$$

for some $g \in H^\infty$. Functional calculus then yields

$$-m_T(\lambda_0)I = m_T(T) - m_T(\lambda_0)I = (T - \lambda_0 I)g(T) = g(T)(T - \lambda_0 I)$$

so that $\lambda_0 I - T$ has an inverse given by $(1/m_T(\lambda_0))g(T)$. Thus we proved that $\sigma(T) \subset \operatorname{supp}(m_T)$. Assume that the inclusion $\sigma(T) \subset \operatorname{supp}(m_T)$ is strict. Then there exists an open set ω such that $\overline{\omega} \cap \sigma(T) = \varnothing$ but $\omega \cap \operatorname{supp}(m_T) \neq \varnothing$. Let θ be a localization of m_T to ω; we know that $\theta \not\equiv 1$ by Remark 2.18 and hence $\ker \theta(T) \neq \{0\}$ by Proposition 4.4. Let $T = \begin{bmatrix} T' & X \\ 0 & T'' \end{bmatrix}$ be the triangularization of T with respect to the decomposition $\mathscr{H} = (\ker \theta(T)) \oplus \mathscr{H}''$. Since $\ker \theta(T)$ is hyperinvariant for T we have $\sigma(T') \subset \sigma(T) \subset \mathbf{C} \setminus \overline{\omega}$. On the other hand we have $m_{T'} = \theta$ so that $\sigma(T') \subset \operatorname{supp}(\theta) \subset \overline{\omega}$ because of the relation already proved between spectrum and support. These two inclusions for $\sigma(T')$ are contradictory and so we conclude that $\sigma(T) = \operatorname{supp}(m_T)$.

For the point spectrum we know that T is completely nonunitary and thus does not have eigenvalues of absolute value one. Hence $\sigma_p(T) \subset \sigma(T) \cap \mathbf{D} \subset \operatorname{supp}(m_T) \cap \mathbf{D}$. For the opposite inclusion we note that each $\lambda_0 \in \operatorname{supp}(m_T) \cap \mathbf{D}$ is a zero of m_T and hence $\ker(T - \lambda_0 I) = \ker(T - \lambda_0 I)(I - \overline{\lambda_0} T)^{-1} = \ker b(T) \neq \{0\}$, by Proposition 4.4, where $b(\lambda) = (\lambda - \lambda_0)/(1 - \overline{\lambda_0}\lambda)$. The theorem is proved.

We conclude this section with some results about the localization of the spectrum of an operator of class C_0. Let $T \in \mathscr{L}(\mathscr{H})$ be an operator on the Banach space $\mathscr{H}$. We recall that a subspace $\mathscr{M} \subset \mathscr{H}$ is a *maximal spectral subspace* for T if $T\mathscr{M} \subset \mathscr{M}$ and for every invariant subspace $\mathscr{N}$ of T such that $\sigma(T \mid \mathscr{N}) \subset \sigma(T \mid \mathscr{M})$ we have $\mathscr{N} \subset \mathscr{M}$. An operator $T \in \mathscr{L}(\mathscr{H})$ is *decomposable* if for every finite open covering $\{G_1, G_2, \dots, G_n\}$ of $\sigma(T)$ there exist maximal spectral subspaces $\mathscr{M}_1, \mathscr{M}_2, \dots, \mathscr{M}_n$ for T such that $\sigma(T \mid \mathscr{M}_j) \subset G_j$, $j = 1, 2, \dots, n$ and $\mathscr{H} = \mathscr{M}_1 + \mathscr{M}_2 + \cdots + \mathscr{M}_n$.

4.12. PROPOSITION. *Let $T \in \mathscr{L}(\mathscr{H})$ be an operator of class C_0, $A \subset \mathbf{C}$, and denote by θ the maximal localization of m_T to A. Then the space $\ker \theta(T)$ is a maximal spectral subspace for T and $\sigma(T \mid \ker \theta(T)) \subseteq \overline{A}$.*

PROOF. By Proposition 4.4 we have $\sigma(T \mid \ker\theta(T)) \subseteq \operatorname{supp}(\theta) \subseteq \overline{A}$. If $\mathscr{M}$ is invariant for T and $\sigma(T \mid \mathscr{M}) \subset \sigma(T \mid \ker\theta(T))$ we have $\operatorname{supp}(m_{T|\mathscr{M}}) \subseteq \operatorname{supp}(\theta) \subseteq \overline{A}$ and hence $m_{T|\mathscr{M}} \mid \theta$ by Remark 2.19. Then we have $\theta(T \mid \mathscr{M}) = 0$ and we conclude that $\mathscr{M} \subset \ker\theta(T)$. The proposition follows.

4.13. PROPOSITION. *Let $T \in \mathscr{L}(\mathscr{H})$ be an operator of class C_0 and let $\{G_1, G_2, \dots, G_n\}$ be a finite open covering of $\sigma(T)$. For each j, $1 \le j \le n$, choose a localization θ_j of m_T to G_j. Then we have*

$$\ker\theta_1(T) + \ker\theta_2(T) + \cdots + \ker\theta_n(T) = \mathscr{H}.$$

PROOF. Let $\{D_1, D_2, \dots, D_n\}$ be an open covering of $\sigma(T)$ such that $\overline{D}_j \subset G_j$, $1 \le j \le n$. The quotient m_T/θ_j has the property that $\operatorname{supp}(m_T/\theta_j) \cap G_j = \varnothing$ from which we conclude that

$$\inf\{|(m_T/\theta_j)(\lambda)| \colon \lambda \in D_j \cap \mathbf{D}\} > 0, \quad 1 \le j \le n. \tag{4.14}$$

On the other hand we clearly have

$$\inf\left\{|m_T(\lambda)| \colon \lambda \in \mathbf{D} \setminus \left(\bigcup_{i=1}^{n} D_i\right)\right\} > 0$$

because $\bigcup_{i=1}^{n} D_i$ covers the support of m_T. Therefore

$$\inf\left\{|(m_T/\theta_j)(\lambda)| \colon \lambda \in \mathbf{D} \setminus \left(\bigcup_{i=1}^{n} D_i\right)\right\} > 0, \quad 1 \le j \le n. \tag{4.15}$$

Combining (4.14) and (4.15) we see that

$$\inf\left\{\sum_{j=1}^{n} |(m_T/\theta_j)(\lambda)| \colon \lambda \in \mathbf{D}\right\} > 0$$

and by Carleson's Corona theorem there exist functions $u_1, u_2, \dots, u_n \in H^\infty$ satisfying the relation

$$\sum_{j=1}^{n} u_j m_T/\theta_j = 1.$$

Functional calculus then yields

$$\sum_{j=1}^{n} (m_T/\theta_j)(T) u_j(T) h = h, \quad h \in \mathscr{H},$$

and to conclude the proof it suffices to show that $(m_T/\theta_j)(T)u_j(T)h \in \ker\theta_j(T)$, $1 \le j \le n$, and this is obvious:

$$\theta_j(T)(m_T/\theta_j)(T)u_j(T) = m_T(T)u_j(T) = 0.$$

4.16. COROLLARY. *Every operator of class C_0 is decomposable.*

PROOF. Let $\{G_1, G_2, \dots, G_n\}$ be an open covering of $\sigma(T)$, and choose an open covering $\{D_1, D_2, \dots, D_n\}$ of $\sigma(T)$ such that $\overline{D}_j \subseteq G_j$, $1 \le j \le n$. For each j denote by θ_j the maximal localization of m_T to D_j. Then the spaces $\ker\theta_j(T)$, $1 \le j \le n$, are maximal spectral by Proposition 4.12, $\sigma(T \mid \ker\theta_j(T)) = \operatorname{supp}(\theta_j) \subseteq \overline{D}_j \subseteq G_j$, and $\ker\theta_1(T) + \ker\theta_2(T) + \cdots + \ker\theta_n(T) = \mathscr{H}$ by Proposition 4.13. The corollary follows.

Exercises

1. Assume that $T \in \mathscr{L}(\mathscr{H})$ is an operator of class C_0 without cyclic vectors. Prove that there exists an invariant subspace $\mathscr{H}'$ for T such that, with the notation of Proposition 4.3, $m_T \not\equiv m_{T'} m_{T''}$.

2. For certain operators T of class C_0 the only hyperinvariant subspaces have the form $\ker\theta(T)$; a trivial example is $T = 0$. Is it always true that every hyperinvariant subspace of T has the form $\ker\theta(T)$?

3. Let $T \in \mathscr{L}(\mathscr{H})$ be an operator of class C_0. Show that there exist hyperinvariant subspaces $\mathscr{H}'$ and $\mathscr{H}''$ for T such that $\mathscr{H}' \cap \mathscr{H}'' = \{0\}$, $\mathscr{H}' \vee \mathscr{H}'' = \mathscr{H}$, $m_{T|\mathscr{H}'}$ is a Blaschke product, and $m_{T|\mathscr{H}''}$ is a singular inner function.

4. Show that the spaces $\mathscr{H}'$ and $\mathscr{H}''$ in the preceding problem are uniquely determined.

5. Let T, $\mathscr{H}$, $\mathscr{H}'$, $\mathscr{H}''$ be as in Exercise 3. Are $\mathscr{H}'$ and $\mathscr{H}''$ maximal spectral spaces?

6. Show that every maximal spectral space $\mathscr{M}$ of an operator T of class C_0 has the form $\ker\theta(T)$ for some maximal localization θ of m_T.

7. A vector $h \in \mathscr{H}$ is a *root vector* for $T \in \mathscr{L}(\mathscr{H})$ if $(\lambda I - T)^n h = 0$ for some $\lambda \in \mathbf{C}$ and $n \in \mathbf{N}$. Prove that the root vectors of an operator T of class C_0 span $\mathscr{H}$ if and only if m_T is a Blaschke product.

CHAPTER 3

Classification Theory

A Jordan block is a compression of the unilateral shift S of multiplicity one to a proper invariant subspace of S^*. A Jordan operator is a direct sum of Jordan blocks with certain additional properties. In this chapter we achieve a complete classification of operators of class C_0 into quasisimilarity classes by showing that each such class contains a unique Jordan operator. We begin with a detailed study of Jordan blocks in Section 1. An abstract characterization of Jordan blocks in terms of defect spaces is given, and the adjoint and the invariant and hyperinvariant subspaces are described. The existence of many maximal vectors for Jordan blocks is shown to have an important arithmetical consequence for H^∞ (see Theorem 1.14). Finally, the commutant lifting theorem is reinterpreted for the case of Jordan blocks, and it is shown to yield an identification of various related algebras. In Section 2 we study operators of class C_0 that have a cyclic vector, also called multiplicity-free operators. It turns out that each such operator is quasisimilar to a unique Jordan block, and this is a first step in the classification theorem. We also show that many of the properties of Jordan blocks, related to commutants and invariant subspaces, are inherited in some form by multiplicity-free operators. The main step in extending the classification results to operators with higher multiplicity is performed in Section 3. The main result is that a maximal cyclic invariant subspace for an operator of class C_0 has an approximate complement which is also invariant. An alternate characterization of multiplicity-free operators is obtained as a consequence. Jordan operators are introduced in Section 4. We show that two quasisimilar Jordan operators must coincide, and this is the uniqueness part of the classification theorem. The main part of the argument is linear algebraic in nature, and it uses the result in Section 1.3. An iterated application of the splitting procedure from Section 3 and a density argument yield a proof of the classification theorem for operators of class C_0 on separable spaces (Theorem 5.1). The case of operators acting on nonseparable spaces is treated next. This requires more detail about the separable case and some set-theoretical maneuvering. Some easy consequences of the classification theorem conclude Section 5. In Section 6 we show that the space on which an operator of class C_0 acts can be decomposed into cyclic invariant subspaces. The kind of decomposition that can be obtained is somewhat weaker than a direct sum decomposition. This parallels the fact that quasisimilarity is a weaker equivalence relation than similarity.

1. A case study: Jordan blocks. We recall the definition (2.1.17) of Jordan blocks. The *Jordan block* $S(\theta)$ associated with the inner function $\theta \in H^\infty$ acts on the space $\mathscr{H}(\theta) = H^2 \ominus \theta H^2$ as follows:

$$S(\theta) = P_{\mathscr{H}(\theta)} S \,|\, \mathscr{H}(\theta) \tag{1.1}$$

or, equivalently,

$$S(\theta)^* = S^* \,|\, \mathscr{H}(\theta). \tag{1.2}$$

Here, as before, the letter S is reserved for the usual unilateral shift on H^2. We clearly have $S(\theta) = S(\theta')$ whenever $\theta \equiv \theta'$. It is also clear that $S(\theta)$ acts on the trivial space $\{0\}$ if $\theta \equiv 1$, that is, if θ is constant. By Proposition 2.2.18 the Jordan block $S(\theta)$ is an operator of class C_0 and

$$m_{S(\theta)} \equiv \theta. \tag{1.3}$$

We recall the fact that $\mathscr{H}(\theta)$ is finite-dimensional if and only if θ is a finite Blaschke product. If θ is a finite Blaschke product then $\dim \mathscr{H}(\theta)$ is given by the number of Blaschke factors of θ.

1.4. LEMMA. *For every nonconstant inner function θ, S is the minimal isometric dilation of $S(\theta)$.*

PROOF. Since S is clearly an isometric dilation of $S(\theta)$, it suffices to show that $\bigvee_{n\geq 0} S^n \mathscr{H}(\theta) = H^2$. Now, the space $\mathscr{H} = \bigvee_{n\geq 0} S^n \mathscr{H}(\theta)$ is invariant for S and hence for $\theta(S)$. We conclude that

$$\mathscr{H} \supset \bigvee_{n\geq 0} \theta(S)^n \mathscr{H}(\theta) = \bigoplus_{n=0}^{\infty} (\theta^{n+1} H^2 \ominus \theta^n H^2) = H^2$$

and the lemma follows.

1.5. COROLLARY. *For every nonconstant inner function θ we have* $\dim(\mathscr{D}_{S(\theta)}) = \dim(\mathscr{D}_{S(\theta)^*}) = 1$.

PROOF. If $U_+ \in \mathscr{L}(\mathscr{K}_+)$ is the minimal isometric dilation of the contraction $T \in \mathscr{L}(\mathscr{H})$, it follows from the proof of Theorem 1.14 that $U_+ | \mathscr{K}_+ \ominus \mathscr{H}$ is a unilateral shift of multiplicity $\dim(\mathscr{D}_T)$. Now, S is a minimal unitary dilation of $S(\theta)$, and $S|H^2 \ominus \mathscr{H}(\theta) = S|\theta H^2$ is clearly a unilateral shift of multiplicity one. We conclude that $\dim(\mathscr{D}_{S(\theta)}) = 1$. Finally, $D^2_{S(\theta)^*} = I - S(\theta)S(\theta)^* = P_{\mathscr{H}(\theta)}(I - SS^*)|\mathscr{H}(\theta)$, so that $\dim(\mathscr{D}_{S(\theta)^*}) = \operatorname{rank}(D^2_{S(\theta)^*}) \leq \operatorname{rank}(I - SS^*) = 1$. Thus, it suffices to show that $D_{S(\theta)^*} \neq 0$. If we had $D_{S(\theta)^*} = 0$, $S(\theta)^*$ would be an isometry, and this contradicts the fact that $S(\theta)$ is of class C_0 and hence of class C_{00} by Corollary 2.4.2.

The preceding corollary has the following converse.

1.6. PROPOSITION. *Let $T \in \mathscr{L}(\mathscr{H})$ be an operator of class $C_{\cdot 0}$ such that $\dim(\mathscr{D}_{T^*}) = 1$. Then either*

(i) *T is unitarily equivalent to the shift S; or*

(ii) *T is unitarily equivalent to $S(\theta)$ for some nonconstant inner function θ.*

PROOF. As shown in §1.2, the minimal isometric dilation of T is unitarily equivalent to S. Thus we may assume that $\mathscr{H} \subset H^2$ is invariant for S^* and $T^* = S^*|\mathscr{H}$. The proposition obviously follows now from Beurling's classification of invariant subspaces for S. Indeed, either $\mathscr{H} = H^2$ or $\mathscr{H} \neq H^2$ and $H^2 \ominus \mathscr{H} = \theta H^2$ for some inner function θ. In this last case, $\mathscr{H} = \mathscr{H}(\theta)$ and $T = S(\theta)$. The function θ cannot be constant because $\dim \mathscr{H} \geq \dim \mathscr{D}_{T^*} = 1$.

1.7. COROLLARY. *For every inner function θ the adjoint $S(\theta)^*$ is unitarily equivalent to $S(\theta^\sim)$.*

PROOF. The operator $T = S(\theta)^*$ satisfies the conditions of Proposition 1.6 and T is not a shift because it is of class C_{00}. Therefore T is unitarily equivalent to $S(\theta')$ for some inner function θ'. We have then

$$\theta' \equiv m_{S(\theta')} = m_T \equiv \theta^\sim$$

by Proposition 2.4.1; the corollary follows.

We recall that a mapping A of a Hilbert space $\mathscr{H}$ into itself is antilinear if it is additive and $A(\lambda x) = \bar{\lambda} A(x)$ for $\lambda \in \mathbf{C}$ and $x \in \mathscr{H}$. If, in addition, A is isometric and onto, then A is called an antiunitary operator. Jordan blocks have the following interesting property.

1.8. PROPOSITION. *For every inner function $\theta \in H^\infty$ there exists an antiunitary operator J on $\mathscr{H}(\theta)$ such that $S(\theta)^* J = J S(\theta)$.*

PROOF. The antiunitary operator $J_0 : H^2 \to H^2$ defined by $J_0 u = u^\sim$, $u \in H^2$, obviously satisfies the relations $SJ_0 = J_0 S$, $J_0(\theta H^2) = \theta^\sim H^2$, and $J_0 \mathscr{H}(\theta) = \mathscr{H}(\theta^\sim)$. Consequently we have $S(\theta^\sim) J_1 = J_1 S(\theta)$ where $J_1 : \mathscr{H}(\theta) \to \mathscr{H}(\theta^\sim)$ is given by $J_0|\mathscr{H}(\theta)$. It suffices then to define J by the formula $J = RJ_1$ where R is any unitary operator such that $S(\theta)^* R = RS(\theta^\sim)$; the existence of such operators R is guaranteed by Corollary 1.7.

We know from §1.3 that $\mu_{S(\theta)} \leq \dim(\mathscr{D}_{S(\theta)^*}) = 1$ and thus $S(\theta)$ must have a cyclic vector. In fact, if we denote, somewhat ambiguously, by 1 the constant function in H^2 whose values are 1 everywhere, the vector $P_{\mathscr{H}(\theta)} 1$ is cyclic for $S(\theta)$. Indeed, 1 is cyclic for S and $S(\theta) P_{\mathscr{H}(\theta)} = P_{\mathscr{H}(\theta)} S$. We note the following formula:

$$P_{\mathscr{H}(\theta)} 1 = 1 - \overline{\theta(0)}\theta \tag{1.9}$$

whose proof we leave as an exercise. The operator $S(\theta)$ has many other cyclic vectors as we shall see shortly.

1.10. PROPOSITION. *Let θ be a nonconstant inner function.*

(i) *For every* $h \in \mathscr{H}(\theta)$ *we have* $m_h \equiv \theta/h \wedge \theta$.

(ii) *Every invariant subspace* $\mathscr{M}$ *of* $S(\theta)$ *has the form* $\phi H^2 \ominus \theta H^2$ *for some inner divisor* ϕ *of* θ. *We have* $\phi H^2 \ominus \theta H^2 = \ker(\theta/\phi)(S(\theta)) = \operatorname{ran}\phi(S(\theta))$.

(iii) *If* $\mathscr{M} = \phi H^2 \ominus \theta H^2$ *is an invariant subspace for* $S(\theta)$ *then* $S(\theta)|\mathscr{M}$ *is unitarily equivalent to* $S(\theta/\phi)$ *and the compression of* $S(\theta)$ *to* $\mathscr{H}(\theta) \ominus \mathscr{M} = \mathscr{H}(\phi)$ *coincides with* $S(\phi)$.

(iv) *A vector* $h \in \mathscr{H}(\theta)$ *is cyclic for* $S(\theta)$ *if and only if* $\theta \wedge h \equiv 1$.

PROOF. (i) Set $u = m_h$ and $v = \theta/h \wedge \theta$. We have $v(S(\theta))h = P_{\mathscr{H}(\theta)}v(S)h = P_{\mathscr{H}(\theta)}vh = P_{\mathscr{H}(\theta)}\theta(h/h \wedge \theta) = 0$ and consequently $u|v$. Conversely, we know that $u(S(\theta))h = 0$ so that $uh = \theta g$ for some $g \in H^2$. Since u divides θ it follows that $h = (\theta/u)g$ and so $(\theta/u)|h$. Since $(\theta/u)|\theta$ obviously, we have $(\theta/u)|h \wedge \theta$ or, equivalently, $v = (\theta/h \wedge \theta)|u$. We conclude that $v \equiv u$.

(ii) If $\mathscr{M}$ is invariant for $S(\theta)$ then $\mathscr{M} \oplus \theta H^2$ is invariant for S; indeed, if $x \in \mathscr{M}$, $Sx = S(\theta)x + P_{\theta H^2}Sx$. By Beurling's theorem there exists an inner function ϕ such that $\mathscr{M} \oplus \theta H^2 = \phi H^2$ and hence $\mathscr{M} = \phi H^2 \ominus \theta H^2$. Clearly ϕ divides θ since $\phi H^2 \supset \theta H^2$. Now, if $h \in H^2 \ominus \phi H^2$ clearly $\phi|h$ so that $\theta/h \wedge \theta|\theta/\phi$ and $(\theta/\phi)(S(\theta))h = 0$ by part (i). Conversely, if $(\theta/\phi)(S(\theta))h = 0$ we have $\theta/h \wedge \theta|\theta/\phi$, which implies that $\phi|h \wedge \theta$ and hence $\phi|h$. Thus $h \in \phi H^2 \cap \mathscr{H}(\theta) = \phi H^2 \ominus \theta H^2$. We have thus proved the equality $\phi H^2 \ominus \theta H^2 = \ker(\theta/\phi)(S(\theta))$. For the second equality we note that $\phi(S(\theta))\mathscr{H}(\theta) = P_{\mathscr{H}(\theta)}\phi(S)\mathscr{H}(\theta) = P_{\mathscr{H}(\theta)}\phi(S)H^2 = P_{\mathscr{H}(\theta)}\phi H^2 = \phi H^2 \ominus \theta H^2$ if $\phi|\theta$.

(iii) The space $\mathscr{H}(\phi) = \mathscr{H}(\theta) \ominus \mathscr{M}$ is invariant for $S(\theta)^*$ and $S(\theta)^*|\mathscr{H}(\phi) = (S^*|\mathscr{H}(\theta))|\mathscr{H}(\phi) = S^*|\mathscr{H}(\phi) = S(\phi)^*$. Thus the compression of $S(\theta)$ to $\mathscr{H}(\phi)$ is $S(\phi)$. Let us denote by T the restriction $S(\theta)|\mathscr{M}$. By Proposition 2.4.4, T is of class C_{00} and $m_T \equiv \theta/\phi$. Then Proposition 1.6 above (applied to T^*) shows that, in order to conclude that T is unitarily equivalent to $S(\theta/\phi)$, it suffices to show that $\dim \mathscr{D}_T = 1$. But we have

$$I - T^*T = P_{\mathscr{M}}(I - S(\theta)^*S(\theta))|\mathscr{M}$$

so that $\dim \mathscr{D}_T = \operatorname{rank}(I - T^*T) \leq \operatorname{rank}(I - S(\theta)^*S(\theta)) = 1$. The inequality $\dim \mathscr{D}_T \geq 1$ is clear if $\mathscr{M} \neq \{0\}$.

(iv) If h is cyclic then we must have $m_h \equiv m_{S(\theta)}$ so that we have $h \wedge \theta \equiv 1$ by (i). Conversely, if $h \wedge \theta \equiv 1$, (ii) shows that h does not belong to any proper invariant subspace of $S(\theta)$ and hence h is a cyclic vector. The proposition is proved.

1.11. COROLLARY. *Every invariant subspace of* $S(\theta)$ *is hyperinvariant.*

PROOF. This follows from the equality $\phi H^2 \ominus \theta H^2 = \ker(\theta/\phi)(S(\theta))$.

1.12. COROLLARY. *Let* θ *be an inner function and* $u \in H^\infty$. *Then* $\ker u(S(\theta)) = (\theta/u \wedge \theta)H^2 \ominus \theta H^2$, $(\operatorname{ran} u(S(\theta)))^- = (u \wedge \theta)H^2 \ominus \theta H^2$, $S(\theta)|\ker u(S(\theta))$ *is unitarily equivalent to* $S(u \wedge \theta)$, *and* $S(\theta)|(\operatorname{ran} u(S(\theta)))^-$ *is unitarily equivalent to* $S(\theta/u \wedge \theta)$.

PROOF. Remark 2.4.10 shows that $\ker u(S(\theta)) = \ker(u \wedge \theta)(S(\theta))$ and then Proposition 1.10(ii) shows that $\ker(u \wedge \theta)(S(\theta)) = (\theta/u \wedge \theta)H^2 \ominus \theta H^2$. By the remark quoted above we have

$$\begin{aligned}\mathscr{H}(\theta) \ominus (\operatorname{ran} u(S(\theta)))^- &= \ker u^\sim(S(\theta)^*) = \ker(u \wedge \theta)^\sim(S(\theta)^*) \\ &= \mathscr{H}(\theta) \ominus (\operatorname{ran}(u \wedge \theta)(S(\theta))).\end{aligned}$$

Thus Proposition 1.10(ii) implies that

$$(\operatorname{ran} u(S(\theta)))^- = \operatorname{ran}(u \wedge \theta)(S(\theta)) = (u \wedge \theta)H^2 \ominus \theta H^2.$$

The last two statements of the corollary obviously follow from Proposition 1.10(iii).

1.13. COROLLARY. *Let θ be an inner function. The set of cyclic vectors for $S(\theta)$ is a dense G_δ in $\mathscr{H}(\theta)$.*

PROOF. Proposition 1.10 implies that a vector $h \in \mathscr{H}(\theta)$ is cyclic if and only if $m_h \equiv \theta$. The corollary follows from Theorem 2.3.6.

Quite interestingly, Theorem 2.3.7 has the following consequence for the arithmetic of Hardy spaces. We will denote by l^1 the space of absolutely summable sequences of complex numbers.

1.14. THEOREM. *Let $\{f_j : j \geq 0\}$ be a bounded sequence of functions in H^2, and let θ be an inner function. The set of those sequences $\{a_j\}$ in l^1 satisfying the relation*

$$\left(\sum_{j=0}^{\infty} a_j f_j\right) \wedge \theta \equiv \left(\bigwedge_{j=0}^{\infty} f_j\right) \wedge \theta$$

is a dense G_δ in l^1.

PROOF. We may assume with no loss of generality that $(\bigwedge_{j=0}^\infty f_j) \wedge \theta \equiv 1$. Indeed, we may replace θ by θ/ϕ and each f_j by f_j/ϕ, where $\phi = (\bigwedge_{j=0}^\infty f_j) \wedge \theta$. Under this additional assumption, the invariant subspace for $S(\theta)$ generated by the vectors $\{P_{\mathscr{H}(\theta)} f_j : j \geq 0\}$ is $\mathscr{H}(\theta)$. Indeed, if the invariant subspace generated by $\{P_{\mathscr{H}(\theta)} f_j : j \geq 0\}$ is $\phi H^2 \ominus \theta H^2$ it follows that $\phi | (\bigwedge_{j=0}^\infty f_j) \wedge \theta$ so $\phi \equiv 1$. We can therefore apply Theorem 2.3.7 to the space $\mathscr{X} = l^1$ and the mapping $X : \mathscr{X} \to \mathscr{H}(\theta)$ defined by

$$X(\{a_j\}) = P_{\mathscr{H}(\theta)}\left(\sum_{j=0}^{\infty} a_j f_j\right), \quad \{a_j\} \in l^1.$$

Consequently the set of sequences $a \in l^1$ for which $m_{Xa} \equiv \theta$ is a dense G_δ in l^1. Finally, the condition $m_{Xa} \equiv \theta$ is equivalent to $Xa \wedge \theta \equiv 1$ and $Xa \wedge \theta \equiv 1$ if and only if $(\sum_{j=0}^\infty a_j f_j) \wedge \theta \equiv 1$, $a = \{a_j\}$.

We remark that a direct proof of the existence of one sequence $\{a_j\}$, satisfying the conditions of the preceding theorem, is quite tedious even in the case of a finite sequence $\{f_j : 1 \leq j \leq n\}$ of functions in H^2.

We conclude this section with a few facts describing various algebras of operators related to a Jordan block. We begin with a result about the operators intertwining two Jordan blocks. In order to simplify the statement we introduce the following notation. If $T \in \mathscr{L}(\mathscr{H})$ and $T' \in \mathscr{L}(\mathscr{H}')$, we denote by $\mathscr{I}(T', T)$ the set of all operators in $\mathscr{L}(\mathscr{H}, \mathscr{H}')$ intertwining T' and T:

$$\mathscr{I}(T, T') = \{X \in \mathscr{L}(\mathscr{H}, \mathscr{H}') : T'X = XT\}. \tag{1.15}$$

If $T' = T$ then $\mathscr{I}(T, T')$ coincides with the commutant $\{T\}'$.

1.16. THEOREM. *Let θ and θ' be two inner functions. For every operator $X \in \mathscr{I}(S(\theta), S(\theta'))$ there exists a function $u \in H^\infty$ such that $\theta'|u\theta$, $||u|| = ||X||$, and*

$$X = P_{\mathscr{H}(\theta')} u(S)|\mathscr{H}(\theta). \tag{1.17}$$

Conversely, every function $u \in H^\infty$ such that $\theta'|u\theta$ determines an operator $X \in \mathscr{I}(S(\theta), S(\theta'))$ defined by (1.17); we have $X = 0$ if and only if $\theta'|u$.

PROOF. The theorem is trivial if either θ or θ' is a constant inner function. We may then assume that θ or θ' are nonconstant inner functions and in this case S is a minimal isometric dilation for both $S(\theta)$ and $S(\theta')$. If $X \in \mathscr{I}(S(\theta'), S(\theta))$ we can apply the lifting Theorem 1.1.10 and thereby find an operator $Y \in \{S\}'$ such that $||Y|| = ||X||$,

$$Y(\theta H^2) \subset \theta' H^2, \tag{1.18}$$

and

$$X = P_{\mathscr{H}(\theta')} Y|\mathscr{H}(\theta). \tag{1.19}$$

The condition $Y \in \{S\}'$ implies that Y is an analytic Toeplitz operator, i.e., $Y = u(S)$ for some $u \in H^\infty$ with $||u|| = ||Y|| = ||X||$. In particular, (1.19) is seen to imply (1.17). Finally, the relation (1.18) implies that $u\theta \in \theta' H^2$ so that $\theta'|u\theta$.

Conversely, the condition $\theta'|u\theta$ implies that $u(S)(\theta H^2) \subset \theta' H^2$ and consequently $u(S)^*\mathscr{H}(\theta') \subset \mathscr{H}(\theta)$. Thus the operator X defined by (1.17) satisfies the relation $X^* = u(S)^*|\mathscr{H}(\theta')$, from which we infer

$$\begin{aligned} S(\theta)^* X^* &= (S^*|\mathscr{H}(\theta))(u(S)^*|\mathscr{H}(\theta')) = (S^* u(S)^*)|\mathscr{H}(\theta') \\ &= (u(S)^* S^*)|\mathscr{H}(\theta') = (u(S)^*|\mathscr{H}(\theta'))S(\theta')^* = X^* S(\theta')^* \end{aligned}$$

or, equivalently, $X \in \mathscr{I}(S(\theta), S(\theta'))$. Finally, the condition $X = 0$ means that $u(S)\mathscr{H}(\theta) \subset \theta' H^2$ and hence $uH^2 = u(S)H^2 = u(S)\mathscr{H}(\theta) + u(S)(\theta H^2) \subset \theta' H^2$. This inclusion implies that $\theta'|u$. That $X = 0$ when $\theta'|u$ is obvious and the proof is complete.

1.20. COROLLARY. *Let θ be an inner function. For every operator $X \in \{S(\theta)\}'$ there exists $u \in H^\infty$, $||u|| = ||X||$, such that $X = u(S(\theta))$. In particular, the commutant $\{S(\theta)\}'$ is isometrically isomorphic to the quotient algebra $H^\infty/\theta H^\infty$.*

PROOF. The existence of u clearly follows from Theorem 1.16; note that the condition $\theta|u\theta$ is satisfied trivially for every $u \in H^\infty$. The functional calculus determines an operator $\Phi : H^\infty/\theta H^\infty \to \{S(\theta)\}'$ of norm ≤ 1. The existence for each $X \in \{S(\theta)\}'$ of a function u with $||u|| = ||X||$ and $X = \Phi(u)$ implies that Φ is an isometry.

A consequence of Corollary 1.20 is that every element $\psi \in H^\infty/\theta H^\infty$ has a representative $u \in H^\infty$ such that $||u|| = ||\psi||$. In other words, every function $v \in H^\infty$ admits a best approximant from θH^∞.

Let us remark that θH^∞ is a weak*-closed subspace of H^∞. Indeed, it is easy to see that $u \in \theta H^\infty$ if and only if $\int_0^{2\pi} e^{int}\overline{\theta(e^{it})}u(e^{it})\,dt = 0$ for $n \geq 1$. It follows that $H^\infty/\theta H^\infty$ is itself the dual of some Banach space and the weak* topology on $H^\infty/\theta H^\infty$ is given by the quotient of the weak*-topology on H^∞. We conclude that the functional calculus $\Phi : H^\infty/\theta H^\infty \to \{S(\theta)\}'$ is continuous when $H^\infty/\theta H^\infty$ is given this weak* topology and $\{S(\theta)\}'$ the relative weak* topology (or ultraweak topology). It is easy to identify the space whose dual is $H^\infty/\theta H^\infty$. First, H^∞ is determined as a subspace of L^∞ by the equations $\int_0^{2\pi} e^{int}u(e^{it})\,dt = 0$, $n \geq 1$. Let H_0^1 denote the closed linear span in L^1 of the functions $\{e^{int} : n \geq 1\}$. Equivalently, $v \in H_0^1$ if and only if $v \in L^1$ and $\int_0^{2\pi} e^{int}v(e^{it})\,dt$, $n \geq 0$. Thus, in the duality between L^1 and L^∞, we have $H^\infty = (H_0^1)^\perp$ and similarly $\theta H^\infty = (\overline{\theta} H_0^1)^\perp$. It follows from the Hahn–Banach theorem that the space $H^\infty/\theta H^\infty$ is isometrically isomorphic to the dual of $\overline{\theta} H_0^1/H_0^1$. More precisely, the duality is given by the bilinear form

$$\langle u + \theta H^\infty, f + H_0^1\rangle = \frac{1}{2\pi}\int_0^{2\pi} u(e^{it})f(e^{it})\,dt; \quad u \in H^\infty, \ f \in \overline{\theta} H_0^1.$$

In order to state the next result we introduce some additional notation. If $T \in \mathscr{L}(\mathscr{H})$, we denote by $\mathscr{A}_T$ the weak*-closed subalgebra of $\mathscr{L}(\mathscr{H})$ generated by T and I, and we denote by $\mathscr{W}_T$ the weakly closed subalgebra of $\mathscr{L}(\mathscr{H})$ generated by T and I. Since $\{T\}'$ is always a weakly closed algebra we have $\mathscr{A}_T \subset \mathscr{W}_T \subset \{T\}'$.

1.21. PROPOSITION. *Let θ be an inner function.*

(i) *We have $\mathscr{A}_{S(\theta)} = \mathscr{W}_{S(\theta)} = \{S(\theta)\}' = \{u(S(\theta)) : u \in H^\infty\}$.*

(ii) *The functional calculus $\Phi : H^\infty/\theta H^\infty \to \mathscr{A}_{S(\theta)}$ is a homeomorphism with respect to the corresponding weak* topologies.*

(iii) *The weak* topology and the weak operator topology coincide on $\mathscr{A}_{S(\theta)}$.*

(iv) *For every weak* continuous functional ϕ on $\mathscr{A}_{S(\theta)}$ and every $\varepsilon > 0$ there exist vectors $x, y \in \mathscr{H}(\theta)$ such that $||x|| \leq (1+\varepsilon)||\phi||^{1/2}$, $||y|| \leq (1+\varepsilon)||\phi||^{1/2}$, and $\phi(A) = (Ax, y)$ for $A \in \mathscr{A}_{S(\theta)}$.*

PROOF. We already know that the functional calculus is continuous in the weak* topologies. Since H^∞ is the weak* closure of polynomials we clearly have

$$\{u(S(\theta)) : u \in H^\infty\} \subset \mathscr{A}_{S(\theta)}.$$

On the other hand we have $\{S(\theta)\}' \subset \{u(S(\theta)) : u \in H^\infty\}$ by the preceding corollary and this, combined with the trivial inclusions $\mathscr{W}_{S(\theta)} \subset \mathscr{A}_{S(\theta)} \subset \{S(\theta)\}'$, gives (i).

We proceed now to proving (iv). If ϕ is a weak* continuous functional on $\mathscr{A}_{S(\theta)}$ then $\phi \circ \Phi$ is weak* continuous on $H^\infty/\theta H^\infty$. We deduce the existence of a function $f \in \overline{\theta} H_0^1$ such that

$$\phi(u(S(\theta))) = \frac{1}{2\pi}\int_0^{2\pi} u(e^{it})f(e^{it})\,dt, \quad u \in H^\infty, \tag{1.22}$$

and $\|f\|_1 < (1+\varepsilon)^2\|\phi\|$. The factorization of the function $\theta f \in H_0^1$ yields functions $g \in H^2$ and $h \in H_0^2$ (i.e., $h(0) = 0$) such that $\theta f = gh$, or $f = g(\overline{\theta} h)$, and allows us to write (1.22) as

$$\phi(u(S(\theta))) = (ug, \theta\bar{h}) = (ug, P_{H^2}(\theta\bar{h})), \tag{1.23}$$

where the scalar product is computed in L^2; indeed, $ug \in H^2$. Now, $h \in H_0^2$ and therefore $\theta\bar{h}$ is orthogonal onto θH^2. It follows that the vector $y = P_{H^2}(\theta\bar{h})$ belongs to $\mathscr{H}(\theta)$ and (1.23) can be transformed into

$$\begin{aligned}\phi(u(S(\theta))) &= (ug, y) = (u(S)g, y) = (g, u(S)^*y)\\ &= (g, u(S(\theta))^*y) = (x, u(S(\theta))^*y)\\ &= (u(S(\theta))x, y),\end{aligned}$$

where $x = P_{\mathscr{H}(\theta)}g$. Thus (iv) is proved and (iii) follows at once. A quick look at the above proof of (iv) shows that the construction can be carried out starting with an arbitrary function $f \in \overline{\theta} H_0^1$. This clearly shows that the mapping $u(S(\theta)) \to u + \theta H^\infty$ is continuous from the weak topology of $\mathscr{A}_{S(\theta)}$ to the weak* topology of $H^\infty/\theta H^\infty$. This observation, combined with (iii) and the weak* continuity of Φ, shows that (ii) is true and thus concludes the proof.

Exercises

1. Show that, for every $\mu \in \mathbf{D}$, the operator $(\mu I - S)(I - \bar{\mu}S)^{-1}$ is unitarily equivalent to S.

2. Show that, for every $\mu \in \mathbf{D}$, $(\mu I - S(\theta))(I - \bar{\mu}S(\theta))^{-1}$ is unitarily equivalent to $S(\theta')$, where $\theta'(\lambda) = \theta((\mu+\lambda)/(1+\bar{\mu}\lambda))$, $\lambda \in \mathbf{D}$.

3. Consider the inner function $\theta(\lambda) = \lambda^n$, $n \geq 1$. Find an orthonormal basis of $\mathscr{H}(\theta)$ in which the matrix of $S(\theta)$ is lower triangular. What is the matrix of $S(\theta)$ in this basis?

4. Use Exercises 2 and 3 to find the matrix of $S(\theta)$, $\theta(\lambda) = ((\lambda+\mu)/(1+\bar{\mu}\lambda))^n$, in a certain orthonormal basis.

5. Let θ be an inner function and define $R \in \mathscr{L}(L^2)$ by $(Rf)(e^{it}) = \theta^{\sim}(e^{it})f(e^{-it})e^{-it}$, $f \in L^2$. Prove that R is unitary, $R\mathscr{H}(\theta) = \mathscr{H}(\theta^{\sim})$, and $S(\theta^{\sim})U = US(\theta)^*$, where $U \in \mathscr{L}(\mathscr{H}(\theta), \mathscr{H}(\theta^{\sim}))$ is defined by $U = R|\mathscr{H}(\theta)$.

6. Assume that $X \in \{S(\theta)\}'$, $||X|| = 1$, and X attains its norm. Show that there exists a unique function $u \in H^\infty$ such that $u(S(\theta)) = X$ and $||u|| = 1$; in addition, u is an inner function. More precisely, if $||Xg|| = ||g|| = 1$, $g \in \mathscr{H}(\theta)$, then $u = Xg/g$.

7. Prove that the only isometries in $\{S(\theta)\}'$ are scalar multiples of I.

8. Let $V \in \mathscr{L}(\mathscr{H}(\theta), \mathscr{H}(\theta^\sim))$ be a unitary operator satisfying the relation $S(\theta^\sim)V = VS(\theta)^*$. Show that V is a scalar multiple of the operator U constructed in Exercise 5.

9. Prove that the antiunitary operator J constructed in Proposition 1.8 is uniquely determined up to a scalar multiple of absolute value one.

10. Prove formula (1.9).

11. Let ϕ be an inner divisor of θ. Show that the operator $\phi(S(\theta))$ is a partial isometry with initial space $\mathscr{H}(\theta/\phi)$ and final space $\phi H^2 \ominus \theta H^2$. This partial isometry provides a unitary equivalence between $S(\theta/\phi)$ and $S(\theta)|\phi H^2 \ominus \theta H^2$.

12. Prove that $u(S(\theta))$ is an invertible operator if and only if $uH^\infty + \theta H^\infty = H^\infty$.

13. Show that $S(\theta)$ is an irreducible operator.

14. Show that $u(S(\theta))$ has closed range if and only if $(u/\phi)H^\infty + (\theta/\phi)H^\infty = H^\infty$, where $\phi \equiv u \wedge \theta$.

15. Let ϕ be an inner divisor of θ, and $\phi' = \theta/\phi$. Assume that $\phi H^\infty + \phi' H^\infty = H^\infty$ so that there are $u, u' \in H^\infty$ satisfying $\phi u + \phi' u' = 1$. Show that the operator $P = (\phi u)(S(\theta))$ is a (nonselfadjoint) projection.

16. Show that every projection $P \in \{S(\theta)\}'$ has the form $P = (\phi u)(S(\theta))$ where ϕ and u are as in Exercise 15.

17. Prove that $S(\theta\theta')$ is similar to $S(\theta) \oplus S(\theta')$ if and only if $\theta H^\infty + \theta' H^\infty = H^\infty$.

18. Show that $\mathscr{H}(\theta)$ and $\mathscr{H}(\theta')$ have positive angle (i.e., there exists a projection P such that $\operatorname{ran} P = \mathscr{H}(\theta)$ and $\ker P \supset \mathscr{H}(\theta')$) if and only if $\theta H^\infty + \theta' H^\infty = H^\infty$.

19. Use Jordan blocks to show that $\mathscr{H}(\theta)$ is finite-dimensional if and only if θ is a finite Blaschke product.

2. Multiplicity-free operators. We recall that an operator T is said to be multiplicity-free if $\mu_T = 1$, i.e., if T has a cyclic vector. The adjoint of a multiplicity-free operator T is not usually multiplicity-free; our first task will be to show that T^* is multiplicity-free if T is of class C_0 and $\mu_T = 1$. We start with two preliminary results.

2.1. LEMMA. *Let $T \in \mathscr{L}(\mathscr{H})$ and $T' \in \mathscr{L}(\mathscr{H}')$ be two completely nonunitary contractions such that $T \prec T'$. Then T is of class C_0 if and only if T' is of class C_0. Moreover, $m_T \equiv m_{T'}$ if T and T' are of class C_0.*

PROOF. If $X \in \mathscr{I}(T,T')$ then $u(T')X = Xu(T)$ for every $u \in H^\infty$. If, in addition, X is a quasiaffinity, it follows that $u(T) = 0$ if and only if $u(T') = 0$ and the lemma obviously follows from these observations.

2.2. PROPOSITION. *Let T be an operator of class C_0. If T is multiplicity-free then $S(m_T) \prec T$; if T^* is multiplicity-free then $T \prec S(m_T)$.*

PROOF. Assume first that T is multiplicity-free. It follows from §1.3 that there exists an operator T' of class $C_{\cdot 0}$ such that $T' \prec T$ and $\dim(\mathscr{D}_{T'^*}) = \mu_T = 1$. The operator T' is of class C_0 by Lemma 2.1 and therefore it cannot be unitarily equivalent to S. Then Proposition 1.6 shows that T' is unitarily equivalent to $S(\theta)$ for some inner function θ; thus $S(\theta) \prec T$. Finally we have $\theta \equiv m_{S(\theta)} \equiv m_T$ and so $S(m_T) \prec T$. If T^* is multiplicity-free it follows by what we have just proved that $S(m_{\tilde{T}}) = S(m_{T^*}) \prec T^*$ and therefore $T \prec S(m_{\tilde{T}})^*$. The operator $S(m_{\tilde{T}})^*$ is unitarily equivalent to $S(m_T)$ by Corollary 1.7. The proposition is proved.

2.3. THEOREM. *For every operator T of class C_0, the following conditions are equivalent:*

(i) *T is multiplicity-free;*
(ii) *T^* is multiplicity-free; and*
(iii) *T is quasisimilar to $S(m_T)$.*

PROOF. It will suffice to prove that (ii)⇒(i). Indeed, it would then follow that (i)⇒(ii) by symmetry. Further, if (i) and (ii) are satisfied, $T \sim S(m_T)$ by Proposition 2.2. Vice versa, if $S(m_T) \prec T$ then $\mu_T \le \mu_{S(m_T)} = 1$ and (i) follows.

Assume therefore that $T \in \mathscr{L}(\mathscr{H})$ and T^* is multiplicity-free. By virtue of Proposition 2.2 we can choose a quasiaffinity X such that

$$S(m_T)X = XT. \tag{2.4}$$

Let $h \in \mathscr{H}$ be a T-maximal vector for T, i.e., $m_h \equiv m_T$; such vectors exist by Theorem 2.3.6. If we denote by $\mathscr{K}$ the cyclic space $\bigvee_{n=0}^{\infty} T^n h$ generated by h then $T\mathscr{K} \subset \mathscr{K}$ and $m_{T|\mathscr{K}} \equiv m_T$. Thus the operator $T|\mathscr{K}$ is multiplicity-free and a second application of Proposition 2.2 yields an injective operator $Y : \mathscr{H}(m_T) \to \mathscr{K}$ such that $Y\mathscr{H}(\theta)$ is dense in $\mathscr{K}$ and

$$TY = YS(m_T). \tag{2.5}$$

Relations (2.4) and (2.5) show that $XY \in \{S(m_T)\}'$ and, of course, XY is injective. We can then apply Corollary 1.20 and Proposition 2.4.9 to deduce the existence of a function u in H^∞ such that

$$XY = u(S(m_T)) \tag{2.6}$$

and $u \wedge m_T \equiv 1$. A calculation based on (2.4) and (2.6) shows that

$$\begin{aligned} X(YX - u(T)) = XYX - Xu(T) &= XYX - u(S(m_T))X \\ &= (XY - u(S(m_T)))X = 0 \end{aligned}$$

and since X is one-to-one we must have $YX = u(T)$. A second application of Proposition 2.4.9 shows that $u(T)$ is a quasiaffinity; indeed, $u \wedge m_T \equiv 1$. In particular, $\mathscr{H} = (u(T)\mathscr{H})^- \subset (Y\mathscr{H}(m_T))^- \subset \mathscr{K}$ so that $\mathscr{K} = \mathscr{H}$ and consequently h is a cyclic vector for T. The theorem is proved.

The above argument actually proves more. The following result is an immediate consequence of this argument.

2.7. COROLLARY. *Let $T \in \mathscr{L}(\mathscr{H})$ be a multiplicity-free operator of class C_0. Then a vector $h \in \mathscr{H}$ is cyclic for T if and only if h is T-maximal. In particular, the set of cyclic vectors for T is a dense G_δ in $\mathscr{H}$.*

2.8. COROLLARY. *Every restriction of a multiplicity-free operator of class C_0 to an invariant subspace is multiplicity-free.*

PROOF. Let T be a multiplicity-free operator and $\mathscr{K}$ an invariant subspace for T. If h is cyclic for T^* then $P_{\mathscr{K}}h$ is cyclic for $(T|\mathscr{K})^*$. Thus $(T|\mathscr{K})^*$, and hence $T|\mathscr{K}$, is multiplicity-free.

In fact the invariant subspaces of a multiplicity-free operator of class C_0 have a classification, analogous to that of invariant subspaces for Jordan blocks. We need first some facts about intertwinings between multiplicity-free operators of class C_0.

2.9. PROPOSITION. *Let T and T' be two multiplicity-free operators of class C_0 such that $m_T \equiv m_{T'}$. Then an operator in $\mathscr{I}(T,T')$ is one-to-one if and only if it has dense range.*

PROOF. If $\theta \equiv m_T$ it follows from Theorem 2.3 that we can find quasiaffinities $X \in \mathscr{I}(T', S(\theta))$ and $Y \in \mathscr{I}(S(\theta), T)$. If A is any operator in $\mathscr{I}(T,T')$ the product XAY commutes with $S(\theta)$ and hence Corollary 1.20 yields a function $u \in H^\infty$ such that

$$XAY = u(S(\theta)).$$

If A is either one-to-one or has dense range then $u(S(\theta))$ has the same property and therefore $u \wedge \theta \equiv 1$ by Proposition 2.4.9. The relations

$$AYX = u(T'), \qquad YXA = u(T) \tag{2.10}$$

are then proved as in the proof of Theorem 2.3. For example,

$$\begin{aligned} X(AYX - u(T')) = XAYX - Xu(T') &= XAYX - u(S(\theta))X \\ &= (XAY - u(S(\theta)))X = 0 \end{aligned}$$

and the first relation in (2.10) follows because X is one-to-one. Finally, we infer from (2.10) that $\operatorname{ran} A \supset \operatorname{ran} u(T')$ and $\ker A \subset \ker u(T)$. The condition $u \wedge \theta \equiv 1$ implies that $u(T)$ and $u(T')$ are quasiaffinities by Proposition 2.4.9 and hence A is necessarily a quasiaffinity. The proposition follows.

An interesting case of Proposition 2.9 occurs when $T = T'$. We recall that for an arbitrary completely nonunitary contraction T the closed operator

$$(u/v)(T) = v(T)^{-1}u(T)$$

is defined whenever $u \in H^\infty$ and $v \in K_T^\infty$. We will denote by $\mathscr{F}_T$ the set of bounded operators of the form $(u/v)(T)$.

2.11. PROPOSITION. *If T is a multiplicity-free operator of class C_0 then $\{T\}' = \mathscr{F}_T$.*

PROOF. Let θ, X, and Y be as in the proof of Proposition 2.9, with $T' = T$. If $A = I$, that proof implies the existence of $v \in H^\infty$ such that $v \wedge \theta \equiv 1$ and $YX = v(T)$; indeed we have only to apply relation (2.10) for $A = I$. Now, if A is arbitrary, we deduce the existence of $u \in H^\infty$ such that $YXA = u(T)$ or $v(T)A = u(T)$. This relation means that $A = (u/v)(T)$ and the proposition is proved.

2.12. REMARK. The preceding proof shows more than the equality $\{T\}' = \mathscr{F}_T$. It shows that there exists a function $v \in H^\infty$ such that $v \wedge m_T \equiv 1$ and every $A \in \{T\}'$ can be written as $A = (u/v)(T)$ for some $u \in H^\infty$. We will show later that the converse of Proposition 2.11 is also true.

We can now give the promised description of invariant subspaces for multiplicity-free operators.

2.13. THEOREM. *For every operator T of class C_0 the following assertions are equivalent:*

(i) *T is multiplicity-free;*
(ii) *For every inner divisor θ of m_T there exists a unique invariant subspace $\mathscr{K}$ for T such that $m_{T|\mathscr{K}} \equiv \theta$;*
(iii) *There do not exist distinct invariant subspaces $\mathscr{K}$ and $\mathscr{K}'$ for T such that $T|\mathscr{K} \prec T|\mathscr{K}'$; and*
(iv) *There do not exist proper invariant subspaces $\mathscr{K}$ for T such that $m_{T|\mathscr{K}} \equiv m_T$.*

If T is multiplicity-free then the unique invariant subspace in (ii) *is given by $\mathscr{K} = \ker \theta(T) = (\operatorname{ran}(m_T/\theta)(T))^-$.*

PROOF. Assume that $T \in \mathscr{L}(\mathscr{H})$ is multiplicity-free, $\mathscr{K}$ is an invariant subspace for T, and $\theta \equiv m_{T|\mathscr{K}}$. The operators $T' = T|\mathscr{K}$ and $T'' = T|\ker\theta(T)$ are multiplicity-free by Corollary 2.8 and they satisfy the relation $T''J = JT'$, where $J : \mathscr{K} \to \ker\theta(T)$ is the inclusion operator. By Proposition 2.9, J must have dense range so that $\mathscr{K} = J\mathscr{K} = \ker\theta(T)$. Thus we proved that (i)$\Rightarrow$(ii). The implication (ii)$\Rightarrow$(iv) is obvious. Assume now that (iv) holds and h is a T-maximal vector. If we set $\mathscr{K} = \bigvee_{n=0}^\infty T^n h$ then $m_{T|\mathscr{K}} = m_T$ and therefore $\mathscr{K} = \mathscr{H}$ by (iv). We conclude that h is a cyclic vector and thus (iv)$\Rightarrow$(i). The implication (iii)$\Rightarrow$(i) is proved in a similar fashion. Indeed, if h and h' are T-maximal vectors, $\mathscr{K} = \bigvee_{n=0}^\infty T^n h$, $\mathscr{K}' = \bigvee_{n=0}^\infty T^n h'$ then $T|\mathscr{K}$ and $T|\mathscr{K}'$ are multiplicity-free and have the same minimal function. By Theorem 2.3 we deduce, using the transitivity of quasisimilarity, then $T|\mathscr{K} \prec T|\mathscr{K}'$. If (iii) holds then we must have $\mathscr{K}' = \mathscr{K}$ and, in particular, $h' \in \mathscr{K}$. We conclude that $\mathscr{K}$ contains every T-maximal vector and hence $\mathscr{K} = \mathscr{H}$ since the set of

T-maximal vectors is dense. Thus (iii) implies that h is a cyclic vector. Finally, the implication (ii)$\Rightarrow$(iii) is obvious because $T|\mathscr{K} \prec T|\mathscr{K}'$ implies, in particular, that $m_{T|\mathscr{K}} \equiv m_{T|\mathscr{K}'}$. The last assertion of the theorem follows because both $T|\ker\theta(T)$ and $T|(\operatorname{ran}(m_T/\theta)(T))^-$ have minimal function θ. The theorem is proved.

2.14. COROLLARY. *Every invariant subspace of a multiplicity-free operator of class C_0 is hyperinvariant.*

Exercises

1. Let $\{\theta_j : j \geq 0\}$ be a sequence of inner functions. Prove that $\bigoplus_{j=0}^{\infty} S(\theta_j)$ is an operator of class C_0 if and only if $\{\theta_j : j \geq 0\}$ has a common (nonzero) multiple in H^∞.

2. Assume that $T = \bigoplus_{j=0}^{\infty} S(\theta_j)$ is of class C_0. Show that T is multiplicity-free if and only if $\theta_i \wedge \theta_j \equiv 1$ for all $i, j \geq 0$, $i \neq j$.

3. Exercise 2 implies that $S(\theta) \oplus S(\theta')$ is quasisimilar to $S(\theta\theta')$ if and only if $\theta \wedge \theta' \equiv 1$. Show that quasisimilarity cannot generally be replaced by the stronger relation of similarity.

4. Assume that $\theta \wedge \theta' \equiv 1$ and construct explicit quasiaffinities intertwining the operators $S(\theta) \oplus S(\theta')$ and $S(\theta\theta')$.

5. Let T be a multiplicity-free operator of class C_0 and $X \in \{T\}'$. Prove that $T|\ker X \sim (T^*|\ker X^*)^*$.

6. Let T be a multiplicity-free operator of class C_0. Show that there exists a function $v \in H^\infty$, $v \wedge m_T \equiv 1$, with the following property. For every weak* continuous functional ϕ on $\mathscr{A}_T$ there exist vectors h, g such that $\phi(v(T)A) = (Ah, g)$ for every $A \in \mathscr{A}_T$.

7. Is it true for every C_0 operator T and every divisor θ of m_T that $\ker\theta(T) = (\operatorname{ran}(m_T/\theta)(T))^-$?

8. Let T be a multiplicity-free operator. Is it always true that every invariant subspace for T is hyperinvariant?

9. Let T be a multiplicity-free operator of class C_0 and $u \in H^\infty$. Prove that $T|\ker u(T)$ is quasisimilar to $S(u \wedge m_T)$.

3. The splitting principle. In this section we start the study of operators of class C_0 with arbitrary multiplicity. The splitting theorem below already suggests how the classification of general operators of class C_0 will look.

3.1. THEOREM. *Let $T \in \mathscr{L}(\mathscr{H})$ be an operator of class C_0, $h \in \mathscr{H}$ a T-maximal vector, $\mathscr{K} = \bigvee_{n=0}^{\infty} T^n h$. Then there exists an invariant subspace $\mathscr{M}$ for T such that $\mathscr{K} \vee \mathscr{M} = \mathscr{H}$ and $\mathscr{K} \cap \mathscr{M} = \{0\}$.*

PROOF. The operator $T_1 = T|\mathscr{K}$ is multiplicity-free and by virtue of Theorem 2.3 there exists a vector $k \in \mathscr{K}$ cyclic for T_1^*. Let us set

$$\mathscr{K}' = \bigvee_{n=0}^{\infty} T^{*^n} k, \qquad \mathscr{M} = \mathscr{H} \ominus \mathscr{K}', \tag{3.2}$$

and define $T_2 \in \mathscr{L}(\mathscr{K}')$ by $T_2^* = T^*|\mathscr{K}'$. Since $\mathscr{K}'$ is invariant for T^*, we have $T_2 P_{\mathscr{K}'} = P_{\mathscr{K}'} T$, and hence the operator $X \in \mathscr{L}(\mathscr{K}, \mathscr{K}')$ defined by $X = P_{\mathscr{K}'}|\mathscr{K}$ satisfies the relation

$$T_2 X = X T_1. \tag{3.3}$$

The idea is to apply Proposition 2.9 to the multiplicity-free operators T_1 and T_2, and therefore it is important to know that $m_{T_1} \equiv m_{T_2}$. We first note that $(\operatorname{ran} X^*)^- = \bigvee_{n=0}^{\infty} X^* T_2^{*^n} k = \bigvee_{n=0}^{\infty} T_1^{*^n} X^* k = \bigvee_{n=0}^{\infty} T_1^{*^n} k = \mathscr{K}$, which implies that X^* has dense range. Thus, if $\theta \equiv m_{T_2}$, we have

$$(\theta(T_1))^* X^* = X^* (\theta(T_2))^* = 0,$$

from which it follows that $\theta(T_1) = 0$. Thus $m_T = m_{T_1}|\theta$ and since $\theta | m_T$ by Proposition 2.4.3 we deduce that $\theta \equiv m_{T_1}$. Proposition 2.9 now implies that X must be a quasiaffinity and the theorem follows from the relations

$$\mathscr{K} \cap \mathscr{M} = \mathscr{K} \cap \ker P_{\mathscr{K}'} = \ker X,$$

and

$$\mathscr{H} \ominus (\mathscr{K} \vee \mathscr{M}) = (\mathscr{H} \ominus \mathscr{K}) \cap \mathscr{K}' = \mathscr{K}' \cap \ker P_{\mathscr{K}} = \ker X^*.$$

3.4. REMARK. If $\mathscr{M}' = \mathscr{H} \ominus \mathscr{K}$ then $\mathscr{M}'$ is invariant for T^*, $\mathscr{K}' \vee \mathscr{M}' = \mathscr{H}$, and $\mathscr{K}' \cap \mathscr{M}' = \{0\}$. This simple observation is refined in Theorem 3.6 below.

3.5. COROLLARY. *Let T be an operator of class C_0. There exists another operator T' of class C_0 such that $S(m_T) \oplus T' \prec T$.*

PROOF. Let $\mathscr{K}$ and $\mathscr{M}$ be as in Theorem 3.1. Then we have $(T|\mathscr{K}) \oplus (T|\mathscr{M}) \prec T$ where the quasiaffinity $X : \mathscr{K} \oplus \mathscr{M} \to \mathscr{H}$ is defined by $X(u \oplus v) = u + v$. The corollary follows now because $T|\mathscr{K}$ is quasisimilar to $S(m_T)$.

Of course this corollary can be applied to T^* and this yields an operator T'' of class C_0 such that $T \prec S(m_T) \oplus T''$. A careful look at Remark 3.4 will show that T'' can be chosen so that $T' \prec T''$ (take $T'' = (T^*|\mathscr{M}')^*$). It is not clear now (but it is true) that T' and T'' are quasisimilar so that we may replace quasiaffine transform by quasisimilarity in Corollary 3.5.

The following refinement of Theorem 3.1 is useful when dealing with quasiaffine transforms.

3.6. THEOREM. *Let $T \in \mathscr{L}(\mathscr{H})$ and $T' \in \mathscr{L}(\mathscr{H}')$ be operators of class C_0, $X \in \mathscr{I}(T, T')$ a quasiaffinity, h a T-maximal vector, $\varepsilon > 0$, $\mathscr{K} = \bigvee_{n=0}^{\infty} T^n h$, and $\mathscr{M}' = \mathscr{H}' \ominus (X\mathscr{K})^-$. There exist subspaces $\mathscr{M}$ of $\mathscr{H}$ and $\mathscr{K}'$ of $\mathscr{H}'$ such that:*

(i) *$\mathscr{M}$ is invariant for T and $\mathscr{K}'$ is invariant for T'^*;*

(ii) $||P_{\mathcal{K}'}Xh - Xh|| < \varepsilon$;
(iii) $\mathcal{M} = \mathcal{H} \ominus (X^*\mathcal{K}')^-$;
(iv) $\mathcal{K} \vee \mathcal{M} = \mathcal{H}$, $\mathcal{K} \cap \mathcal{M} = \{0\}$, $\mathcal{K}' \vee \mathcal{M}' = \mathcal{H}'$, $\mathcal{K}' \cap \mathcal{M}' = \{0\}$; *and*
(v) $P_{\mathcal{K}'}X|\mathcal{K}$ *and* $P_{\mathcal{M}'}X|\mathcal{M}$ *are quasiaffinities.*

PROOF. The operator $T_1' = T'|(X\mathcal{K})^-$ is multiplicity-free and $m_{T_1'} \equiv m_{T'} \equiv m_T$. Indeed, if $\theta \equiv m_{T_1'}$ then $X\theta(T)h = \theta(T_1')Xh = 0$ so $\theta(T)h = 0$ because X is one-to-one. We conclude that $m_T = m_h|\theta$ and the divisibility of $m_T \equiv m_{T'}$ by θ is obvious. Thus the construction in Theorem 3.1 can be done with T_1' in the place of T_1. This time we will take advantage of the density of cyclic vectors for $T_1'^*$ and choose such a cyclic vector k that

$$||Xh - k|| < \varepsilon. \tag{3.7}$$

We can then define $\mathcal{K}' = \bigvee_{n=0}^{\infty} T'^{*^n} k$ and $\mathcal{M} = \mathcal{H} \ominus (X^*\mathcal{K}')^-$; Remark 3.4 shows then that $\mathcal{K}' \vee \mathcal{M}' = \mathcal{H}'$ and $\mathcal{K}' \cap \mathcal{M}' = \{0\}$. As in the proof of Theorem 3.1 the operator T_2' defined by $T_2'^* = T'^*|\mathcal{K}'$ is multiplicity-free, has minimal function m_T, and $P_{\mathcal{K}'}|(X\mathcal{K})^-$ is a quasiaffinity. Since $X|\mathcal{K} : \mathcal{K} \to (X\mathcal{K})^-$ is obviously a quasiaffinity, it follows that

$$P_{\mathcal{K}'}X|\mathcal{K} = (P_{\mathcal{K}'}|(X\mathcal{K})^-)(X|\mathcal{K})$$

is also a quasiaffinity.

By (3.7) we certainly have

$$||P_{\mathcal{K}'}Xh - Xh|| \leq ||k - Xh|| < \varepsilon.$$

We have already verified (i), (ii), (iii), the second half of (iv), and the first half of (v). In order to conclude the proof we need to play a little more with Proposition 2.9. The operators T_1 and T_2 defined by

$$T_1 = T|\mathcal{K}, \qquad T_2^* = T^*|(X^*\mathcal{K}')^-$$

are multiplicity-free and have minimal function m_T. For T_1 this is obvious, while for T_2 we have $T_2'^* \prec T_2^*$ via the quasiaffinity $X^*|\mathcal{K}' : \mathcal{K}' \to (X^*\mathcal{K}')^-$. We already know that the operator

$$P_{\mathcal{K}}X^*|\mathcal{K}' = (P_{\mathcal{K}'}X|\mathcal{K})^*$$

is a quasiaffinity. Since

$$P_{\mathcal{K}}X^*|\mathcal{K}' = (P_{\mathcal{K}}|(X^*\mathcal{K}')^-)(X^*|\mathcal{K}')$$

we deduce that $P_{\mathcal{K}}|(X^*\mathcal{K}')^-$ has dense range in $\mathcal{K}$. Since we also have $P_{\mathcal{K}}|(X^*\mathcal{K}')^- \in \mathcal{I}(T_2^*, T_1^*)$, Proposition 2.9 implies that $Y = P_{\mathcal{K}}|(X^*\mathcal{K}')^-$ is a quasiaffinity and hence

$$\mathcal{K} \cap \mathcal{M} = \ker Y^* = \{0\},$$

and

$$\begin{aligned}\mathcal{H} \ominus (\mathcal{K} \vee \mathcal{M}) = (\mathcal{H} \ominus \mathcal{K}) \cap (\mathcal{H} \ominus \mathcal{M}) &= (\mathcal{H} \ominus \mathcal{K}) \cap (X^*\mathcal{K}')^- \\ &= \ker Y = \{0\}.\end{aligned}$$

It is easy now to show that $P_{\mathcal{M}'}X|\mathcal{M}$ is a quasiaffinity:

$$((P_{\mathcal{M}'}X)\mathcal{M})^- = ((P_{\mathcal{M}'}X)(\mathcal{M}+\mathcal{K}))^- = ((P_{\mathcal{M}'}X)(\mathcal{M}\vee\mathcal{K}))^-$$
$$= (P_{\mathcal{M}'}X\mathcal{H})^- = P_{\mathcal{M}'}\mathcal{H}' = \mathcal{M}'$$

and, analogously,

$$((P_{\mathcal{M}}X^*)\mathcal{M}')^- = ((P_{\mathcal{M}}X^*)(\mathcal{M}'\vee\mathcal{K}'))^- = (P_{\mathcal{M}}X^*\mathcal{H}')^- = \mathcal{M},$$

where we used the facts that $P_{\mathcal{M}'}X|\mathcal{K} = 0$ and $P_{\mathcal{M}}X^*|\mathcal{K}' = 0$, which follow from the definitions of $\mathcal{M}$ and $\mathcal{M}'$. The proof is now complete.

We give a first use of the splitting principle by proving the converse to Proposition 2.11.

3.8. THEOREM. *For every operator T of class C_0 the following assertions are equivalent:*

(i) *T is multiplicity-free;*

(ii) *$\{T\}'$ is commutative; and*

(iii) *$\{T\}' = \mathcal{F}_T$.*

PROOF. We already know from Proposition 2.11 that (i)$\Rightarrow$(iii), and (iii) trivially implies (ii). It remains to show that the commutant of an operator T of class C_0 with $\mu_T \geq 2$ cannot be commutative. Let T, $\mathcal{K}$, and $\mathcal{M}$ be as in Theorem 3.1; if $\mu_T \geq 2$ we must have $\mathcal{K} \neq \mathcal{H}$ and hence $\mathcal{M} \neq \{0\}$. Define $\mathcal{K}'$, T_1, and T_2 by

$$\mathcal{K}' = \mathcal{H} \ominus \mathcal{M}, \qquad T_1 = T|\mathcal{K}, \qquad T_2^* = T^*|\mathcal{K}'.$$

The operator $X = P_{\mathcal{K}'}|\mathcal{K}$ is a quasiaffinity and $X \in \mathcal{I}(T_1, T_2)$. Since both T_1 and T_2 are multiplicity-free we deduce that T_1 and T_2 are quasisimilar (they are both quasisimilar to $S(m_T)$ by Theorem 2.3(iii)). Let $Y \in \mathcal{I}(T_2, T_1)$ be a quasiaffinity and define $A \in \{T\}'$ by $A = YP_{\mathcal{K}'}$. We clearly have

$$\ker A = \ker P_{\mathcal{K}'} = \mathcal{M}, \qquad (A\mathcal{H})^- = \mathcal{K}. \tag{3.9}$$

Assume that we can find a nonzero operator $Z \in \mathcal{I}(T_2, T|\mathcal{M})$. Then the operator $B \in \{T\}'$ defined by $B = ZP_{\mathcal{K}'}$ is such that $AB = 0$ and $((BA)\mathcal{H})^- = (ZP_{\mathcal{K}'}Y\mathcal{K}')^- = (ZP_{\mathcal{K}'}\mathcal{K})^- = (Z\mathcal{K}')^- \neq \{0\}$ so that A and B do not commute. Thus, in order to show that $\{T\}'$ is not commutative, it suffices to show the existence of such a Z. Since $\mathcal{M} \neq \{0\}$, $T|\mathcal{M}$ has a nonzero cyclic space $\mathcal{M}_1$ and it would suffice to find $Z \in \mathcal{I}(T_2, T|\mathcal{M}_1)$, $Z \neq 0$. Finally, if $\theta \equiv m_T \equiv m_{T_2}$ and $\theta' \equiv m_{T|\mathcal{M}_1}$, then $T|\mathcal{M}_1 \sim S(\theta')$, $T_2 \sim S(\theta)$, and $\theta'|\theta$ and therefore we can as well prove that there are nonzero operators in $\mathcal{I}(S(\theta), S(\theta'))$ whenever $\theta'|\theta$ and θ' is nonconstant. The existence of such operators follows from Theorem 1.16 and thus we conclude our proof.

Exercises

1. Assume that the operator $T \in \mathcal{L}(\mathcal{H})$ of class C_0 has the following splitting property. For every cyclic subspace $\mathcal{K}$ for T there exists an invariant subspace

$\mathscr{M}$ such that $\mathscr{K} \cap \mathscr{M} = \{0\}$ and $\mathscr{K} \vee \mathscr{M} = \mathscr{H}$. Prove that m_T is a Blaschke product with simple zeros.

2. Is the converse to the previous exercise true?

3. Show that there exist operators T (not of class C_0) that have the splitting property in Exercise 1 and $\sigma_p(T) = \varnothing$.

4. Let T, T', and T'' be operators of class C_0 such that $S(m_T) \oplus T' \prec T \prec S(m_T) \oplus T''$. Prove that $m_{T'} \equiv m_{T''}$. (Note that the relation $T' \prec T''$ is not part of the assumption.)

5. Construct an operator T (on a possibly nonseparable Hilbert space) such that $\{T\}'$ is commutative but T is not multiplicity-free.

6. Construct a multiplicity-free operator T such that $\{T\}'$ is not commutative.

4. Jordan operators. The splitting principle in §3, and particularly Corollary 3.5, shows that for a given operator T of class C_0 we can find inner functions $\theta_0, \theta_1, \ldots, \theta_n$ and another operator T' of class C_0 such that $\theta_{j+1}|\theta_j$, $0 \le j \le n-1$, and

$$S(\theta_0) \oplus S(\theta_1) \oplus \cdots \oplus S(\theta_n) \oplus T' \prec T.$$

An inductive application of this observation (transfinite if the space is not separable) yields a direct sum of Jordan blocks that is actually quasisimilar to T. For reasons of uniqueness we allow only certain kinds of direct sums of Jordan blocks that we call Jordan operators. In order to describe Jordan operators we will need some facts about ordinals. We adopt the view that an *ordinal number* is a set α that is transitive with respect to "$\in$," i.e., $\beta \in \gamma \in \alpha$ implies $\beta \in \alpha$, and is totally ordered by "$\in$." Thus $0 = \varnothing$, $1 = \{\varnothing\}$, $2 = \{\varnothing, \{\varnothing\}\}, \ldots$ and, in general, $\alpha = \{\beta : \beta \in \alpha\}$. We will write, as usual, $\alpha < \beta$ if α and β are ordinals and $\alpha \in \beta$. An ordinal α is a *cardinal number* if it is not equipotent with any smaller ordinal. Thus $0, 1, 2, \ldots$ are cardinal numbers and ω, the first infinite ordinal, is also a cardinal number and is sometimes denoted $\aleph_0$. Ordinals are well ordered and so the definition

$$\operatorname{card}(\alpha) = \text{first } \beta \text{ such that } \beta \le \alpha \text{ and } \beta \text{ is equipotent to } \alpha$$

makes sense; it associates a cardinal number with every ordinal. The axiom of choice implies that every set M is equipotent to some cardinal α; we write $\alpha = \operatorname{card}(M)$ in this case.

4.1. DEFINITION. Let γ be a cardinal number and $\Theta = \{\theta_\alpha : \alpha < \gamma\}$ a family of inner functions. Then Θ is called a *model function* if $\theta_\alpha|\theta_\beta$ whenever $\operatorname{card}(\beta) \le \operatorname{card}(\alpha) < \gamma$. The *Jordan operator* $S(\Theta)$ determined by the model function Θ is the C_0 operator defined as

$$S(\Theta) = \bigoplus_{\alpha < \gamma'} S(\theta_\alpha), \qquad \gamma' = \min\{\beta : \theta_\beta \equiv 1\}.$$

We will denote by $\mathscr{H}(\Theta)$ the space of $S(\Theta)$.

Some comments are in order here. The condition that Θ be a model function implies that $m_{S(\Theta)} = \theta_0$ and $\theta_\alpha \equiv \theta_\beta$ if $\operatorname{card}(\alpha) = \operatorname{card}(\beta)$. Thus the operator $S(\Theta)$ contains many identical summands; for example, it contains at least $\aleph_1$ summands of the form $S(\theta_\omega)$ if $\gamma' > \omega$. Thus, $S(\Theta)$ acts on a separable space if and only if $\theta_\omega \equiv 1$; in this case all summands $S(\theta_\alpha)$ with $\alpha \geq \omega$ act on the trivial space $\{0\}$. The fact that some θ_α may be constant suggests the following definition.

4.2. DEFINITION. Let $\Theta = \{\theta_\alpha : \alpha < \gamma\}$ and $\Theta' = \{\theta'_\alpha : \alpha < \gamma'\}$ be two model functions. We say that Θ and Θ' are *equivalent*, and we write $\Theta \equiv \Theta'$, if $\theta_\alpha \equiv \theta'_\alpha$ for $\alpha < \min\{\gamma, \gamma'\}$ and $\theta_\alpha \equiv 1$ [resp. $\theta'_\alpha \equiv 1$] if $\gamma' \leq \alpha < \gamma$ [resp. $\gamma \leq \alpha < \gamma'$].

It is clear that the relation $\Theta \equiv \Theta'$ implies that $S(\Theta) = S(\Theta')$. Thus the number γ is not particularly relevant in the definition of Jordan operators—it plays only a set-theoretical role. It may be easier to think that θ_α is defined for every ordinal α, $\theta_\alpha \equiv 1$ if α is large enough, and $S(\Theta) = \bigoplus_{\alpha<\gamma} S(\theta_\alpha)$ where γ is the first ordinal such that $\theta_\gamma \equiv 1$.

The main goal of this section is to prove that Jordan operators have an important uniqueness property: each quasisimilarity class in $\mathscr{L}(\mathscr{H})$ contains, up to unitary equivalence, at most one Jordan operator. Then we will show in §5 that the quasisimilarity class of every operator of class C_0 contains a Jordan operator.

We recall that we denoted by μ_T the multiplicity of an operator T (§1.3).

4.3. DEFINITION. Let T be an operator of class C_0. The *multiplicity function* ν_T associates with each inner function θ the cardinal number

$$\nu_T(\theta) = \mu_{T|(\operatorname{ran}\theta(T))^-}.$$

If $T \in \mathscr{L}(\mathscr{H})$ we obviously have

$$\mu_T \leq \dim \mathscr{H} \leq \aleph_0 \mu_T$$

and we deduce the following inequalities for the multiplicity function:

$$\nu_T(\theta) \leq \dim(\operatorname{ran}\theta(T))^- \leq \aleph_0 \nu_T(\theta). \tag{4.4}$$

In particular, we have

$$\nu_T(\theta) = \dim(\operatorname{ran}\theta(T))^- \text{ if } \nu_T(\theta) \geq \aleph_0 \text{ or } \dim(\operatorname{ran}\theta(T))^- > \aleph_0. \tag{4.5}$$

4.6. LEMMA. *Let T and T' be two operators of class C_0. If $T \prec T'$ then $\nu_{T'}(\theta) \leq \nu_T(\theta)$ for every inner function θ. In particular, the multiplicity function ν_T is a quasisimilarity invariant of T.*

PROOF. If $X \in \mathscr{I}(T, T')$ is any operator, then we also have $\theta(T')X = X\theta(T)$ for every inner function θ. If, in addition, X is a quasiaffinity, we deduce that $T|(\operatorname{ran}\theta(T))^- \prec T'|(\operatorname{ran}\theta(T'))^-$ via the quasiaffinity $X|(\operatorname{ran}\theta(T))^-$. By §1.3 we have $\nu_{T'}(\theta) \leq \nu_T(\theta)$ and the lemma is proved.

Our next task is to determine the function ν_T corresponding with a Jordan operator T. If $T \in \mathscr{L}(\mathscr{H})$ is an operator and γ a cardinal number we will denote by $T^{(\gamma)}$ the direct sum of γ copies of T acting on the direct sum $\mathscr{H}^{(\gamma)}$ of γ copies of $\mathscr{H}$.

4.7. LEMMA. *Let n and k be natural numbers and θ a nonconstant inner function. If there exists an injective operator $X \in \mathscr{I}(S(\theta)^{(k)}, S(\theta)^{(n)})$ then $k \le n$.*

PROOF. Let $P_i : \mathscr{H}(\theta)^{(n)} \to \mathscr{H}(\theta)$ and $Q_j : \mathscr{H}(\theta)^{(k)} \to \mathscr{H}(\theta)$, $1 \le i \le n$, $1 \le j \le k$, be the natural projections, and assume that $X \in \mathscr{I}(S(\theta)^{(k)}, S(\theta)^{(n)})$. The operators $P_i X Q_j^*$ commute with $S(\theta)$ and by virtue of Corollary 1.20 we can find $a_{ij} \in H^\infty$ such that $P_i X Q_j^* = a_{ij}(S(\theta))$, $1 \le i \le n$, $1 \le j \le k$. Thus the operator X can be written as

$$X\left(\bigoplus_{j=1}^{k} h_j\right) = \bigoplus_{i=1}^{n} P_{\mathscr{H}(\theta)}\left(\sum_{j=1}^{k} a_{ij} h_j\right), \quad h_j \in \mathscr{H}(\theta),\ 1 \le j \le k. \tag{4.8}$$

Assume now that X is one-to-one. In this case not all functions a_{ij} can be divisible by θ; in fact, if all a_{ij} are divisible by θ then $X = 0$ and this is not one-to-one unless $\mathscr{H}(\theta) = \{0\}$. Therefore there exists a minor of maximum rank of the matrix $[a_{ij}]_{i,j}$ that is not divisible by θ, and there is no loss of generality in assuming that this minor is $|a_{ij}|_{1\le i,j\le r}$, where $r \le \min(k, n)$. Assume that $k > n$ and write the determinant

$$\begin{vmatrix} a_{11} & a_{12} & \cdots & a_{1,r+1} \\ a_{21} & a_{22} & \cdots & a_{2,r+1} \\ \cdot & \cdot & \cdots & \cdot \\ a_{r1} & a_{r2} & \cdots & a_{r,r+1} \\ x_1 & x_2 & \cdots & x_{r+1} \end{vmatrix} = \sum_{j=1}^{r+1} x_j u_j$$

that we developed following the last row. The sum $\sum_{j=1}^{r+1} a_{ij} u_j$ is zero if $1 \le i \le r$ and equals a minor of order $r+1$ (hence divisible by θ) if $i > r$. Then the vector $h = \bigoplus_{j=1}^{k} h_j \in \mathscr{H}(\theta)^{(k)}$ defined by $h_j = P_{\mathscr{H}(\theta)} u_j$, $1 \le j \le r+1$, $h_j = 0$ otherwise, satisfies the relation $Xh = 0$ and hence $h = 0$. We conclude that $P_{\mathscr{H}(\theta)} u_{r+1} = 0$ or $u_{r+1} \in \theta H^2$. But $u_{r+1} = |a_{ij}|_{1\le i,j\le r}$ was chosen not divisible by θ, a contradiction. We conclude that we necessarily have $k \le n$, thus finishing the proof.

4.9. COROLLARY. *Let n and k be natural numbers and θ a nonconstant inner function. If there exists an operator $X \in \mathscr{I}(S(\theta)^{(k)}, S(\theta)^{(n)})$ with dense range then $k \ge n$.*

PROOF. We can apply Lemma 4.8 to X^* since $S(\theta)^*$ is unitarily equivalent to $S(\theta^\sim)$.

4.10. LEMMA. *Let T be an operator of class C_0, $n = \dim(\mathscr{D}_{T^*})$, and $\theta \equiv m_T$. There exists a surjective operator $X \in \mathscr{I}(S(\theta)^{(n)}, T)$.*

PROOF. The minimal isometric dilation of T is a unilateral shift of multiplicity n. We may assume with no loss of generality that the space $\mathscr{H}$ of T is contained in $(H^2)^{(n)}$ and $T^* = S^{*^{(n)}}|\mathscr{H}$. Now we have

$$0 = \theta^{\sim}(T^*) = \theta^{\sim}(S^{*^{(n)}})|\mathscr{H}$$

so that $\mathscr{H} \subset \ker \theta^{\sim}(S^{*^{(n)}}) = \mathscr{H}(\theta)^{(n)}$. The operator X is defined then as $P_{\mathscr{H}}|\mathscr{H}(\theta)^{(n)}$.

4.11. LEMMA. *Let θ be a nonconstant inner function and n a natural number. The multiplicity of $S(\theta)^{(n)}$ equals n.*

PROOF. The inequality $\mu_{S(\theta)^{(n)}} \leq n$ is clear since $S(\theta)$ is multiplicity-free. Set $k = \mu_{S(\theta)^{(n)}}$ and apply §1.3 to find an operator T of class C_0 such that $T \prec S(\theta)^{(n)}$ and $\dim(\mathscr{D}_{T^*}) = k$. If $Y \in \mathscr{I}(T, S(\theta)^{(n)})$ is a quasiaffinity, and $X \in \mathscr{I}(S(\theta)^{(k)}, T)$ is surjective, then $YX \in \mathscr{I}(S(\theta)^{(k)}, S(\theta)^{(n)})$ has dense range. We conclude that $k \geq n$ by Corollary 4.9. The lemma is proved.

We are now ready to treat the multiplicity function of a Jordan operator.

4.12. THEOREM. *Let $\Theta = \{\theta_\alpha : \alpha < \gamma\}$ be a model function.*
(i) $\mu_{S(\Theta)} = \min\{\alpha : \theta_\alpha \equiv 1\}$.
(ii) $\nu_{S(\Theta)}(\theta) = \min\{\alpha : \theta_\alpha | \theta\}$ *for every inner function* θ.

PROOF. Note first that the two minima in the statement are actually cardinal numbers by the definition of model functions. An easy argument shows that the more general statement (ii) is implied by (i). Indeed, for each α the restriction $S(\theta_\alpha)|(\operatorname{ran} \theta(S(\theta_\alpha)))^-$ is unitarily equivalent to $S(\theta_\alpha/\theta \wedge \theta_\alpha)$ by Corollary 1.12 so that $S(\Theta)|(\operatorname{ran} \theta(S(\Theta)))^-$ is unitarily equivalent to $\bigoplus_{\alpha<\gamma} S(\theta_\alpha/\theta_\alpha \wedge \theta)$. Now (ii) follows from (i) if we note that $\{\theta_\alpha/\theta_\alpha \wedge \theta : \alpha < \gamma\}$ is also a model function and $\theta_\alpha | \theta$ if and only if $\theta_\alpha/\theta_\alpha \wedge \theta \equiv 1$.

In order to prove (i) we set $\beta = \min\{\alpha : \theta_\alpha \equiv 1\}$. The direct sum $S(\Theta) = \bigoplus_{\alpha<\beta} S(\theta_\alpha)$ contains exactly β multiplicity-free summands and the inequality

$$\mu_{S(\Theta)} \leq \beta$$

follows at once. If $\beta > \aleph_0$ we have

$$\beta \leq \dim \mathscr{H}(\Theta) \leq \aleph_0 \beta = \beta$$

and hence $\dim(\mathscr{H}(\Theta)) = \beta$. The equality $\mu_{S(\Theta)} = \beta$ follows now from the inequalities

$$\mu_{S(\Theta)} \leq \beta \leq \aleph_0 \mu_{S(\Theta)}.$$

We still have to treat the case in which $\beta \leq \omega$. First consider the subcase $\beta < \omega$. Then $\theta_{\beta-1}$ is a nonconstant divisor of $\theta_1, \theta_2, \ldots, \theta_{\beta-2}$ and therefore by Proposition 1.10(iii), $S(\theta_{\beta-1})^* = S(\theta_j)^*|\mathscr{H}(\theta_{\beta-1})$, $j = 0, 2, \ldots, \beta-1$, from which we deduce

$$(S(\theta_{\beta-1})^{(\beta)})^* = S(\Theta)^*|\left(\bigoplus_{j=0}^{\beta-1} \mathscr{H}(\theta_{\beta-1})\right).$$

This relation implies that

$$\beta = \mu_{S(\theta_{\beta-1})^{(\beta)}} \leq \mu_{S(\Theta)}$$

and (i) is proved if $\beta < \omega$. Finally, if $\beta = \omega$, then we clearly have

$$\mu_{S(\Theta)} \geq \mu_{\bigoplus_{j=0}^{n-1} S(\theta_j)} = n$$

for every $n < \omega$. Indeed, all functions θ_j, $j < \omega$, are nonconstant by the definition of β. We conclude that in this case we have $\mu_{S(\Theta)} \geq \omega = \beta$. The theorem is proved.

We have thus shown how to compute $\mu_{S(\Theta)}$ given the model function Θ. It is an easy task to reverse this process.

4.13. COROLLARY. *If $\Theta = \{\theta_\alpha : \alpha < \gamma\}$ is a model function then*

$$\theta_\alpha \equiv \bigwedge\{\theta : \nu_{S(\Theta)}(\theta) \leq \mathrm{card}(\alpha)\}$$

for every ordinal α.

PROOF. We have $\nu_{S(\Theta)}(\theta_\alpha) = \min\{\beta : \theta_\alpha | \theta_\beta\} \leq \alpha$ so that $\nu_{S(\Theta)}(\theta_\alpha) \leq \mathrm{card}(\alpha)$ ($\nu_{S(\Theta)}(\theta_\alpha)$ is a cardinal number!) and we conclude that the function

$$\phi = \bigwedge\{\theta : \nu_{S(\Theta)}(\theta) \leq \mathrm{card}(\alpha)\}$$

divides θ_α. Now let θ be an inner function such that $\nu_{S(\Theta)}(\theta) \leq \mathrm{card}(\alpha) \leq \alpha$. Since $\nu_{S(\Theta)}(\theta) = \min\{\beta : \theta_\beta | \theta\}$ we deduce that $\theta_\alpha | \theta$. Because θ was arbitrary such that $\nu_{S(\Theta)}(\theta) \leq \mathrm{card}(\alpha)$ we must have $\theta_\alpha | \phi$. The corollary follows.

The uniqueness property of Jordan operators is now quite transparent.

4.14. THEOREM. *If Θ and Θ' are two model functions and $S(\Theta) \prec S(\Theta')$ then $\Theta \equiv \Theta'$ and hence $S(\Theta) = S(\Theta')$.*

PROOF. Assume that $\Theta = \{\theta_\alpha\}$ and $\Theta' = \{\theta'_\alpha\}$. If $S(\Theta) \prec S(\Theta')$, Lemma 4.6 implies that $\nu_{S(\Theta')}(\theta) \leq \nu_{S(\Theta)}(\theta)$ for every inner function θ. Thus, for every ordinal α,

$$\{\theta : \nu_{S(\Theta)}(\theta) \leq \mathrm{card}(\alpha)\} \subset \{\theta : \nu_{S(\Theta')}(\theta) \leq \mathrm{card}(\alpha)\}$$

so that $\theta'_\alpha = \bigwedge\{\theta : \nu_{S(\Theta')}(\theta) \leq \mathrm{card}(\alpha)\}$ divides $\theta_\alpha = \bigwedge\{\theta : \nu_{S(\Theta)}(\theta) \leq \mathrm{card}(\alpha)\}$. Finally, we have $S(\Theta')^* \prec S(\Theta)^*$ and these two operators are also (unitarily equivalent to) Jordan operators, determined by the model functions $\{\theta'^{\sim}_\alpha\}$ and $\{\theta^{\sim}_\alpha\}$, respectively. By the first part of the proof we have $\theta^{\sim}_\alpha | \theta'^{\sim}_\alpha$ so that $\theta_\alpha | \theta'_\alpha$. We conclude that $\theta_\alpha \equiv \theta'_\alpha$ and the theorem follows.

Exercises

1. Let $\mathscr{F}$ be a family of inner functions such that if $\theta, \theta' \in \mathscr{F}$ then either $\theta | \theta'$ or $\theta' | \theta$. Associate a cardinal number $\gamma(\theta)$ with each function $\theta \in \mathscr{F}$ and set $T = \bigoplus_{\theta \in \mathscr{F}} S(\theta)^{(\gamma(\theta))}$. Under what condition is T unitarily equivalent to a Jordan operator?

2. Compute the function ν_T if $T = S(\theta_1) \oplus S(\theta_2)$, where θ_1 and θ_2 are two inner functions. (It is not assumed here that $\theta_2 | \theta_1$.)

3. Find a Jordan operator $S(\Theta)$ such that $\nu_T = \nu_{S(\Theta)}$, where T is as in Exercise 2.

4. Let $\theta_1(\lambda) = \lambda^2$, $\theta_2(\lambda) = \lambda$, $\lambda \in \mathbf{D}$, and define $T_1 = S(\theta_1)^{(\omega)}$, $T_2 = S(\theta_1)^{(\omega)} \oplus S(\theta_2)^{(\omega)}$. Find a quasiaffinity in $\mathscr{I}(T_2, T_1)$. Does this contradict Theorem 4.14?

5. Show that the commutant of $S(\theta)^{(n)}$ $(n < \omega)$ is isomorphic to the algebra of $n \times n$ matrices over $H^\infty / \theta H^\infty$.

5. The classification theorem. We are about to show that operators of class C_0 can be completely classified using the relation of quasisimilarity. We begin with the particular case of separably acting operators.

5.1. THEOREM. *Let T be an operator of class C_0 acting on a separable space. There exists a Jordan operator $S(\Theta)$ that is quasisimilar to T. Moreover, $S(\Theta)$ is uniquely determined by either of the relations $S(\Theta) \prec T$ and $T \prec S(\Theta)$.*

PROOF. Assume $T \in \mathscr{L}(\mathscr{H})$ and $\mathscr{H}$ is separable. Let $\{h_n : n \geq 0\}$ be a dense sequence in $\mathscr{H}$, and let $\{k_n : n \geq 0\}$ be a sequence in which each h_n is repeated infinitely often. We construct inductively vectors $f_0, f_1, f_2, \ldots$ in $\mathscr{H}$ and invariant subspaces $\mathscr{M}_{-1}, \mathscr{M}_0, \mathscr{M}_1, \ldots$ for T such that $\mathscr{M}_{-1} = \mathscr{H}$, and

$$f_j \in \mathscr{M}_{j-1}, \qquad m_{f_j} = m_{T|\mathscr{M}_{j-1}}; \tag{5.2}$$

$$\mathscr{K}_j \vee \mathscr{M}_j = \mathscr{M}_{j-1}, \qquad \mathscr{K}_j \cap \mathscr{M}_j = \{0\}, \quad \text{where } \mathscr{K}_j = \bigvee_{n=0}^{\infty} T^n f_j; \tag{5.3}$$

$$\|k_j - P_{\mathscr{K}_0 \vee \mathscr{K}_1 \vee \cdots \vee \mathscr{K}_j} k_j\| \leq 2^{-j} \tag{5.4}$$

for $j = 0, 1, 2, \ldots$. Assume that f_j and $\mathscr{M}_j$ have already been defined for $j < n$ and let us construct f_n and $\mathscr{M}_n$; note that if $n = 0$ only $\mathscr{M}_{-1}$ has been constructed and there is no f_{-1}. A repeated application of (5.3) yields

$$\mathscr{H} = \mathscr{M}_{-1} = \mathscr{K}_0 \vee \mathscr{M}_0 = \mathscr{K}_0 \vee \mathscr{K}_1 \vee \mathscr{M}_1 = \cdots = \mathscr{K}_0 \vee \mathscr{K}_1 \vee \cdots \vee \mathscr{K}_{n-1} \vee \mathscr{M}_{n-1}$$

so that we can find vectors $u_n \in \mathscr{K}_0 \vee \mathscr{K}_1 \vee \cdots \vee \mathscr{K}_{n-1}$ and $v_n \in \mathscr{M}_{n-1}$ such that

$$\|k_n - u_n - v_n\| \leq 2^{-n-1}. \tag{5.5}$$

We can then find a vector $f_n \in \mathscr{M}_{n-1}$ such that $m_{f_n} = m_{T|\mathscr{M}_{n-1}}$ (f_n is a $(T|\mathscr{M}_{n-1})$-maximal vector) and

$$\|v_n - f_n\| \leq 2^{-n-1}. \tag{5.6}$$

An application of the splitting principle (Theorem 3.1) to the operator $T|\mathscr{M}_{n-1}$ shows the existence of an invariant subspace $\mathscr{M}_n$ satisfying (5.3) for $j = n$. Since (5.2) is satisfied for $j = n$ by the choice of f_n, it remains to verify (5.4). But it is clear that relations (5.5) and (5.6) imply that

$$\|k_n - P_{\mathscr{K}_0 \vee \mathscr{K}_1 \vee \cdots \vee \mathscr{K}_n} k_n\| \leq \|k_n - u_n - f_n\| \leq \|k_n - u_n - v_n\| + \|v_n - f_n\| \leq 2^{-n}.$$

Thus the existence of $\{f_j : j \geq 0\}$ and $\{\mathscr{M}_j : j \geq 0\}$ is proved by induction. An important consequence of (5.4) is that

$$\mathscr{H} = \bigvee_{j<\omega} \mathscr{K}_j. \tag{5.7}$$

Indeed, (5.4) implies that

$$\operatorname{dist}\left(k_n, \bigvee_{j<\omega} \mathscr{K}_j\right) \to 0 \quad \text{as } n \to \infty$$

and, since each h_i is repeated infinitely often among the k_n, this implies $h_i \in \bigvee_{j<\omega} \mathscr{K}_j$ for all i. We now define a model function $\Theta = \{\theta_j : j < \omega\}$ by setting $\theta_j \equiv m_{f_j}$ for $j < \omega$. Relation (5.2) and the fact that $\mathscr{M}_{j+1} \subset \mathscr{M}_j$ easily imply that $\theta_{j+1}|\theta_j$ for $j < \omega$ and hence Θ is indeed a model function. The operator $T|\mathscr{K}_j$ is a multiplicity-free operator and has minimal function θ_j so that Proposition 2.2 implies the existence of a quasiaffinity $X_j \in \mathscr{I}(S(\theta_j), T|\mathscr{K}_j)$, $j < \omega$. We can then define an operator $X \in \mathscr{I}(S(\Theta), T)$ by

$$X\left(\bigoplus_{j<\omega} g_j\right) = \sum_{j<\omega} (2^{-j}/\|X_j\|) X_j g_j, \qquad \bigoplus_{j<\omega} g_j \in \mathscr{H}(\Theta) = \bigoplus_{j<\omega} \mathscr{H}(\theta_j).$$

The reader will verify without difficulty that this formula defines a bounded operator X. Relation (5.7) implies that X must have dense range so that X is a quasiaffinity if we can prove that $\ker X = \{0\}$. Assume that the vector $g = \bigoplus_{j<\omega} g_j \in \ker X$, $g \neq 0$, and g_n is the first nonzero component of g. By the definition of X we have

$$X_n g_n = -\sum_{1 \leq j < \omega} (2^{-j}/\|X_{n+j}\|) X_{n+j} g_{n+j}. \tag{5.8}$$

The left-hand side of (5.8) belongs to $\mathscr{K}_n$ while the right-hand side belongs to $\bigvee_{1 \leq j < \omega} \mathscr{K}_{n+j} \subset \mathscr{M}_n$. By (5.3) we have $\mathscr{K}_n \cap \mathscr{M}_n = \{0\}$ so that $X_n g_n = 0$ and $g_n = 0$ by the fact that X_n is one-to-one. This contradiction clearly implies that $g = 0$ so that $\ker X = \{0\}$ and X is a quasiaffinity.

So far we proved the existence of a Jordan operator $S(\Theta)$ such that $S(\Theta) \prec T$. We can apply the same argument to T^*, and, in view of the fact that adjoints of Jordan operators are also Jordan operators, we deduce the existence of a Jordan operator $S(\Theta')$ such that $T \prec S(\Theta')$. The theorem now follows easily from the uniqueness property of Jordan operators. Indeed, if Θ and Θ' are any model functions such that $S(\Theta) \prec T \prec S(\Theta')$ we deduce by transitivity that $S(\Theta) \prec S(\Theta')$ and hence $S(\Theta) = S(\Theta')$ by Theorem 4.14. This proves that $S(\Theta) \sim T$ as well as the uniqueness assertion.

Before extending Theorem 5.1 to nonseparably acting operators we note two interesting properties specific to the separable case. In the sequel we will call $S(\Theta)$ the *Jordan model* of the operator T if $S(\Theta) \sim T$.

5.9. COROLLARY. *Let T be an operator of class C_0 acting on the separable space $\mathscr{H}$. Assume that the vectors $\{f_j : j < \omega\}$ and the invariant subspaces $\{\mathscr{M}_j : j < \omega\}$ satisfy relations* (5.2) *and* (5.3) *with $\mathscr{M}_{-1} = \mathscr{H}$. If $\theta_j = m_{f_j}$, $j < \omega$, then $\bigoplus_{j<\omega} S(\theta_j)$ is the Jordan model of T.*

PROOF. Fix a natural number n and set $f'_j = f_j$ and $\mathscr{M}'_j = \mathscr{M}_j$ for $0 \leq j < n$. Following the argument in the proof of Theorem 5.1 we can extend these finite sequences to infinite sequences $\{f'_j : j < \omega\}$ and $\{\mathscr{M}'_j : j < \omega\}$ satisfying (5.2) and (5.3) for all j and (5.4) for $j \geq n$ (with f_j and $\mathscr{M}_j$ replaced by f'_j and $\mathscr{M}'_j$, of course). Then the relation $\mathscr{H} = \bigvee_{j<\omega} \mathscr{K}'_j$, $\mathscr{K}'_j = \bigvee_{n=0}^{\infty} T^n f'_j$ follows at once and this is what is needed to prove that $S(\Theta) \prec T$, where $\Theta = \{\theta'_j : j < \omega\}$ is the model function defined by $\theta'_j \equiv m_{f'_j}$, $j < \omega$. Recalling now that $f_j = f'_j$ for $j < n$ we conclude that the Jordan model $S(\Theta)$ of T satisfies the relations $\theta'_j \equiv m_{f_j}$ for $j < n$. Since n is arbitrary, the corollary follows.

5.10. COROLLARY. *Let T and T' be separably acting operators of class C_0, and let $S(\Theta) = \bigoplus_{j<\omega} S(\theta_j)$ be the Jordan model of T. If $m_{T'}|\theta_j$ for all $j < \omega$ then $S(\Theta)$ is also the Jordan model of $T \oplus T'$.*

PROOF. Assume that $T \in \mathscr{L}(\mathscr{H})$, $T' \in \mathscr{L}(\mathscr{H}')$, and the vectors $\{f_j : j < \omega\}$ and the invariant subspaces $\{\mathscr{M}_j : j < \omega\}$ satisfy the relations (5.2)–(5.4) with $\mathscr{M}_{-1} = \mathscr{H}$. If we set $\mathscr{M}'_j = \mathscr{M}_j \oplus \mathscr{H}'$ then we have

$$m_{(T\oplus T')|\mathscr{M}'_j} \equiv m_{T|\mathscr{M}_j} \vee m_{T'} \equiv \theta_j \vee m_{T'} \equiv \theta_j$$

for $j < \omega$ and consequently the vectors $\{f_j \oplus 0 : j < \omega\}$ and subspaces $\{\mathscr{M}'_j : j < \omega\}$ satisfy relations (5.2) and (5.3) with $\mathscr{M}'_{-1} = \mathscr{H} \oplus \mathscr{H}'$, $\mathscr{M}_j$ replaced by $\mathscr{M}'_j$, and T replaced by $T \oplus T'$. It suffices then to apply Corollary 5.9 to $T \oplus T'$.

This last corollary sheds some light on the reason behind the definition of Jordan operators. To see this, consider two inner functions θ and θ' such that $\theta'|\theta$ and define $\theta_j \equiv \theta$, $\theta_{\omega+j} \equiv \theta'$ for $j < \omega$. By Corollary 5.10, the Jordan model of $\bigoplus_{j<2\omega} S(\theta_j)$ is $\bigoplus_{j<\omega} S(\theta_j)$. Thus adding countably many copies of $S(\theta')$ to $S(\theta)^{(\omega)}$ will not change the quasisimilarity class—we have to add at least $\aleph_1$ summands of the form $S(\theta')$. Now, $S(\theta)^{(\omega)} \oplus S(\theta')^{(\aleph_1)}$ is unitarily equivalent to the Jordan operator $S(\Theta)$, where $\Theta = \{\theta_\alpha : \alpha < \aleph_1\}$ is defined by $\theta_\alpha \equiv \theta$ for $\alpha < \omega$, and $\theta_\alpha \equiv \theta'$ for $\omega \leq \alpha < \aleph_1$.

Let us consider now the case of nonseparably acting operators.

5.11. LEMMA. *Let $T \in \mathscr{L}(\mathscr{H})$ be an operator of class C_0. There exists a separable reducing subspace $\mathscr{H}_0$ for T with the following property: if $\bigoplus_{j<\omega} S(\theta_j)$ is the Jordan model of $T|\mathscr{H}_0$ then $m_{T|\mathscr{H}\ominus\mathscr{H}_0}|\theta_j$ for all $j < \omega$.*

PROOF. We apply once again the technique in the proof of Theorem 5.1 to construct sequences $\{f_j : j < \omega\}$ and $\{\mathscr{M}_j : j < \omega\}$ satisfying relations (5.2) and (5.3), with $\mathscr{M}_{-1} = \mathscr{H}$. We now define $\mathscr{H}_0$ as the smallest reducing subspace for T containing all the vectors f_j, $j < \omega$. An application of Corollary 5.9 to the operator $T|\mathscr{H}_0$, the vectors f_j, and the subspaces

$$\mathscr{M}'_j = \mathscr{M}_j \cap \mathscr{H}_0 = \mathscr{M}_j \ominus (\mathscr{H} \ominus \mathscr{H}_0)$$

shows that the Jordan model of $T|\mathscr{H}_0$ is $\bigoplus_{j<\omega} S(\theta_j)$, where $\theta_j \equiv m_{f_j}$ for $j < \omega$. Finally, we have $\mathscr{H} \ominus \mathscr{H}_0 \subset \mathscr{M}_j$ and therefore

$$m_{T|\mathscr{H}\ominus\mathscr{H}_0} | m_{T|\mathscr{M}_j} \equiv \theta_j$$

for $j < \omega$. The lemma is proved.

We recall that an ordinal α is a *limit ordinal* if it is not the immediate successor of another ordinal. Sample limit ordinals are 0, ω, 2ω, and every uncountable cardinal. Every ordinal α can be uniquely written as $\alpha = \beta + k$ with β a limit ordinal and $k < \omega$.

5.12. PROPOSITION. *Let $T \in \mathscr{L}(\mathscr{H})$ be an operator of class C_0. We can associate with each limit ordinal α a reducing subspace $\mathscr{H}_\alpha$ for T such that:*

(i) *$\mathscr{H}_\alpha \perp \mathscr{H}_\beta$ if $\alpha \neq \beta$, and $\mathscr{H} = \bigoplus_\alpha \mathscr{H}_\alpha$; and*
(ii) *if the Jordan model of $T|\mathscr{H}_\alpha$ is $\bigoplus_{i<\omega} S(\theta_{\alpha+i})$ then $\theta_{\alpha+i}|\theta_{\beta+j}$ whenever $\alpha > \beta$ and $i, j < \omega$.*

PROOF. We use transfinite induction to associate with each limit ordinal γ reducing spaces $\mathscr{H}_\gamma$ and $\mathscr{M}_\gamma$ for T with the following properties:

$$\mathscr{H}_\gamma \text{ is separable,} \tag{5.13}$$

$$\left(\bigoplus_{\beta\leq\gamma} \mathscr{H}_\beta\right) \oplus \mathscr{M}_\gamma = \mathscr{H}; \tag{5.14}$$

$$m_{T|\mathscr{M}_\gamma} | \theta_{\gamma+k} \quad \text{for } k < \omega; \tag{5.15}$$

and

$$\theta_{\gamma+k}|\theta_{\beta+j} \quad \text{for } \beta < \gamma \text{ and } j, k < \omega, \tag{5.16}$$

where we have denoted by $\bigoplus_{j<\omega} S(\theta_{\beta+j})$ the Jordan model of $T|\mathscr{H}_\beta$. For $\gamma = 0$ let $\mathscr{H}_0$ be the space given by Lemma 5.11 and set $\mathscr{M}_0 = \mathscr{H} \ominus \mathscr{H}_0$. Relations (5.13), (5.14), and (5.15) are then satisfied (for $\beta = \gamma = 0$) while (5.16) is vacuous for $\gamma = 0$. Now let α be a limit ordinal and assume that the spaces $\mathscr{H}_\gamma$ and $\mathscr{M}_\gamma$ have been constructed to satisfy (5.13)–(5.16) for all $\gamma < \alpha$. In order to construct $\mathscr{H}_\alpha$ and $\mathscr{M}_\alpha$ we note that (5.14) implies the equality

$$\left(\bigoplus_{\beta<\alpha} \mathscr{H}_\beta\right) \oplus \left(\bigcap_{\beta<\alpha} \mathscr{M}_\beta\right) = \mathscr{H}.$$

We can then apply Lemma 5.11 once more to the operator $T|(\bigcap_{\beta<\alpha} \mathscr{M}_\beta)$ to find a separable reducing subspace $\mathscr{H}_\alpha \subset \bigcap_{\beta<\alpha} \mathscr{M}_\beta$ such that setting $\mathscr{M}_\alpha = (\bigcap_{\beta<\alpha} \mathscr{M}_\beta) \ominus \mathscr{H}_\alpha$ condition (5.15) is satisfied for $\gamma = \alpha$. Now, properties (5.13) and (5.14) are clearly satisfied for $\gamma = \alpha$ so we have only to verify (5.16) for $\gamma = \alpha$. If $\beta < \alpha$ we have $\mathscr{H}_\alpha \subset \bigcap_{\alpha'<\alpha} \mathscr{M}_{\alpha'} \subset \mathscr{M}_\beta$ and therefore $\theta_{\alpha+i}|m_{T|\mathscr{M}_\beta}$, $i < \omega$. But then we apply (5.15) for $\gamma = \beta$ to conclude that $\theta_{\alpha+i}|\theta_{\beta+j}$ for

$i, j < \omega$. The existence of $\mathscr{H}_\alpha$ and $\mathscr{M}_\alpha$ then follows by induction; the reader will have noticed that (5.16) is in fact a consequence of the previous two conditions.

Property (5.14) shows that $\mathscr{H}_{\gamma+\omega} \subset \mathscr{M}_\gamma$ and therefore $\mathscr{M}_\gamma \neq 0$ if $\mathscr{H}_{\gamma+\omega} \neq \{0\}$. Since the number of nonzero subspaces $\mathscr{H}_\alpha$ cannot exceed the dimension of $\mathscr{H}$, we conclude that $\mathscr{M}_\gamma = 0$ for some γ, and for such γ relation (5.14) implies that $\bigoplus_\alpha \mathscr{H}_\alpha = \mathscr{H}$. The proposition follows.

5.17. COROLLARY. *Let $T \in \mathscr{L}(\mathscr{H})$ be an operator of class C_0. There exists a family of inner functions $\{\theta_\alpha : \alpha < \gamma\}$ indexed by a segment of the ordinal numbers such that:*

(i) *T is quasisimilar to $\bigoplus_\alpha S(\theta_\alpha)$; and*

(ii) *$\theta_\alpha | \theta_\beta$ whenever $\alpha \geq \beta$.*

PROOF. If $\mathscr{H}_\alpha$ and $\theta_{\alpha+j}$ are given by Proposition 5.12, the operator T is clearly quasisimilar to $\bigoplus_\beta S(\theta_\beta)$, where the sum is extended over all ordinals β for which θ_β is not constant–these ordinals form a segment of the ordinals by property (ii) of that proposition. The corollary now becomes obvious.

The operator $\bigoplus_\alpha S(\theta_\alpha)$ in Corollary 5.17 is not necessarily a Jordan operator and hence it does not deserve to be called the Jordan model of T. Moreover, this operator is not uniquely determined by T, and this can be seen from the example following Corollary 5.10. Our task is to perform some surgery on the operator $\bigoplus_\alpha S(\theta_\alpha)$ given by Corollary 5.17 and show that this operator is quasisimilar to a Jordan operator.

Let us fix an operator $T_0 = \bigoplus_\alpha S(\theta_\alpha)$, where $\{\theta_\alpha : \alpha < \gamma\}$ are such that $\theta_\beta | \theta_\alpha$ whenever $\alpha \leq \beta$.

5.18. DEFINITION. An ordinal number α is said to be *good* if we have $\theta_\beta \equiv \theta_\alpha$ whenever $\beta \geq \alpha$ and $\text{card}(\beta) = \text{card}(\alpha)$. The remaining ordinals will be called *bad*.

Of course, the notion of good and bad is relative to the operator T_0. It is also quite clear that T_0 is a Jordan operator if and only if all ordinals are good. Thus we must try to eliminate the direct summands of T_0 corresponding with bad ordinals. The finite ordinals are good because there is only one ordinal with given finite cardinal, and it is not a priori obvious that other good ordinals exist.

5.19. LEMMA. *The set A of (classes of) inner functions defined by $A = \{\theta : \theta \equiv \theta_\alpha$ for some $\alpha\}$ is at most countable.*

PROOF. Choose a point $\lambda_0 \in \mathbf{D}$ such that $\theta_0(\lambda_0) \neq 0$. If $\alpha < \beta$ then $\theta_\beta | \theta_\alpha$ and therefore $0 < |\theta_\alpha(\lambda_0)| \leq |\theta_\beta(\lambda_0)|$. Moreover, if $|\theta_\alpha(\lambda_0)| = |\theta_\beta(\lambda_0)|$ it follows from the maximum modulus principle that $\theta_\alpha \equiv \theta_\beta$. It follows that the set $\{\theta : \theta \equiv \theta_\alpha \text{ for some } \alpha\}$ has the same cardinality as $K = \{|\theta_\alpha(\lambda_0)|\}_\alpha$. Now we saw that the mapping $\lambda_0 \to |\theta_\alpha(\lambda_0)|$ is increasing and therefore K is a well-ordered set. The lemma follows now from the fact that every well-ordered set of real numbers is at most countable. This last assertion is proved as follows. For every element $x \in K$ we choose a rational number $f(x)$ such that $x < f(x) < x'$

if x has an immediate successor x', and $x < f(x)$ if x is the last element of K. Thus K is equipotent with a set of rational numbers.

We can now show that there is a plentiful supply of good ordinals. We remind the reader that θ_α are defined only for $\alpha < \gamma$, where γ can be taken to be a sufficiently large cardinal. Thus no set-theoretical difficulties will arise.

5.20. LEMMA. *For each transfinite cardinal* $\aleph$ *there exist good ordinals* α *with* $\mathrm{card}(\alpha) = \aleph$. *Moreover, if* $g(\aleph) = \min\{\alpha : \mathrm{card}(\alpha) = \aleph,\ \alpha \text{ is good}\}$, *then the set* $\{\aleph : g(\aleph) \neq \aleph\}$ *is at most countable.*

PROOF. Let $\aleph$ be a transfinite cardinal and denote by $\aleph'$ the successor of $\aleph$ in the series of cardinals; of course $\aleph' > \omega$. The set of ordinals $\{\alpha : \mathrm{card}(\alpha) = \aleph\}$ has cardinality $\aleph'$ and we have

$$\{\alpha : \mathrm{card}(\alpha) = \aleph\} = \bigcup_{\theta \in A} \{\alpha : \mathrm{card}(\alpha) = \aleph,\ \theta_\alpha \equiv \theta\},$$

where A is the countable set in Lemma 5.19. The sets $\{\alpha : \mathrm{card}(\alpha) = \aleph,\ \theta_\alpha \equiv \theta\}$ cannot all have cardinality $\leq \aleph$ because $\aleph_0 \cdot \aleph = \aleph < \aleph'$. Thus there must exist a function $\theta \in A$ such that the set

$$\Sigma = \{\alpha : \mathrm{card}(\alpha) = \aleph,\ \theta_\alpha \equiv \theta\}$$

has cardinality $\aleph'$. We now show that every element α of Σ is a good ordinal. Assume to the contrary that there exists β such that $\mathrm{card}(\beta) = \aleph$ but $\theta \not\equiv \theta_\beta$. It follows then that θ_β does not divide θ_α for $\alpha \in \Sigma$ and hence $\alpha < \beta$. Thus $\Sigma \subset \{\alpha : \alpha < \beta\}$ and therefore $\mathrm{card}(\Sigma) \leq \mathrm{card}(\beta) = \aleph$, a contradiction. Thus Σ consists of good ordinals.

For the last assertion we note that, if $\aleph_1$ and $\aleph_2$ are two distinct cardinals such that $g(\aleph_1) > \aleph_1$ and $g(\aleph_2) > \aleph_2$, then $\theta_{\aleph_1} \not\equiv \theta_{\aleph_2}$. Indeed, assume for example that $\aleph_1 < \aleph_2$. Then we have $\aleph_1 < g(\aleph_1) < \aleph_2$ and $\theta_{\aleph_2} | \theta_{g(\aleph_1)}$ while $\theta_{g(\aleph_1)} \not\equiv \theta_{\aleph_1}$ by the definition of $g(\aleph_1)$. Thus the mapping $\aleph \to \theta_\aleph$ is one-to-one from the set $\{\aleph : \aleph \neq g(\aleph)\}$ to the set A in Lemma 5.19. Our result now obviously follows from that lemma.

5.21. REMARK. The preceding proof shows actually that the set $\{\alpha : \mathrm{card}(\alpha) = \aleph,\ \alpha \text{ is good}\} = \{\alpha : g(\aleph) \leq \alpha < \aleph'\}$ has cardinality $\aleph'$ while the set of bad ordinals

$$\{\alpha : \mathrm{card}(\alpha) = \aleph,\ \aleph \leq \alpha < g(\aleph)\}$$

has cardinality $\leq \aleph$. We can in fact construct a bijection of all ordinals onto the good ordinals. Indeed, we recall that if two ordinals α and β satisfy the relation $\alpha < \beta$ then β can be written uniquely as $\beta = \alpha + \gamma$. The number γ thus defined will be denoted $\beta - \alpha$. Then the bijection mentioned above can be obtained by associating with each ordinal α the good ordinal $g(\mathrm{card}(\alpha)) + (\alpha - \mathrm{card}(\alpha))$. Consequently, the operator $\bigoplus_{\alpha \text{ is good}} S(\theta_\alpha)$ is unitarily equivalent to the Jordan operator $S(\Theta')$ where $\Theta' = \{\theta'_\alpha\}$ is given by $\theta'_\alpha \equiv \theta_{g(\mathrm{card}(\alpha))}$. The following result gives the last tool that is needed in eliminating all summands corresponding with bad ordinals.

5.22. LEMMA. *There exists a function f defined on the set of all bad ordinal numbers such that*

(i) $f(\alpha) < \operatorname{card}(\alpha)$ *is a good limit ordinal; and*

(ii) $f(\alpha) \neq f(\beta)$ *if* $\alpha \neq \beta$ *and* $\operatorname{card}(\alpha) = \operatorname{card}(\beta) > \aleph_0$.

If f is any such function then $f(\alpha) = 0$ whenever $\aleph_0 \leq \alpha < g(\aleph_0)$ and the set $\{\alpha : f(\alpha) = \beta\}$ is at most countable for every limit ordinal β.

PROOF. If $\aleph_0 \leq \alpha < g(\aleph_0)$ then 0 is the only good limit ordinal that is $\leq \alpha$ and therefore we must define $f(\alpha) = 0$. Now, if $\aleph$ is an uncountable ordinal then it follows from the preceding remark that

$$\{\alpha : \alpha < \aleph,\ \alpha \text{ is good}\}$$

has cardinality $\aleph$ and consequently

$$\Sigma_\aleph = \{\alpha : \alpha < \aleph,\ \alpha \text{ is a good limit ordinal}\}$$

also has cardinality $\aleph$. The set of bad ordinals

$$\Omega_\aleph = \{\alpha : \aleph \leq \alpha < g(\aleph)\}$$

has cardinality $\leq \aleph$ and therefore one can define f such that $f(\Omega_\aleph) \subset \Sigma_\aleph$ and $f|\Omega_\aleph$ is one-to-one. This finishes the proof of the existence of f. We need only to check now that $\{\alpha : f(\alpha) = \beta\}$ is at most countable. By (ii), for each uncountable $\aleph$ such that $\aleph \neq g(\aleph)$ there exists at most one α such that $\operatorname{card}(\alpha) = \aleph$ and $f(\alpha) = \beta$. Thus the lemma follows from the fact the sets $\{\aleph : \aleph > \aleph_0,\ g(\aleph) \neq \aleph\}$ and $\{\alpha : \aleph_0 \leq \alpha < g(\aleph_0)\}$ are at most countable.

5.23. THEOREM. *Every operator T of class C_0 is quasisimilar to a Jordan operator $S(\Theta)$. Moreover, $S(\Theta)$ is uniquely determined by either of the relations $S(\Theta) \prec T$ and $T \prec S(\Theta)$.*

PROOF. The uniqueness part follows as in the proof of Theorem 5.1. For the existence part we may assume that $T = \bigoplus_\alpha S(\theta_\alpha)$, where the inner functions $\{\theta_\alpha\}$ are as in Corollary 5.17. Now, as seen in Remark 5.21, the operator $T' = \bigoplus_{\alpha \text{ is good}} S(\theta_\alpha)$ is unitarily equivalent to a Jordan operator. Therefore it will suffice to show that T and T' are quasisimilar. Let f be the function on bad ordinals given by Lemma 2.22, and note that T is unitarily equivalent to

$$T'' = T''' \oplus \left\{ \bigoplus_{\substack{\alpha \text{ is a good} \\ \text{limit ordinal}}} \left[\left(\bigoplus_{j<\omega} S(\theta_{\alpha+j}) \right) \oplus \left(\bigoplus_{f(\beta)=\alpha} S(\theta_\beta) \right) \right] \right\},$$

where T''' collects all good ordinals that do not have the form $\alpha + j$ with a good limit ordinal α and $j < \omega$. Of course $\alpha + j$ is good wherever α is good and $j < \omega$. Now, we know that the relation $f(\beta) = \alpha$ implies that $\operatorname{card}(\beta) \geq \aleph_0$ and $\alpha \leq \operatorname{card} \beta$. Consequently $\alpha + j \leq \beta$ for all $j < \omega$ and this implies that $\theta_\beta | \theta_{\alpha+j}$

if $f(\beta) = \alpha$ and $j < \omega$. Therefore the minimal function of the separably acting operator $\bigoplus_{f(\beta)=\alpha} S(\theta_\beta)$ divides $\theta_{\alpha+j}$ for all $j < \omega$ and Corollary 5.10 implies that

$$\left(\bigoplus_{j<\omega} S(\theta_{\alpha+j})\right) \oplus \left(\bigoplus_{f(\beta)=\alpha} S(\theta_\beta)\right)$$

is quasisimilar to $\bigoplus_{j<\omega} S(\theta_{\alpha+j})$. We conclude that T'' is quasisimilar to

$$T''' \oplus \left\{ \bigoplus_{\substack{\alpha \text{ is a good} \\ \text{limit ordinal}}} \left(\bigoplus_{j<\omega} S(\theta_{\alpha+j})\right)\right\}$$

while this last operator is unitarily equivalent to T'. The theorem follows.

The operator $S(\Theta)$ whose existence has just been proved will be called the *Jordan model* of T. We conclude this section with a few useful consequences of the existence of Jordan models.

5.24. DEFINITION. Let T be an operator of class C_0. The model function $M_T = \{M_T(\alpha)\}$ is defined by

$$M_T(\alpha) = \bigwedge\{\theta : \nu_T(\theta) \leq \text{card}(\alpha)\}.$$

It is easy to check that M_T is indeed a model function and $M_T(0) \equiv m_T$.

5.25. COROLLARY. *Every operator T of class C_0 is quasisimilar to the Jordan operator $S(M_T)$.*

PROOF. Let $S(\Theta)$ be the Jordan model of the operator T of class C_0, where $\Theta = \{\theta_\alpha\}$ is a model function. Our corollary now obviously follows from Corollary 4.13 combined with Lemma 4.6.

5.26. COROLLARY. *For every operator T of class C_0 we have $\mu_T = \mu_{T^*}$.*

PROOF. Let $S(\Theta)$ be the Jordan model of T. By Lemma 1.3.3 we must have $\mu_T = \mu_{S(\Theta)}$ and $\mu_{T^*} = \mu_{S(\Theta)^*}$ and the equality $\mu_{S(\Theta)} = \mu_{S(\Theta)^*}$ follows by Theorem 4.12.

5.27. COROLLARY. *If T is an operator of class C_0 and $\mathscr{M}$ is an invariant subspace for T, then $\mu_{T|\mathscr{M}} \leq \mu_T$.*

PROOF. We have $(T|\mathscr{M})^* P_{\mathscr{M}} = P_{\mathscr{M}} T^*$ and Lemma 1.3.3 implies that $\mu_{(T|\mathscr{M})^*} \leq \mu_{T^*}$. The result follows now from the preceding corollary because $T|\mathscr{M}$ is also an operator of class C_0.

5.28. DEFINITION. An operator T can be *injected in* T' if there exists an injective operator $X \in \mathscr{I}(T, T')$. We indicate by $T \prec^i T'$ the fact that T can be injected in T'.

5.29. COROLLARY. *If T and T' are operators of class C_0 and $T \prec^i T'$ then $\mu_T \leq \mu_{T'}$.*

PROOF. Let $X \in \mathcal{I}(T',T)$ be one-to-one. Then $X^* \in \mathcal{I}(T^*,T'^*)$ has dense range and the inequality $\mu_{T^*} \le \mu_{T'^*}$ follows from Lemma 1.3.3. It suffices now to apply Corollary 5.26.

5.30. PROPOSITION. *Let $T \in \mathcal{L}(\mathcal{H})$ be an operator of class C_0. There exists a bounded, one-to-one, with dense range, conjugate linear operator $J : \mathcal{H} \to \mathcal{H}$ such that $T^*J = JT$.*

PROOF. Let $S(\Theta)$ be the Jordan model of T, where $\Theta = \{\theta_\alpha\}$ is a model function. By Proposition 1.8 we can find antiunitary operators J_α on $\mathcal{H}(\theta_\alpha)$ such that $S(\theta_\alpha)^*J_\alpha = J_\alpha S(\theta_\alpha)$. Now choose a quasiaffinity $X \in \mathcal{I}(T, S(\Theta))$ and define $J = X^*(\bigoplus_\alpha J_\alpha)X$. It is clear that J is antilinear, one-to-one, and has dense range. The relation $T^*J = JT$ follows easily from the fact that $X^* \in \mathcal{I}(S(\Theta)^*, T^*)$.

5.31. PROPOSITION. *Let T and T' be two operators of class C_0. The following assertions are equivalent:*

(i) $T \prec^i T'$;
(ii) $T^* \prec^i T'^*$;
(iii) $\nu_T(\theta) \le \nu_{T'}(\theta)$ *for every inner function* θ; *and*
(iv) $M_T(\alpha)|M_{T'}(\alpha)$ *for every ordinal* α.

PROOF. Let J and J' be antilinear operators as in Proposition 5.30 such that $TJ = JT^*$ and $T'^*J' = J'T'$. If $X \in \mathcal{I}(T,T')$ is one-to-one then $J'XJ \in \mathcal{I}(T^*,T'^*)$ is one-to-one and we see that (i)⇒(ii). For reasons of symmetry, (i) and (ii) are equivalent. Assume that $T \prec^i T'$. Then

$$T|(\operatorname{ran}\theta(T))^- \prec^i T'|(\operatorname{ran}\theta(T'))^-,$$

the intertwining being realized by $X|(\operatorname{ran}\theta(T))^-$, where X is any injection in $\mathcal{I}(T,T')$. Therefore the implication (i)⇒(iii) follows from Corollary 5.29.

Now assume that (iii) holds. Then we clearly have

$$\{\theta : \nu_T(\theta) \le \operatorname{card}(\alpha)\} \supset \{\theta : \nu_{T'}(\theta) \le \operatorname{card}(\alpha)\}$$

so that clearly $M_T(\alpha) = \bigwedge\{\theta : \nu_T(\theta) \le \operatorname{card}(\alpha)\}$ divides $M_{T'}(\alpha)$ for every ordinal α.

Finally, assuming (iv), we infer from Proposition 1.10 that $S(M_T(\alpha))$ is unitarily equivalent to some restriction of $S(M_{T'}(\alpha))$ to an invariant subspace. Consequently $S(M_T)$ is unitarily equivalent to the restriction of $S(M_{T'})$ to some invariant subspace, and this clearly implies that $S(M_T) \prec^i S(M_{T'})$. The relation $T \prec^i T'$ now follows from the transitivity of "$\prec^i$" and therefore (iv)⇒(i). The proof is complete.

5.32. PROPOSITION. *Let T and T' be two operators of class C_0. The following assertions are equivalent:*

(i) $T \prec T'$;
(ii) $T \prec^i T'$ *and* $T' \prec^i T$;
(iii) $\nu_T(\theta) = \nu_{T'}(\theta)$ *for every inner function* θ; *and*

(iv) $T \sim T'$.

PROOF. Assume that $T \prec T'$ and $S(\Theta)$ is the Jordan model of T. By transitivity we have $S(\Theta) \prec T'$ and hence $S(\Theta)$ is also the Jordan model of T'. Consequently, T and T' are quasisimilar. We have thus established that (i)⇒(iv). That (iv) implies (ii) is trivial and the implication (ii)⇒(iii) obviously follows from Proposition 5.31. Assuming (iii), the same proposition implies that $M_T(\alpha) \equiv M_{T'}(\alpha)$ for every ordinal α and thus T and T' have the same Jordan model, in particular $T \prec T'$. The proposition follows.

We finally note a description of the invariant subspaces of an operator of class C_0.

5.33. PROPOSITION. *Let T be an operator of class C_0 and $\mathscr{M}$ an invariant subspace for T. There exist operators X and Y in $\{T\}'$ such that*

$$\mathscr{M} = (\operatorname{ran} X)^- = \ker Y.$$

PROOF. We obviously have $T|\mathscr{M} \prec^i T$ and therefore we deduce $(T|\mathscr{M})^* \prec^i T^*$. Let $X^* : \mathscr{M} \to \mathscr{H}$ be one-to-one with the property that $T^*X^* = (T|\mathscr{M})^*X^*$. Then $X : \mathscr{H} \to \mathscr{M}$ has dense range and, when regarded as an operator on $\mathscr{H}$, $X \in \{T\}'$. The same argument, applied to T^* and $\mathscr{H} \ominus \mathscr{M}$, shows the existence of $Y \in \{T\}'$ such that $(\operatorname{ran} Y^*)^- = \mathscr{H} \ominus \mathscr{M}$. Clearly then $\mathscr{M} = \ker Y^*$, thus concluding the proof.

Exercises

1. Show that every operator T of class C_0 has invariant subspaces that are not hyperinvariant if $\mu_T \geq 2$.

2. Let θ and θ' be inner functions. Find the Jordan model of $S(\theta) \oplus S(\theta')$.

3. Let θ, θ', and θ'' be inner functions. Find the Jordan model of $S(\theta) \oplus S(\theta') \oplus S(\theta'')$.

4. Let $\{\theta_j : j < \omega\}$ be a sequence of inner functions such that $\theta_j | \theta_{j+1}$, $j < \omega$, and $T = \bigoplus_{j<\omega} S(\theta_j)$ is an operator of class C_0. Find the Jordan model of T.

5. For $t > 0$ define $\theta_t(\lambda) = \exp(t(\lambda+1)/(\lambda-1))$, $\lambda \in D$. Find the Jordan model of $\bigoplus_{t\in[0,1]} S(\theta_t)$.

6. Let $S(\Theta)$ be a Jordan operator and $\aleph$ a cardinal number. Find the Jordan model of $S(\Theta)^{(\aleph)}$.

7. Let $T \in \mathscr{L}(\mathscr{H})$ be an operator of class C_0 with Jordan model $S(\Theta)$, $\Theta = \{\theta_\alpha\}$. Assume that $\mathscr{H}_0$ is a separable reducing subspace for T and $T|\mathscr{H}_0$ has Jordan model $\bigoplus_{j<\omega} S(\theta_j)$. Prove that $m_{T|\mathscr{H}\ominus\mathscr{H}_0} | \theta_j$ for all $j < \omega$. (Compare this result with Lemma 5.11.)

8. Generalize Exercise 7 to nonseparable reducing spaces.

9. Extend Corollary 5.10 to nonseparably acting operators.

10. Find an operator T and an invariant subspace $\mathcal{M}$ for T such that $\mathcal{M}$ cannot be written as $(\operatorname{ran} X)^-$ or $\ker X$ with $X \in \{T\}'$.

11. Find operators T and T' such that $T \prec^i T'$ and $T' \prec^i T$ but T and T' are not quasisimilar.

12. Let $T \in \mathcal{L}(\mathcal{H})$ satisfy the relation $T^2 = 0$. Show that T can be written as

$$T = \begin{bmatrix} 0 & A & 0 \\ 0 & 0 & 0 \\ 0 & 0 & 0 \end{bmatrix},$$

with respect to a certain decomposition $\mathcal{H} = \mathcal{H}' \oplus \mathcal{H}'' \oplus \mathcal{H}'''$, such that A is a quasiaffinity.

13. Let T and A be as above. Show that T has closed range if and only if A is invertible. If A is invertible, show that T is similar to

$$\begin{bmatrix} 0 & I & 0 \\ 0 & 0 & 0 \\ 0 & 0 & 0 \end{bmatrix}$$

on $\mathcal{H}' \oplus \mathcal{H}' \oplus \mathcal{H}'''$.

14. Deduce from Exercise 13 that any operator T such that $T^2 = 0$ and $\operatorname{ran} T$ is closed is similar to an operator T' of the form $T' = J_1^{(\alpha)} \oplus J_2^{(\beta)}$ where $J_1 = 0 \in \mathcal{L}(\mathbf{C})$ and $J_2 = \left[\begin{smallmatrix} 0 & 1 \\ 0 & 0 \end{smallmatrix}\right] \in \mathcal{L}(\mathbf{C}^2)$ and α, β are cardinals. Note that T' is not unitarily equivalent to the Jordan model of T unless either $\beta < \omega$ or $\beta < \alpha$.

15. Generalize Exercise 14. More precisely, assume $T \in \mathcal{L}(\mathcal{H})$ is such that $T^n = 0 \neq T^{n-1}$ and $T\mathcal{H}, T^2\mathcal{H}, \dots, T^{n-1}\mathcal{H}$ are closed. Prove that T is similar with an operator T' of the form $J_1^{(\alpha_1)} \oplus J_2^{(\alpha_2)} \oplus \cdots \oplus J_n^{(\alpha_n)}$, where $J_1, J_2, \dots, J_n$ are Jordan cells and $\alpha_1, \alpha_2, \dots, \alpha_n$ are cardinals.

16. In order to extend Exercise 15 we call Jordan cells the operators $J_{\lambda,k} \in \mathcal{L}(\mathbf{C}^k)$ defined for $\lambda \in \mathbf{C}$ and $k \geq 1$ by $J_{\lambda,k} e_j = \lambda e_j + e_{j-1}$, $j \geq 2$, $J_{\lambda,k} e_1 = \lambda e_1$. Prove that an algebraic operator $T \in \mathcal{L}(\mathcal{H})$ is similar to an orthogonal sum of Jordan cells if and only if $q(T)\mathcal{H}$ is closed for every polynomial q.

17. Does Exercise 16 have an analogue for operators of class C_0?

6. Approximate decompositions. If T is a linear operator on a finite-dimensional space $\mathcal{H}$, then the classical theorem of Jordan shows that T is similar to a direct sum of Jordan cells. In particular, $\mathcal{H}$ can be decomposed into a direct (not orthogonal) sum of invariant subspaces for T such that the restriction of T to each of these subspaces is similar to some Jordan cell. We cannot expect this kind of decomposition to hold for arbitrary operators of class C_0 because quasisimilarity does not usually imply similarity. In this section we show that a weaker notion of decomposition allows us to prove analogues of the finite-dimensional result.

6.1. DEFINITION. Let $\mathscr{H}$ be a Hilbert space and $\{\mathscr{H}_j : j \in J\}$ a family of closed subspaces of $\mathscr{H}$ satisfying the relation $\mathscr{H} = \bigvee_{j\in J} \mathscr{H}_j$.

(i) We say that $\mathscr{H}$ is the *approximate sum* of the family $\{\mathscr{H}_j : j \in J\}$ if $\mathscr{H}_i \cap (\bigvee_{j\neq i} \mathscr{H}_j) = \{0\}$ for every $i \in J$.

(ii) We say that $\mathscr{H}$ is the *almost direct sum* of the family $\{\mathscr{H}_j : j \in J\}$ if for every family $\{K_\alpha : \alpha \in A\}$ of subsets of J such that $\bigcap_{\alpha\in A} K_\alpha = \varnothing$ we have $\bigcap_{\alpha\in A}(\bigvee_{j\in K_\alpha} \mathscr{H}_j) = \{0\}$.

(iii) We say that $\mathscr{H}$ is the *quasidirect sum* of the family $\{\mathscr{H}_j : j \in J\}$ if for every family $\{K_\alpha : \alpha \in A\}$ of subsets of J we have $\bigcap_{\alpha\in A}(\bigvee_{j\in K_\alpha} \mathscr{H}_j) = \bigvee_{j\in K} \mathscr{H}_j$, where $K = \bigcap_{\alpha\in A} K_\alpha$.

It is quite obvious that every quasidirect sum is an almost direct sum, and every almost direct sum is an approximate sum. The almost direct sums seem to be the easiest to deal with because of the following simple result.

6.2. LEMMA. *Let $\{\mathscr{H}_j : j \in J\}$ be a family of subspaces of $\mathscr{H}$, and set $\mathscr{K}_i = \mathscr{H} \ominus (\bigvee_{j\neq i} \mathscr{H}_j)$ for $i \in I$. Then $\mathscr{H}$ is the almost direct sum of the family $\{\mathscr{H}_j : j \in J\}$ if and only if*

$$\bigvee_{j\in J} \mathscr{H}_j = \mathscr{H} = \bigvee_{j\in J} \mathscr{K}_j.$$

PROOF. Assume that $\mathscr{H}$ is the almost direct sum of $\{\mathscr{H}_j : j \in J\}$. Since $\bigcap_{i\in J}\{j \in J : j \neq i\} = \varnothing$, we must have

$$\bigvee_{i\in J} \mathscr{K}_i = \mathscr{H} \ominus \left(\bigcap_{i\in J}\left(\bigvee_{j\neq i} \mathscr{H}_j\right)\right) = \mathscr{H} \ominus \{0\} = \mathscr{H}$$

and, of course, $\bigvee_{j\in J} \mathscr{H}_j = \mathscr{H}$. Conversely, assume that $\bigvee_{j\in J} \mathscr{H}_j = \mathscr{H} = \bigvee_{j\in J} \mathscr{K}_j$ and $\{K_\alpha : \alpha \in A\}$ is a family of subsets of J such that $\bigcap_{\alpha\in A} K_\alpha = \varnothing$. Then we must have $\bigcup_{\alpha\in A}(J \backslash K_\alpha) = J$ and hence

$$\begin{aligned} \mathscr{H} \ominus \left(\bigcap_{\alpha\in A}\left(\bigvee_{j\in K_\alpha} \mathscr{H}_j\right)\right) &= \bigvee_{\alpha\in A}\left(\mathscr{H} \ominus \bigvee_{j\in K_\alpha} \mathscr{H}_j\right) \\ &\supset \bigvee_{\alpha\in A}\left(\bigvee_{i\in J\backslash K_\alpha} \mathscr{K}_i\right) \\ &= \bigvee_{i\in J} \mathscr{K}_i = \mathscr{H}, \end{aligned}$$

where the inclusion holds because $\mathscr{K}_i \perp \mathscr{H}_j$ for $i \neq j$. We conclude that $\bigcap_{\alpha\in A}(\bigvee_{j\in K_\alpha} \mathscr{H}_j) = \{0\}$. The lemma follows.

An obvious consequence of Lemma 6.2 is that, if $\mathscr{H}$ is the almost direct sum of $\{\mathscr{H}_j : j \in J\}$, then $\mathscr{H}$ is also the almost direct sum of $\{\mathscr{K}_j : j \in J\}$.

6.3. LEMMA. *Let $T \in \mathscr{L}(\mathscr{H})$ be an operator of class C_0, and let $S(\Theta)$, $\Theta = \{\theta_\alpha : \alpha < \gamma\}$, be the Jordan model of T. If $\mathscr{H}'$ is an arbitrary separable* *subspace of $\mathscr{H}$, there exists a reducing separable subspace $\mathscr{K}_0$ for T such that $\mathscr{K}_0 \supset \mathscr{H}'$ and $T|\mathscr{K}_0$ is quasisimilar to $\bigoplus_{j<\omega} S(\theta_j)$.*

PROOF. Let X be any quasiaffinity satisfying the relation $TX = XS(\Theta)$ and denote by $\mathscr{K}_0$ the smallest reducing subspace for T containing $\mathscr{H}'$ and $X(\bigoplus_{j<\omega} \mathscr{H}(\theta_j))$. The subspace $\mathscr{K}_0$ is separable so that the Jordan model of $T|\mathscr{K}_0$ must have the form $\bigoplus_{j<\omega} S(\theta'_j)$. We obviously have $T|\mathscr{K}_0 \prec^i T$ and hence $\theta'_j|\theta_j$ for $j < \omega$ by Proposition 5.31. On the other hand we have $X(\bigoplus_{j<\omega} \mathscr{H}(\theta_j)) \subset \mathscr{K}_0$ and hence $X|\bigoplus_{j<\omega} \mathscr{H}(\theta_j)$ is an injection in $\mathscr{I}(\bigoplus_{j<\omega} S(\theta_j), T|\mathscr{K}_0)$. We conclude that $\theta_j|\theta'_j$ by Proposition 5.31 and thus that $\theta_j \equiv \theta'_j$, $j < \omega$.

The following result shows that, in our search for approximate decompositions, we can restrict ourselves to separably acting operators. The reader may want to compare this result with Proposition 5.12.

6.4. THEOREM. *Let $T \in \mathscr{L}(\mathscr{H})$ be an operator of class C_0, and let $S(\Theta)$, $\Theta = \{\theta_\alpha\}$, be the Jordan model of T. We can associate with each limit ordinal α a reducing subspace $\mathscr{K}_\alpha$ for T such that*

(i) *$\mathscr{K}_\alpha \perp \mathscr{K}_\beta$ for $\alpha \neq \beta$;*

(ii) *$\mathscr{H} = \bigoplus_\alpha \mathscr{K}_\alpha$; and*

(iii) *$T|\mathscr{K}_\alpha$ is quasisimilar to $\bigoplus_{j<\omega} S(\theta_{\alpha+j})$.*

PROOF. Fix a quasiaffinity $X \in \mathscr{I}(S(\Theta), T)$. We construct by transfinite induction reducing subspaces $\mathscr{K}_\alpha$ for each limit ordinal α with the following properties:

$$\mathscr{K}_\alpha \perp \mathscr{K}_\gamma \quad \text{for } \gamma < \alpha; \tag{6.5}$$

$$\bigoplus_{\gamma\le\alpha} \mathscr{K}_\gamma \supset X\left(\bigoplus_{\gamma\le\alpha}\bigoplus_{j<\omega} \mathscr{H}(\theta_{\gamma+j})\right); \tag{6.6}$$

and

$$T|\mathscr{K}_\alpha \quad \text{is quasisimilar to } \bigoplus_{j<\omega} S(\theta_{\alpha+j}). \tag{6.7}$$

For $\alpha = 0$ let $\mathscr{K}_0$ be given by Lemma 6.3 with $\mathscr{H}' = X(\bigoplus_{j<\omega} \mathscr{H}(\theta_j))$. Conditions (6.6) and (6.7) are satisfied by that lemma, while (6.5) is vacuous for $\alpha = 0$. Now assume that β is a limit ordinal and the spaces $\mathscr{K}_\alpha$ have already been defined for every limit ordinal $\alpha < \beta$; note that β must be transfinite. The spaces

$$\mathscr{L} = \bigoplus_{\alpha<\beta} \mathscr{K}_\alpha, \qquad \mathscr{K} = \mathscr{H} \ominus \mathscr{L}$$

obviously reduce T and (6.6) implies that

$$\mathscr{L} \supset X\left(\bigoplus_{\gamma<\beta}\bigoplus_{j<\omega} \mathscr{H}(\theta_{\gamma+j})\right). \tag{6.8}$$

The idea is now to obtain $\mathscr{K}_\beta$ by applying Lemma 6.3 to $T|\mathscr{K}$ with $\mathscr{H}' = P_{\mathscr{K}}X(\bigoplus_{j<\omega}\mathscr{H}(\theta_{\beta+j}))$. The space $\mathscr{K}_\beta$ thus obtained obviously satisfies (6.5) for $\alpha = \beta$. Relation (6.6) with $\alpha = \beta$ is also easy to deduce from (6.8) and the fact that $\mathscr{K}_\beta \supset \mathscr{H}'$. Finally, Lemma 6.3 implies that $T|\mathscr{K}_\beta$ is quasisimilar to $\bigoplus_{j<\omega} S(\theta_j')$, where $S(\Theta')$, $\Theta' = \{\theta_\alpha'\}$, is the Jordan model of $T|\mathscr{K}$. Thus (6.6) will be proved if we show that $\theta_\gamma' \equiv \theta_{\beta+\gamma}$ for every ordinal γ. To do this we note that (6.8) implies the inclusion

$$X^*\mathscr{K} \subset \bigoplus_\gamma \mathscr{H}(\theta_{\beta+\gamma}),$$

from which it follows that

$$T^*|\mathscr{K} \prec^i \bigoplus_\gamma S(\theta_{\beta+\gamma}).$$

Now we remark that the family $\{\theta_{\beta+\gamma}\}$ is a model function; indeed, $\text{card}(\gamma) = \text{card}(\gamma')$ implies $\text{card}(\beta+\gamma) = \text{card}(\beta+\gamma')$ and hence $\theta_{\beta+\gamma} \equiv \theta_{\beta+\gamma'}$. Consequently, Proposition 5.31 implies

$$\theta_\gamma' | \theta_{\beta+\gamma}, \qquad \gamma \geq 0. \tag{6.9}$$

On the other hand, Corollary 5.25 shows that

$$\begin{aligned}\theta_{\beta+\gamma} &\equiv \bigwedge\{\theta : \nu_T(\theta) \leq \text{card}(\beta+\gamma)\}\\ &= \bigwedge\{\theta : \nu_{(T|\mathscr{K})\oplus(T|\mathscr{L})}(\theta) \leq \text{card}(\beta+\gamma)\}\end{aligned}$$

and we clearly have

$$\begin{aligned}\nu_{(T|\mathscr{K})\oplus(T|\mathscr{L})}(\theta) &\leq \nu_{T|\mathscr{K}}(\theta) + \mu_{T|\mathscr{L}} \leq \nu_{T|\mathscr{K}}(\theta) + \aleph_0\,\text{card}(\beta)\\ &\leq \nu_{T|\mathscr{K}}(\theta) + \text{card}(\beta),\end{aligned}$$

where we used the fact that $\text{card}(\beta) \geq \aleph_0$. Thus the inequality $\nu_{T|\mathscr{K}}(\theta) \leq \text{card}(\gamma)$ implies $\nu_T(\theta) \leq \text{card}(\beta+\gamma)$ so that

$$\theta_{\beta+\gamma} | \bigwedge\{\theta : \nu_{T|\mathscr{K}}(\theta) \leq \text{card}(\gamma)\} \equiv \theta_\gamma'.$$

This relation combined with (6.9) shows that $\theta_\gamma' \equiv \theta_{\beta+\gamma}$ and thus finishes the inductive construction of the reducing spaces $\mathscr{K}_\alpha$. This also concludes the proof of the theorem because relation (6.6) shows that $\bigoplus_\alpha \mathscr{K}_\alpha = \mathscr{H}$.

The proof of the following result is a refinement of the proof of Theorem 5.1 using the more complicated splitting principle in Theorem 3.6.

6.10. THEOREM. *Let $T \in \mathscr{L}(\mathscr{H})$ be an operator of class C_0, and let $S(\Theta)$, $\Theta = \{\theta_\alpha\}$, be the Jordan model of T. We can associate with each ordinal α an invariant subspace $\mathscr{H}_\alpha$ for T with the following properties*:

(i) *$\mathscr{H}$ is the almost direct sum of $\{\mathscr{H}_\alpha\}$*;

(ii) *$T|\mathscr{H}_\alpha$ is quasisimilar to $S(\theta_\alpha)$ for each α; and*

(iii) *$\mathscr{H}_{\alpha+i} \perp \mathscr{H}_{\beta+j}$ if α and β are different limit ordinals and $i, j < \omega$.*

PROOF. By virtue of Theorem 6.4 it suffices to consider the case in which $\mathscr{H}$ is separable. Indeed, assume that the spaces $\mathscr{K}_\alpha$ are provided by Theorem 6.4, and for each α we write $\mathscr{K}_\alpha$ as the almost direct sum of invariant subspaces $\{\mathscr{H}_{\alpha+j} : j < \omega\}$ such that $T|\mathscr{H}_{\alpha+j}$ is quasisimilar to $S(\theta_{\alpha+j})$. Then the spaces $\{\mathscr{H}_{\alpha+j} : \alpha = \text{limit ordinal}, j < \omega\}$ satisfy conditions (i)–(iii) of the present theorem.

Assume thus that $\mathscr{H}$ is separable and, as in the proof of Theorem 5.1, choose dense sequences $\{h_n : n \geq 0\}$ and $\{k_n : n \geq 0\}$ in $\mathscr{H}$ such that each vector h_i appears infinitely often among the k_n. We will construct inductively vectors $f_0, f_1, f_2, \ldots$ in $\mathscr{H}$, subspaces $\mathscr{H}_0, \mathscr{H}_1, \ldots$ and $\mathscr{M}_0, \mathscr{M}_1, \ldots$ invariant for T, and subspaces $\mathscr{H}_0', \mathscr{H}_1', \ldots$ and $\mathscr{M}_0', \mathscr{M}_1', \ldots$ invariant for T^* with the following properties:

$$f_j \in \mathscr{M}_{j-1} \quad \text{and} \quad m_{f_j} \equiv m_{T|\mathscr{M}_{j-1}}, \quad \text{where } \mathscr{M}_{-1} = \mathscr{H}; \tag{6.11}$$

$$\mathscr{H}_j = \bigvee_{k=0}^{\infty} T^k f_j; \tag{6.12}$$

$$\mathscr{M}_j' = \mathscr{H} \ominus (\mathscr{H}_0 \vee \mathscr{H}_1 \vee \cdots \vee \mathscr{H}_j), \qquad \mathscr{M}_j = \mathscr{H} \ominus (\mathscr{H}_0' \vee \mathscr{H}_1' \vee \cdots \vee \mathscr{H}_j'); \tag{6.13}$$

$$P_{\mathscr{H}_j'}|\mathscr{H}_j \text{ and } P_{\mathscr{M}_j'}|\mathscr{M}_j \text{ are quasiaffinities;} \tag{6.14}$$

and

$$\begin{cases} \|k_i - P_{\mathscr{H}_0 \vee \mathscr{H}_1 \vee \cdots \vee \mathscr{H}_j} k_i\| < 2^{-j}, & \text{if } j \text{ is even and } i = j/2, \\ \|k_i - P_{\mathscr{H}_0' \vee \mathscr{H}_1' \vee \cdots \vee \mathscr{H}_j'} k_i\| < 2^{-j}, & \text{if } j \text{ is odd and } i = (j-1)/2. \end{cases} \tag{6.15}$$

Assume that n is a natural number such that vectors f_j and subspaces $\mathscr{H}_j$, $\mathscr{M}_j$, $\mathscr{H}_j'$, $\mathscr{M}_j'$ satisfying (6.11)–(6.15) for $j < n$ have been constructed (note that this assumption is vacuously satisfied for $n = 0$). In order to choose the vector f_n we distinguish three cases. If $n = 0$ we choose $f_0 \in \mathscr{M}_{-1} = \mathscr{H}$ satisfying (6.11) for $j = n$ and such that

$$\|f_0 - k_0\| < 1; \tag{6.16}$$

the existence of such a vector f_0 follows from the density of T-maximal vectors (Theorem 2.3.6). If $n > 0$ and $n = 2i$ we remark that by (6.13) and (6.14) we have

$$\begin{aligned} \mathscr{H} &= (\mathscr{H}_0 \vee \mathscr{H}_1 \vee \cdots \vee \mathscr{H}_{n-1}) \oplus \mathscr{M}_{n-1}' = (\mathscr{H}_0 \vee \mathscr{H}_1 \vee \cdots \vee \mathscr{H}_{n-1}) \vee P_{\mathscr{M}_{n-1}'} \mathscr{M}_{n-1} \\ &= \mathscr{H}_0 \vee \mathscr{H}_1 \vee \cdots \vee \mathscr{H}_{n-1} \vee \mathscr{M}_{n-1} \end{aligned}$$

and consequently we can choose $u \in \mathscr{H}_0 \vee \mathscr{H}_1 \vee \cdots \vee \mathscr{H}_{n-1}$ and $v \in \mathscr{M}_{n-1}$ such that

$$\|k_i - u - v\| < 2^{-n-1}. \tag{6.17}$$

We can then choose $f_n \in \mathscr{M}_{n-1}$ satisfying (6.11) for $j = n$ and, in addition,

$$\|f_n - v\| < 2^{-n-1}. \tag{6.18}$$

Finally, if $n > 0$ and $n = 2i + 1$, analogous calculations show that $\mathscr{H} = \mathscr{H}_0' \vee \mathscr{H}_1' \vee \cdots \vee \mathscr{H}_{n-1}' \vee \mathscr{M}_n'$ so that we can find $u' \in \mathscr{H}_0' \vee \mathscr{H}_1' \vee \cdots \vee \mathscr{H}_{n-1}'$ and $v' \in \mathscr{M}_{n-1}'$ satisfying

$$(6.19) \qquad ||k_i - u' - v'|| < 2^{-n-1}.$$

One more use of Theorem 2.3.6 allows us to find f_n satisfying (6.11) for $j = n$ and such that

$$(6.20) \qquad ||P_{\mathscr{M}_{n-1}'} f_n - v'|| < 2^{-n-2};$$

here, of course, we use the fact that $P_{\mathscr{M}_{n-1}'}\mathscr{M}_{n-1}$ is dense in $\mathscr{M}_{n-1}'$. The vector f_n being chosen, we can define $\mathscr{H}_n = \bigvee_{k=0}^{\infty} T^k f_n$ and $\mathscr{M}_n' = \mathscr{M}_{n-1}' \ominus (P_{\mathscr{M}_{n-1}'}\mathscr{H}_n)^-$. Theorem 3.6 applied to the quasiaffinity $X = P_{\mathscr{M}_{n-1}'}|\mathscr{M}_{n-1} \in \mathscr{I}(T|\mathscr{M}_{n-1}, (T^*|\mathscr{M}_{n-1}')^*)$ with $\varepsilon = 2^{-n-2}$ provides a subspace $\mathscr{H}_n'$ invariant for $T^*|\mathscr{M}_{n-1}'$ and a subspace $\mathscr{M}_n$ invariant for $T|\mathscr{M}_{n-1}$ such that

$$(6.21) \qquad ||P_{\mathscr{H}_n'} P_{\mathscr{M}_{n-1}'} f_n - P_{\mathscr{M}_{n-1}'} f_n|| < 2^{-n-2},$$

$\mathscr{M}_n = \mathscr{M}_{n-1} \ominus (P_{\mathscr{M}_{n-1}}\mathscr{H}_n')^-$, and $P_{\mathscr{H}_n'}X|\mathscr{H}_n$ and $P_{\mathscr{M}_n'}X|\mathscr{M}_n$ are quasiaffinities. Since $P_{\mathscr{H}_n'}X|\mathscr{H}_n = P_{\mathscr{H}_n'}|\mathscr{H}_n$ and $P_{\mathscr{M}_n'}X|\mathscr{M}_n = P_{\mathscr{M}_n'}|\mathscr{M}_n$ it follows that (6.14) is satisfied for $j = n$. Condition (6.12) is verified by the definition of $\mathscr{H}_n$. Next observe that

$$\begin{aligned}
\mathscr{M}_n &= \mathscr{M}_{n-1} \ominus (P_{\mathscr{M}_{n-1}}\mathscr{H}_n')^- = \mathscr{M}_{n-1} \cap (\mathscr{H} \ominus \mathscr{H}_n') \\
&= [\mathscr{H} \ominus (\mathscr{H}_1' \vee \mathscr{H}_2' \vee \cdots \vee \mathscr{H}_{n-1}')] \cap (\mathscr{H} \ominus \mathscr{H}_n') \\
&= \mathscr{H} \ominus (\mathscr{H}_1' \vee \mathscr{H}_2' \vee \cdots \vee \mathscr{H}_n'),
\end{aligned}$$

and an analogous calculation for $\mathscr{M}_n'$ shows that (6.13) is also true for $j = n$. Property (6.15) is verified for $n = 0$ because

$$||k_0 - P_{\mathscr{H}_0} k_0|| \le ||k_0 - f_0|| < 1 = 2^0$$

by (6.16). If $n = 2i > 0$ then

$$\begin{aligned}
||k_i - P_{\mathscr{H}_0 \vee \mathscr{H}_1 \vee \cdots \vee \mathscr{H}_n} k_i|| &\le ||k_i - u - f_n|| \le ||k_i - u - v|| + ||v - f_n|| \\
&< 2^{-n-1} + 2^{-n-1} = 2^{-n}
\end{aligned}$$

by (6.17) and (6.18). If $n = 2i + 1$ we have

$$\begin{aligned}
||k_i - P_{\mathscr{H}_0' \vee \mathscr{H}_1' \vee \cdots \vee \mathscr{H}_n'} k_i|| &\le ||k_i - u' - P_{\mathscr{H}_n'} P_{\mathscr{M}_{n-1}'} f_n|| \\
&\le ||k_i - u' - v'|| + ||v' - P_{\mathscr{M}_{n-1}'} f_n|| + ||P_{\mathscr{M}_{n-1}'} f_n - P_{\mathscr{H}_n'} P_{\mathscr{M}_{n-1}'} f_n|| \\
&< 2^{-n-1} + 2^{-n-2} + 2^{-n-2} = 2^{-n}
\end{aligned}$$

by (6.19), (6.20), and (6.21). Thus in all cases condition (6.15) is satisfied for $j = n$. The existence of the objects f_j, $\mathscr{H}_j$, $\mathscr{M}_j$, $\mathscr{H}_j'$, $\mathscr{M}_j'$ for all j follows by induction.

Now, as in the proof of Theorem 5.1, the inequalities (6.15) imply that

$$\bigvee_{j=0}^{\infty} \mathscr{H}_j = \bigvee_{j=0}^{\infty} \mathscr{H}_j' = \mathscr{H}.$$

Furthermore, if $i \neq j$ we have $\mathscr{H}_i \perp \mathscr{H}_j'$. Indeed, assume for example that $i < j$, and note that $\mathscr{H}_i \perp \mathscr{H}_i'$ and $\mathscr{H}_j' \subset \mathscr{M}_i'$ by (6.13). Therefore we have $\mathscr{H}_j' \subset \mathscr{H} \ominus (\bigvee_{i\neq j} \mathscr{H}_i)$ and hence

$$\bigvee_{j=0}^{\infty} \left(\mathscr{H} \ominus \left(\bigvee_{i\neq j} \mathscr{H}_i \right)\right) = \mathscr{H}.$$

By Lemma 6.2, $\mathscr{H}$ is the almost direct sum of the space $\{\mathscr{H}_j : j < \omega\}$. The theorem follows now because $T|\mathscr{H}_j$ is quasisimilar to $S(\theta_j)$, with $\theta_j \equiv m_{f_j}$, and $\bigoplus_{j<\omega} S(\theta_j)$ is the Jordan model of T by Corollary 5.9.

Exercises

1. Let $T \in \mathscr{L}(\mathscr{H})$ be an operator of class C_0, and $\{\theta_i : i \in I\}$ a family of pairwise relatively prime inner divisors of m_T such that $\bigvee\{\theta_i : i \in I\} \equiv m_T$ (note that I is at most countable by Proposition 2.2.14). Show that $\mathscr{H}$ is the quasidirect sum of the family $\{\mathscr{H}_i : i \in I\}$ defined by $\mathscr{H}_i = \ker \theta_i(T)$.

2. Let $\mathscr{H}$ be a Hilbert space with an orthonormal basis $\{e_n : n \geq 0\}$. For $n \geq 1$ denote by $\mathscr{H}_n$ the one-dimensional subspace generated by $e_n + ne_0$. Show that $\mathscr{H}$ is the approximate sum, but not the almost direct sum, of the spaces $\{\mathscr{H}_n : n \geq 1\}$.

3. Let $\mathscr{H}$ be a separable infinite-dimensional space, and let $T \in \mathscr{L}(\mathscr{H})$ be such that $(T\mathscr{H})^- = \ker T$. Show that there exists a quasidirect decomposition $\mathscr{H} = \bigvee_{n=1}^{\infty} \mathscr{H}_n$ such that each $\mathscr{H}_n$ is invariant for T and $T|\mathscr{H}_n$ is similar to $\left[\begin{smallmatrix} 0 & 1 \\ 0 & 0 \end{smallmatrix}\right] \in \mathscr{L}(\mathbf{C}^2)$. A proof can be based on the following exercise.

4. Let T be as in Exercise 3, and let ε be an arbitrary positive number. Show that there exists a bounded sequence $\{\alpha_n : n \geq 0\}$ of positive numbers and an invertible operator $X \in \mathscr{L}(\mathscr{H})$ such that $||X||\,||X^{-1}|| \leq 1 + \varepsilon$ and $X^{-1}TX$ is unitarily equivalent to $\bigoplus_{n=0}^{\infty} \left[\begin{smallmatrix} 0 & \alpha_n \\ 0 & 0 \end{smallmatrix}\right]$.

CHAPTER 4

Applications of Jordan Models

We have seen in Chapter 3 that every operator T of class C_0 is quasisimilar to a unique Jordan operator T', called the Jordan model of T. In the first two sections of this chapter we are concerned with those properties that T may share with T'. Thus, in Section 1 we show that T (resp., the commutant $\{T\}'$) is reflexive if and only if T' (resp., $\{T'\}'$) is reflexive. Moreover, the reflexivity of T can be shown to depend on the reflexivity of a single Jordan block, which can easily be calculated from the Jordan model of T. In Section 2 we relate the lattices of hyperinvariant subspaces of T and T'. It is shown that generally the lattice corresponding with T' can be identified with a part (a retract) of that corresponding with T. Thus the Jordan model has the smallest lattice of hyperinvariant subspaces in a given quasisimilarity class. In Section 3 we give an application to the classification of operator semigroups. We begin with some generalities about semigroups, and then we concentrate on those semigroups that vanish for large times. These latter semigroups are closely related with a special subclass of C_0 operators, and they are classified accordingly into quasisimilarity classes. Semigroups of operators are also related with the Hilbert space representations of the convolution algebra $L^1(0,1)$. These representations are classified up to quasisimilarity in Section 4.

1. Algebras generated by T, and reflexivity. We recall that for an arbitrary operator T in $\mathscr{L}(\mathscr{H})$ we denote by $\mathscr{A}_T$ [resp., $\mathscr{W}_T$] the weak* [resp., weakly] closed algebra generated by T and I. If T is a completely nonunitary contraction then $\mathscr{F}_T$ denotes the set of all operators $X \in \mathscr{L}(\mathscr{H})$ such that $X = v(T)^{-1}u(T)$ for some $v \in K_T^\infty$ and $u \in H^\infty$.

Other algebras can be associated with T as follows. If $\mathscr{A}$ is an arbitrary subalgebra of $\mathscr{L}(\mathscr{H})$ then $\mathrm{Lat}(\mathscr{A})$ denotes the collection of all closed invariant subspaces for $\mathscr{A}$, i.e., $\mathscr{M} \in \mathrm{Lat}(\mathscr{A})$ if $X\mathscr{M} \subset \mathscr{M}$ for every $X \in \mathscr{A}$. If $\mathscr{L}$ is a collection of closed subspaces of $\mathscr{H}$ we denote by $\mathrm{Alg}(\mathscr{L})$ the set of those X in $\mathscr{L}(\mathscr{H})$ such that $X\mathscr{M} \subset \mathscr{M}$ for every $\mathscr{M}$ in $\mathscr{L}$. $\mathrm{Alg}(\mathscr{L})$ is always a weakly closed subalgebra of $\mathscr{L}(\mathscr{H})$. We can thus form for every algebra $\mathscr{A}$ the larger algebra $\mathrm{Alg\,Lat}(\mathscr{A})$.

1.1. DEFINITION. An algebra $\mathscr{A} \subseteq \mathscr{L}(\mathscr{H})$ is said to be *reflexive* if $\mathscr{A} = \mathrm{Alg\,Lat}(\mathscr{A})$. An operator T is said to be *reflexive* [resp., *hyperreflexive*] if $\mathscr{W}_T$ [resp., $\{T\}'$] is reflexive.

We clearly have $\mathrm{Lat}(T) = \mathrm{Lat}(\mathscr{W}_T) = \mathrm{Lat}(\mathscr{A}_T)$ so that T is reflexive if and only if $\mathrm{Alg\,Lat}(T) = \mathscr{W}_T$. The following result plays a basic role in the study of reflexive operators of class C_0, and it is interesting on its own.

1.2. THEOREM. *For every operator T of class C_0 we have*

$$\{T\}'' = \{T\}' \cap \mathrm{Alg\,Lat}(T) = \mathscr{A}_T = \mathscr{W}_T = \mathscr{F}_T.$$

PROOF. It clearly suffices to verify the following six inclusions:

$$\mathscr{W}_T \subseteq \{T\}'' \subseteq \{T\}' \cap \mathrm{Alg\,Lat}(T) \subseteq \mathscr{F}_T \subseteq \{T\}'' \subseteq \mathscr{A}_T \subseteq \mathscr{W}_T. \tag{1.3}$$

The first of these inclusions is true for arbitrary operators T. The inclusion $\{T\}'' \subset \{T\}'$ is also true for arbitrary T. Now, since T is of class C_0, every invariant subspace $\mathscr{M}$ for T has the form $\mathscr{M} = \ker Y$ for some $Y \in \{T\}'$ by Proposition 3.5.33. If $X \in \{T\}''$ we must have $X\mathscr{M} \subseteq \mathscr{M}$ because X and Y commute. We conclude that $\{T\}'' \subseteq \mathrm{Alg\,Lat}(T)$ and hence the second inclusion in (1.3) is proved.

For the proof of the third inclusion we will use the splitting principle. Assume that $T \in \mathscr{L}(\mathscr{H})$, $X \in \{T\}' \cap \mathrm{Alg\,Lat}(T)$, $f \in \mathscr{H}$, and $\mathscr{K} = \bigvee_{n=0}^{\infty} T^n f$. Then $X\mathscr{K} \subseteq \mathscr{K}$ and $X|\mathscr{K} \in \{T|\mathscr{K}\}'$. Since $T|\mathscr{K}$ is multiplicity-free and $m_{T|\mathscr{K}} = m_f$, it follows from Theorem 3.3.8 that there exist functions $u_f, v_f \in H^\infty$ such that

$$v_f \wedge m_f \equiv 1, \tag{1.4}$$

and $v_f(T|\mathscr{K})(X|\mathscr{K}) = u_f(T|\mathscr{K})$, in particular,

$$v_f(T)Xf = u_f(T)f. \tag{1.5}$$

Assume now that h is a T-maximal vector, $\mathscr{K}_0 = \bigvee_{n=0}^{\infty} T^n h$, and $\mathscr{M}_0 \in \mathrm{Lat}(T)$ is chosen according to Theorem 3.3.1: $\mathscr{K}_0 \cap \mathscr{M}_0 = \{0\}$ and $\mathscr{K}_0 \vee \mathscr{M}_0 = \mathscr{H}$. We claim that $h+g$ is also a T-maximal vector for every $g \in \mathscr{M}_0$. Indeed, the relation $u(T)(h+g) = 0$ implies that

$$u(T)h = -u(T)g \in \mathscr{K}_0 \cap \mathscr{M}_0$$

and therefore $u(T)h = 0$. Thus m_T divides u because h is T-maximal. Therefore we will have

$$v_h \wedge m_T \equiv v_{h+g} \wedge m_T \equiv 1$$

for every $g \in \mathscr{M}$. We want to show that $v_h(T)X = u_h(T)$ so that $X = (u_h/v_h)(T)$. To do this we play an elementary game with relation (1.5) applied to $f = h$ and $f = h + g$. Namely, we have

$$(v_{h+g}(T)X - u_{h+g}(T))h = -(v_{h+g}(T)X - u_{h+g}(T))g \in \mathscr{K}_0 \cap \mathscr{M}_0 = \{0\}$$

by (1.5) applied to $f = h + g$, and we conclude that

$$v_{h+g}(T)Xh = u_{h+g}(T)h.$$

Combining this with (1.5) for $f = h$ we get

$$\begin{aligned} &v_h(T)u_{h+g}(T)h - v_{h+g}(T)u_h(T)h \\ &\qquad = v_h(T)v_{h+g}(T)Xh - v_{h+g}(T)v_h(T)Xh = 0 \end{aligned}$$

so that $m_T \equiv m_h$ divides $v_h u_{h+g} - v_{h+g} u_h$. Therefore

$$v_h(T)u_{h+g}(T) = v_{h+g}(T)u_h(T)$$

and hence

$$\begin{aligned} v_{h+g}(T)v_h(T)X(h+g) &= v_h(T)v_{h+g}(T)X(h+g) \\ &= v_h(T)u_{h+g}(T)(h+g) \\ &= v_{h+g}(T)u_h(T)(h+g). \end{aligned}$$

Since $v_{h+g} \wedge m_T \equiv 1$, the operator $v_{h+g}(T)$ is a quasiaffinity and the last equality above implies

$$v_h(T)X(h+g) = u_h(T)(h+g)$$

which, in combination with (1.5) with $f = h$, yields $v_h(T)Xg = u_h(T)g$. We conclude that $v_h(T)X \mid \mathscr{M}_0 = u_h(T) \mid \mathscr{M}_0$ and, since $v_h(T)X \mid \mathscr{K}_0 = u_h(T) \mid \mathscr{K}_0$ by the definition of u_h and v_h, we must have $v_h(T)X = v_h(T)X \mid \mathscr{K}_0 \vee \mathscr{M}_0 = u_h(T)$. Thus $X \in \mathscr{F}_T$, and the third inclusion in (1.3) is proved.

The inclusion $\mathscr{F}_T \subseteq \{T\}''$ is true for every completely nonunitary contraction T. Indeed, if $X = (u/v)(T) \in \mathscr{F}_T$ and $Y \in \{T\}'$, we must have

$$v(T)XY = u(T)Y = Yu(T) = Yv(T)X = v(T)YX,$$

which implies $XY = YX$ because $v \in K_T^\infty$ and hence $v(T)$ is one-to-one.

The proof of the inclusion $\{T\}'' \subseteq \mathscr{A}_T$ is based on a classical argument, essentially due to von Neumann. Let $X \in \{T\}''$, and denote by T' (resp., X') the direct sum of infinitely many copies of T (resp., X). It is easy to check that T' is an operator of class C_0 with $m_{T'} = m_T$, and $X' \in \{T'\}''$. By the second inclusion in (1.3), which has already been proved for all operators of class C_0, we must have $X' \in \operatorname{Alg}\operatorname{Lat}(T')$. Let $V = \{Y : \sum_{j=0}^\infty \|Yh_j - Xh_j\|^2 < \varepsilon^2\}$ be an arbitrary ultrastrong neighborhood of X, and set $h = \bigoplus_{j=0}^\infty h_j$. The cyclic space $\mathscr{K} = \bigvee_{n=0}^\infty T'^n h$ is then invariant for X' so that there must exist a polynomial p satisfying the inequality $\|X'h - p(T')h\| < \varepsilon$. But this means that $p(T) \in V$ and we conclude that $X \in \mathscr{A}_T$ (of course, $\mathscr{A}_T$ coincides with the ultrastrong closure of polynomials in T).

Finally, the inclusion $\mathscr{A}_T \subseteq \mathscr{W}_T$ is true for any operator T, and this concludes the proof of the theorem.

1.6. COROLLARY. *For every operator T of class C_0 there exists a function v in H^∞, $v \wedge m_T \equiv 1$, with the following property. Every operator $X \in \mathscr{A}_T$ can be written as $X = (u/v)(T)$ for some $u \in H^\infty$.*

PROOF. This follows from the fact that the function v_h in the preceding proof can be chosen independently of X (cf. Remark 3.2.12).

There are a few immediate consequences of Theorem 1.2 for the reflexivity of operators of class C_0.

1.7. COROLLARY. *An operator T of class C_0 is reflexive if and only if* $\operatorname{Alg}\operatorname{Lat}(T) \subseteq \{T\}'$.

PROOF. The inclusion $\operatorname{Alg}\operatorname{Lat}(T) \subseteq \{T\}'$ is clearly true if T is reflexive. Conversely, if T is of class C_0 and $\operatorname{Alg}\operatorname{Lat}(T) \subset \{T\}'$, then

$$\operatorname{Alg}\operatorname{Lat}(T) = \operatorname{Alg}\operatorname{Lat}(T) \cap \{T\}' = \mathscr{W}_T$$

by Theorem 1.2.

1.8. COROLLARY. *Let $T \in \mathscr{L}(\mathscr{H})$ be an operator of class C_0, and let $\{\mathscr{M}_j : j \in J\} \subseteq \operatorname{Lat}(T)$ be such that $\bigvee_{j\in J} \mathscr{M}_j = \mathscr{H}$. If $T \,|\, \mathscr{M}_j$ is reflexive for every $j \in J$ then T is reflexive.*

PROOF. If $X \in \operatorname{Alg}\operatorname{Lat}(T)$ then clearly $X\mathscr{M}_j \subseteq \mathscr{M}_j$ and $X \,|\, \mathscr{M}_j$ belongs to $\operatorname{Alg}\operatorname{Lat}(T \,|\, \mathscr{M}_j)$ for $j \in J$. If each $T \,|\, \mathscr{M}_j$ is reflexive, we must have $X \,|\, \mathscr{M}_j \in \{T \,|\, \mathscr{M}_j\}'$, and consequently

$$\ker(XT - TX) \supseteq \bigvee_{j\in J} \mathscr{M}_j = \mathscr{H}.$$

Thus $X \in \{T\}'$, and the reflexivity of T follows from the preceding corollary.

1.9. COROLLARY. *Assume that T is a reflexive operator of class C_0 and $X \in \mathscr{W}_T$. Then $T \,|\, (\operatorname{ran} X)^-$ is also reflexive.*

PROOF. If $T \in \mathscr{L}(\mathscr{H})$ and $Y \in \operatorname{Alg}\operatorname{Lat}(T \,|\, (X\mathscr{H})^-)$ then the product YX can be regarded as an operator acting on $\mathscr{H}$, with range contained in $(X\mathscr{H})^-$. By Theorem 1.2 we have $X \in \operatorname{Alg}\operatorname{Lat}(T)$. If $\mathscr{M} \in \operatorname{Lat}(T)$ then $(X\mathscr{M})^- \in \operatorname{Lat}(T \,|\, (X\mathscr{M})^-)$ so that $Y(X\mathscr{M})^- \subseteq (X\mathscr{M})^-$. Thus

$$YX\mathscr{M} \subseteq Y(X\mathscr{M})^- \subseteq (X\mathscr{M})^- \subseteq \mathscr{M}$$

for every $\mathscr{M} \in \operatorname{Lat}(T)$. Thus $YX \in \operatorname{Alg}\operatorname{Lat}(T)$ and hence $YX \in \{T\}'$ by the reflexivity of T. Now remark that

$$\begin{aligned}[][Y(T \,|\, (X\mathscr{H})^-) - (T \,|\, (X\mathscr{H})^-)Y]X &= YTX - TYX \\ &= YTX - YXT \\ &= TYX - YTX = 0\end{aligned}$$

and we conclude that $Y \in \{T \,|\, (X\mathscr{H})^-\}'$. The reflexivity of $T \,|\, (X\mathscr{H})^-$ follows now from Corollary 1.7.

Our main problem in the rest of this section will be to prove that the reflexivity and hyperreflexivity of operators of class C_0 is a quasisimilarity invariant. We will then be able to characterize reflexive operators in terms of their Jordan models. In doing so we need an auxiliary property, which we call property $(*)$.

1.10. DEFINITION. A completely nonunitary contraction T is said to have *property* $(*)$ if for any quasiaffinity $X \in \{T\}'$ there exist a quasiaffinity $Y \in \{T\}'$ and a function u in H^∞ such that $XY = YX = u(T)$.

Of course the product XY is a quasiaffinity so that the function u in the above definition must belong to K_T^∞.

1.11. LEMMA. *Let T and T' be two quasisimilar completely nonunitary contractions.*

(i) *T has property $(*)$ if and only if T' has property $(*)$.*
(ii) *If T has property $(*)$ then we can find quasiaffinities $A \in \mathcal{I}(T',T)$ and $B \in \mathcal{I}(T,T')$, and a function $u \in H^\infty$, such that $AB = u(T)$ and $BA = u(T')$.*

PROOF. Let us assume that T has property $(*)$ and $B' \in \mathcal{I}(T,T')$ and $A \in \mathcal{I}(T',T)$ are two quasiaffinities. If $X \in \{T'\}'$ is an arbitrary quasiaffinity then $AXB' \in \{T\}'$ is also a quasiaffinity and hence we can find a quasiaffinity $Y' \in \{T\}'$ and u in H^∞ such that

$$Y'AXB' = AXB'Y' = u(T).$$

Now remark that

$$A(XB'Y'A - u(T')) = (AXB'Y' - u(T))A = 0,$$
$$(B'Y'AX - u(T'))B' = B'(Y'AXB' - u(T)) = 0,$$

and hence $XB'Y'A = B'Y'AX = u(T')$ because A and B' are quasiaffinities. Since $B'Y'A$ is clearly a quasiaffinity, we see that T' has property $(*)$. By symmetry, (i) is proved.

To prove (ii) we specialize the previous calculations by taking $X = I \in \{T'\}'$. Then we obtain Y' and u satisfying $Y'AB' = AB'Y' = u(T)$. If we set $B = B'Y'$ then clearly $AB = u(T)$. Finally,

$$(BA - u(T'))B = B(AB - u(T)) = 0$$

and we conclude that $BA = u(T')$. The proof is complete.

The following result shows that property $(*)$ is important in the study of reflexivity.

1.12. LEMMA. *Let T and T' be two quasisimilar operators of class C_0. If T has property $(*)$ then T is reflexive if and only if T' is reflexive.*

PROOF. Choose quasiaffinities A, B and $u \in H^\infty$ as in Lemma 2.11(ii). Assume that T is reflexive, $X \in \operatorname{Alg}\operatorname{Lat}(T')$, and $\mathcal{M} \in \operatorname{Lat}(T)$. Then we have $(B\mathcal{M})^- \in \operatorname{Lat}(T')$ so that

$$AXB\mathcal{M} \subseteq A(B\mathcal{M})^- \subseteq (AB\mathcal{M})^- = (u(T)\mathcal{M})^-$$

and we conclude that $AXB \in \operatorname{Alg}\operatorname{Lat}(T)$. Now, T is reflexive, so that $AXB \in \{T\}'$, and the relation

$$A(XT' - T'X)B = AXBT - TAXB = 0$$

shows that $X \in \{T'\}'$ because A and B are quasiaffinities. The operator X was arbitrary in $\operatorname{Alg}\operatorname{Lat}(T')$ and therefore T' is reflexive by Corollary 1.7. The lemma follows obviously by reasons of symmetry.

Unfortunately, not every operator of class C_0 has property $(*)$. We can, however, produce a family of operators with property $(*)$ that will suffice for our purposes.

1.13. PROPOSITION. *Let θ_0 and θ_1 be two inner functions such that $\theta_1 \mid \theta_0$. Then the operator $T = S(\theta_0) \oplus S(\theta_1)$ has property $(*)$.*

PROOF. Let P_0 and P_1 denote the projections of $\mathscr{H} = \mathscr{H}(\theta_0) \oplus \mathscr{H}(\theta_1)$ onto $\mathscr{H}(\theta_0)$ and $\mathscr{H}(\theta_1)$, respectively. If $X \in \{T\}'$ then $P_i^* X P_j \in \mathscr{I}(S(\theta_j), S(\theta_i))$ for $0 \le i, j \le 1$, and by virtue of Theorem 3.1.6 we can find functions $a_{ij} \in H^\infty$ such that

$$\theta_0 \mid a_{01}\theta_1 \tag{1.14}$$

and

$$P_i^* X P_j h = P_{\mathscr{H}(\theta_i)}(a_{ij}h), \quad h \in \mathscr{H}(\theta_j),\ 0 \le i, j \le 1. \tag{1.15}$$

Vice versa, if $A = \left[\begin{smallmatrix} a_{00} & a_{01} \\ a_{10} & a_{11} \end{smallmatrix}\right]$ is a matrix of functions in H^∞ for which (1.14) holds, then there exists an operator $X \in \{T\}'$ satisfying (1.15). Of course, the matrix A is not uniquely determined by X. We can always change a_{ij} into $a_{ij} + u_{ij}\theta_j$, where u_{ij} are arbitrary functions in H^∞.

Assume for the moment that $\theta_0 \wedge \det(A) \equiv 1$, where $\det(A) = a_{00}a_{11} - a_{01}a_{10}$. Then the matrix $B = \left[\begin{smallmatrix} a_{11} & -a_{01} \\ -a_{10} & a_{00} \end{smallmatrix}\right]$ determines an operator $Y \in \{T\}'$, and the easily verified relations $AB = BA = uI$, $u = \det(A)$, imply that $XY = YX = u(T)$. Moreover, the relation $\theta_0 \wedge u \equiv 1$ implies that $u \in K_T^\infty$ because $m_T \equiv \theta_0$, and hence $u(T)$ is a quasiaffinity. We deduce that the operator Y is also a quasiaffinity.

The considerations above indicate that, in order to show that T has property $(*)$, it suffices to prove that for every quasiaffinity $X \in \{T\}'$ we can find a matrix A satisfying (1.14) and (1.15) and such that $\theta_0 \wedge \det(A) \equiv 1$. Assume therefore that X is a quasiaffinity, and the matrix A satisfies (1.14) and (1.15). We note first that

$$a_{00} \wedge a_{01} \wedge \theta_0 \equiv 1. \tag{1.16}$$

Indeed, if $q \equiv a_{00} \wedge a_{01} \wedge \theta_0$ then we see from (1.15) that $P_{\mathscr{H}(\theta_0)} X \mathscr{H} \subseteq qH^2 \ominus \theta_0 H^2$, and hence $q \equiv 1$ because X has dense range. Moreover, we have

$$\theta_1 \wedge \det(A) \equiv 1. \tag{1.17}$$

Indeed, if $p \equiv \theta_1 \wedge \det(A)$ and we define

$$h = P_{\mathscr{H}(\theta_0)}(-a_{01}\theta_1/p) \oplus P_{\mathscr{H}(\theta_1)}(a_{00}\theta_1/p),$$

an easy calculation (using the fact that $P_{\mathscr{H}(\theta)}(aP_{\mathscr{H}(\theta)}f) = P_{\mathscr{H}(\theta)}(af)$ if $a \in H^\infty$, $f \in H^2$, and θ is inner) shows that $P_0 X h = 0$ and

$$P_1 X h = P_{\mathscr{H}(\theta_1)}(\theta_1 \det(A)/p) = 0.$$

By the injectivity of X we must have $h = 0$ and therefore $\theta_0 \mid (-a_{01}\theta_1/p)$ and $\theta_1 \mid (a_{00}\theta_1/p)$. We deduce that $p \mid (a_{01}\theta_1/\theta_0)$ and $p \mid a_{00}$. Since $(a_{01}\theta_1/\theta_0) \mid a_{01}$ and $p \mid \theta_1$ by the definition of p, we easily deduce that $p \mid (\theta_1 \wedge a_{01} \wedge a_{00})$ and hence $p \equiv 1$ by (1.16). It is easy to see now that (1.16) and (1.17) imply

$$(\theta_1 a_{00} \wedge \theta_1 a_{01} \wedge \det(A)) \wedge \theta_0 \equiv 1. \tag{1.18}$$

Indeed, if r denotes the left-hand side of (1.18), then $r \mid \det(A)$ and hence $r \wedge \theta_1 \equiv 1$ by (1.17). Then we see that the relation $r \mid \theta_1 a_{00}$ [resp., $r \mid \theta_1 a_{01}$] implies $r \mid a_{00}$ [resp., $r \mid a_{01}$] and hence $r \mid a_{00} \wedge a_{01} \wedge \theta_0$. The relation $r \equiv 1$ now follows from (1.16).

An easy application of Theorem 3.1.14 implies the existence of scalars x and y such that

$$(\det(A) + x\theta_1 a_{00} + y\theta_1 a_{01}) \wedge \theta_0 \equiv 1$$

(we may assume that the coefficient of $\det(A)$ is one because of the density assertion in the theorem cited above). We now define

$$A' = \begin{bmatrix} a_{00} & a_{01} \\ a_{10} - y\theta_1 & a_{11} + x\theta_1 \end{bmatrix}$$

and note that, by the remarks above, A' also determines X. Finally, we have

$$\det(A') = \det(A) + x\theta_1 a_{00} + y\theta_1 a_{01}$$

and hence $\theta_0 \wedge \det(A') \equiv 1$. The proposition is proved.

The proposition just proved certainly applies to $T = S(\theta_0)$ since we are allowed to take $\theta_1 \equiv 1$. It is clear however that for this particular case the proposition follows immediately from Theorem 3.1.6.

Lemma 1.12 and Proposition 1.13 already show that reflexivity is a quasisimilarity invariant for operators of class C_0 with multiplicity ≤ 2. It is therefore interesting to know which Jordan operators with multiplicity ≤ 2 are reflexive. The proof of the following result is based on our very explicit knowledge of the invariant subspaces of a Jordan block.

1.19. PROPOSITION. *Let θ_0 and θ_1 be two inner functions such that $\theta_1 \mid \theta_0$. The operator $T = S(\theta_0) \oplus S(\theta_1)$ is reflexive if and only if the Jordan block $S(\theta_0/\theta_1)$ is reflexive.*

PROOF. An easy application of Corollary 3.1.12 shows that $\operatorname{ran} \theta_1(T) = (\theta_1 H^2 \ominus \theta_0 H^2) \oplus \{0\}$ and thus $T \mid \operatorname{ran} \theta_1(T)$ is unitarily equivalent to $S(\theta_0/\theta_1)$ by Proposition 3.1.10(iii). If T is reflexive then $S(\theta_0/\theta_1)$ is reflexive by Corollary 1.9.

Assume now that $X \in \operatorname{Alg}\operatorname{Lat}(T)$. The subspaces $\mathscr{H}(\theta_0) \oplus \{0\}$ and $\{0\} \oplus \mathscr{H}(\theta_1)$ are invariant for T, hence for X, and therefore X can be written as $X = X_0 \oplus X_1$ with $X_j \in \operatorname{Alg}\operatorname{Lat}(S(\theta_j))$, $j = 0, 1$. Now consider the subspaces $\mathscr{M}_0, \mathscr{M}_1 \in \operatorname{Alg}\operatorname{Lat}(T)$ described by

$$\mathscr{M}_0 = \{(\theta_0/\theta_1)h \oplus h : h \in \mathscr{H}(\theta_1)\}$$

and

$$\mathscr{M}_1 = \{(\theta_0/\theta_1)S(\theta_1)h \oplus h : h \in \mathscr{H}(\theta_1)\}.$$

The inclusion $X\mathscr{M}_0 \subseteq \mathscr{M}_0$ yields

$$X_0(\theta_0/\theta_1)h = (\theta_0/\theta_1)X_1 h, \quad h \in \mathscr{H}(\theta_1),$$

and the inclusion $X\mathscr{M}_1 \subseteq \mathscr{M}_1$ yields

$$X_0(\theta_0/\theta_1)S(\theta_1)h = (\theta_0/\theta_1)S(\theta_1)X_1h, \qquad h \in \mathscr{H}(\theta_1).$$

We combine the second equality above with the first in which h is replaced by $S(\theta_1)h$ to obtain

$$(\theta_0/\theta_1)X_1S(\theta_1)h = (\theta_0/\theta_1)S(\theta_1)X_1h, \qquad h \in \mathscr{H}(\theta_1).$$

Since θ_0/θ_1 is an inner function, this last equality shows that $X_1 \in \{S(\theta_1)\}'$ and hence there exists $u \in H^\infty$ such that $X_1 = u(S(\theta_1))$ by Corollary 3.1.20. Thus we deduce the existence of an operator $Y_0 \in \operatorname{Alg}\operatorname{Lat}(S(\theta_0))$ such that

$$X - u(T) = Y_0 \oplus 0 \in \operatorname{Alg}\operatorname{Lat}(T). \tag{1.20}$$

For every inner divisor q of θ_0/θ_1 we now consider the subspace $\mathscr{N}_q \in \operatorname{Lat}(T)$ defined by

$$\mathscr{N}_q = \{qh \oplus P_{\mathscr{H}(\theta_1)}h : h \in \mathscr{H}(\theta_0/q)\}.$$

The inclusion $(Y_0 \oplus 0)\mathscr{N}_q \subseteq \mathscr{N}_q$ means that for every $h \in \mathscr{H}(\theta_0/q)$ we can find $h' \in \mathscr{H}(\theta_0/q)$ such that $Y_0(qh) = qh'$ and $P_{\mathscr{H}(\theta_1)}h' = 0$. The second equality implies, of course, that $h' \in \theta_1H^2$ so that $h' \in \theta_1H^2 \cap \mathscr{H}(\theta_0/q) = \theta_1H^2 \ominus (\theta_0/q)H^2$ and hence $qh' \in q\theta_1H^2 \ominus \theta_0H^2$. We have thus proved that

$$Y_0(qH^2 \ominus \theta_0H^2) = Y_0(q\mathscr{H}(\theta_0/q)) \subseteq q\theta_1H^2 \ominus \theta_0H^2 \tag{1.21}$$

for every inner divisor q of θ_0/θ_1. If $q = 1$ and $q = \theta_0/\theta_1$ we obtain the following particular cases of (1.21):

$$\operatorname{ran}(Y_0) \subseteq \theta_1H^2 \ominus \theta_0H^2, \qquad \ker(Y_0) \supseteq (\theta_0/\theta_1)H^2 \ominus \theta_0H^2. \tag{1.22}$$

Relations (1.21) and (1.22) can be used to produce an operator in $\operatorname{Alg}\operatorname{Lat}(S(\theta_0/\theta_1))$. Indeed, as shown in Proposition 3.1.10, $S(\theta_0) \mid \theta_1H^2 \ominus \theta_0H^2$ is unitarily equivalent to $S(\theta_0/\theta_1)$. In fact, the unitary operator $V : \theta_1H^2 \ominus \theta_0H^2 \to \mathscr{H}(\theta_0/\theta_1)$ given by $Vh = h/\theta_1$ satisfies the relation

$$VS(\theta_0) \mid (\theta_1H^2 \ominus \theta_0H^2) = S(\theta_0/\theta_1)V = P_{\mathscr{H}(\theta_0/\theta_1)}S(\theta_0)V.$$

The operator $Z = VY_0 \mid \mathscr{H}(\theta_0/\theta_1)$ belongs to $\operatorname{Alg}\operatorname{Lat}(S(\theta_0/\theta_1))$. Indeed, if $\mathscr{M}$ is an invariant subspace for $S(\theta_0/\theta_1)$, we must have $\mathscr{M} = qH^2 \ominus (\theta_0/\theta_1)H^2$ for some inner divisor q of θ_0/θ_1, and (1.21) implies that

$$\begin{aligned} Z\mathscr{M} &\subseteq VY_0(qH^2 \ominus (\theta_0/\theta_1)H^2) = VY_0(qH^2 \ominus \theta_0H^2) \\ &\subseteq V(q\theta_1H^2 \ominus \theta_0H^2) = qH^2 \ominus (\theta_0/\theta_1)H^2 = \mathscr{M}. \end{aligned}$$

Assuming now that $S(\theta_0/\theta_1)$ is reflexive, it follows that Z commutes with $S(\theta_0/\theta_1)$ and hence

$$\begin{aligned} &V(Y_0S(\theta_0) - S(\theta_0)Y_0) \mid \mathscr{H}(\theta_0/\theta_1) \\ &= [VY_0P_{\mathscr{H}(\theta_0/\theta_1)}S(\theta_0) - V(S(\theta_0) \mid (\theta_1H^2 \ominus \theta_0H^2))Y_0] \mid \mathscr{H}(\theta_0/\theta_1) \\ &= ZS(\theta_0/\theta_1) - S(\theta_0/\theta_1)Z = 0, \end{aligned}$$

where we have used the equalities $Y_0S(\theta_0) = Y_0P_{\mathscr{H}(\theta_0/\theta_1)}S(\theta_0)$ and $S(\theta_0)Y_0 = (S(\theta_0) \mid (\theta_1H^2 \ominus \theta_0H^2))Y_0$, which follow from (1.22). Thus $Y_0S(\theta_0) = S(\theta_0)Y_0$

on the space $\mathscr{H}(\theta_0/\theta_1)$. On the orthogonal complement $(\theta_0/\theta_1)H^2 \ominus \theta_0 H^2$ of this space we have $Y_0 S(\theta_0) = S(\theta_0) Y_0 = 0$ by (1.22). We conclude that $Y_0 \in \{S(\theta_0)\}'$. Thus, if $S(\theta_0/\theta_1)$ is reflexive, it follows from the above argument and (1.20) that every operator $X \in \operatorname{Alg}\operatorname{Lat}(T)$ commutes with T. Our proposition follows by Corollary 1.7.

The characterization of reflexive operators of class C_0 now becomes quite an easy task.

1.23. THEOREM. *Let T be an operator of class C_0, and let $S(\Theta)$, $\Theta = \{\theta_\alpha\}$, be the Jordan model of T. Then T is reflexive if and only if $S(\theta_0/\theta_1)$ is reflexive.*

PROOF. Assume that $T \in \mathscr{L}(\mathscr{H})$ and $X \in \mathscr{I}(S(\Theta), T)$ is a quasiaffinity. The operators $T \mid (\operatorname{ran}(\theta_1(T)))^-$ and $S(\Theta) \mid \operatorname{ran}\theta_1(S(\Theta))$ are quasisimilar since $X \mid \operatorname{ran}\theta_1(S(\Theta))$ is a quasiaffinity intertwining them. Thus $T \mid (\operatorname{ran}\theta_1(T))^-$ is quasisimilar to $S(\theta_0/\theta_1)$. If T is reflexive, it follows from Corollary 1.9, Lemma 1.12, and Proposition 1.13 that $S(\theta_0/\theta_1)$ is reflexive.

Conversely, assume that $S(\theta_0/\theta_1)$ is reflexive, and for each ordinal α consider the subspaces $\mathscr{H}_\alpha$, $\mathscr{K}_\alpha \in \operatorname{Lat}(T)$ defined by

$$\mathscr{H}_\alpha = \left\{X\left(\bigoplus f_\beta\right) : f_\beta = 0 \quad \text{for } \beta \neq \alpha\right\}^-$$

and

$$\mathscr{K}_\alpha = \left\{X\left(\bigoplus f_\beta\right) : f_0 \in (\theta_0/\theta_\alpha)H^2 \ominus \theta_0 H^2, f_\alpha = 0 \quad \text{for } \alpha \neq 0\right\}^- .$$

The restriction $T \mid \mathscr{H}_0 \vee \mathscr{H}_1$ is quasisimilar to $S(\theta_0) \oplus S(\theta_1)$, while $T \mid \mathscr{H}_\alpha \vee \mathscr{K}_\alpha$ is quasisimilar to $S(\theta_\alpha) \oplus S(\theta_\alpha)$ for $\alpha > 0$; this follows from Proposition 3.5.32 since a suitable restriction of X provides one of the needed intertwining operators. All these restrictions are then reflexive by Lemma 1.12, Proposition 1.13, and Proposition 1.19. Finally we note that

$$(\mathscr{H}_0 \vee \mathscr{H}_1) \vee \left(\bigvee_{\alpha \geq 1} (\mathscr{H}_\alpha \vee \mathscr{K}_\alpha)\right) = \bigvee_{\alpha \geq 0} \mathscr{H}_\alpha = \mathscr{H}$$

and the reflexivity of T follows from Corollary 1.8. The theorem is proved.

We turn now to the question of hyperreflexivity for operators of class C_0. The quasisimilarity invariance is an easier problem now, as seen from the following result.

1.24. PROPOSITION. *If the operators T and T' are quasisimilar, and one of them is hyperreflexive, then so is the other.*

PROOF. Let $Y \in \mathscr{I}(T', T)$ and $X \in \mathscr{I}(T, T')$ be two quasiaffinities, and let $A \in \operatorname{Alg}\operatorname{Lat}(\{T\}')$. We claim that $XAY \in \operatorname{Alg}\operatorname{Lat}(\{T'\}')$. Indeed, for each $\mathscr{M}$ in $\operatorname{Lat}(\{T'\}')$, the subspace

$$\mathscr{N} = \bigvee_{Z \in \{T\}'} ZY\mathscr{M}$$

clearly belongs to $\operatorname{Lat}(\{T\}')$ and

$$X\mathscr{N} = \bigvee_{Z \in \{T\}'} XZY\mathscr{M} \subset \bigvee_{Z' \in \{T'\}'} Z'\mathscr{M} = \mathscr{M}.$$

Hence we have $XAY\mathscr{M} \subset XA\mathscr{N} \subset X\mathscr{N} \subset \mathscr{M}$ so that XAY leaves each $\mathscr{M} \in \operatorname{Lat}(\{T'\}')$ invariant, as claimed. If T' is hyperreflexive we must have $XAY \in \{T'\}'$ for every $A \in \operatorname{Alg}\operatorname{Lat}(\{T\}')$ so that

$$X(AT - TA)Y = XAYT - TXAY = 0,$$

and hence $A \in \{T\}'$ because X and Y are quasiaffinities. Thus the hyperreflexivity of T follows from the hyperreflexivity of T', and the proposition follows for reasons of symmetry.

1.25. THEOREM. *Assume that T is an operator of class C_0. Then T is hyperreflexive if and only if $S(m_T)$ is reflexive.*

PROOF. Proposition 1.24 shows that we can restrict ourselves to operators T of the from $S(\Theta)$, where $\Theta = \{\theta_\alpha\}$ is a model function. Assume first that $S(\Theta)$ is hyperreflexive and $X \in \operatorname{Alg}\operatorname{Lat}(S(\theta_0))$. We claim that the operator $Y = \bigoplus_\alpha Y_\alpha$, where $Y_0 = X$ and $Y_\alpha = 0$ for $\alpha \neq 0$, belongs to $\operatorname{Alg}\operatorname{Lat}(\{S(\Theta)\}')$. Indeed, a subspace $\mathscr{M} \in \operatorname{Lat}\{S(\Theta)\}'$ must have the form $\mathscr{M} = \bigoplus_\alpha \mathscr{M}_\alpha$, with $\mathscr{M}_\alpha \in \operatorname{Lat}(S(\theta_\alpha))$, and this clearly implies that $Y\mathscr{M} \subset \mathscr{M}$. Thus $Y \in \{S(\Theta)\}'$ by the assumption that $S(\Theta)$ is hyperreflexive, and hence $X \in \{S(\theta_0)\}'$. The reflexivity of $S(\theta_0)$ follows from Corollary 1.7.

Conversely, assume that $S(\theta_0)$ is reflexive. By Proposition 3.1.10, $S(\theta_\alpha)$ is unitarily equivalent to

$$S(\theta_0) \mid \operatorname{ran} u_\alpha(S(\theta_0)), \qquad u_\alpha = \theta_0/\theta_\alpha,$$

and therefore $S(\theta_\alpha)$ is reflexive for every ordinal α by Corollary 1.9. Let us define operators $R_{\alpha\beta}$ in $\{S(\Theta)\}'$ as follows: $R_{\alpha\beta}(\bigoplus_\gamma h_\gamma) = \bigoplus_\gamma k_\gamma$, where $k_\gamma = 0$ for $\gamma \neq \alpha$, and

$$\begin{aligned} k_\alpha &= P_{\mathscr{H}(\theta_\alpha)} h_\beta \quad \text{whenever } \alpha > \beta, \\ &= (\theta_\alpha/\theta_\beta) h_\beta \quad \text{whenever } \alpha \le \beta; \end{aligned} \tag{1.26}$$

the fact that $R_{\alpha\beta}$ commutes with $S(\Theta)$ follows from Theorem 3.1.16. Clearly $P_\alpha = R_{\alpha\alpha}$ coincides with the orthogonal projection of $\mathscr{H}(\Theta)$ onto its α-component space.

For every A in $\operatorname{Alg}\operatorname{Lat}(\{S(\Theta)\}')$ we have $P_\alpha A P_\beta \in \operatorname{Alg}\operatorname{Lat}(\{S(\Theta)\}')$ and $A = \sum_{\alpha,\beta} P_\alpha A P_\beta$ unconditionally in the strong operator topology. To conclude the proof it will suffice to show that each $P_\alpha A P_\beta$ commutes with $S(\Theta)$. Now, the operators $R_{\beta\alpha} P_\alpha A P_\beta$ and $P_\alpha A P_\beta R_{\beta\alpha}$ also belong to $\operatorname{Alg}\operatorname{Lat}(\{S(\Theta)\}')$ and have the form $\bigoplus_\gamma T_\gamma$, with $T_\gamma = 0$ for $\gamma \neq \beta$ and $\gamma \neq \alpha$, respectively. Considering hyperinvariant subspaces of the form $\ker \theta(S(\Theta))$, with θ a divisor of θ_0, it is easy to see that $T_\gamma \in \operatorname{Alg}\operatorname{Lat}(S(\theta_\gamma))$ for each γ, so that T_γ commutes with $S(\theta_\gamma)$ by the reflexivity of $S(\theta_\gamma)$. Thus $R_{\beta\alpha}P_\alpha AP_\beta$ and $P_\alpha AP_\beta R_{\beta\alpha}$ commute with $S(\Theta)$, hence

$$R_{\beta\alpha}(P_\alpha AP_\beta S(\Theta) - S(\Theta)P_\alpha AP_\beta) = (P_\alpha AP_\beta S(\Theta) - S(\Theta)P_\alpha AP_\beta)R_{\beta\alpha} = 0.$$

Now, if the range of $R_{\beta\alpha}$ does not contain $\operatorname{ran} P_\alpha$, it follows that $\beta < \alpha$ and therefore $R_{\beta\alpha}$ is one-to-one on $\operatorname{ran} P_\alpha$. In either case the last equality shows that $P_\alpha AP_\beta \in \{S(\Theta)\}'$, and the theorem is proved.

Theorems 1.23 and 1.25 show that the characterization of those inner functions θ for which $S(\theta)$ is reflexive is quite important. Unfortunately, such a characterization is unknown at the time of this writing. There are several partial results, some of which are outlined in the exercises.

Exercises

1. Show that there exist operators of class C_0 that do not have property $(*)$.

2. Prove that $S^{(n)}$ and $S(\theta)^{(n)}$ have property $(*)$ if $n < \omega$, but $S^{(\omega)}$ does not have property $(*)$. Here, as usual, S denotes the unilateral shift on H^2.

3. Show that $S \oplus S(\theta)$ has property $(*)$ for every inner function θ.

4. Find a completely nonunitary contraction T for which $\mathscr{W}_T \neq \mathscr{F}_T$.

5. Find an operator T such that $\{T\}'' = \{T\}' \neq \mathscr{W}_T$.

6. Assume that T is an operator for which $\{T\}' \cap \operatorname{Alg}\operatorname{Lat}(T) = \mathscr{W}_T$. Show that $T \oplus T$ is reflexive.

7. Find a completely nonunitary contraction T such that $\mathscr{F}_T$ is not contained in $\mathscr{W}_T$.

8. Prove the following generalization of Lemma 1.12. Let T and T' be two quasisimilar completely nonunitary contractions. Assume that $\mathscr{F}_T = \mathscr{W}_T$, $\mathscr{F}_{T'} = \mathscr{W}_{T'}$, and T has property $(*)$. Prove that T is reflexive if and only if T' is reflexive.

9. Let T and T' be two completely nonunitary contractions. Assume that there exist quasiaffinities $B \in \mathscr{I}(T,T')$ and $A \in \mathscr{I}(T',T)$ and an outer function $u \in H^\infty$ such that $AB = u(T)$. Prove that T is reflexive if and only if T' is reflexive.

10. Let θ be an inner function. Prove that $S \oplus S(\theta)$ is hyperreflexive if and only if $S(\theta)$ is reflexive.

11. Let T be an operator of class C_0, and let $\{\theta_j : j \in J\}$ be a family of inner divisors of m_T such that $\bigvee_{j\in J} \theta_j \equiv m_T$. Prove that T is reflexive if and only if $T \mid \ker \theta_j(T)$ is reflexive for every $j \in J$.

12. An operator T is said to be *unicellular* if $\operatorname{Lat}(T)$ is totally ordered (by inclusion). Show that a unicellular operator acting on a space of dimension > 1 cannot be reflexive.

13. Show that a unicellular operator of class C_0 must be multiplicity-free. Moreover, a multiplicity-free operator T of class C_0 is unicellular if and only if $S(m_T)$ is unicellular.

14. Show that $S(\theta)$ is unicellular if and only if θ has one of the following forms: (i) $\theta(\lambda) = ((\mu - \lambda)/(1 - \bar{\mu}\lambda))^n$, $\lambda \in \mathbf{D}$, for some $\mu \in \mathbf{D}$ and $n \in \mathbf{N}$; or (ii) $\theta(\lambda) = \exp(t(\lambda + \varsigma)/(\lambda - \varsigma))$, $\lambda \in \mathbf{D}$, where $t > 0$ and $|\varsigma| = 1$.

15. Let $\{\theta_j : j \in J\}$ be a family of inner functions such that $\theta_i \wedge \theta_j \equiv 1$ for $i, j \in J$, $i \neq j$. Assuming that $\theta = \bigvee_{j \in J} \theta_j$ exists, show that $S(\theta)$ is reflexive if and only if $S(\theta_j)$ is reflexive for every j.

16. Let θ be an inner function such that $S(\theta)$ is reflexive. Show that all the zeros of θ are simple and the singular measure defining the singular part of θ has no atoms. Conversely, if θ is a Blaschke product with simple zeros then $S(\theta)$ is reflexive.

17. Let T be an algebraic operator. Find necessary and sufficient conditions for the reflexivity of T.

18. Let T and T' be two operators of class C_0. Prove that the following assertions are equivalent:

(i) $m_T \wedge m_{T'} \equiv 1$;
(ii) $\operatorname{Lat}(T \oplus T') = \{\mathscr{M} \oplus \mathscr{N} : \mathscr{M} \in \operatorname{Lat}(T),\ \mathscr{N} \in \operatorname{Lat}(T')\}$;
(iii) $\mathscr{W}_{T \oplus T'} = \{A \oplus A' : A \in \mathscr{W}_T,\ A' \in \mathscr{W}_{T'}\}$;
(iv) $\{T \oplus T'\}' = \{A \oplus A' : A \in \{T\}',\ A' \in \{T'\}'\}$; and
(v) P, $P' \in \mathscr{W}_{T \oplus T'}$, where P and P' denote the projections onto the components of T and T', respectively.

2. Hyperinvariant subspaces. We recall that the members of $\operatorname{Lat}(\{T\}')$, $T \in \mathscr{L}(\mathscr{H})$, are also called hyperinvariant subspaces for T. If T and T' are quasisimilar operators, it does not usually follow that there exists a natural isomorphism (or any isomorphism whatsoever) between $\operatorname{Lat}(\{T\}')$ and $\operatorname{Lat}(\{T'\}')$. However, some information about $\operatorname{Lat}(\{T\}')$ can be obtained by exploiting the Jordan model of T in case T is an operator of class C_0. We start with a study of $\operatorname{Lat}(\{T\}')$ in case T is a Jordan operator.

2.1. PROPOSITION. *Let $T = S(\Theta)$, $\Theta = \{\theta_\alpha\}$, be a Jordan operator.*

(i) *A subspace $\mathscr{K} \subset \mathscr{H}(\Theta)$ is hyperinvariant for T if and only if it has the form*

$$\mathscr{K} = \bigoplus_{\alpha} (\theta''_\alpha H^2 \ominus \theta_\alpha H^2), \tag{2.2}$$

where $\theta''_\alpha \mid \theta_\alpha$ for all α, and $\Theta' = \{\theta'_\alpha = \theta_\alpha/\theta''_\alpha\}$, $\Theta'' = \{\theta''_\alpha\}$ are model functions.

(ii) *If* $\mathscr{K}$ *is given by* (2.2), *and* $T = \begin{bmatrix} T' & X \\ 0 & T'' \end{bmatrix}$ *is the triangularization of* T *with respect to the decomposition* $\mathscr{H}(\Theta) = \mathscr{K} \oplus (\mathscr{H}(\Theta) \ominus \mathscr{K})$, *then the Jordan models of* T' *and* T'' *are* $S(\Theta')$ *and* $S(\Theta'')$, *respectively. In particular, we have* $m_T = m_{T'}m_{T''}$.

(iii) *If* $\mathscr{K}_1, \mathscr{K}_2 \in \operatorname{Lat}(\{T\}')$ *are such that* $T \,|\, \mathscr{K}_1 \sim T \,|\, \mathscr{K}_2$, *then* $\mathscr{K}_1 = \mathscr{K}_2$.

PROOF. Assertion (ii) clearly follows from formula (2.2) and Proposition 3.1.10. If $\mathscr{K}_1, \mathscr{K}_2 \in \operatorname{Lat}(\{T\}')$ and $T \,|\, \mathscr{K}_1 \sim T \,|\, \mathscr{K}_2$, then $T \,|\, \mathscr{K}_1$ and $T \,|\, \mathscr{K}_2$ must have the same Jordan model. By (ii), $\mathscr{K}_1$ is completely determined by the Jordan model if $T \,|\, \mathscr{K}_1$, and hence we must have $\mathscr{K}_1 = \mathscr{K}_2$. Thus we need only to prove (i).

Let the operators $R_{\alpha\beta}$ and $P_\alpha = R_{\alpha\alpha}$ be defined as in the proof of Theorem 1.25 (cf. relation (1.26)). We claim that $\{T\}'$ is the strongly closed operator algebra generated by $\{R_{\alpha\beta}\}$ and the operators $\{u(T) : u \in H^\infty\}$. Indeed, as in the proof of Theorem 1.25, it suffices to show that for every $A \in \{T\}'$ and all ordinals α, β, there exists a function $u \in H^\infty$ such that $P_\alpha A P_\beta = P_\alpha u(T) P_\beta$. But an application of Theorem 3.1.16 shows exactly that.

The preceding remark implies that a subspace $\mathscr{K} \subset \mathscr{H}(\Theta)$ is hyperinvariant for T if and only if it is invariant for T and for the operators $\{R_{\alpha\beta}\}$. Assume thus that $\mathscr{K} \in \operatorname{Lat}(\{T\}')$; we see immediately that $P_\alpha \mathscr{K} \subset \mathscr{K}$, so that $\mathscr{K}$ can be written as $\mathscr{K} = \bigoplus_\alpha \mathscr{K}_\alpha$, where each $\mathscr{K}_\alpha$ is invariant for $S(\theta_\alpha)$. By Proposition 3.1.10 there exist inner divisors θ''_α of θ_α such that $\mathscr{K}_\alpha = \theta''_\alpha H^2 \ominus \theta_\alpha H^2$. If $\alpha < \beta$ are two ordinals, the conditions $R_{\alpha\beta}\mathscr{K} \subset \mathscr{K}$ and $R_{\beta\alpha}\mathscr{K} \subset \mathscr{K}$ are easily translated into

$$P_{\mathscr{H}(\theta_\beta)}\mathscr{K}_\alpha \subset \mathscr{K}_\beta \quad \text{and} \quad (\theta_\alpha/\theta_\beta)\mathscr{K}_\beta \subset \mathscr{K}_\alpha, \tag{2.3}$$

respectively. The first of these relations implies that $\theta''_\alpha H^2 \ominus \theta_\alpha H^2 = \mathscr{K}_\alpha \subset \mathscr{K}_\beta \oplus \theta_\beta H^2 = \theta''_\beta H^2$, and consequently $\theta''_\beta \,|\, \theta''_\alpha$. The second relation in (2.3) implies that $(\theta_\alpha/\theta_\beta)\theta''_\beta H^2 \subset \theta''_\alpha H^2$ and hence $\theta''_\alpha \,|\, (\theta_\alpha\theta''_\beta/\theta_\beta)$ or, equivalently, $(\theta_\beta/\theta''_\beta) \,|\, (\theta_\alpha/\theta''_\alpha)$. These relations easily imply that Θ' and Θ'' are model functions. Conversely, if $\mathscr{K}$ is given by (2.2) and Θ' and Θ'' are model functions, the considerations above can be reversed to yield the inclusions $R_{\alpha\beta}\mathscr{K} \subset \mathscr{K}$ for all α, β. Since $\mathscr{K}$ is clearly invariant for T, it follows that $\mathscr{K} \in \operatorname{Lat}(\{T\}')$. The proof is now complete.

2.4. REMARK. Let $\mathscr{K}_1, \mathscr{K}_2 \in \operatorname{Lat}(\{T\}')$ be given by $\mathscr{K}_j = \bigoplus_\alpha (\theta_\alpha^{(j)} H^2 \ominus \theta_\alpha H^2)$, $j = 1, 2$. Then we can easily check the following formulas:

$$\mathscr{K}_1 \cap \mathscr{K}_2 = \bigoplus_\alpha ((\theta_\alpha^{(1)} \vee \theta_\alpha^{(2)})H^2 \ominus \theta_\alpha H^2),$$

$$\mathscr{K}_1 \vee \mathscr{K}_2 = \bigoplus_\alpha ((\theta_\alpha^{(1)} \wedge \theta_\alpha^{(2)})H^2 \ominus \theta_\alpha H^2).$$

In particular, $\mathscr{K}_1 \subset \mathscr{K}_2$ if and only if $\theta_\alpha^{(2)} \,|\, \theta_\alpha^{(1)}$ for every ordinal α.

2.5. COROLLARY. *If* T *is a Jordan operator then* $\operatorname{Lat}(\{T\}')$ *is the complete lattice generated by subspaces of the form* $\ker u(T)$ *and* $(\operatorname{ran} u(T))^-$ *for* u *in* H^∞.

PROOF. Let $\mathscr{K}$ be the hyperinvariant subspace for $T = S(\Theta)$ given by (2.2). By Corollary 3.1.12 we have

$$\ker \theta'_\gamma(T) = \bigoplus_\alpha ((\theta_\alpha/\theta_\alpha \wedge \theta'_\gamma)H^2 \ominus \theta_\alpha H^2) = \bigoplus_\alpha \mathscr{M}_\alpha$$

and

$$(\operatorname{ran} \theta''_{\gamma+1}(T))^- = \bigoplus_\alpha ((\theta_\alpha \wedge \theta''_{\gamma+1})H^2 \ominus \theta_\alpha H^2) = \bigoplus_\alpha \mathscr{N}_\alpha$$

for every ordinal γ. If $\alpha < \gamma$ then $\theta'_\gamma \,|\, \theta'_\alpha$ so that $\theta''_\alpha = \theta_\alpha/\theta'_\alpha \,|\, \theta_\alpha/\theta'_\gamma$ and hence $\mathscr{M}_\alpha \subset \theta''_\alpha H^2 \ominus \theta_\alpha H^2$ for such α. If $\alpha = \gamma$ then $\mathscr{M}_\alpha = \theta''_\alpha H^2 \ominus \theta_\alpha H^2$, and if $\alpha > \gamma$ then $\theta'_\alpha \,|\, \theta'_\gamma$ so that $\theta''_\alpha H^2 \ominus \theta_\alpha H^2 = \ker \theta'_\alpha(S(\theta_\alpha)) \subset \mathscr{M}_\alpha$. Analogously we have, if $\alpha < \gamma + 1$, $\theta''_{\gamma+1} \,|\, \theta''_\alpha$, and hence $\mathscr{N}_\alpha \supset \theta''_\alpha H^2 \ominus \theta_\alpha H^2$, $\mathscr{N}_{\gamma+1} = \theta''_{\gamma+1} H^2 \ominus \theta_{\gamma+1} H^2$ and, if $\alpha > \gamma+1$, then $\theta''_\alpha \,|\, \theta''_{\gamma+1}$ and hence $\mathscr{N}_\alpha \subset \theta''_\alpha H^2 \ominus \theta_\alpha H^2$. These calculations show that we have

$$\begin{aligned}
&(\operatorname{ran} \theta''_{\gamma+1}(T))^- \cap \ker \theta'_\gamma(T) \\
&= \left(\bigoplus_{\alpha<\gamma} \mathscr{M}_\alpha\right) \oplus (\theta''_\gamma H^2 \ominus \theta_\gamma H^2) \oplus (\theta''_{\gamma+1} H^2 \ominus \theta_{\gamma+1} H^2) \oplus \left(\bigoplus_{\alpha>\gamma+1} \mathscr{N}_\alpha\right),
\end{aligned}$$

and this space is contained in $\mathscr{K}$. Therefore we have

$$\mathscr{K} = \bigvee_\gamma ((\operatorname{ran} \theta''_{\gamma+1}(T))^- \cap \ker \theta'_\gamma(T)),$$

and this concludes the proof.

It is easy to see that $\operatorname{Lat}(\{T\}')$ is not usually generated by $\{\ker u(T), (\operatorname{ran} u(T))^- : u \in H^\infty\}$ as a lattice (i.e., infinite lattice operations are really needed in the above proof).

2.6. PROPOSITION. *The lattice* $\operatorname{Lat}(\{T\}')$ *of the Jordan operator* $T = S(\Theta)$ *is totally ordered by inclusion if and only if one of the two following situations occurs:*

(i) $\theta_0(\lambda) \equiv ((\lambda - \mu)/(1 - \bar{\mu}\lambda))^n$, $\lambda \in \mathbf{D}$, *for some* $\mu \in \mathbf{D}$ *and some natural number* n, *and* θ_α *equals* 1, $((\lambda-\mu)/(1-\bar{\mu}\lambda))^{n-1}$, *or* θ_0 *for every ordinal* α.

(ii) $\theta_0(\lambda) \equiv \exp(t(\lambda + \varsigma)/(\lambda - \varsigma))$, $\lambda \in \mathbf{D}$, *for some* $t > 0$ *and* ς *with* $|\varsigma| = 1$, *and* θ_α *equals* 1 *or* θ_0 *for every ordinal* α.

PROOF. If θ and θ' are two inner divisors of m_T then $\ker \theta(T) \subset \ker \theta'(T)$ if and only if $\theta \,|\, \theta'$. This follows, for example, from Theorem 2.4.6. If $\operatorname{Lat}(\{T\}')$ is totally ordered, it follows that the divisors of m_T are totally ordered by divisibility, and this already implies that $\theta_0 \equiv m_T$ must have one of the forms described in (i) and (ii).

Let us consider case (i) first. Since $\theta_\alpha \,|\, \theta_0$, we must have $\theta_\alpha(\lambda) = ((\lambda-\mu)/(1-\bar{\mu}\lambda))^{n(\alpha)}$, $\lambda \in \mathbf{D}$, where $n(\alpha)$ is a nonincreasing function of α and $n(0) = n$. An easy translation of Proposition 2.1(i) shows that $\operatorname{Lat}(\{T\}')$ is isomorphic with the collection of all nonincreasing functions $k(\alpha)$ with the property that the function $n(\alpha) - k(\alpha)$ is also nonincreasing in α. The order relation on this lattice is "pointwise inequality" by Remark 2.4.

Assume that there exists an ordinal number β such that $n(\beta) \notin \{n, n-1, 0\}$. We then define the functions $k_1(\alpha) = \max\{n(\alpha) - 1, 0\}$ and $k_2(\alpha) = \min\{n(\alpha), n(\beta)\}$. These two functions define incomparable hyperinvariant subspaces because

$$k_1(0) = n - 1 > k_2(0) = n(\beta),$$

and

$$k_1(\beta) = n(\beta) - 1 < k_2(\beta) = n(\beta).$$

Thus we must have $n(\alpha) \in \{n, n-1, 0\}$ if $\mathrm{Lat}(\{T\}')$ is totally ordered. Conversely, if $n(\alpha) \in \{n, n-1, 0\}$ for every α, $\mathrm{Lat}(\{T\}')$ is totally ordered. Indeed, let k_1 and k_2 be two arbitrary nonincreasing functions such that $n(\alpha) - k_1(\alpha)$ and $n(\alpha) - k_2(\alpha)$ are also nonincreasing. If k_1 and k_2 are not comparable, there exist ordinals $\alpha < \beta$ such that, for example, $k_1(\alpha) < k_2(\alpha)$ and $k_1(\beta) > k_2(\beta)$. Since $n(\alpha) \leq n(\beta) + 1$ by the assumption, we must have $n(\beta) - k_2(\beta) \leq n(\alpha) - k_2(\alpha) \leq n(\beta) + 1 - k_2(\alpha)$. Then we obtain $k_1(\beta) \leq k_1(\alpha) \leq k_2(\alpha) \leq k_2(\beta)$, a contradiction. Thus $\mathrm{Lat}(\{T\}')$ must be totally ordered.

Now consider case (ii). Then we have $\theta_\alpha(\lambda) = \exp(t(\alpha)(\lambda + \varsigma)/(\lambda - \varsigma))$, $\lambda \in \mathbf{D}$, where $t(\alpha)$ is a nonincreasing function and $t(\alpha) \in [0, t]$. As in the preceding argument, $\mathrm{Lat}(\{T\}')$ is isomorphic to the pointwise ordered lattice of all nonincreasing functions $s(\alpha)$ such that $s(\alpha) \in [0, t(\alpha)]$ and $t(\alpha) - s(\alpha)$ is also nonincreasing. Assume that there exists an ordinal β such that $t(\beta) \notin \{t, 0\}$, and choose a positive number $\varepsilon < \min\{t(\beta), t - t(\beta)\}$. Define the functions $s_1(\alpha) = \max\{t(\alpha) - \varepsilon, 0\}$ and $s_2(\alpha) = \min\{t(\alpha), t(\beta)\}$. These two functions correspond to noncomparable hyperinvariant subspaces because $s_1(\beta) = t(\beta) - \varepsilon < s_2(\beta) = t(\beta)$, while $s_1(0) = t(0) > s_2(0) = t(\beta)$. Thus $t(\alpha) \in \{0, t\}$ if $\mathrm{Lat}(\{T\}')$ is to be totally ordered. Conversely, if $t(\alpha) \in \{0, t\}$ for all α, and $s(\alpha) \in [0, t]$ is a nonincreasing function such that $t - s(\alpha)$ is also nonincreasing, it clearly follows that s must be constant on the set $\{\alpha : t(\alpha) \neq 0\}$. Thus $\mathrm{Lat}(\{T\}')$ is isomorphic with the segment $[0, t]$ in this case and the proposition follows.

If T and T' are two arbitrary operators one can always construct a map $\phi : \mathrm{Lat}(\{T\}') \to \mathrm{Lat}(\{T'\}')$ by setting

$$\phi(\mathscr{M}) = \bigvee\{X\mathscr{M} : X \in \mathscr{I}(T, T')\} \tag{2.7}$$

where, as usual, $X \in \mathscr{I}(T, T')$ means that $T'X = XT$. This map may be trivial, i.e., $\phi(\mathscr{M}) = \{0\}$ for every $\mathscr{M} \in \mathrm{Lat}(\{T\}')$, if there is no special relationship between T and T'. If we define $\psi : \mathrm{Lat}(\{T'\}') \to \mathrm{Lat}(\{T\}')$ by a formula analogous to (2.7), we must have $\psi(\phi(\mathscr{M})) \subset \mathscr{M}$ and $\phi(\psi(\mathscr{N})) \subset \mathscr{N}$ for $\mathscr{M} \in \mathrm{Lat}(\{T\}')$ and $\mathscr{N} \in \mathrm{Lat}(\{T'\}')$. Indeed, we have

$$\begin{aligned}\psi(\phi(\mathscr{M})) &= \bigvee\{Y\phi(\mathscr{M}) : Y \in \mathscr{I}(T', T)\}\\ &= \bigvee\{YX\mathscr{M} : Y \in \mathscr{I}(T', T),\ X \in \mathscr{I}(T, T')\}\\ &\subset \bigvee\{Z\mathscr{M} : Z \in \{T\}'\} \subset \mathscr{M}\end{aligned}$$

since $\mathscr{M} \in \mathrm{Lat}(\{T\}')$. The maps ϕ and ψ described above are particularly interesting if T is the Jordan model of the operator T' of class C_0. The following result shows in particular that, in a family of quasisimilar operators of class C_0, the Jordan model has the smallest lattice of hyperinvariant subspaces.

2.8. THEOREM. *Let $T \in \mathscr{L}(\mathscr{H})$ be an operator of class C_0 with Jordan model $S(\Theta) \in \mathscr{L}(\mathscr{H}(\Theta))$, $\Theta = \{\theta_\alpha\}$. Define maps $\phi : \mathrm{Lat}(\{T\}') \to \mathrm{Lat}(\{S(\Theta)\}')$, $\psi : \mathrm{Lat}(\{S(\Theta)\}') \to \mathrm{Lat}(\{T\}')$, and $\psi_* : \mathrm{Lat}(\{T^*\}') \to \mathrm{Lat}(\{S(\Theta)^*\}')$ by formulas analogous to* (2.7).

(i) *There exist $X \in \mathscr{I}(S(\Theta), T)$ and $Y \in \mathscr{I}(T, S(\Theta))$ such that $\psi(\mathscr{M}) = (Y\mathscr{M})^- = X^{-1}(\mathscr{M})$ for every $\mathscr{M} \in \mathrm{Lat}(\{T\}')$. In particular, $S(\Theta)\,|\,\psi(\mathscr{M})$ is unitarily equivalent to the Jordan model of $T\,|\,\mathscr{M}$.*

(ii) *$\psi \circ \phi$ is the identity mapping on* $\mathrm{Lat}(\{S(\Theta)\}')$.

(iii) *$\psi_*(\mathscr{H} \ominus \mathscr{M}) = \mathscr{H}(\Theta) \ominus \psi(\mathscr{M})$ for $\mathscr{M} \in \mathrm{Lat}(\{T\}')$.*

PROOF. Let $\mathscr{H} = \bigvee_\alpha \mathscr{H}_\alpha$ be the almost direct decomposition provided by Theorem 3.6.10. Thus $\mathscr{H}_\alpha \in \mathrm{Lat}(T)$, $T\,|\,\mathscr{H}_\alpha \sim S(\theta_\alpha)$, $\mathscr{H}_{\alpha+i} \perp \mathscr{H}_{\beta+j}$ if α and β are different limit ordinals and $i, j < \omega$. Let us set $\mathscr{H}'_\alpha = \mathscr{H} \ominus (\bigvee_{\beta\neq\alpha} \mathscr{H}_\beta) \in \mathrm{Lat}(T^*)$. By Lemma 3.6.2 we must have $\mathscr{H} = \bigvee_\alpha \mathscr{H}'_\alpha$, and this implies that $P_{\mathscr{H}'_\alpha}|\mathscr{H}_\alpha$ is a quasiaffinity. Indeed, $\mathscr{H}_\beta \perp \mathscr{H}'_\alpha$ if $\alpha \neq \beta$, so that $(P_{\mathscr{H}'_\alpha}\mathscr{H}_\alpha)^- = (P_{\mathscr{H}'_\alpha}(\bigvee_\alpha \mathscr{H}_\alpha))^- = P_{\mathscr{H}'_\alpha}\mathscr{H} = \mathscr{H}'_\alpha$. Thus $P_{\mathscr{H}'_\alpha}|\mathscr{H}_\alpha$ has dense range and, in an analogous manner, the adjoint $(P_{\mathscr{H}'_\alpha}|\mathscr{H}_\alpha)^* = P_{\mathscr{H}_\alpha}|\mathscr{H}'_\alpha$ also has dense range. Furthermore, if we denote $T_{\mathscr{H}'_\alpha} = (T^*\,|\,\mathscr{H}'_\alpha)^*$, we have $T_{\mathscr{H}'_\alpha}(P_{\mathscr{H}'_\alpha}|\mathscr{H}_\alpha) = (P_{\mathscr{H}'_\alpha}|\mathscr{H}_\alpha)(T\,|\,\mathscr{H}_\alpha)$ so that $T\,|\,\mathscr{H}_\alpha \prec T_{\mathscr{H}'_\alpha}$. By Proposition 3.5.32, $T_{\mathscr{H}'_\alpha}$ must be quasisimilar to $S(\theta_\alpha)$, hence we can choose quasiaffinities $X_\alpha \in \mathscr{I}(S(\theta_\alpha), T\,|\,\mathscr{H}_\alpha)$ and $Y_\alpha \in \mathscr{I}(T_{\mathscr{H}'_\alpha}, S(\theta_\alpha))$. We may clearly assume that $\sum_{n<\omega} \|Y_{\alpha+n}\| \leq 1$ and $\sum_{n<\omega} \|X_{\alpha+n}\| \leq 1$ for every limit ordinal α. Now define quasiaffinities $Y \in \mathscr{I}(T, S(\Theta))$ and $X \in \mathscr{I}(S(\Theta), T)$ as follows:

$$Xh = \sum_\alpha X_\alpha h_\alpha,\ h = \bigoplus_\alpha h_\alpha \in \mathscr{H}(\Theta);\quad Yh = \bigoplus_\alpha Y_\alpha P_{\mathscr{H}'_\alpha} h,\ h \in \mathscr{H}. \tag{2.9}$$

The operators X and Y are indeed bounded and, in fact, $\|X\| \leq 1$ and $\|Y\| \leq 1$. The reader will have no difficulty showing that X and Y are indeed quasiaffinities.

Now we remark that the operator $Y_\alpha(P_{\mathscr{H}'_\alpha}|\mathscr{H}_\alpha)X_\alpha \in \{S(\theta_\alpha)\}'$ is a quasiaffinity and hence Corollary 3.1.20 implies the existence of $u_\alpha \in H^\infty$ such that

$$Y_\alpha(P_{\mathscr{H}'_\alpha}|\mathscr{H}_\alpha)X_\alpha = u_\alpha(S(\theta_\alpha)), \qquad u_\alpha \wedge \theta_\alpha \equiv 1. \tag{2.10}$$

Note in addition that $Y_\alpha(P_{\mathscr{H}'_\alpha}|\mathscr{H}_\beta)X_\beta = 0$ if $\alpha \neq \beta$, and therefore formulas (2.9) and (2.10) imply that

$$YX = \bigoplus_\alpha u_\alpha(S(\theta_\alpha)). \tag{2.11}$$

If $\mathscr{K} \in \mathrm{Lat}(\{S(\Theta)\}')$ is given by formula (2.2), we see that $u_\alpha(S(\theta_\alpha))\,|\,\theta''_\alpha H^2 \ominus \theta_\alpha H^2$ is a quasiaffinity and therefore (2.11) implies the equality $(YX\mathscr{K})^- = \mathscr{K}$. Since we clearly have $(YX\mathscr{K})^- \subset \psi(\phi(\mathscr{K}))$, and $\psi(\phi(\mathscr{K})) \subset \mathscr{K}$ by the remarks preceding the theorem, we conclude that $\psi(\phi(\mathscr{K})) = \mathscr{K}$, $\mathscr{K} \in \mathrm{Lat}(\{S(\Theta)\}')$, and this proves assertion (ii).

Next we show that $(Y\mathscr{M})^- \in \mathrm{Lat}(\{S(\Theta)\}')$ whenever $\mathscr{M} \in \mathrm{Lat}(\{T\}')$. Since $(Y\mathscr{M})^-$ is clearly invariant for $S(\Theta)$, it suffices to show that $(Y\mathscr{M})^-$ is invariant under the operators $R_{\beta\alpha}$ defined in the proof of Theorem 1.25. In order to do this we construct some imperfect replicas $Q_{\beta\alpha} \in \{T\}'$ of $R_{\beta\alpha}$ as follows:

$$\begin{aligned} Q_{\beta\alpha} &= X_\beta P_{\mathscr{H}(\theta_\beta)} Y_\alpha P_{\mathscr{H}'_\alpha} \quad \text{if } \alpha \leq \beta, \\ &= X_\beta (\theta_\beta/\theta_\alpha) Y_\alpha P_{\mathscr{H}'_\alpha} \quad \text{if } \alpha > \beta, \end{aligned} \tag{2.12}$$

where $(\theta_\beta/\theta_\alpha)$ designates a multiplication operator. We claim that

$$YQ_{\beta\alpha} = u_\beta(S(\Theta))R_{\beta\alpha}Y \tag{2.13}$$

for all ordinals α and β. Indeed, assume for example that $\alpha \leq \beta$. Keeping in mind the fact that $P_{\mathscr{H}'_\gamma} \mid \mathscr{H}_\beta = 0$ if $\gamma \neq \beta$, we see that

$$\begin{aligned} YQ_{\beta\alpha} &= \bigoplus_\gamma Y_\gamma P_{\mathscr{H}'_\gamma} X_\beta P_{\mathscr{H}(\theta_\beta)} Y_\alpha P_{\mathscr{H}'_\alpha} \\ &= \left(\bigoplus_{\gamma\neq\beta} 0\right) \oplus (Y_\beta P_{\mathscr{H}'_\beta} X_\beta P_{\mathscr{H}(\theta_\beta)} Y_\alpha P_{\mathscr{H}'_\alpha}) \\ &= \left(\bigoplus_{\gamma\neq\beta} 0\right) \oplus (u_\beta(S(\theta_\beta)) P_{\mathscr{H}(\theta_\beta)} Y_\alpha P_{\mathscr{H}'_\alpha}) \\ &= \left(\bigoplus_{\gamma\neq\beta} 0\right) \oplus (P_{\mathscr{H}(\theta_\beta)} u_\beta(S(\theta_\beta)) Y_\alpha P_{\mathscr{H}'_\alpha}) \\ &= R_{\beta\alpha} u_\beta(S(\Theta))Y \\ &= u_\beta(S(\Theta))R_{\beta\alpha}Y, \end{aligned}$$

where we have used (2.9), (2.10), and the fact that $P_{\mathscr{H}(\theta_\beta)} \in \mathscr{I}(S(\theta_\alpha), S(\theta_\beta))$. The equality (2.13) is proved analogously for the case in which $\beta < \alpha$.

Assume now that $\mathscr{M} \in \mathrm{Lat}(\{T\}')$ and note that (2.13) implies the inclusions

$$u_\beta(S(\Theta))R_{\beta\alpha}(Y\mathscr{M})^- \subset (YQ_{\beta\alpha}\mathscr{M})^- \subset (Y\mathscr{M})^-.$$

Moreover, we know from (2.10) that $u_\beta \wedge \theta_\beta \equiv 1$ so that $u_\beta(S(\Theta))$ is a quasi-affinity on the range of $R_{\beta\alpha}$. We conclude that

$$R_{\beta\alpha}(Y\mathscr{M})^- = (u_\beta(S(\Theta))(R_{\beta\alpha}(Y\mathscr{M})^-))^- \subset (Y\mathscr{M})^-,$$

and therefore $(Y\mathscr{M})^-$ is hyperinvariant for $S(\Theta)$. A completely analogous argument shows that $(X^*\mathscr{N})^- \in \mathrm{Lat}(\{S(\Theta)^*\}')$ whenever $\mathscr{N} \in \mathrm{Lat}(\{T^*\}')$.

Let $\mathscr{M} \in \mathrm{Lat}(\{T\}')$ and note that the fact that $(Y\mathscr{M})^-$ is hyperinvariant for $S(\Theta)$ does not allow us to conclude that $(Y\mathscr{M})^- = \psi(\mathscr{M})$. To prove this equality we note that $X^*(\mathscr{H} \ominus \mathscr{M}) \subset \mathscr{H}(\Theta) \ominus (Y\mathscr{M})^-$, as is readily seen from the equality $(Yh, X^*g) = (XYh, g) = 0$, which is verified for $h \in \mathscr{M}$ and $g \in \mathscr{H} \ominus \mathscr{M}$ because $XYh \in \mathscr{M}$. In fact this argument does not depend on the particular form of

X and Y, but only on the fact that $Y \in \mathcal{I}(T, S(\Theta))$ and $X \in \mathcal{I}(S(\Theta), T)$. Therefore we must have

$$\psi_*(\mathcal{H} \ominus \mathcal{M}) \subset \mathcal{H}(\Theta) \ominus \psi(\mathcal{M}).$$

Taking into account the obvious inclusions $Y\mathcal{M} \subset \psi(\mathcal{M})$ and $X^*(\mathcal{H} \ominus \mathcal{M}) \subset \psi_*(\mathcal{H} \ominus \mathcal{M})$ we deduce the following relations:

$$\begin{aligned}(2.14)\quad T^* \,|\, \mathcal{H} \ominus \mathcal{M} &\prec S(\Theta)^* \,|\, (X^*(\mathcal{H} \ominus \mathcal{M}))^- \overset{i}{\prec} S(\Theta)^* \,|\, \psi_*(\mathcal{H} \ominus \mathcal{M}) \\ &\overset{i}{\prec} S(\Theta)^* \,|\, \mathcal{H}(\Theta) \ominus \psi(\mathcal{M}) \overset{i}{\prec} S(\Theta)^* \,|\, \mathcal{H}(\Theta) \ominus (Y\mathcal{M})^-.\end{aligned}$$

Finally, the operator $Z = P_{\mathcal{H}(\Theta) \ominus (Y\mathcal{M})^-} Y \,|\, \mathcal{H} \ominus \mathcal{M}$ clearly has dense range and

$$(S(\Theta)^* \,|\, \mathcal{H}(\Theta) \ominus (Y\mathcal{M})^-)^* Z = Z(T^* \,|\, \mathcal{H} \ominus \mathcal{M})^*,$$

thus showing that $S(\Theta)^* \,|\, \mathcal{H}(\Theta) \ominus (Y\mathcal{M})^- \overset{i}{\prec} T^* \,|\, \mathcal{H} \ominus \mathcal{M}$. Combining this with (2.14) and Proposition 3.5.32 we see that all five operators in (2.14) are quasisimilar and, since all the subspaces appearing in (2.14) are hyperinvariant, we conclude that

$$(X^*(\mathcal{H} \ominus \mathcal{M}))^- = \psi_*(\mathcal{H} \ominus \mathcal{M}) = \mathcal{H}(\Theta) \ominus \psi(\mathcal{M}) = \mathcal{H}(\Theta) \ominus (Y\mathcal{M})^-$$

by Proposition 2.1(iii). This proves both assertions (i) and (iii) of the present theorem, thus concluding the proof.

2.15. COROLLARY. *Let T be an operator of class C_0, and let $\mathcal{M}$ be a hyperinvariant subspace for T. If $T = \begin{bmatrix} T' & X \\ 0 & T'' \end{bmatrix}$ is the triangularization of T with respect to the decomposition $\mathcal{H} = \mathcal{M} \oplus (\mathcal{H} \ominus \mathcal{M})$, then $m_T \equiv m_{T'} m_{T''}$.*

PROOF. If ψ and $S(\Theta)$ are as in the preceding theorem, then T' is quasisimilar to $S(\Theta) \,|\, \psi(\mathcal{M})$ and T''^* is quasisimilar to $S(\Theta)^* \,|\, \psi_*(\mathcal{H} \ominus \mathcal{M}) = S(\Theta)^* \,|\, \mathcal{H}(\Theta) \ominus \psi(\mathcal{M})$. Now the equality $m_T \equiv m_{T'} m_{T''}$ clearly follows from Proposition 2.1(ii).

If $T \in \mathcal{L}(\mathcal{H})$ and $\mathcal{M}$ is a subspace of $\mathcal{H}$, we will denote by $T_{\mathcal{M}} \in \mathcal{L}(\mathcal{M})$ the compression of T to $\mathcal{M}$: $T_{\mathcal{M}} = P_{\mathcal{M}} T \,|\, \mathcal{M}$. If $\mathcal{M} \in \mathrm{Lat}(T)$ we have $T_{\mathcal{M}} = T \,|\, \mathcal{M}$, and if $\mathcal{M} \in \mathrm{Lat}(T^*)$ then $T_{\mathcal{M}} = (T^* \,|\, \mathcal{M})^*$. Clearly $(T^*_{\mathcal{M}}) = (T_{\mathcal{M}})^*$ for every T and $\mathcal{M}$.

2.16. COROLLARY. *Let $T \in \mathcal{L}(\mathcal{H})$ and $T' \in \mathcal{L}(\mathcal{H}')$ be two quasisimilar operators of class C_0, let $S(\Theta)$ be their common Jordan model, and let $\eta : \mathrm{Lat}(\{T\}') \to \mathrm{Lat}(\{T'\}')$, $\psi : \mathrm{Lat}(\{T\}') \to \mathrm{Lat}(\{S(\Theta)\}')$, $\psi' : \mathrm{Lat}(\{T'\}') \to \mathrm{Lat}(\{S(\Theta)\}')$ be defined by formulas analogous to* (2.7).

(i) *$\psi'(\eta(\mathcal{M})) = \psi(\mathcal{M})$ for $\mathcal{M} \in \mathrm{Lat}(\{T\}')$. In particular, $T \,|\, \mathcal{M} \sim T' \,|\, \eta(\mathcal{M})$, $\mathcal{M} \in \mathrm{Lat}(\{T\}')$.*

(ii) *If $\mathcal{M} \in \mathrm{Lat}(\{T\}')$, $\mathcal{M}' \in \mathrm{Lat}(\{T'\}')$, and $T \,|\, \mathcal{M} \sim T' \,|\, \mathcal{M}'$ then $T_{\mathcal{H} \ominus \mathcal{M}} \sim T'_{\mathcal{H}' \ominus \mathcal{M}'}$.*

PROOF. The inclusion $\psi'(\eta(\mathcal{M})) \subset \psi(\mathcal{M})$ is obvious. Hence we have

$$T \,|\, \mathcal{M} \overset{i}{\prec} S(\Theta) \,|\, \psi'(\eta(\mathcal{M})) \overset{i}{\prec} S(\Theta) \,|\, \psi(\mathcal{M}) \prec T \,|\, \mathcal{M},$$

where the first injection follows because $\psi'(\eta(\mathcal{M}))$ must contain $BA\mathcal{M}$ if $B \in \mathcal{I}(T', S(\Theta))$ and $A \in \mathcal{I}(T, T')$ are quasiaffinities, and the last relation follows from Theorem 2.8. We conclude by Proposition 3.5.32 that $S(\Theta)\,|\,\psi'(\eta(\mathscr{M})) \sim S(\Theta)\,|\,\psi(\mathscr{M})$, and hence $\psi'(\eta(\mathscr{M})) = \psi(\mathscr{M})$ by Proposition 2.1(iii).

To prove (ii) we note that $S(\Theta)\,|\,\psi(\mathscr{M})$ and $S(\Theta)\,|\,\psi'(\mathscr{M}')$ must be quasisimilar, and hence $\psi(\mathscr{M}) = \psi'(\mathscr{M}')$ by Proposition 2.1(iii). Theorem 2.8 now implies that both $T_{\mathscr{H}\ominus\mathscr{M}}$ and $T'_{\mathscr{H}'\ominus\mathscr{M}'}$ are quasisimilar to $S(\Theta)_{\mathscr{N}}$, where $\mathscr{N} = \mathscr{H}(\Theta)\ominus\psi(\mathscr{M}) = \mathscr{H}(\Theta)\ominus\psi'(\mathscr{M}')$.

2.17. COROLLARY. *Let $T \in \mathscr{L}(\mathscr{H})$ be an operator of class C_0 with Jordan model $S(\Theta)$, let $\phi : \operatorname{Lat}(\{S(\Theta)\}') \to \operatorname{Lat}(\{T\}')$, $\phi_* : \operatorname{Lat}(\{S(\Theta)^*\}') \to \operatorname{Lat}(\{T^*\}')$ be defined by formulas analogous to (2.7), and let $\mathscr{M}$ be a hyperinvariant subspace for $S(\Theta)$. Among the spaces $\mathscr{N} \in \operatorname{Lat}(\{T\}')$, with the property that $T\,|\,\mathscr{N} \sim S(\Theta)\,|\,\mathscr{M}$, $\phi(\mathscr{M})$ is the smallest one and $\mathscr{H}\ominus\phi_*(\mathscr{H}(\Theta)\ominus\mathscr{M})$ is the largest one.*

PROOF. Let ψ be as in Theorem 2.8. By that theorem, $T\,|\,\mathscr{N} \sim S(\Theta)\,|\,\mathscr{M}$ implies $\psi(\mathscr{N}) = \mathscr{M}$ and hence $\phi(\mathscr{M}) = \phi(\psi(\mathscr{N})) \subset \mathscr{N}$. Also, $T\,|\,\mathscr{N} \sim S(\Theta)\,|\,\mathscr{M}$ implies that $T^*\,|\,\mathscr{H}\ominus\mathscr{N} \sim S(\Theta)^*\,|\,\mathscr{H}(\Theta)\ominus\mathscr{M}$ and hence the inclusion $\phi_*(\mathscr{H}(\Theta)\ominus\mathscr{M}) \subset \mathscr{H}\ominus\mathscr{N}$ follows as above. Thus $\mathscr{N} \subset \mathscr{H}\ominus\phi_*(\mathscr{H}(\Theta)\ominus\mathscr{M})$, as desired.

2.18. COROLLARY. *Under the conditions of the preceding corollary, the following assertions are equivalent:*

(i) *ϕ is a bijection;*

(ii) *ϕ_* is a bijection;*

(iii) *$\mathscr{H}\ominus\phi(\mathscr{M}) = \phi_*(\mathscr{H}(\Theta)\ominus\mathscr{M})$ for $\mathscr{M} \in \operatorname{Lat}(\{S(\Theta)\}')$; and*

(iv) *if $\mathscr{N}_1, \mathscr{N}_2 \in \operatorname{Lat}(\{T\}')$ and $T\,|\,\mathscr{N}_1 \sim T\,|\,\mathscr{N}_2$, then $\mathscr{N}_1 = \mathscr{N}_2$.*

PROOF. According to Theorem 2.8, ϕ has the left inverse ψ, and hence (i) holds if and only if ψ is one-to-one. We know from Proposition 2.1 that (iv) holds for $T = S(\Theta)$. Thus ψ is one-to-one if and only if (iv) holds, and thus the equivalence of (i) and (iv) is established. By Theorem 2.8 we also have $\mathscr{H}\ominus\psi(\mathscr{M}) = \psi_*(\mathscr{H}(\Theta)\ominus\mathscr{M})$, $\mathscr{M} \in \operatorname{Lat}(\{S(\Theta)\}')$. We conclude that ψ is one-to-one if and only if ψ_* is one-to-one. But ψ_* is the left inverse of ϕ_* and the equivalence of (i) and (ii) follows at once. Finally we establish the equivalence of (iii) and (iv). If $\mathscr{M} \in \operatorname{Lat}(\{S(\Theta)\}')$, we have $T\,|\,\phi(\mathscr{M}) \sim T\,|\,\mathscr{H}\ominus\phi_*(\mathscr{H}(\Theta)\ominus\mathscr{M}) \sim S(\Theta)\,|\,\mathscr{M}$ and therefore the equality in (iii) holds if (iv) is true. Conversely, if (iii) holds and $\mathscr{N}_1, \mathscr{N}_2$ are as in (iv), then Corollary 2.17 implies the following inclusions:

$$\phi(\mathscr{M}) \subset \mathscr{N}_j \subset \mathscr{H}\ominus\phi_*(\mathscr{H}(\Theta)\ominus\mathscr{M}) = \phi(\mathscr{M}), \qquad j = 1, 2,$$

where $\mathscr{M} = \psi(\mathscr{N}_1) = \psi(\mathscr{N}_2)$. Thus $\mathscr{N}_1 = \mathscr{N}_2 = \phi(\mathscr{M})$ and the corollary is proved.

Exercises

1. Let $\{\theta_j : j \in J\}$ be a family of inner functions. Characterize the lattice of hyperinvariant subspaces of the operator $T = \bigoplus_{j\in J} S(\theta_j)$.

2. Let T be the operator in Exercise 1. Show that $\mathrm{Lat}(\{T\}')$ is distributive (i.e., $\mathcal{M}_1 \vee (\mathcal{M}_2 \cap \mathcal{M}_3) = (\mathcal{M}_1 \vee \mathcal{M}_2) \cap (\mathcal{M}_1 \vee \mathcal{M}_3)$ and $\mathcal{M}_1 \cap (\mathcal{M}_2 \vee \mathcal{M}_3) = (\mathcal{M}_1 \cap \mathcal{M}_2) \vee (\mathcal{M}_1 \cap \mathcal{M}_3)$ for $\mathcal{M}_1, \mathcal{M}_2, \mathcal{M}_3 \in \mathrm{Lat}(\{T\}')$) and selfdual (i.e., the ordered set $(\mathrm{Lat}(\{T\}'), \subset)$ is isomorphic to the ordered set $(\mathrm{Lat}(\{T\}'), \supset)$).

3. Let T be as above. Show that $\mathrm{Lat}(\{T\}')$ is complemented if and only if each θ_j is a Blaschke product with simple zeros.

4. Let $T = \bigoplus_{j<\omega} S(\theta_j)$ be a Jordan operator. Show that in general $\mathrm{Lat}(\{T\}')$ is not generated as a lattice by the subset $\{(\mathrm{ran}\, u(T))^-, \ker u(T) : u \in H^\infty\}$.

5. Find an operator T of class C_0 with Jordan model $S(\Theta)$ such that:

(i) $\mathrm{Lat}(\{S(\Theta)\}')$ is finite and totally ordered; and

(ii) $\mathrm{Lat}(\{T\}')$ is not totally ordered.

6. Let $\{t_n : n > 0\}$ be a dense sequence in the interval $(0,1)$ and set $\theta_n(\lambda) = \exp(t_n(\lambda+1)/(\lambda-1))$, $\lambda \in \mathbf{D}$, $n > 0$. Show that $\mathrm{Lat}(\{T\}')$, where $T = \bigoplus_{n=1}^{\infty} S(\theta_n)$, is isomorphic to the lattice $\mathcal{L}$ of classes of equivalence of measurable functions $f : [0,1] \to [0,1]$, where two functions are equivalent if they coincide almost everywhere.

7. Let T be as in Exercise 6. Find the Jordan model $S(\Theta)$ of T, and show that $\mathrm{Lat}(\{S(\Theta)\}')$ is isomorphic to the segment $[0,1]$. Find mappings $[0,1] \to \mathcal{L}$ and $\mathcal{L} \to [0,1]$ that correspond with the maps $\mathrm{Lat}(\{S(\Theta)\}') \to \mathrm{Lat}(\{T\}')$ and $\mathrm{Lat}(\{T\}') \to \mathrm{Lat}(\{S(\Theta)\}')$ given by $\mathcal{M} \to \phi(\mathcal{M})$, $\mathcal{M} \to \mathcal{H} \ominus \phi_*(\mathcal{H}(\Theta) \ominus \mathcal{M})$, $\mathcal{N} \to \psi(\mathcal{N})$ and $\mathcal{N} \to \mathcal{H}(\Theta) \ominus \psi_*(\mathcal{H} \ominus \mathcal{N})$, $\mathcal{M} \in \mathrm{Lat}(\{S(\Theta)\}')$, $\mathcal{N} \in \mathrm{Lat}(\{T\}')$.

8. Let T, T', $S(\Theta)$, and η, ψ, ψ' be as in Corollary 2.16. Consider in addition maps $\phi : \mathrm{Lat}(\{S(\Theta)\}') \to \mathrm{Lat}(\{T\}')$ and $\phi' : \mathrm{Lat}(\{S(\Theta)\}') \to \mathrm{Lat}(\{T'\}')$ defined in the usual manner. Show that $\phi' = \eta \circ \phi$, but η may be different from $\phi' \circ \psi$.

9. Find an operator T for which $\mathrm{Lat}(\{T\}')$ is not distributive.

10. Find an operator T for which $\mathrm{Lat}(\{T\}')$ is not selfdual.

11. What can be said about the lattice $\mathrm{Lat}(\{T\}')$ if T is an algebraic operator?

3. Semigroups and Volterra operators. We will need a few definitions and results from the theory of strongly continuous semigroups of operators. For the reader's convenience, we recall the pertinent facts.

3.1. DEFINITION. A family $\{T(t) : t \geq 0\}$ of operators acting on the Hilbert space $\mathcal{H}$ is said to be a *strongly continuous semigroup of operators* if the following conditions are satisfied:

(i) $T(0) = I$, $T(t+s) = T(t)T(s)$, $t, s \geq 0$;

(ii) $\lim_{t \downarrow 0} \|T(t)h - h\| = 0$ for all $h \in \mathcal{H}$.

If, in addition, $\|T(t)\| \leq 1$ for all $t \geq 0$, then $\{T(t) : t \geq 0\}$ is called a *strongly continuous semigroup of contractions.*

3.2. DEFINITION. Let $\{T(t) : t \geq 0\} \subset \mathscr{L}(\mathscr{H})$ be a strongly continuous semigroup of operators. The *generator* A of $\{T(t) : t \geq 0\}$ is the (generally unbounded) operator defined by $Ah = \lim_{t\downarrow 0}(T(t)h - h)/t$; the domain of A consists of those vectors h for which the limit exists.

We recall that an unbounded operator A on $\mathscr{H}$ is said to be invertible if there exists $B \in \mathscr{L}(\mathscr{H})$ such that $AB = I$ and $BAh = h$ for h in the domain of A.

3.3. THEOREM. *Assume that A is the generator of the strongly continuous semigroup of contractions $\{T(t) : t \geq 0\} \subset \mathscr{L}(\mathscr{H})$. Then*

(i) *$-A$ is accretive, i.e.,* $\mathrm{Re}(Ah, h) \leq 0$ *for h in the domain of A;*
(ii) *$\lambda I - A$ is invertible whenever* $\mathrm{Re}\,\lambda > 0$ *and, for such λ, $\|(\lambda I - A)^{-1}\| \leq 1/\mathrm{Re}\,\lambda$ and $(\lambda I - A)^{-1}h = \int_0^\infty e^{-\lambda t}T(t)h\,dt$, $h \in \mathscr{H}$. Conversely, if A is an operator satisfying the inequalities $\|(\lambda I - A)^{-1}\| \leq 1/\lambda$ for $\lambda > 0$, there exists a unique semigroup of contractions whose generator is A.*

The last statement in the preceding theorem is a particular case of the Hille–Yosida theorem, which gives a characterization of generators for arbitrary strongly continuous semigroups.

3.4. LEMMA. *Assume that A is the generator of the strongly continuous semigroup $\{T(t)\colon t \geq 0\}$ of contractions acting on $\mathscr{H}$. Then the operator $T = (A + I)(A - I)^{-1} = I + 2(A - I)^{-1}$ is a contraction, 1 is not an eigenvalue of T, and*

$$A = (T + I)(T - I)^{-1} = I + 2(T - I)^{-1}. \tag{3.5}$$

Conversely, if $T \in \mathscr{L}(\mathscr{H})$ is a contraction and 1 is not an eigenvalue of T, then the operator A defined by (3.5) *is the generator of a semigroup of contractions.*

PROOF. It is clear that T is a bounded operator since $(A - I)^{-1}$ is bounded. Moreover, if h is in the domain of A we have by Theorem 3.3(i)

$$\begin{aligned}\|(A + I)h\|^2 &= \|Ah\|^2 + \|h\|^2 + 2\,\mathrm{Re}(Ah, h)\\ &\leq \|Ah\|^2 + \|h\|^2 - 2\,\mathrm{Re}(Ah, h)\\ &= \|(A - I)h\|^2,\end{aligned}$$

and this inequality yields $\|Ty\| \leq \|y\|$, $y \in \mathscr{H}$, upon substituting $(A - I)h$ by y. The equality $T - I = 2(A - I)^{-1}$ shows that $T - I$ is one-to-one and $2(T - I)^{-1} = A - I$. Relation (3.5) is an immediate consequence of the last equality.

Assume now that A is defined by (3.5), where T is a contraction and $I - T$ is one-to-one. An easy calculation shows that $\lambda I - A = ((\lambda - 1)T - (\lambda + 1)I)(T - I)^{-1}$, and hence $(\lambda I - A)^{-1} = (I - T)((\lambda - 1)T - (\lambda + 1)I)^{-1}$ exists for $\mathrm{Re}\,\lambda > 0$.

Moreover, we deduce from the properties of the functional calculus that

$$\begin{aligned}\|(\lambda I - A)^{-1}\| &\le \sup\{|(\mu-1)/((\lambda-1)\mu-(\lambda+1))| : |\mu|<1\}\\ &= \sup\{|(\mu-1)/((\lambda-1)\mu-(\lambda+1))| : \mu=(\varsigma+1)/(\varsigma-1),\\ &\qquad\qquad \operatorname{Re}\varsigma<0\}\\ &= \sup\{1/|\lambda-\varsigma| : \operatorname{Re}\varsigma<0\}\\ &= 1/\operatorname{Re}\lambda.\end{aligned}$$

Theorem 3.3 implies now that A is the generator of a uniquely determined strongly continuous semigroup of contractions. The lemma is proved.

3.6. DEFINITION. Let A be the generator of the strongly continuous semigroup of contractions $\{T(t) : t \ge 0\}$. The *cogenerator* of $\{T(t) : t \ge 0\}$ is the contraction T defined by $T = (A+I)(A-I)^{-1}$.

In what follows we will use the notation e_t for the singular inner function defined for $t \ge 0$ by

$$e_t(\lambda) = \exp\left(t\frac{\lambda+1}{\lambda-1}\right), \quad \lambda \in \mathbf{D}. \tag{3.7}$$

3.8. LEMMA. *Assume that $U \in \mathscr{L}(\mathscr{H})$ is an absolutely continuous unitary operator. Then U is the cogenerator of a strongly continuous semigroup $\{U(t) : t \ge 0\}$ of unitary operators, where*

$$U(t) = e_t(U), \quad t \ge 0. \tag{3.9}$$

PROOF. The fact that 1 is not an eigenvalue of U is obvious since U is absolutely continuous. The lemma is obvious if 1 does not belong to the spectrum $\sigma(U)$ of U. Indeed, in that case

$$\frac{d}{dt}(e_t(U)) = (U+I)(U-I)^{-1}e_t(U) \tag{3.10}$$

in the norm topology because

$$\frac{d}{dt}(e_t(\lambda)) = \frac{\lambda+1}{\lambda-1}e_t(\lambda)$$

uniformly for $\lambda \in \sigma(U)$. The general case now follows easily: write $U = \bigoplus_{n=1}^{\infty} U_n$ such that $1 \notin \sigma(U_n)$ for $n \ge 1$; this can be achieved by using the spectral theorem. We conclude that $\{e_t(U) : t \ge 0\} = \{\bigoplus_{n=1}^{\infty} e_t(U_n) : t \ge 0\}$ is a strongly continuous semigroup and, if A denotes the generator of this semigroup, (3.10) implies the equality $A(U-I)k = (U+I)k$ for a dense set of vectors k (namely, the vectors with finitely many nonzero components in the decomposition $U = \bigoplus_{n=1}^{\infty} U_n$). We easily infer that $A = (U+I)(U-I)^{-1}$, and hence $U = (A+I)(A-I)^{-1}$ is the cogenerator of $\{e_t(U) : t \ge 0\}$.

3.11. PROPOSITION. *Assume that $T \in \mathscr{L}(\mathscr{H})$ is a completely nonunitary contraction. Then T is the cogenerator of a strongly continuous semigroup $\{T(t) : t \ge 0\}$, where*

$$T(t) = e_t(T), \quad t \ge 0. \tag{3.12}$$

PROOF. The minimal unitary dilation $U \in \mathcal{L}(\mathcal{K})$ of T is absolutely continuous by Theorem 1.2.13, and hence U is the cogenerator of $\{e_t(U) : t \geq 0\}$ by the preceding lemma. Consequently the operators $e_t(T) = P_{\mathcal{H}} e_t(U) | \mathcal{H}$ form a strongly continuous semigroup of contractions (multiplicativity follows from the equalities $e_s e_t = e_{s+t}$ in H^∞). Denote by A and B the generators of the semigroups $\{e_t(T) : t \geq 0\}$ and $\{e_t(U) : t \geq 0\}$, respectively. We have $I - U = 2(I - B)^{-1}$ so that

$$(I - U)k = 2\int_0^\infty e^{-t} e_t(U)k\,dt, \quad k \in \mathcal{K}. \tag{3.13}$$

Assume that $k \in \mathcal{H}$, and apply P to both sides of (3.13) to get

$$(I - T)k = 2\int_0^\infty e^{-t} e_t(T)k\,dt = 2(I - A)^{-1}k, \quad k \in \mathcal{H}.$$

We thus have $I - T = 2(I - A)^{-1}$, so that $T = (A + I)(A - I)^{-1}$ is the cogenerator of $\{e_t(T),\, t \geq 0\}$, as desired.

The reader will certainly note that the proof of Lemma 3.8 really applies to all unitary operators U for which 1 is not an eigenvalue. Proposition 3.11 also works for arbitrary contractions T for which 1 is not an eigenvalue. The problem is that we defined the functional calculus only for completely nonunitary contractions. This restriction will not affect us here because our main object of study in this section will be a semigroup $\{T(t) : t \geq 0\}$ such that $T(\tau) = 0$ for sufficiently large τ.

3.14. PROPOSITION. *Let $\{T(t) : t \geq 0\} \subset \mathcal{L}(\mathcal{H})$ be a strongly continuous semigroup of contractions with cogenerator T. Assume that there exists a number $\tau > 0$ such that $T(\tau) = 0$ and $T(t) \neq 0$ for $t < \tau$. Then T is an operator of class C_0 and $m_T \equiv e_\tau$.*

PROOF. The proposition obviously follows from (3.12) once we know that T is completely nonunitary. Assume that $T = U \oplus T'$, with U unitary. If $\{U(t) : t \geq 0\}$ and $\{T'(t) : t \geq 0\}$ are the semigroups with cogenerators U and T', respectively, then $T(t) = U(t) \oplus T'(t)$, $t \geq 0$. We have seen that $U(t)$ is unitary for all $t \geq 0$, and this contradicts the equality $T(\tau) = 0$, unless U acts on the trivial space $\{0\}$. The proposition follows.

The preceding proposition clearly hints at a possible classification of all semigroups $\{T(t) : t \geq 0\}$ with $T(\tau) = 0$. In analogy with the case of single operators, we give the following definition.

3.15. DEFINITION. The semigroups of operators $\{T(t) : t \geq 0\} \subset \mathcal{L}(\mathcal{H})$ and $\{T'(t) : t \geq 0\} \subset \mathcal{L}(\mathcal{H}')$ are said to be *quasisimilar* if there exist quasi-affinities $X \in \mathcal{L}(\mathcal{H}, \mathcal{H}')$ and $Y \in \mathcal{L}(\mathcal{H}', \mathcal{H})$ satisfying the equalities $T'(t)X = XT(t)$, $T(t)Y = YT'(t)$, $t \geq 0$.

3.16. LEMMA. *Assume that $\{T(t) : t \geq 0\} \subset \mathcal{L}(\mathcal{H})$ and $\{T'(t) : t \geq 0\} \subset \mathcal{L}(\mathcal{H}')$ are strongly continuous semigroups of contractions with completely* *nonunitary cogenerators T and T', respectively. Then the two semigroups are quasisimilar if and only if T and T' are quasisimilar.*

PROOF. It suffices to show that an operator $X \in \mathscr{L}(\mathscr{H}, \mathscr{H}')$ satisfies the relation $T'X = XT$ if and only if $T'(t)X = XT(t)$ for all $t \geq 0$. But this clearly follows from the formulas $T(t) = e_t(T)$, $T'(t) = e_t(T')$, $t \geq 0$, and $(I - T)h = 2\int_0^\infty e^{-t}T(t)h$, $h \in \mathscr{H}$, $(I - T')h' = 2\int_0^\infty e^{-t}T'(t)h'\,dt$, $h' \in \mathscr{H}'$.

The classification theorem for operators of class C_0 now implies that every semigroup of the kind considered in Proposition 3.14 is quasisimilar to a "Jordan model," i.e., to a semigroup whose cogenerator is a Jordan operator. It is interesting therefore to find nice semigroups whose cogenerators are unitarily equivalent to $S(e_\alpha)$, $\alpha > 0$.

3.17. LEMMA. *Let $\{T(t) : t \geq 0\} \subset \mathscr{L}(\mathscr{H})$ be a strongly continuous semigroup of contractions, and let $\tau > 0$ be such that $T(\tau) = 0$. Denote by A and T the generator and cogenerator of $\{T(t) : t \geq 0\}$, respectively.*

(i) *$(\lambda I - A)^{-1}$ exists for all $\lambda \in \mathbf{C}$, and*

$$(\lambda I - A)^{-1}h = \int_0^\infty e^{-\lambda t}T(t)h\,dt = \int_0^\tau e^{-\lambda t}T(t)\,dt, \quad \lambda \in \mathbf{C}, \quad h \in \mathscr{H}.$$

(ii) *The bounded operator $V = -A^{-1}$ is accretive and $\sigma(V) = \{0\}$.*

(iii) *We have $T = (I - V)(I + V)^{-1}$ and $\dim(\mathscr{D}_T) = \dim((V + V^*)\mathscr{H})$.*

PROOF. Define $R(\lambda) \in \mathscr{L}(\mathscr{H})$, $\lambda \in \mathbf{C}$, by $R(\lambda)h = \int_0^\tau e^{-\lambda t}T(t)h\,dt$, $h \in \mathscr{H}$. Clearly $R(\lambda)$ is analytic in λ. The equations $(\lambda I - A)R(\lambda)h = h$, $R(\lambda)(\lambda I - A)k = k$ hold for $\operatorname{Re}\lambda > 0$, $h \in \mathscr{H}$, and k in the domain of A. Now (i) follows obviously from these identities and the unique extension principle for analytic maps.

We already know from Theorem 3.3 that $\operatorname{Re}(Ah, h) \leq 0$ for h in the domain of A. The inequality $\operatorname{Re}(Vk, k) \geq 0$ follows at once if we set $h = Vk$, $k \in \mathscr{H}$. If $\lambda \neq 0$ then the operator $\lambda^{-1}(I - (I + \lambda A)^{-1})$ is the inverse of $\lambda I - V$. Thus $\sigma(V) = \{0\}$ since $\sigma(V)$ cannot be empty. Finally, the relation $T = (I - V)(V + I)^{-1}$ is immediate from $T = (A + I)(A - I)^{-1}$, and an easy calculation shows that $I - T^*T = 2(V^* + I)^{-1}(V^* + V)(V + I)^{-1}$, thus completing the proof of (iii).

We now introduce some examples of semigroups. If a, b are real numbers and $a < b$ we will denote by $L^2(a, b)$ the subspace of $L^2(-\infty, +\infty)$ consisting of those (classes of) functions vanishing almost everywhere on $(-\infty, a) \cup (b, +\infty)$. Now fix a positive number τ and define a semigroup of contractions $\{T_\tau(t) : t \geq 0\} \subset L^2(0, \tau)$ by the formula

$$\begin{aligned}(T_\tau(t)f)(x) &= f(x - t) \quad && \text{if } t \leq x \leq \tau,\\ &= 0 && \text{if } 0 \leq x < t, \quad f \in L^2(0, \tau).\end{aligned}$$

It is easy to check that this is a strongly continuous semigroup of contractions, $T_\tau(\tau) = 0$, and $T_\tau(t) \neq 0$ for $t < \tau$. We denote by T_τ and A_τ the cogenerator and generator of $\{T_\tau(t) : t \geq 0\}$, and set $V_\tau = -A_\tau^{-1}$.

3.18. LEMMA. *Assume that* τ *is a positive real number.*

(i) $(V_\tau f)(x) = \int_0^x f(t)\,dt, \quad 0 \le x \le \tau,\ f \in L^2(0,\tau)$.

(ii) T_τ *is unitarily equivalent to* $S(e_\tau)$.

PROOF. An easily justifiable calculation, based on Lemma 3.17(i), shows that

$$\begin{aligned}(V_\tau f)(x) &= (-A_\tau^{-1} f)(x)\\ &= -\int_0^\tau (T_\tau(t)f)(x)\,dt\\ &= -\int_0^x f(x-t)\,dt\\ &= \int_0^x f(t)\,dt\end{aligned}$$

for $x \in (0,\tau)$ and $f \in L^2(0,\tau)$. To prove (ii) we observe that T_τ is an operator of class C_0 with minimal function e_τ. By Proposition 3.1.6, T_τ is unitarily equivalent to $S(e_\tau)$ if we can prove that $\dim(\mathscr{D}_{T_\tau}) = 1$. Now, by Lemma 3.17(iii), we must show that $V_\tau + V_\tau^*$ has rank one. A calculation shows that

$$(V_\tau^* f)(x) = \int_x^\tau f(t)\,dt, \quad 0 \le x \le \tau,\ f \in L^2(0,\tau),$$

and hence

$$((V_\tau + V_\tau^*)f)(x) = \int_0^\tau f(t)\,dt, \quad 0 \le x \le \tau,\ f \in L^2(0,\tau).$$

We conclude that the range of $V_\tau + V_\tau^*$ consists of the constant functions, and therefore (ii) follows by the remarks above.

3.19. REMARK. We could define a semigroup $\{T(t) : t \ge 0\} \subset \mathscr{L}(L^2(a,b))$ by $(T(t)f)(x) = f(x-t)$, $t \le x \le b$, $(T(t)f)(x) = 0$, $a \le x < t$, $f \in L^2(a,b)$. It is clear, however, that this semigroup is unitarily equivalent to $\{T_\tau(t) : t \ge 0\}$ if $\tau = b - a$.

3.20. DEFINITION. Assume that β is a given cardinal number, and $\tau : \{\alpha : \alpha < \beta\} \to (0, +\infty)$ is a function such that $\tau(\alpha) \le \tau(\alpha')$ whenever $\alpha \ge \alpha'$, and $\tau(\alpha) = \tau(\alpha')$ whenever $\operatorname{card}(\alpha) = \operatorname{card}(\alpha')$. Then the semigroup $\{T(t) : t \ge 0\}$ defined by $T(t) = \bigoplus_{\alpha<\beta} T_{\tau(\alpha)}(t)$ is called a *canonical semigroup.*

The following result is an easy consequence of Theorem 3.5.23 and of the preceding observations (especially Lemma 3.16).

3.21. THEOREM. *Assume that* $\{T(t) : t \ge 0\}$ *is a strongly continuous semigroup of contractions, and* $T(\tau) = 0$ *for some* $\tau > 0$. *Then* $\{T(t) : t \ge 0\}$ *is quasisimilar to a uniquely determined canonical semigroup.*

It is interesting that Theorem 3.21 extends to general strongly continuous semigroups (i.e., not necessarily contraction semigroups). Assume that $\{T(t) : t \ge 0\} \subset \mathscr{L}(\mathscr{H})$ is a strongly continuous semigroup and $T(\tau) = 0$ for some $\tau > 0$. For fixed $h \in \mathscr{H}$ the function $t \mapsto T(t)h$ is continuous, has compact support, and is therefore bounded. By the uniform boundedness principle we have $\sup\{\|T(t)\| : t \ge 0\} < \infty$. This observation is basic in the proof of the following result.

3.22. THEOREM. *Assume that* $\{T(t) : t \geq 0\} \subset \mathscr{L}(\mathscr{H})$ *is a strongly continuous semigroup, and* $T(\tau) = 0$ *for some* $\tau > 0$. *Then* $\{T(t) : t \geq 0\}$ *is quasisimilar to a uniquely determined canonical semigroup.*

PROOF. As noted above, we have $C = \sup\{\|T(t)\| : t \geq 0\} < \infty$. Let us introduce a new scalar product $(\cdot,\cdot)_1$ on $\mathscr{H}$ as follows:

$$(h,k)_1 = \int_0^\tau (T(t)h, T(t)k)\,dt, \quad h,k \in \mathscr{H}. \tag{3.23}$$

If we denote $\|h\|_1 = (h,h)_1$, we have

$$\|h\|_1 \leq C\|h\|, \quad h \in \mathscr{H}. \tag{3.24}$$

Denote by $\mathscr{H}_1$ the completion of $\mathscr{H}$ under the new scalar product, and denote by $J : \mathscr{H} \to \mathscr{H}_1$ the operator of inclusion. Then J is bounded by (3.24) and hence it is a quasiaffinity. Now, if $0 \leq s \leq \tau$, we have by (3.23)

$$\begin{aligned}
\|T(s)h\|_1^2 &= \int_0^\tau \|T(t)T(s)h\|^2\,dt \\
&= \int_0^\tau \|T(t+s)h\|^2\,dt \\
&= \int_0^{\tau-s} \|T(t+s)h\|^2\,dt \\
&= \int_s^\tau \|T(t)h\|^2\,dt \\
&\leq \int_0^\tau \|T(t)h\|^2\,dt = \|h\|_1^2,
\end{aligned}$$

and this shows that $T(s)$ extends by continuity to a contraction $T_1(s) \in \mathscr{L}(\mathscr{H}_1)$. If $h \in \mathscr{H}$, we have

$$\lim_{t\downarrow 0} \|T_1(t)h - h\|_1 \leq C \lim_{t\downarrow 0} \|T(t)h - h\| = 0$$

and, since $\mathscr{H}$ is dense in $\mathscr{H}_1$, we conclude that $\{T_1(t) : t \geq 0\}$ is a strongly continuous semigroup of contractions. Since it is also clear that $T_1(\tau) = 0$, an application of Theorem 3.22 shows the existence of a canonical semigroup $\{T'(t) : t \geq 0\} \subset \mathscr{L}(\mathscr{H}')$ and of a quasiaffinity $X \in \mathscr{L}(\mathscr{H}_1, \mathscr{H}')$ such that $T'(t)X = XT_1(t)$, $t \geq 0$. Since obviously $T_1(t)J = JT(t)$ for $t \geq 0$, the operator XJ is a quasiaffinity and $T'(t)(XJ) = (XJ)T(t)$, $t \geq 0$. This proves one half of the theorem. To prove the other half, we note that the same argument applied to the semigroup $\{T(t)^* : t \geq 0\}$ yields a contraction semigroup $\{T_2(t) : t \geq 0\} \subset \mathscr{L}(\mathscr{H}_2)$, and a quasiaffinity $J_* \in \mathscr{L}(\mathscr{H}_2, \mathscr{H})$, satisfying $T(t)J_* = J_*T_2(t)$, $t \geq 0$. A second application of Theorem 3.22 now gives a canonical semigroup $\{T''(t) : t \geq 0\} \subset \mathscr{L}(\mathscr{H}'')$, and a quasiaffinity $Y \in \mathscr{L}(\mathscr{H}'', \mathscr{H}_2)$ such that $T_2(t)Y = YT''(t)$, $t \geq 0$. We therefore have $T(t)(J_*Y) = (J_*Y)T''(t)$ for $t \geq 0$. The relations $T'(t)(XJJ_*Y) = (XJJ_*Y)T''(t)$, $t \geq 0$, show, via Lemma 3.16 and Theorem 3.5.23, that $T'(t) = T''(t)$, $t \geq 0$. Thus $\{T(t) : t \geq 0\}$ is quasisimilar to $\{T'(t) : t \geq 0\}$, and the above argument also yields the uniqueness of the canonical semigroup. The theorem is proved.

We now turn our attention to the operator V_τ defined above (cf. Lemma 3.18). These operators (especially V_1) are known as the Volterra integral operators, and they have interesting lattices of invariant subspaces. It is clear that $L^2(s,\tau)$ is invariant under V_τ whenever $S \in [0,\tau]$; we will see that these are the only invariant subspaces of V_τ. Let us note that $V \in L^2(s,\tau)$ is unitarily equivalent to $V_{\tau-s}$, and an explicit unitary equivalence $U : L^2(0,\tau-s) \to L^2(s,\tau)$ is given by $(Uf)(x) = f(x-s)$, $x \in [s,\tau]$, $f \in L^2(0,\tau-s)$.

3.25. LEMMA. *Assume that* $V \in \mathscr{L}(\mathscr{H})$, $\sigma(V) = \{0\}$, *and* $T = (I-V)(I+V)^{-1}$. *Then* T *and* V *have the same invariant subspaces.*

PROOF. It suffices to show that each of the operators T and V can be approximated in norm by a sequence of polynomials in the other. Indeed, since $\sigma(V) = \{0\}$ and $\sigma(T) = \{1\}$, the following series converge in norm:

$$T = -I + 2(I+V)^{-1} = -I + 2\sum_{n=0}^{\infty}(-1)^n V^n,$$
$$V = I - (I - \tfrac{1}{2}(I-T))^{-1} = -\sum_{n=1}^{\infty}(\tfrac{1}{2})^n (I-T)^n.$$

3.26. THEOREM. *The only invariant subspaces of* V_τ *are the spaces* $L^2(s,\tau)$, $s \in [0,\tau]$.

PROOF. By Lemma 3.26, V_τ and $T_\tau = (I - V_\tau)(I + V_\tau)^{-1}$ have the same invariant subspaces. The operator T_τ is unitarily equivalent to $S(e_\tau)$ by Lemma 3.18. The inner divisors of the function e_τ are (equivalent to) the functions e_s, $s \in [0,\tau]$, and we deduce from Proposition 3.1.10 that the lattice of invariant subspaces of T_τ (and hence of V_τ) is totally ordered by inclusion. Now let $\mathscr{M}$ be an invariant subspace for V_τ and set

$$A = \{s \in [0,\tau] : L^2(s,\tau) \subset \mathscr{M}\},$$
$$B = \{s \in [0,\tau] : L^2(s,\tau) \supset \mathscr{M}\}.$$

A moment's thought shows that $0 \in A$, $\tau \in B$, A and B are closed intervals, and $A \cup B = [0,\tau]$. Since $[0,\tau]$ is connected, $A \cap B$ is not empty, and clearly $\mathscr{M} = L^2(s,\tau)$ if $s \in A \cap B$. The theorem follows.

In view of Proposition 3.14 and Lemma 3.17, it is an interesting problem to give a characterization of those accretive operators V such that $\sigma(V) = \{0\}$, and the contraction $T = (I-V)(I+V)^{-1}$ is of class C_0. Note that $\sigma(T) = \{1\}$, and therefore the minimal function of T (in case T is of class C_0) must have the form e_τ for some $\tau > 0$.

3.27. LEMMA. *Assume that* $T \in \mathscr{L}(\mathscr{H})$ *is a completely nonunitary contraction and* $\sigma(T) = \{1\}$. *Then* T *is of class* C_{00}.

PROOF. Since T^* is also completely nonunitary and $\sigma(T^*) = \{1\}$, it will suffice to show that T is of class $C_{\cdot 0}$. To do this we consider the minimal isometric dilation $U_+ \in \mathscr{L}(\mathscr{K}_+)$ of T. By Proposition 1.2.7 we must show that the residual part $\mathscr{R}$ of U_+ is $\{0\}$. Now, the operator $R = U_+ \mid \mathscr{R}$ is absolutely continuous by Theorem 2.2.13. If E denotes the spectral measure of R, i.e., $R = \int_{\mathbf{T}} \varsigma E(d\varsigma)$, then we have, in particular, $E(\{1\}) = 0$.

Now denote $X = P_{\mathscr{R}} \mid \mathscr{H}$ and observe that

$$RX = XT. \tag{3.28}$$

If σ is a closed subset of $\mathbf{T}$, $1 \notin \sigma$, and $h \in \mathscr{H}$, we can now define an entire function f as follows:

$$\begin{aligned} f(\lambda) &= (\lambda I - R \mid E(\sigma)\mathscr{R})^{-1} E(\sigma) X h \quad && \text{if } \lambda \notin \sigma, \\ &= E(\sigma) X (\lambda I - T)^{-1} h && \text{if } \lambda \neq 1. \end{aligned}$$

The fact that f is well defined for $\lambda \in \mathbf{C}$ follows from the equality $(\lambda I - R)^{-1} X = X(\lambda I - T)^{-1}$, $\lambda \notin \mathbf{T}$, which is a direct consequence of (3.28). Since $\lim_{|\lambda| \to \infty} f(\lambda) = 0$, we deduce from Liouville's theorem that $f(\lambda) \equiv 0$ and, in particular, $E(\sigma) X h = -R f(0) = 0$. Now, σ was an arbitrary closed set with $1 \notin \sigma$, and the relation $E(\{1\}) = 0$ implies that $Xh = 0$, $h \in \mathscr{H}$. But relation (1.2.8) implies that

$$\lim_{n \to \infty} \|T^{*^n} h\| = \|Xh\| = 0, \quad h \in \mathscr{H},$$

and hence T is of class $C_{\cdot 0}$, as claimed.

We do not give a complete answer to the problem stated before Lemma 3.27, but we use this lemma to give an abstract characterization of the Volterra operators.

3.29. THEOREM. *Assume that* $V \in \mathscr{L}(\mathscr{H})$ *is an accretive operator,* $\sigma(V) = \{0\}$, *and* $\ker V = \{0\}$. *If* $V + V^*$ *has rank one then* V *is unitarily equivalent to* V_τ, *where* $\tau = \|V + V^*\|$.

PROOF. Note that the operator $T = (I - V)(I + V)^{-1}$ is a contraction, $\sigma(T) = \{1\}$, and $\ker(I - T) = \ker V = \{0\}$. It follows that T is completely nonunitary. Indeed, if $T = U \oplus T'$, with U a unitary operator acting on a nonzero space, we would have $\sigma(U) = \{1\}$, and hence $U = I$, contradicting the equality $\ker(I - T) = \{0\}$. By Lemma 3.27, T is of class C_{00}. Furthermore, a calculation similar to the proof of Lemma 3.17(iii) shows that $\dim(\mathscr{D}_T) = 1$, and hence T is unitarily equivalent to $S(\theta)$ for some inner function θ by Proposition 3.1.6. The equality $\sigma(T) = \sigma(S(\theta)) = \{1\}$ now implies that $\theta \equiv e_\tau$ for some $\tau > 0$, and hence $V = (I - T)(I + T)^{-1}$ is unitarily equivalent to $(I - S(e_\tau))(I + S(e_\tau))^{-1}$. This last operator is unitarily equivalent to V_τ by Lemma 3.18. To calculate τ we simply note that $\|V + V^*\| = \|V_\tau + V_\tau^*\| = \tau$.

Exercises

1. Provide the details for the extension of Lemma 3.8 [resp., Proposition 3.11] to the case in which U [resp., T] is an arbitrary unitary operator [resp., contraction] such that $\ker(I-U)=\{0\}$ [resp., $\ker(I-T)=\{0\}$].

2. Let $\{T(t): t\geq 0\}$, A, and T be as in Lemma 3.17, and $V=-A^{-1}$. Prove that $\|(\lambda I-A)^{-1}\|\leq(1-\exp(-\tau\operatorname{Re}\lambda))/\operatorname{Re}\lambda$, $\operatorname{Re}\lambda\neq 0$, and $\limsup_{\mu\downarrow 0}(\mu\log\|(\mu I-V)^{-1}\|)\leq\tau$. The estimate for $\|(\lambda I-A)^{-1}\|$ also holds when $\operatorname{Re}\lambda=0$ if we understand the right-hand side as a limit.

3. Verify the strong continuity of the semigroup $\{T_\tau(t): t\geq 0\}$.

4. Verify the following fact, used in the proof of Theorem 3.22: $\{T(t)^*: t\geq 0\}$ is strongly continuous whenever $\{T(t): t\geq 0\}$ is a strongly continuous semigroup.

5. Assume that T is a completely nonunitary contraction, and $\sigma(T)\subset\mathbf{T}$ has Lebesgue measure zero. Prove that T is of class C_{00}.

6. Let s and t be two positive numbers. Theorem 3.29 implies that the operators sV_t and tV_s are unitarily equivalent. Find an explicit unitary equivalence between these two operators.

7. Find an explicit unitary equivalence between V_t and V_t^*, $t\geq 0$.

8. Determine the semigroup generated by $-(V_t^*)^{-1}$, $t\geq 0$.

4. Representations of the convolution algebra $L^1(0,1)$. It is well known that $L^1(0,1)$ is a commutative Banach algebra under the convolution "$*$" defined by

$$(4.1)\qquad (f*g)(x)=\int_0^x f(t)g(x-t)\,dt,\quad x\in(0,1),\ f,g\in L^1(0,1).$$

In this section we will give a classification of the Hilbert space representations of this algebra.

4.2. DEFINITION. A *representation* of the algebra $L^1(0,1)$ on the Hilbert space $\mathscr{H}$ is a continuous linear map $\rho: L^1(0,1)\to\mathscr{L}(\mathscr{H})$ such that $\rho(f*g)=\rho(f)\rho(g)$, $f,g\in L^1(0,1)$. If ρ and ρ' are representations of $L^1(0,1)$ on $\mathscr{H}$ and $\mathscr{H}'$, respectively, then $\rho\oplus\rho'$ is the representation of $L^1(0,1)$ on $\mathscr{H}\oplus\mathscr{H}'$ defined by $(\rho\oplus\rho')(f)=\rho(f)\oplus\rho'(f)$, $f\in L^1(0,1)$.

The simplest representation of $L^1(0,1)$ on $\mathscr{H}$ is the *null representation* defined by $\rho(f)=0$ for all f. It is clear that a null representation is uniquely determined, up to unitary equivalence, by the cardinal number $\dim(\mathscr{H})$. For the sake of uniqueness we will fix once and for all a Hilbert space $\mathscr{H}_\aleph$ of dimension $\aleph$ for each cardinal $\aleph$, and call the null representation on $\mathscr{H}_\aleph$ *the canonical null representation of multiplicity* $\aleph$. A more interesting class of representations is provided by the following result.

4.3. LEMMA. *Let $\{T(t) : t \geq 0\}$ be a strongly continuous semigroup of operators on $\mathscr{H}$, and assume that $T(1) = 0$. Then there exists a representation ρ of $L^1(0,1)$ on $\mathscr{H}$ such that*

$$\rho(f)h = \int_0^1 f(t)T(t)h\,dt, \qquad h \in \mathscr{H},\ f \in L^1(0,1).$$

PROOF. It is clear that the integral defining $\rho(f)h$ makes sense and it is linear in h and f. Moreover,

$$\|\rho(f)h\| \leq \int_0^1 |f(t)|\,\|T(t)\|h\,dt \leq C\|f\|_1\|h\|,$$

for all h and f, where $C = \sup\{\|T(t)\| : t \geq 0\} < \infty$. Thus ρ is a continuous linear mapping from $L^1(0,1)$ to $\mathscr{L}(\mathscr{H})$. In order to verify the multiplicativity of ρ, fix $f, g \in L^1(0,1)$ and $h \in \mathscr{H}$ and compute:

$$\begin{aligned}
(f * g)h &= \int_0^1 (f * g)(t)T(t)h\,dt \\
&= \int_0^1 \left[\int_0^t f(s)g(t-s)\,ds\right] T(t)h\,dt \\
&= \int_0^1 f(s)T(s)\left[\int_s^1 g(t-s)T(t-s)h\,dt\right] ds \\
&= \int_0^1 f(s)T(s)\left[\int_0^{1-s} g(t)T(t)h\,dt\right] ds \\
&= \int_0^1 f(s)T(s)\left[\int_0^1 g(t)T(t)h\,dt\right] ds \\
&= \int_0^1 f(s)T(s)\rho(g)h\,ds \\
&= \rho(f)\rho(g)h.
\end{aligned}$$

Here we have used the fact that $T(t)T(s) = T(t+s) = 0$ if $t > 1 - s$. The lemma is proved.

The representation ρ constructed in Lemma 4.3 will be called the representation determined by the semigroup $\{T(t) : t \geq 0\}$.

4.4. DEFINITION. A *canonical representation* of $L^1(0,1)$ is a representation of the form $\rho \oplus \rho'$, where ρ is a canonical null representation and ρ' is the representation determined by a canonical semigroup $\{T(t) : t \geq 0\}$ with $T(1) = 0$.

We now introduce the equivalence relation that we use in our classification result.

4.5. DEFINITION. Let ρ and ρ' be representations of $L^1(0,1)$ on the Hilbert spaces $\mathscr{H}$ and $\mathscr{H}'$, respectively. We say that ρ is a *quasiaffine transform* of ρ' if there exists a quasiaffinity $X \in \mathscr{L}(\mathscr{H}, \mathscr{H}')$ such that $\rho'(f)X = X\rho(f)$, $f \in L^1(0,1)$. The representations ρ and ρ' are *quasisimilar* if each is a quasiaffine transform of the other.

As in the case of single operators, we use the symbols $\rho \prec \rho'$ [resp., $\rho \sim \rho'$] to indicate that ρ is a quasiaffine transform of ρ' [resp., ρ and ρ' are quasisimilar]. We want to prove first that canonical representations have a uniqueness property with respect to quasiaffine transforms.

4.6. LEMMA. *Let ρ and ρ' be the representations determined by the semigroups $\{T(t) : t \geq 0\} \subset \mathscr{L}(\mathscr{H})$ and $\{T'(t) : t \geq 0\} \subset \mathscr{L}(\mathscr{H}')$, respectively. An operator $X \in \mathscr{L}(\mathscr{H}, \mathscr{H}')$ satisfies the conditions $\rho'(f)X = X\rho(f)$, $f \in L^1(0,1)$, if and only if $T'(t)X = XT(t)$, $t \geq 0$.*

PROOF. Assume first that $T'(t)X = XT(t)$, $t \geq 0$, $f \in L^1(0,1)$, and $h \in \mathscr{H}$. Then clearly

$$\begin{aligned} \rho'(f)Xh &= \int_0^1 f(t)T'(t)Xh\,dt \\ &= \int_0^1 f(t)XT(t)h\,dt \\ &= X\int_0^1 f(t)T(t)h\,dt \\ &= X\rho(f)h. \end{aligned}$$

Conversely, assume that $\rho'(f)X = X\rho(f)$, $f \in L^1(0,1)$. If we fix h, this equality applied to the characteristic function of $(0,s)$ yields $X\int_0^s T(t)h\,dt = \int_0^s T'(t)Xh\,dt$, $0 < s < 1$. From the strong continuity of the two semigroups we deduce that

$$T(s)h = \frac{d}{ds}\int_0^s T(t)h\,dt \quad \text{and} \quad T'(s)Xh = \frac{d}{ds}\int_0^s T'(t)Xh\,dt,$$

so we immediately infer the equality $T'(s)Xh = XT(s)h$, $0 < s < 1$. The lemma follows at once.

The preceding proof also has the following consequence.

4.7. COROLLARY. *If ρ is the representation determined by the semigroup $\{T(t) : t \geq 0\} \subset \mathscr{L}(\mathscr{H})$, then $\mathscr{H} = \bigvee\{\rho(f)\mathscr{H} : f \in L^1(0,1)\}$ and $\bigcap\{\ker \rho(f) : f \in L^1(0,1)\} = \{0\}$.*

PROOF. We have $h = \lim_{s\downarrow 0} s^{-1}\int_0^s T(t)h\,dt$, $h \in \mathscr{H}$, and the integrals belong to $\bigvee\{\rho(f)\mathscr{H} : f \in L^1(0,1)\}$. Moreover, if $h \in \bigcap\{\ker \rho(f) : f \in L^1(0,1)\}$ then clearly $h = 0$ by the same formula. The corollary is proved.

We are now ready for the uniqueness result.

4.8. PROPOSITION. *If ρ and ρ' are two canonical representations of $L^1(0,1)$ and $\rho \prec \rho'$, then $\rho = \rho'$.*

PROOF. By Definition 4.4 we have $\rho = \rho_1 \oplus \rho_2$ and $\rho' = \rho_1' \oplus \rho_2'$, where ρ_1 and ρ_1' are null representations on $\mathscr{H}_1$ and $\mathscr{H}_1'$, respectively, while ρ_2 and ρ_2' are determined by the canonical semigroups $\{T(t) : t \geq 0\} \subset \mathscr{L}(\mathscr{H}_2)$ and $\{T'(t) : t \geq 0\} \subset \mathscr{L}(\mathscr{H}_2')$. Assume that $X \in \mathscr{L}(\mathscr{H}_1 \oplus \mathscr{H}_2, \mathscr{H}_1' \oplus \mathscr{H}_2')$ is a quasiaffinity satisfying the relations $\rho'(f)X = X\rho(f)$, $f \in L^1(0,1)$. We clearly have

$$X\left(\bigvee\{\rho(f)\mathscr{H} : f \in L^1(0,1)\}\right) \subset \bigvee\{\rho'(f)\mathscr{H}' : f \in L^1(0,1)\},$$

and

$$X\left(\bigcap\{\ker\rho(f) : f \in L^1(0,1)\}\right) \subset \bigcap\{\ker\rho'(f) : f \in L^1(0,1)\},$$

where we have set $\mathscr{H} = \mathscr{H}_1 \oplus \mathscr{H}_2$ and $\mathscr{H}' = \mathscr{H}_1' \oplus \mathscr{H}_2'$. By Corollary 4.7, these inclusions are equivalent to $X(\{0\}\oplus\mathscr{H}_2) \subset \{0\}\oplus\mathscr{H}_2'$ and $X(\mathscr{H}_1\oplus\{0\}) \subset \mathscr{H}_1'\oplus\{0\}$, respectively. Thus $X = X_1 \oplus X_2$, where $X_j \in \mathscr{L}(\mathscr{H}_j, \mathscr{H}_j')$ are quasiaffinities for $j = 1, 2$. We see already that $\mathscr{H}_1$ and $\mathscr{H}_1'$ have the same dimension and hence $\rho_1 = \rho_1'$ by the definition of canonical null representations. Next, we infer the relations $T'(t)X_2 = X_2T(t)$, $t \geq 0$, which imply that $\rho_2 = \rho_2'$ by Lemma 3.16 and Theorem 3.5.23. The proposition follows.

Our next task is to show that every representation of $L^1(0,1)$ is quasisimilar to a canonical representation, which will be uniquely determined by virtue of Proposition 4.8. The first step will be to show that every representation ρ is similar to a sum $\rho_1 \oplus \rho_2$, where ρ_1 is a null representation and ρ_2 is determined by a strongly continuous semigroup of operators.

We begin with a few facts about $L^1(0,1)$. For fixed $a \geq 0$ and $\varepsilon > 0$ we define functions $f_{a,\varepsilon} \in L^1(0,1)$ as follows:

$$\begin{aligned} f_{a,\varepsilon} &= \varepsilon^{-1}\chi_{(a,a+\varepsilon)} && \text{if } a+\varepsilon \leq 1, \\ &= \varepsilon^{-1}\chi_{(a,1)} && \text{if } a < 1 < a+\varepsilon, \\ &= 0 && \text{if } a \geq 1, \end{aligned}$$

where χ_I denotes the characteristic function of the set I. We also define the translation operators $\tau_a : L^1(0,1) \to L^1(0,1)$ by

$$\begin{aligned} (\tau_a f)(x) &= 0 && \text{if } x \leq a, \\ &= f(x-a) && \text{if } a < x \leq 1. \end{aligned}$$

Note that $\tau_a\tau_b = \tau_{a+b}$, and $\tau_a = 0$ if $a \geq 1$.

4.9. LEMMA. *For every $g \in L^1(0,1)$ we have*
(i) $\lim_{\varepsilon\downarrow 0} \|f_{a,\varepsilon} * g - \tau_a g\|_1 = 0$, $a \geq 0$, *and*
(ii) $\lim_{a\downarrow 0} \|\tau_a g - g\|_1 = 0$.

PROOF. We have

$$\|f_{a,\varepsilon} * g - \tau_a g\|_1 \leq \|f_{a,\varepsilon}\|_1\|g\|_1 + \|\tau_a g\|_1 \leq 2\|g\|_1,$$

and, similarly, $\|\tau_a g - g\|_1 \leq 2\|g\|_1$. Therefore it suffices to prove the lemma for a dense collection of functions $g \in L^1(0,1)$. Assume that g is continuous and has compact support in $(0,1)$. Then, for $a<1$ and $\varepsilon<1-a$, we have

$$\begin{aligned}|(f_{a,\varepsilon}*g)(x)-(\tau_a g)(x)| &= \left|\varepsilon^{-1}\int_a^{a+\varepsilon} g(x-t)\,dt - g(x-a)\right| \\ &= \left|\varepsilon^{-1}\int_a^{a+\varepsilon}(g(x-t)-g(x-a))\,dt\right| \\ &\le \sup\{|g(y)-g(z)|\colon y,z\in(0,1),\ |y-z|\le\varepsilon\},\end{aligned}$$

and we see by the uniform continuity of g that $f_{a,\varepsilon}*g$ converges uniformly to $\tau_a g$ as $\varepsilon\to 0$. Thus (i) is proved because the case in which $a\ge 1$ is trivial. Similarly, if g is continuous with compact support in $(0,1)$ then $\tau_a g$ converges uniformly to g as $a\to 0$. The lemma follows.

We can now prove a converse to Corollary 4.7.

4.10. PROPOSITION. *Assume that ρ is a representation of $L^1(0,1)$ on the Hilbert space $\mathscr{H}$. If $\mathscr{H}=\bigvee\{\rho(f)\mathscr{H} : f\in L^1(0,1)\}$, then ρ is induced by a strongly continuous semigroup of operators $\{T(t) : t\ge 0\}\subset\mathscr{L}(\mathscr{H})$ such that $T(1)=0$.*

PROOF. It follows from Lemma 4.9(i) that

$$\lim_{\varepsilon\downarrow 0}\rho(f_{a,\varepsilon})\rho(g)=\rho(\tau_a g) \tag{4.11}$$

for all $a\ge 0$ and $g\in L^1(0,1)$. Since we have

$$\|\rho(f_{a,\varepsilon})\|\le\|\rho\|\,\|f_{a,\varepsilon}\|\le\|\rho\|,$$

and the ranges of $\rho(g)$ generate $\mathscr{H}$, (4.11) implies the existence of an operator $T(a)\in\mathscr{L}(\mathscr{H})$ such that $\|T(a)\|\le\|\rho\|$ and

$$T(a)\rho(g)=\rho(\tau_a g),\qquad g\in L^1(0,1),\ a\ge 0. \tag{4.12}$$

It is now easy to show that $\{T(t) : t\ge 0\}$ is a semigroup of operators. Indeed, if $a,b\ge 0$, (4.12) implies

$$\begin{aligned}T(a)T(b)\rho(g) &= T(a)\rho(\tau_b g) \\ &= \rho(\tau_a\tau_b g) \\ &= \rho(\tau_{a+b}g) \\ &= T(a+b)\rho(g),\qquad g\in L^1(0,1),\end{aligned}$$

and the equality $T(a+b)=T(a)T(b)$ follows from the hypothesis. Similarly, $T(1)\rho(g)=\rho(\tau_1 g)=\rho(0)=0$, $g\in L^1(0,1)$, so that $T(1)=0$. To verify the strong continuity, we use Lemma 4.9(ii):

$$\lim_{a\downarrow 0}\|T(a)\rho(g)-\rho(g)\|=\lim_{a\downarrow 0}\|\rho(\tau_a g-g)\|=0,\quad g\in L^1(0,1).$$

Thus we have $\lim_{a\downarrow 0}\|T(a)h-h\|=0$ for a dense set of vectors h, and the strong continuity follows because $\{\|T(a)\| : a\ge 0\}$ is uniformly bounded (by $\|\rho\|$).

Now, the continuity of ρ and the equality $f * g = \int_0^1 f(t)(\tau_t g)\,dt$ justify the following calculation:

$$\begin{aligned}\int_0^1 f(t)T(t)\rho(g)h\,dt &= \int_0^1 f(t)\rho(\tau_t g)h\,dt \\ &= \rho\left(\int_0^1 f(t)(\tau_t g)\,dt\right)h \\ &= \rho(f*g)h \\ &= \rho(f)\rho(g)h, \quad h \in \mathscr{H},\ f,g \in L^1(0,1).\end{aligned}$$

We conclude that for $f \in L^1(0,1)$ we have $\rho(f)k = \int_0^1 f(t)T(t)k\,dt$ whenever k has the form $k = \rho(g)h$, $g \in L^1(0,1)$, $h \in \mathscr{H}$. Therefore ρ is the representation determined by the strongly continuous semigroup $\{T(t) : t \geq 0\}$, and this concludes the proof of our proposition.

4.13. PROPOSITION. *Every representation of $L^1(0,1)$ on a Hilbert space is similar to a representation of the form $\rho_1 \oplus \rho_2$, where ρ_1 is a null representation and ρ_2 is determined by a strongly continuous semigroup of operators.*

PROOF. Let ρ be a representation of $L^1(0,1)$ on the Hilbert space $\mathscr{H}$, and denote $\mathscr{H}_2 = \bigvee\{\rho(f)\mathscr{H} : f \in L^1(0,1)\}$, $\mathscr{H}_1 = \mathscr{H} \ominus \mathscr{H}_2$. The family of operators $\{\rho(f_{0,\varepsilon}) : \varepsilon > 0\}$ is uniformly bounded by $\|\rho\|$. Since the unit ball of $\mathscr{L}(\mathscr{H})$ is compact in the weak operator topology, we infer the existence of an operator $P \in \mathscr{L}(\mathscr{H})$ that belongs to the weak closure of $\{\rho(f_{0,\varepsilon}) : 0 < \varepsilon < \alpha\}$ for every $\alpha > 0$. We claim that P is a projection onto $\mathscr{H}_2$, i.e.,

$$P^2 = P \quad \text{and} \quad P\mathscr{H} = \mathscr{H}_2. \tag{4.14}$$

The fact that $P\mathscr{H} \subset \mathscr{H}_2$ is clear because P is a weak limit of the family $\{\rho(f_{0,\varepsilon}) : 0 < \varepsilon < \alpha\}$, $\alpha > 0$. Therefore (4.14) follows if we can prove that $Ph = h$ for a dense family of vectors h in $\mathscr{H}_2$. It suffices thus to prove that $P\rho(g) = \rho(g)$ for $g \in L^1(0,1)$. Given $\delta > 0$ and $g \in L^1(0,1)$ we can choose, by Lemma 4.9(i), a number $\alpha > 0$ such that

$$\|f_{0,\varepsilon} * g - g\| \leq \delta, \quad 0 < \varepsilon < \alpha,$$

and consequently

$$\|\rho(f_{0,\varepsilon})\rho(g) - \rho(g)\| \leq \delta\|\rho\|, \quad 0 < \varepsilon < \alpha.$$

The definition of P now implies that $\|P\rho(g) - \rho(g)\| \leq \delta\|\rho\|$ and, since δ is arbitrary, $P\rho(g) = \rho(g)$. The proof of (4.14) is complete.

Next observe that the algebra $L^1(0,1)$ is commutative and, since the commutant of $\{\rho(f) : f \in L^1(0,1)\}$ is closed in the weak operator topology, we conclude that P commutes with $\rho(f)$, $f \in L^1(0,1)$. Thus we have

$$P\rho(f) = \rho(f)P = \rho(f), \quad f \in L^1(0,1). \tag{4.15}$$

The idea now is to transform P into an orthogonal projection by using a similarity. To do this we denote by Q the orthogonal projection onto $\mathscr{H}_2$, so that $PQ = Q$ and $QP = P$. The operator $X = I + P - Q$ is invertible, $X^{-1} = I - P + Q$, and $Q = XPX^{-1}$. We then define the representation ρ' of $L^1(0,1)$ by $\rho'(f) = X\rho(f)X^{-1}$, $f \in L^1(0,1)$. Relations (4.15) now imply that

$$Q\rho'(f) = \rho'(f)Q = \rho'(f), \qquad f \in L^1(0,1),$$

and hence $\rho'(f)$ can be written as $\rho'(f) = \rho_1(f) \oplus \rho_2(f)$, where $\rho_1(f) = 0 \in \mathcal{L}(\mathcal{H}_1)$ and $\rho_2(f) = \rho'(f) \mid \mathcal{H}_2 \in \mathcal{L}(\mathcal{H}_2)$. By virtue of Proposition 4.10, our proof will be done if we can show that

$$\mathcal{H}_2 = \bigvee\{\rho_2(f)\mathcal{H}_2 : f \in L^1(0,1)\}. \tag{4.16}$$

We have

$$\begin{aligned}\rho_2(f)\mathcal{H}_2 &= \rho'(f)Q\mathcal{H} = X\rho(f)X^{-1}Q\mathcal{H}\\ &= X\rho(f)PX^{-1}\mathcal{H} = X\rho(f)P\mathcal{H}\\ &= XP\rho(f)\mathcal{H} = P\rho(f)\mathcal{H} = \rho(f)\mathcal{H},\end{aligned}$$

where we used the relations $Q\mathcal{H} = \mathcal{H}_2$, $X^{-1}Q = PX^{-1}$, $X^{-1}\mathcal{H} = \mathcal{H}$, (4.15), and the fact that $Xh = h$ if $h \in \mathcal{H}_2$. Now (4.16) follows at once from the definition of $\mathcal{H}_2$, thus concluding our proof.

The classification theorem becomes now almost obvious.

4.17. THEOREM. *Every representation ρ of $L^1(0,1)$ on a Hilbert space is quasisimilar to a canonical representation ρ'. The canonical representation ρ' is uniquely determined by the relation $\rho' \prec \rho$.*

PROOF. By Proposition 4.13 we may assume that $\rho = \rho_1 \oplus \rho_2$, where ρ_1 is a null representation and ρ_2 is determined by a strongly continuous semigroup. Clearly ρ_1 is unitarily equivalent to a canonical null representation. For ρ_2 we apply Lemma 4.6 and Theorem 3.22 to conclude that it is quasisimilar to the representation determined by a certain canonical semigroup. The existence assertion of the theorem is obtained by summing up these observations. The uniqueness assertion is an immediate consequence of Proposition 4.8.

Exercises

1. Prove that the Banach algebras $L^1(0,1)$ and $L^1(0,a)$ are isometrically isomorphic for $a > 0$.

2. Find an explicit formula for the representation of $L^1(0,1)$ determined by the semigroup $\{T_\tau(t) : t \geq 0\} \subset \mathcal{L}(L^2(0,1))$.

3. Show that $f * g \in L^2(0,1)$ and $\|f * g\|_2 \leq \|f\|_1\|g\|_2$ whenever $f \in L^1(0,1)$ and $g \in L^2(0,1)$.

4. Prove analogues of Propositions 4.10 and 4.13 for representations of the convolution algebras $L^1(0,+\infty)$ and $L^1(-\infty,+\infty)$.

CHAPTER 5

Characteristic Functions and the Class C_0

In order to go more deeply in the study of operators of class C_0 we need to use functional models. An example of a functional model that has occurred in previous chapters is provided by the Jordan blocks. Functional models are introduced in Section 1, after a discussion of Hilbert space-valued and operator-valued functions on the unit disc or circle. This discussion is self-contained, and we made an effort not to use radial limits of analytic functions. After the introduction of inner functions, we define functional models and show how isometric dilations help identify every operator (in a certain class) with a functional model. We conclude the section with a description of the invariant subspaces of a functional model (in terms of factorizations) and with a translation of the commutant lifting theorem into the language of functional models. Section 2 is dedicated to a presentation of the exterior powers and other tensor operations applied to a Hilbert space, and to Hilbert space operators. Since we require some properties of the group C^*-algebra of the symmetric group, we felt it was appropriate to include a discussion of finite-dimensional C^*-algebras. This section requires as a prerequisite only basic facts about C^*-algebras. We did not find it necessary to refer to the specific structure of projections in the C^*-algebra of the symmetric group (the Young symmetrizers). Section 3 is the most important part of this chapter. Here we establish necessary and sufficient conditions for a functional model to be of class C_0, and we give methods for calculating the inner functions in the Jordan model of an operator by using functional models. The results in this section can be viewed as generalizations of the fact that the Jordan model of a finite matrix can be obtained from its characteristic matrix. The determinants and minors used in linear algebra are replaced here by the tensor operations in Section 2. All Hilbert spaces in this chapter are assumed to be separable.

1. Functional models of contractions. Let $T \in \mathscr{L}(\mathscr{H})$ be a contraction of class $C_{\cdot 0}$, i.e., $\lim_{n\to\infty} \|T^{*^n} x\| = 0$ for all x in $\mathscr{H}$. As seen from Corollary 1.2.11, the minimal isometric dilation $U_+ \in \mathscr{L}(\mathscr{K}_+)$ of T is a unilateral shift of multiplicity $\dim(\mathscr{D}_{T^*})$, and $U_+ \mid \mathscr{K}_+ \ominus \mathscr{H}$ is a unilateral shift of multiplicity $\dim(\mathscr{D}_T)$. If $\dim(\mathscr{D}_{T^*}) = \dim(\mathscr{D}_T) = 1$ then T is seen to be unitarily equivalent to one of the operators $S(\theta)$ studied in §3.1. The functional representation of the space $\mathscr{H}(\theta)$ played an important role in §3.1, and it is this functional representation that we want to extend to arbitrary operators of class $C_{\cdot 0}$. For the sake of simplicity all Hilbert spaces will be supposed to be separable when talking about functional models.

Let $\mathcal{F}$ be a (separable) Hilbert space. We denote by $H^2(\mathcal{F})$ the set of all analytic functions $u : \mathbf{D} \to \mathcal{F}$ with a power series expansion $u(\lambda) = \sum_{n=0}^{\infty} \lambda^n a_n$, $\lambda \in \mathbf{D}$, such that $\sum_{n=0}^{\infty} \|a_n\|^2 < \infty$. The space $H^2(\mathcal{F})$ is a Hilbert space under the norm

$$\|u\|_{H^2(\mathcal{F})}^2 = \sum_{n=0}^{\infty} \|a_n\|^2, \qquad u(\lambda) = \sum_{n=0}^{\infty} \lambda^n a_n \in H^2(\mathcal{F}).$$

A trivial calculation proves the following alternative formula:

$$\|u\|_{H^2(\mathcal{F})}^2 = \sup\left\{\frac{1}{2\pi}\int_0^{2\pi} \|u(re^{it})\|^2\,dt : 0 < r < 1\right\}. \tag{1.1}$$

We denote by $L^2(\mathcal{F})$ the Hilbert space of (classes of) measurable functions $f : \mathbf{T} \to \mathcal{F}$ such that $\|f\|_{L^2(\mathcal{F})}^2 = (1/2\pi)\int_0^{2\pi} \|f(e^{it})\|^2\,dt$ is finite. Note that, since $\mathcal{F}$ is separable, all notions of measurability for $\mathcal{F}$-valued functions coincide. If $\mathcal{F}$ were not separable we should define $L^2(\mathcal{F})$ as a set of separably valued functions.

If $f \in L^2 = L^2(\mathbf{C})$ and $a \in \mathcal{F}$ then the function $fa \in L^2(\mathcal{F})$ is defined by $(fa)(\varsigma) = f(\varsigma)a$, $\varsigma \in \mathbf{T}$. Now, let χ denote the identity function of $\mathbf{T}$ (i.e., $\chi(\varsigma) = \varsigma$ for $\varsigma \in \mathbf{T}$), and let $\{a_n : -\infty < n < +\infty\}$ be a sequence of vectors in $\mathcal{F}$ such that $\sum_{n=-\infty}^{+\infty} \|a_n\|^2 < \infty$. An easy calculation shows that

$$(\chi^n a_n, \chi^m a_m)_{L^2(\mathcal{F})} = 0, \quad n \neq m,$$

so that the series $\sum_{n=-\infty}^{+\infty} \chi^n a_n$ converges unconditionally in $L^2(\mathcal{F})$. As in the scalar case, every $f \in L^2(\mathcal{F})$ can be written as a series $f = \sum_{n=-\infty}^{+\infty} \chi^n a_n$, where a_n are the Fourier coefficients of f defined by

$$a_n = \frac{1}{2\pi}\int_0^{2\pi} e^{-int} f(e^{it})\,dt, \quad -\infty < n < +\infty.$$

We see now that $H^2(\mathcal{F})$ can be identified with a subspace of $L^2(\mathcal{F})$. Indeed, if $u(\lambda) = \sum_{n=0}^{\infty} \lambda^n a_n$, $\lambda \in \mathbf{D}$, is an arbitrary element of $H^2(\mathcal{F})$, the series $\sum_{n=0}^{\infty} \chi^n a_n$ converges in $L^2(\mathcal{F})$. We may therefore regard $H^2(\mathcal{F})$ as a subspace of $L^2(\mathcal{F})$ consisting of those functions $f \in L^2(\mathcal{F})$ whose negative Fourier coefficients vanish. This convention greatly facilitates notation, while introducing practically no ambiguity. Thus, an element u of $H^2(\mathcal{F})$ defines an analytic function in $\mathbf{D}$ as well as an almost everywhere defined function on $\mathbf{T}$; moreover, the relation

$$u(\varsigma) = \lim_{r \uparrow 1} u(r\varsigma) \tag{1.2}$$

holds for almost every $\varsigma \in \mathbf{T}$. Since we do not prove (1.2), we will try not to use this relation. We can, however, use the fact that $\lim_{r\uparrow 1} \|u - u_r\|_{L^2(\mathcal{F})} = 0$, where $u_r(\varsigma) = u(r\varsigma)$, $\varsigma \in \mathbf{T}$. This follows from the easily verified equality

$$\|u - u_r\|^2_{L^2(\mathscr{F})} = \sum_{n=1}^{\infty} (1 - r^n)^2 \|a_n\|^2 \quad \text{if } u(\lambda) = \sum_{n=0}^{\infty} \lambda^n a_n,\ \lambda \in \mathbf{D}.$$

In particular, if u is bounded on $\mathbf{D}$, then it has the same bound almost everywhere on $\mathbf{T}$.

We now define operators $S_{\mathscr{F}} \in \mathscr{L}(H^2(\mathscr{F}))$ and $U_{\mathscr{F}} \in \mathscr{L}(L^2(\mathscr{F}))$ as follows:

$$(1.3)\qquad \begin{aligned} (S_{\mathscr{F}} u)(\lambda) &= \lambda u(\lambda), \quad \lambda \in \mathbf{D},\ u \in H^2(\mathscr{F}), \\ (U_{\mathscr{F}} u)(\varsigma) &= \varsigma u(\varsigma), \quad \varsigma \in \mathbf{T},\ u \in L^2(\mathscr{F}). \end{aligned}$$

A quick look at the Fourier coefficients of $S_{\mathscr{F}} u$ and $U_{\mathscr{F}} u$ shows that $S_{\mathscr{F}}$ is a unilateral shift of multiplicity $\dim(\mathscr{F})$, $U_{\mathscr{F}}$ is a unitary operator, and $U_{\mathscr{F}} \mid H^2(\mathscr{F}) = S_{\mathscr{F}}$. The operator $U_{\mathscr{F}}$ is called a bilateral shift of multiplicity $\dim(\mathscr{F})$; quite clearly $U_{\mathscr{F}}$ is the minimal unitary dilation of $S_{\mathscr{F}}$ because

$$(1.4)\qquad L^2(\mathscr{F}) = \bigvee_{n=0}^{\infty} U^{*^n} H^2(\mathscr{F}).$$

Now assume that $\mathscr{F}$ and $\mathscr{F}'$ are two (separable) Hilbert spaces. We denote by $H^\infty(\mathscr{L}(\mathscr{F}, \mathscr{F}'))$ the Banach space of all bounded analytic functions $\Phi : \mathbf{D} \to \mathscr{L}(\mathscr{F}, \mathscr{F}')$ with the norm

$$\|\Phi\|_\infty = \sup\{\|\Phi(\lambda)\| : |\lambda| < 1\}.$$

For every $\Phi \in H^\infty(\mathscr{L}(\mathscr{F}, \mathscr{F}'))$ we can define the Toeplitz operator $T_\Phi \in \mathscr{L}(H^2(\mathscr{F}), H^2(\mathscr{F}'))$ as follows:

$$(1.5)\qquad (T_\Phi u)(\lambda) = \Phi(\lambda) u(\lambda), \quad \lambda \in \mathbf{D},\ u \in H^2(\mathscr{F}).$$

It will be convenient to write $\Phi u = T_\Phi u$ for $u \in H^2(\mathscr{F})$ and $\Phi \in H^\infty(\mathscr{L}(\mathscr{F}, \mathscr{F}'))$. We see immediately, using formula (1.1), that T_Φ is indeed a bounded operator and $\|T_\Phi\| \le \|\Phi\|_\infty$. It is also clear that we have the following relation: $S_{\mathscr{F}'} T_\Phi = T_\Phi S_{\mathscr{F}}$. This intertwining formula allows us to extend T_Φ to the entire space $L^2(\mathscr{F})$. Indeed, assume that X is an arbitrary operator in $\mathscr{I}(S_{\mathscr{F}}, S_{\mathscr{F}'})$ and note that the operators $X_n \in \mathscr{L}(U^{*^n}_{\mathscr{F}} H^2(\mathscr{F}), U^{*^n}_{\mathscr{F}'} H^2(\mathscr{F}'))$ defined by $X_n = U^{*^n}_{\mathscr{F}'} X U^n_{\mathscr{F}} \mid U^{*^n}_{\mathscr{F}} H^2(\mathscr{F})$ satisfy the relations $X_0 = X$, $X_{n+1} \mid U^{*^n}_{\mathscr{F}} H^2(\mathscr{F}) = X_n$, and $\|X_n\| = \|X\|$, $n > 0$. We conclude that there exists an operator $Y \in \mathscr{L}(L^2(\mathscr{F}), L^2(\mathscr{F}'))$ such that $\|Y\| = \|X\|$ and $Y \mid U^{*^n}_{\mathscr{F}} H^2(\mathscr{F}) = X_n$ for all natural n. The relation $U_{\mathscr{F}'} Y = Y U_{\mathscr{F}}$ is also easy to check.

If, in the construction above, $X = T_\Phi$, the resulting operator Y will be denoted by M_Φ, and we will set

$$\Phi u = M_\Phi u, \quad u \in L^2(\mathscr{F}).$$

To convince ourselves that the above notation is justified, we have to define the boundary values $\Phi(e^{it})$ of a given function $\Phi \in H^\infty(\mathscr{L}(\mathscr{F}, \mathscr{F}'))$. Let Σ be a countable dense set in $\mathscr{F}$, and for each $a \in \Sigma$ choose a function $f_a : \mathbf{T} \to \mathscr{F}'$ such that $(T_\Phi a)(\varsigma) = f_a(\varsigma)$ almost everywhere on $\mathbf{T}$. An easy argument shows that there exists a subset N of $\mathbf{T}$, with Lebesgue measure zero, such that $f_{\alpha a+\beta b}(\varsigma) = \alpha f_a(\varsigma) + \beta f_b(\varsigma)$ and $\|f_a(\varsigma)\| \le \|\Phi\|_\infty \|a\|$ for every $\varsigma \in \mathbf{T}\backslash N$, $a, b \in \Sigma$, and rational scalars α, β (here we use the fact about bounded functions noted after (1.2)). It follows that for every $\varsigma \in \mathbf{T}\backslash N$ we can define an operator $\Phi(\varsigma) \in \mathscr{L}(\mathscr{F}, \mathscr{F}')$ such that $\Phi(\varsigma)a = f_a(\varsigma)$, $a \in \Sigma$, and $\|\Phi(\varsigma)\| \le \|\Phi\|_\infty$. It is now clear that

$$M_\Phi(\chi^n a)(\varsigma) = (M_\Phi U^n_{\mathscr{F}} a)(\varsigma) = (U^n_{\mathscr{F}'} M_\Phi a)(\varsigma) = \varsigma^n \Phi(\varsigma)a = \Phi(\varsigma)(\varsigma^n a)$$

if $-\infty < n < +\infty$, $a \in \Sigma$, and $\varsigma \in \mathbf{T}\backslash N$, and this can be extended to show that

$$(M_\Phi u)(\varsigma) = \Phi(\varsigma)u(\varsigma), \quad u \in L^2(\mathscr{F}), \tag{1.6}$$

for almost every $\varsigma \in \mathbf{T}$. Let us finally note that the above construction (or relation (1.6)) shows that the mapping $\varsigma \to \Phi(\varsigma)$ is strongly measurable in the sense that the function $\Phi(\varsigma)a$ is measurable for every $a \in \mathscr{F}$.

One reason why Toeplitz operators are important is the following result.

1.7. THEOREM. *Let $\mathscr{F}$ and $\mathscr{F}'$ be separable Hilbert spaces. For every operator $X \in \mathscr{I}(S_{\mathscr{F}}, S_{\mathscr{F}'})$ there exists a function $\Phi \in H^\infty(\mathscr{L}(\mathscr{F}, \mathscr{F}'))$ such that $X = T_\Phi$ and $\|\Phi\|_\infty = \|X\|$. In particular, the equalities $\|M_\Phi\| = \|T_\Phi\| = \|\Phi\|_\infty$ hold for every $\Phi \in H^\infty(\mathscr{L}(\mathscr{F}, \mathscr{F}'))$.*

PROOF. For every $a \in \mathscr{F}$ the functions $Xa \in H^2(\mathscr{F}')$ can be written as $(Xa)(\lambda) = \sum_{n=0}^\infty \lambda^n X_n a$, $\lambda \in \mathbf{D}$, where $\|X_n a\| \le \|Xa\| \le \|X\|\,\|a\|$. These inequalities and the linearity of X show that $X_n \in \mathscr{L}(\mathscr{F}, \mathscr{F}')$ and the power series $\Phi(\lambda) = \sum_{n=0}^\infty \lambda^n X_n$ converges for $\lambda \in \mathbf{D}$. In order to prove that Φ is bounded, fix $\mu \in \mathbf{D}$ and observe that $X^*(\ker(S^*_{\mathscr{F}'} - \bar\mu I)) \subset \ker(S^*_{\mathscr{F}} - \bar\mu I)$. Now, $\ker(S^* - \bar\mu I)$ consists of all vectors of the form $(1 - \bar\mu\chi)^{-1}b$, with $b \in \mathscr{F}$, and hence for every b' in $\mathscr{F}'$ there exists b in $\mathscr{F}$ such that $\|b\| \le \|X\|\,\|b'\|$ and $X^*((1-\bar\mu\chi)^{-1}b') = (1-\bar\mu\chi)^{-1}b$. Let b and b' be related as above, and let $a \in \mathscr{F}$. Then we have

$$\begin{aligned}
(\Phi(\mu)a, b')_{\mathscr{F}'} &= \sum_{n=0}^\infty \mu^n (X_n a, b')_{\mathscr{F}'} \\
&= \sum_{n=0}^\infty \mu^n (Xa, \chi^n b')_{H^2(\mathscr{F}')} \\
&= (Xa, (1-\bar\mu\chi)^{-1}b')_{H^2(\mathscr{F}')} \\
&= (a, (1-\bar\mu\chi)^{-1}b))_{H^2(\mathscr{F})} \\
&= (a, b)_{\mathscr{F}},
\end{aligned}$$

and, consequently,

$$|(\Phi(\mu)a, b')_{\mathscr{F}'}| \le \|a\|\,\|b\| \le \|X\|\,\|a\|\,\|b'\|, \quad a \in \mathscr{F},\ b' \in \mathscr{F}',$$

from which it follows that $\|\Phi(\mu)\| \le \|X\|$, $\mu \in \mathbf{D}$. The equality $Xu = T_\Phi u$ is clear now if u is a constant function and, if $u \in \mathscr{F}$ and $n \ge 0$,

$$X(\chi^n u) = X(S^n_{\mathscr{F}} u) = S^n_{\mathscr{F}'} X u = S^n_{\mathscr{F}'} T_\Phi u = T_\Phi(\chi^n u),$$

so that $X = T_\Phi$ on a dense subset of $H^2(\mathscr{F})$. Finally, we proved earlier that $\|M_\Phi\| = \|T_\Phi\| \le \|\Phi\|_\infty$, and the above proof yields the opposite inequality. The theorem follows.

1.8. LEMMA. *A Toeplitz operator T_Φ is an isometry if and only if $\Phi(\varsigma)$ is an isometry for almost every $\varsigma \in \mathbf{T}$.*

PROOF. Assume that $\Phi(\varsigma)$ is an isometry for almost every $\varsigma \in \mathbf{T}$. Since $T_\Phi = M_\Phi \mid H^2(\mathscr{F})$, formula (1.6) implies that

$$\begin{aligned}\|T_\Phi u\|^2 = \|M_\Phi u\|^2 &= \frac{1}{2\pi}\int_0^{2\pi} \|\Phi(e^{it})u(e^{it})\|^2\,dt \\ &= \frac{1}{2\pi}\int_0^{2\pi} \|u(e^{it})\|^2\,dt = \|u\|^2\end{aligned}$$

for every $u \in H^2(\mathscr{F})$; hence T_Φ is an isometry.

Conversely, assume that T_Φ is an isometry, and choose a countable dense set Σ in $\mathscr{F}$. Then

$$0 = \|a\|^2 - \|T_\Phi a\|^2 = \frac{1}{2\pi}\int_0^{2\pi} (\|a\|^2 - \|\Phi(e^{it})a\|^2)\,dt$$

for every $a \in \Sigma$. Now, $\|\Phi\|_\infty = \|T_\Phi\| = 1$, so that $\|\Phi(e^{it})\| \le 1$ almost everywhere. We infer the existence of a subset $N \subset \mathbf{T}$ of Lebesgue measure zero such that $\|\Phi(\varsigma)a\| = \|a\|$ for $\varsigma \in \mathbf{T}\backslash N$ and $a \in \Sigma$. Since Σ is dense, $\Phi(\varsigma)$ is an isometry for $\varsigma \in \mathbf{T}\backslash N$. The lemma is proved.

For every $\Phi \in H^\infty(\mathscr{L}(\mathscr{F},\mathscr{F}'))$ we construct $\Phi^\sim \in H^\infty(\mathscr{L}(\mathscr{F}',\mathscr{F}))$ by the formula $\Phi^\sim(\lambda) = \Phi(\bar\lambda)^*$. It is clear that $\|\Phi^\sim\|_\infty = \|\Phi\|_\infty$.

1.9. DEFINITION. A function $\Phi \in H^\infty(\mathscr{L}(\mathscr{F},\mathscr{F}'))$ is said to be *inner* if T_Φ is an isometry. The function Φ is **-inner* if $T_{\Phi^\sim}$ is an isometry. Finally, Φ is *two-sided inner* if it is both inner and *-inner.

We can now prove a generalization of Beurling's invariant subspace theorem.

1.10. THEOREM. *Let $\mathscr{F}$ be a separable Hilbert space and $\mathscr{M} \subset H^2(\mathscr{F})$ an invariant subspace for $S_\mathscr{F}$. There exists a separable Hilbert space $\mathscr{G}$ and an inner function $\Phi \in H^\infty(\mathscr{L}(\mathscr{G},\mathscr{F}))$ such that $\mathscr{M} = \Phi H^2(\mathscr{G}) = T_\Phi H^2(\mathscr{G})$.*

PROOF. The isometry $S_\mathscr{F} \mid \mathscr{M}$ is completely nonunitary and hence a unilateral shift. Let $\mathscr{G}$ be a Hilbert space whose dimension equals the multiplicity of $S_\mathscr{F} \mid \mathscr{M}$, and let $X : H^2(\mathscr{G}) \to \mathscr{M}$ be a unitary operator such that $(S_\mathscr{F} \mid \mathscr{M})X = XS_\mathscr{G}$. Such a unitary operator exists because all shifts with the same multiplicity are unitarily equivalent. We can then consider X as an operator in $\mathscr{L}(H^2(\mathscr{G}), H^2(\mathscr{F}))$ and, as such, $X \in \mathscr{I}(S_\mathscr{G}, S_\mathscr{F})$. By Theorem 1.7 we have $X = T_\Phi$ for some inner function $\Phi \in H^\infty(\mathscr{L}(\mathscr{G},\mathscr{F}))$. The equality $\mathscr{M} = XH^2(\mathscr{G}) = \Phi H^2(\mathscr{G})$ is clear, and the theorem is proved.

We are now ready to study the functional models associated with contractions of class of $C_{\cdot 0}$.

1.11. DEFINITION. Let $\mathscr{F}$ and $\mathscr{F}'$ be two separable Hilbert spaces, and let $\Theta \in H^\infty(\mathscr{L}(\mathscr{F},\mathscr{F}'))$ be an inner function. The *functional model* associated with Θ is the operator $S(\Theta)$ acting on $\mathscr{H}(\Theta) = H^2(\mathscr{F}')\ominus\Theta H^2(\mathscr{F})$ and defined by $S(\Theta) = P_{\mathscr{H}(\Theta)}S_{\mathscr{F}'} \mid \mathscr{H}(\Theta)$ or, equivalently, $S(\Theta)^* = S^*_{\mathscr{F}'} \mid \mathscr{H}(\Theta)$.

Thus a Jordan block $S(\theta)$ is a special kind of functional model. It is clear that every functional model $S(\Theta)$ is an operator of class $C_{\cdot 0}$; the converse is also true, up to unitary equivalence.

1.12. PROPOSITION. *Every separably acting contraction $T \in \mathscr{L}(\mathscr{H})$ of class $C_{\cdot 0}$ is unitarily equivalent to a functional model.*

PROOF. Let $U_+ \in \mathscr{L}(\mathscr{K}_+)$ be the minimal isometric dilation of T, so that $T^* = U_+^* \mid \mathscr{H}$. By Proposition 1.2.7, U_+ is a unilateral shift. We may therefore assume (up to unitary equivalence) that $U_+ = S_{\mathscr{F}'} \in \mathscr{L}(H^2(\mathscr{F}'))$ for some separable Hilbert space $\mathscr{F}'$. Now, $\mathscr{M} = H^2(\mathscr{F}')\ominus\mathscr{H}$ is invariant under $S_{\mathscr{F}'}$, and hence $\mathscr{M} = \Theta H^2(\mathscr{F})$ for some separable space $\mathscr{F}$ and some inner function $\Theta \in H^\infty(\mathscr{L}(\mathscr{F},\mathscr{F}'))$. We then have $\mathscr{H} = \mathscr{K}_+ \ominus (\mathscr{K}_+ \ominus \mathscr{H}) = H^2(\mathscr{F}') \ominus \Theta H^2(\mathscr{F})$ and $T = S(\Theta)$. The proposition is proved.

The function Θ provided by Proposition 1.12 is not uniquely determined, so that it is desirable to pick in some canonical way a function Θ for every operator T of class $C_{\cdot 0}$. In order to do this we recall from Proposition 1.2.9 and the proof of Theorem 1.1.4 that if $T \in \mathscr{L}(\mathscr{H})$ is of class $C_{\cdot 0}$ and $U_+ \in \mathscr{L}(\mathscr{K}_+)$ is the minimal isometric dilation of T, then U_+ is a unilateral shift of multiplicity $\dim(\mathscr{D}_{T^*})$ and $U_+ \mid \mathscr{K}_+ \ominus \mathscr{H}$ is a unilateral shift of multiplicity $\dim(\mathscr{D}_T)$. More precisely, if $\mathscr{K}_+$ is given by $\mathscr{K}_+ = \mathscr{H} \oplus (\bigoplus_{n=0}^\infty \mathscr{D}_n)$, $\mathscr{D}_n = \mathscr{D}_T$, $n \geq 0$, then the mapping $\phi_* : \mathscr{D}_{T^*} \to \mathscr{K}_+$ defined by $\phi_*(x) = D_{T^*}x \oplus (-T^*x) \oplus 0 \oplus \cdots$, $x \in \mathscr{D}_{T^*}$, is an isometry of $\mathscr{D}_{T^*}$ onto the wandering space of U_+.

Recall now that the coefficients X_n in the proof of Theorem 1.7 were constructed such that

$$(Xa, S^n_{\mathscr{F}'}b)_{H^2(\mathscr{F}')} = (X_n a, b)_{\mathscr{F}'}, \quad a \in \mathscr{F},\ b \in \mathscr{F}'.$$

Keeping this in mind, we calculate for every $a \in \mathscr{D}_T$ and $b \in \mathscr{D}_{T^*}$ the scalar products $(0 \oplus a \oplus 0 \oplus \cdots, U_+^n \phi_*(b))$, $n \geq 0$. For $n = 0$ we have

$$(0 \oplus a \oplus 0 \oplus \cdots, \phi_*(b)) = (0 \oplus a, D_{T^*}b \oplus (-T^*b)) = -(Ta, b).$$

For $n > 0$ we have $U_+^n \phi_*(b) = T^n D_{T^*} b \oplus D_T T^{n-1} D_{T^*} b \oplus \cdots$, so that

$$(0 \oplus a \oplus 0 \oplus \cdots, U_+^n \phi_*(b)) = (a, D_T T^{n-1} D_{T^*} b) = (D_{T^*} T^{*^{n-1}} D_T a, b).$$

Finally, we compute the analytic function corresponding with the operators figuring in these scalar products:

$$-T + \sum_{n=1}^\infty \lambda^n D_{T^*} T^{*^{n-1}} D_T = -T + \lambda D_{T^*}(I - \lambda T^*)^{-1} D_T.$$

1.13. DEFINITION. The function $\Theta_T : \mathbf{D} \to \mathscr{L}(\mathscr{D}_T, \mathscr{D}_{T^*})$ defined by $\Theta_T(\lambda) = [-T + \lambda D_{T^*}(I - \lambda T^*)^{-1} D_T] | \mathscr{D}_T$ is called the *characteristic function* of the contraction T.

1.14. COROLLARY. *Let T be a separably acting operator of class $C_{\cdot 0}$. Then Θ_T is an inner function, and T is unitarily equivalent with $S(\Theta_T)$. In particular, $S_{\mathscr{D}_{T^*}}$ is a minimal isometric dilation of $S(\Theta_T)$.*

PROOF. This obviously follows from the proof of Theorem 1.7 and the calculation preceding Definition 1.13.

In general it is not true that the characteristic function of $S(\Theta)$ equals Θ. For example, assume that $\Theta \in H^\infty(\mathscr{L}(\mathscr{F},\mathscr{F}'))$ is a constant unitary operator, that is, $\Theta(\lambda) = \Theta_0$, $\lambda \in \mathbf{D}$, where Θ_0 is a unitary operator from $\mathscr{F}$ onto $\mathscr{F}'$. Then clearly $\Theta H^2(\mathscr{F}) = H^2(\mathscr{F}')$ so that $T = S(\Theta)$ acts on a trivial space, and consequently $\Theta_T(\lambda)$ acts between two trivial spaces. Thus all constant unitary functions produce the same functional model up to unitary equivalence. In a similar fashion, if $\Theta_j \in H^\infty(\mathscr{L}(\mathscr{F}_j,\mathscr{F}_j')), j = 1,2$, are inner functions such that Θ_2 is a constant unitary operator, and if $\Theta_1 \oplus \Theta_2 \in H^\infty(\mathscr{L}(\mathscr{F}_1 \oplus \mathscr{F}_2, \mathscr{F}_1' \oplus \mathscr{F}_2'))$ is defined by

$$(\Theta_1 \oplus \Theta_2)(\lambda) = \Theta_1(\lambda) \oplus \Theta_2(\lambda), \qquad \lambda \in \mathbf{D}, \tag{1.15}$$

then $S(\Theta_1 \oplus \Theta_2)$ is unitarily equivalent to $S(\Theta_1)$.

Now assume that $\Theta_j \in H^\infty(\mathscr{L}(\mathscr{F}_j,\mathscr{F}_j'))$ are inner functions and $U \in \mathscr{L}(\mathscr{F}_1,\mathscr{F}_2)$, $U' \in \mathscr{L}(\mathscr{F}_1',\mathscr{F}_2')$ are unitary operators such that

$$\Theta_1(\lambda) = U'^*\Theta_2(\lambda)U, \qquad \lambda \in \mathbf{D}. \tag{1.16}$$

Then again $S(\Theta_1)$ and $S(\Theta_2)$ are unitarily equivalent. An explicit unitary equivalence $V : \mathscr{H}(\Theta_1) \to \mathscr{H}(\Theta_2)$ can be defined by $(Vf)(\lambda) = U'f(\lambda)$, $\lambda \in \mathbf{D}$, $f \in \mathscr{H}(\Theta_1)$.

Interestingly enough, the two cases described by (1.15) and (1.16) cover essentially all situations in which two inner functions produce unitarily equivalent functional models. In order to show this, we need a simple observation about unilateral shifts. Assume that $\mathscr{F}$ is a separable Hilbert space and $\mathscr{M},\mathscr{N} \subset \mathscr{F}$ are subspaces such that $\mathscr{F} = \mathscr{M} \oplus \mathscr{N}$. Then we have $H^2(\mathscr{M}), H^2(\mathscr{N}) \subset H^2(\mathscr{F})$ and $H^2(\mathscr{F}) = H^2(\mathscr{M}) \oplus H^2(\mathscr{N})$. Moreover, $H^2(\mathscr{M})$ and $H^2(\mathscr{N})$ are invariant under $S_{\mathscr{F}}$ and hence they are both reducing. Conversely, it is easy to check that every reducing subspace for $S_{\mathscr{F}}$ has the form $H^2(\mathscr{M})$ for some subspace $\mathscr{M}$ of $\mathscr{F}$. We will also need the following simple result.

1.17. LEMMA. *If $\Theta \in H^\infty(\mathscr{L}(\mathscr{F},\mathscr{F}'))$ is an inner function and $\Theta H^2(\mathscr{F}) = H^2(\mathscr{F}')$, then Θ is a unitary constant.*

PROOF. Assume that $\Theta H^2(\mathscr{F}) = H^2(\mathscr{F}')$. Then T_Θ is a unitary operator implementing a unitary equivalence between $S_{\mathscr{F}}$ and $S_{\mathscr{F}'}$. In particular, $T_\Theta(\ker S_{\mathscr{F}}^*) = \ker S_{\mathscr{F}'}^*$. Since $\ker S_{\mathscr{F}}^*$ consists of the constant functions in $H^2(\mathscr{F})$, it follows that for every $a \in \mathscr{F}$ the function $(T_\Theta a)(\lambda) = \Theta(\lambda)a$ is constant. Therefore $\Theta(\lambda) = \Theta(0)$ and $\Theta(0)$ must be an isometry because Θ is an inner function. Moreover, $\Theta(0)\mathscr{F} = T_\Theta(\ker S_{\mathscr{F}}^*) = \ker S_{\mathscr{F}'}^* = \mathscr{F}'$ so that $\Theta(0)$ is a unitary operator. The lemma follows.

1.18. PROPOSITION. *Let $\Theta \in H^\infty(\mathscr{L}(\mathscr{F},\mathscr{F}'))$ be an inner function, and set $T = S(\Theta)$. There exist a separable Hilbert space $\mathscr{N}$ and unitary operators $U \in \mathscr{L}(\mathscr{F},\mathscr{D}_T \oplus \mathscr{N})$, $U' \in \mathscr{L}(\mathscr{F}',\mathscr{D}_{T^*} \oplus \mathscr{N})$ such that $\Theta(\lambda) = U'^*(\Theta_T(\lambda) \oplus I_{\mathscr{N}})U$, $\lambda \in \mathbf{D}$.*

PROOF. We denote $\mathscr{K}_+ = \bigvee_{n=0}^\infty S_{\mathscr{F}'}^n \mathscr{H}(\Theta)$ and note that $S_{\mathscr{F}'} \mid \mathscr{K}_+$ is a minimal isometric dilation of $S(\Theta)$. Moreover, $\mathscr{K}_+$ is a reducing space for $S_{\mathscr{F}'}$. Indeed, keeping in mind the relation $T^* = S(\Theta)^* = S_{\mathscr{F}'}^* \mid \mathscr{H}(\Theta)$, we have

$$\begin{aligned} S_{\mathscr{F}'}^* \mathscr{K}_+ &= \bigvee_{n=0}^\infty S_{\mathscr{F}'}^* S_{\mathscr{F}'}^n \mathscr{H}(\Theta) \\ &= (S_{\mathscr{F}'}^* \mathscr{H}(\Theta)) \vee \left(\bigvee_{n=1}^\infty S_{\mathscr{F}'}^{n-1} \mathscr{H}(\Theta) \right) \\ &= (S(\Theta)^* \mathscr{H}(\Theta)) \vee \left(\bigvee_{n=0}^\infty S_{\mathscr{F}'}^n \mathscr{H}(\Theta) \right) \\ &\subset \mathscr{K}_+. \end{aligned}$$

By the observations preceding Lemma 1.17, $\mathscr{F}'$ can be written as a sum $\mathscr{F}' = \mathscr{M} \oplus \mathscr{N}$ such that $\mathscr{K}_+ = H^2(\mathscr{M})$. We now have

$$\Theta H^2(\mathscr{F}) = (\mathscr{K}_+ \ominus \mathscr{H}(\Theta)) \oplus H^2(\mathscr{N})$$

and, since T_Θ is an isometry, $\mathscr{F}$ must have a decomposition of the form $\mathscr{F} = \mathscr{P} \oplus \mathscr{Q}$ with the property that $\Theta H^2(\mathscr{P}) = \mathscr{K}_+ \ominus \mathscr{H}(\Theta)$ and $\Theta H^2(\mathscr{Q}) = H^2(\mathscr{N})$. It is clear that $\Theta(\lambda)\mathscr{P} \subset \mathscr{M}$ and $\Theta(\lambda)\mathscr{Q} \subset \mathscr{N}$, $\lambda \in \mathbf{D}$, so that Θ can be written as $\Theta = \Theta_1 \oplus \Theta_2$, where $\Theta_1 \in H^\infty(\mathscr{L}(\mathscr{P},\mathscr{M}))$ and $\Theta_2 \in H^\infty(\mathscr{L}(\mathscr{Q},\mathscr{N}))$ are inner functions. In addition, Θ_2 must be a unitary constant by Lemma 1.17.

Since $S_{\mathscr{M}} = S_{\mathscr{F}'} \mid \mathscr{K}_+$ is a minimal isometric dilation of $T = S(\Theta)$ while, by Corollary 1.14, $S_{\mathscr{D}_{T^*}}$ is a minimal isometric dilation of $S(\Theta_T)$, there must exist a function $\Phi \in H^\infty(\mathscr{L}(\mathscr{M},\mathscr{D}_{T^*}))$ such that T_Φ is unitary and $T_\Phi(\Theta_1 H^2(\mathscr{P})) = T_\Phi(\mathscr{K}_+ \ominus \mathscr{H}(\Theta)) = \Theta_T H^2(\mathscr{D}_T)$. Of course, Lemma 1.17 implies that $\Phi(\lambda) = \Phi(0)$, $\lambda \in \mathbf{D}$, is a unitary operator. Moreover, the relation $\Phi\Theta_1 H^2(\mathscr{P}) = \Theta_T H^2(\mathscr{D}_T)$ shows that $S_{\mathscr{P}}$ and $S_{\mathscr{D}_T}$ are unitarily equivalent. Thus there exists a unitary operator $V \in \mathscr{L}(\mathscr{P},\mathscr{D}_T)$ such that $\Phi(\lambda)\Theta_1(\lambda) = \Theta_T(\lambda)V$, $\lambda \in \mathbf{D}$. We can now define $U' : \mathscr{F}' = \mathscr{M} \oplus \mathscr{N} \to \mathscr{D}_{T^*} \oplus \mathscr{N}$ and $U : \mathscr{F} = \mathscr{P} \oplus \mathscr{Q} \to \mathscr{D}_T \oplus \mathscr{N}$ by the formulas $U' = \Phi(0) \oplus I_{\mathscr{N}}$, $U = V \oplus \Theta_2(0)$. It is an easy exercise to verify that U and U' satisfy the desired identities. The proposition is proved.

1.19. DEFINITION. Two inner functions Θ and Θ' are said to be *equivalent* if $S(\Theta)$ and $S(\Theta')$ are unitarily equivalent operators.

Proposition 1.18 gives a more concrete form of the equivalence relation just defined. It is interesting to note that every inner function Θ is equivalent to an inner function Θ' acting between two infinite-dimensional Hilbert spaces. Indeed, it suffices to observe that Θ and $\Theta \oplus \Theta_1$ are equivalent, where $\Theta_1(\lambda) = I_{\mathscr{K}}$ and $\mathscr{K}$ is an infinite-dimensional Hilbert space. This observation will be useful, even though the space $\mathscr{H}(\Theta)$ is easier to understand when Θ acts between finite-dimensional spaces.

1.20. PROPOSITION. *For every two-sided inner function Θ, the operators $S(\Theta)^*$ and $S(\Theta^\sim)$ are unitarily equivalent.*

PROOF. Assume that $\Theta \in H^\infty(\mathscr{L}(\mathscr{F},\mathscr{F}'))$ is two-sided inner, and set $T = S(\Theta)$. It clearly follows from Definition 1.13 that $\Theta_{T^*}(\lambda) = \Theta_T^\sim(\lambda)$, $\lambda \in \mathbf{D}$, so that Θ_T is also two-sided inner by Proposition 1.18. Again by Proposition 1.18, $S(\Theta_T^\sim)$ and $S(\Theta^\sim)$ are unitarily equivalent; consequently our proposition will follow once we know that T^* is unitarily equivalent to $S(\Theta_{T^*})$. Of course, by Corollary 1.14, it suffices to verify that T^* is of class $C_{\cdot 0}$ or, equivalently, that T is of class $C_{0\cdot}$.

Now, the operators $\Theta(\varsigma)$ are unitary for almost every $\varsigma \in \mathbf{T}$, and this clearly implies that M_Θ is a unitary operator of $L^2(\mathscr{F})$ onto $L^2(\mathscr{F}')$. If we denote by U_- the compression of $U_{\mathscr{F}'}$ to $L^2(\mathscr{F}')\ominus\Theta H^2(\mathscr{F})$, then U_- is unitarily equivalent to the compression of $U_{\mathscr{F}}$ to $L^2(\mathscr{F})\ominus H^2(\mathscr{F})$, and hence U_-^* is a unilateral shift. Finally, we have $T = S(\Theta) = U_- \mid \mathscr{H}(\Theta)$ and hence T is of class $C_{0\cdot}$ because U_- is of class $C_{0\cdot}$. The proof is now complete.

As noted in the proof above, if $\Theta \in H^\infty(\mathscr{L}(\mathscr{F},\mathscr{F}'))$ is a two-sided inner function then $\Theta(\varsigma)$ is a unitary operator for almost every $\varsigma \in \mathbf{T}$. Hence Θ is equivalent to a function $\Theta_1 \in H^\infty(\mathscr{L}(\mathscr{F}))$. Indeed, it suffices to define $\Theta_1(\lambda) = \Theta(\lambda)U$, where $U \in \mathscr{L}(\mathscr{F}',\mathscr{F})$ is an arbitrary unitary operator. This observation will be used often in the sequel because it greatly facilitates notation.

We can now give a description, similar to Proposition 3.1.10, of the invariant subspaces of a functional model.

1.21. PROPOSITION. *Let $\Theta \in H^\infty(\mathscr{F},\mathscr{F}')$ be an inner function.*

(i) *If $\mathscr{M} \subset \mathscr{H}(\Theta)$ is an invariant subspace, there exist a Hilbert space $\mathscr{K}$ and inner functions $\Theta_1 \in H^\infty(\mathscr{F},\mathscr{K})$, $\Theta_2 \in H^\infty(\mathscr{K},\mathscr{F}')$ such that $\Theta(\lambda) = \Theta_2(\lambda)\Theta_1(\lambda)$, $\lambda \in \mathbf{D}$, and*

$$\mathscr{M} = \Theta_2 H^2(\mathscr{K}) \ominus \Theta H^2(\mathscr{F}). \tag{1.22}$$

(ii) *Conversely, if $\mathscr{K}$, Θ_1, and Θ_2 are as above, then (1.22) defines an invariant subspace for $S(\Theta)$. Moreover, if $S(\Theta) = \begin{bmatrix} T_1 & X \\ 0 & T_2 \end{bmatrix}$ is the triangularization of $S(\Theta)$ with respect to the decomposition $\mathscr{H}(\Theta) = \mathscr{M} \oplus (\mathscr{H}(\Theta) \ominus \mathscr{M})$, then T_j is unitarily equivalent to $S(\Theta_j)$ for $j = 1, 2$.*

PROOF. Let $\mathscr{M}$ be invariant for $S(\Theta)$. Then the space $\mathscr{M} \oplus \Theta H^2(\mathscr{F})$ is invariant for $S_{\mathscr{F}'}$ and Theorem 1.10 implies the existence of $\mathscr{K}$ and $\Theta_2 \in H^\infty(\mathscr{L}(\mathscr{K},\mathscr{F}'))$ such that (1.22) holds. Furthermore, the inclusion $\Theta H^2(\mathscr{F}) \subset \Theta_2 H^2(\mathscr{K})$ implies the existence of an isometry $V \in I(S_{\mathscr{F}}, S_{\mathscr{K}})$ such that $T_\Theta = T_{\Theta_2} V$. Of course, by Theorem 1.7, we have $V = T_{\Theta_1}$, where $\Theta_1 \in H^\infty(\mathscr{L}(\mathscr{K},\mathscr{F}))$ is an inner function. The relation $T_\Theta = T_{\Theta_2} T_{\Theta_1}$, applied to constant functions, clearly implies $\Theta(\lambda) = \Theta_2(\lambda)\Theta_1(\lambda)$, $\lambda \in \mathbf{D}$.

To prove (ii) we first note that the invariance of the subspace $\mathscr{M}$ described by (1.22) is obvious. It is also clear that $\mathscr{H}(\Theta) \ominus \mathscr{M} = H^2(\mathscr{F}') \ominus \Theta_2 H^2(\mathscr{K}) = \mathscr{H}(\Theta_2)$ and

$$T_2^* = S(\Theta)^* \mid \mathscr{H}(\Theta) \ominus \mathscr{M} = S_{\mathscr{F}'}^* \mid \mathscr{H}(\Theta_2) = S(\Theta_2)^*.$$

It remains to prove the unitary equivalence of T_1 with $S(\Theta_1)$. This unitary equivalence is realized by the operator T_{Θ_2} because

$$\begin{aligned} T_{\Theta_2}\mathscr{H}(\Theta_1) &= \Theta_2(H^2(\mathscr{K}) \ominus \Theta_1 H^2(\mathscr{F})) \\ &= \Theta_2 H^2(\mathscr{K}) \ominus \Theta_2\Theta_1 H^2(\mathscr{F}) \\ &= \Theta_2 H^2(\mathscr{K}) \ominus \Theta H^2(\mathscr{F}) \\ &= \mathscr{M}. \end{aligned}$$

The relation $T_1(T_{\Theta_2} \mid \mathscr{H}(\Theta_1)) = (T_{\Theta_2} \mid \mathscr{H}(\Theta_1))S(\Theta_1)$ is an easy consequence of the fact that $T_{\Theta_2} \in \mathscr{I}(S_{\mathscr{K}}, S_{\mathscr{F}'})$.

We finally turn to the commutant lifting problem in the context of functional models. Let $\Theta \in H^\infty(\mathscr{L}(\mathscr{F}, \mathscr{F}'))$ be an inner function, and consider the Toeplitz operator $T_\Phi \in \{S_{\mathscr{F}'}\}'$, $\Phi \in H^\infty(\mathscr{L}(\mathscr{F}'))$. If we have $\Phi\Theta H^2(\mathscr{F}) \subset \Theta H^2(\mathscr{F})$ then $\mathscr{H}(\Theta)$ is a semi-invariant subspace for T_Φ and hence the operator

$$P_{\mathscr{H}(\Theta)} T_\Phi \mid H(\Theta) \tag{1.23}$$

commutes with $S(\Theta)$. The notation $\Phi(S(\Theta))$ might seem appropriate for the operator in (1.23), but we will abstain from using it. The commutant lifting theorem (Theorem 1.1.10) combined with Theorem 1.7 shows that all operators commuting with $S(\Theta)$ have the form (1.23). More generally, we have the following result.

1.24. PROPOSITION. *Let $\Theta \in H^\infty(\mathscr{L}(\mathscr{F}, \mathscr{F}'))$ and $\Theta_1 \in H^\infty(\mathscr{L}(\mathscr{F}_1, \mathscr{F}_1'))$ be two inner functions, and $X \in \mathscr{I}(S(\Theta), S(\Theta_1))$. There exists a function $\Phi \in H^\infty(\mathscr{L}(\mathscr{F}', \mathscr{F}_1'))$ such that $\|\Phi\|_\infty = \|X\|$, $\Phi\Theta H^2(\mathscr{F}) \subset \Theta_1 H^2(\mathscr{F}_1)$, and $X = P_{\mathscr{H}(\Theta_1)} T_\Phi \mid \mathscr{H}(\Theta)$.*

PROOF. By Theorem 1.1.10, there exists $Y \in \mathscr{I}(S_{\mathscr{F}'}, S_{\mathscr{F}_1'})$ such that $Y\Theta H^2(\mathscr{F}) \subset \Theta_1 H^2(\mathscr{F}_1)$, $\|Y\| = \|X\|$, and $X = P_{\mathscr{H}(\Theta_1)} Y \mid \mathscr{H}(\Theta)$. Now, Theorem 1.7 shows that $Y = T_\Phi$ for some $\Phi \in H^\infty(\mathscr{L}(\mathscr{F}', \mathscr{F}_1'))$ such that $\|\Phi\|_\infty = \|Y\|$. The proposition follows at once.

Let us analyze more closely the condition $\Phi\Theta H^2(\mathscr{F}) \subset \Theta_1 H^2(\mathscr{F}_1)$. This condition shows that for every $u \in H^2(\mathscr{F})$ we can find v in $H^2(\mathscr{F}_1)$ such that $T_\Phi T_\Theta u = T_{\Theta_1} v$. Moreover, since Θ_1 and Θ are inner, we have

$$\|v\| = \|\Theta_1 v\| = \|\Phi\Theta u\| \le \|\Phi\|_\infty \|\Theta u\| = \|\Phi\|_\infty \|u\|.$$

This shows that v is uniquely determined and hence it deserves to be denoted $v = Zu$, where $Z \in \mathscr{L}(H^2(\mathscr{F}), H^2(\mathscr{F}_1))$ and $\|Z\| \le \|\Phi\|_\infty$. The trivial calculation

$$T_{\Theta_1} Z S_{\mathscr{F}} = T_\Phi T_\Theta S_{\mathscr{F}} = S_{\mathscr{F}_1'} T_\Phi T_\Theta = S_{\mathscr{F}_1'} T_{\Theta_1} Z = T_{\Theta_1} S_{\mathscr{F}_1} Z$$

shows that $Z \in \mathcal{I}(S_{\mathcal{F}}, S_{\mathcal{F}_1})$, and hence Z can be written as $Z = T_\Psi$ for some $\Psi \in H^\infty(\mathcal{L}(\mathcal{F}, \mathcal{F}_1))$. Thus we have $\Phi\Theta H^2(\mathcal{F}) \subset \Theta_1 H^2(\mathcal{F}_1)$ if and only if there exists $\Psi \in H^\infty(\mathcal{L}(\mathcal{F}, \mathcal{F}_1))$ satisfying the relation

$$\Phi\Theta = \Theta_1\Psi, \tag{1.25}$$

i.e., $\Phi(\lambda)\Theta(\lambda) = \Theta_1(\lambda)\Psi(\lambda)$, $\lambda \in \mathbf{D}$.

Now let us remark that $P_{\mathcal{H}(\Theta_1)}T_\Phi \mid \mathcal{H}(\Theta) = 0$ if and only if $\Phi H^2(\mathcal{F}') \subset \Theta_1 H^2(\mathcal{F}_1)$, and an argument analogous to that above shows that this happens if and only if

$$\Phi = \Theta_1\Xi \tag{1.26}$$

for some function $\Xi \in H^\infty(\mathcal{L}(\mathcal{F}, \mathcal{F}_1))$. An argument almost identical with the one used in the proof of Corollary 3.1.20 yields the following result.

1.27. COROLLARY. *Let $\Theta \in H^\infty(\mathcal{L}(\mathcal{F}, \mathcal{F}'))$ and $\Theta_1 \in H^\infty(\mathcal{L}(\mathcal{F}_1, \mathcal{F}_1'))$ be two inner functions. If we denote by $H^\infty_{\Theta,\Theta_1}$ the set of those functions $\Phi \in H^\infty(\mathcal{L}(\mathcal{F}, \mathcal{F}'))$ for which $\Phi\Theta H^2(\mathcal{F}) \subset \Theta_1 H^2(\mathcal{F}_1)$, the space $\mathcal{I}(S(\Theta), S(\Theta_1))$ is isometrically isomorphic with the quotient $H^\infty_{\Theta,\Theta_1}/\Theta_1 H^\infty(\mathcal{L}(\mathcal{F}, \mathcal{F}_1))$.*

Of course, in the case in which $\Theta = \Theta_1$, the space $\Theta H^\infty(\mathcal{L}(\mathcal{F}, \mathcal{F}_1))$ is an ideal in $H^\infty_{\Theta,\Theta}$, and the above-mentioned isomorphism is also an algebra isomorphism.

Exercises

1. Assume that $\mathcal{F}$ is a nonseparable Hilbert space and $L^2(\mathcal{F})$ is defined as the set of (classes of) separably valued weakly measurable functions $f : \mathbf{T} \to \mathcal{F}$ such that $\|f\|^2_{L^2(\mathcal{F})} = (1/2\pi)\int_0^{2\pi} \|f(e^{it})\|^2\,dt < \infty$. Show that, as in the separable case, $L^2(\mathcal{F})$ is isomorphic with $\bigoplus_{n=-\infty}^{\infty} \mathcal{F}_n$, $\mathcal{F}_n = \mathcal{F}$, and a unitary operator $U : \bigoplus_{n=-\infty}^{\infty} \mathcal{F}_n \to L^2(\mathcal{F})$ is defined by $U(\bigoplus_{n=-\infty}^{\infty} a_n) = \sum_{n=-\infty}^{\infty} \chi^n a_n$.

2. Show that formula (1.2) still holds if $\mathcal{F}$ is not separable.

3. Assume that at least one of the spaces $\mathcal{F}$ and $\mathcal{F}'$ is nonseparable, and $\Phi \in H^\infty(\mathcal{L}(\mathcal{F}, \mathcal{F}'))$. The operators T_Φ and M_Φ can be defined as in the separable case, and the analogue of Theorem 1.7 holds. Show that operators $\Phi(\varsigma)$, depending strongly measurably on ς and satisfying (1.6), exist. Prove that the $\Phi(\varsigma)$ are not uniquely determined almost everywhere.

4. Show that Lemma 1.8 may fail in the nonseparable case.

5. Check the details of the construction of the operators M_Φ. That is, show that the operators X_n have the stated properties.

6. let $\{\theta_j : j < \omega\}$ be a model function, and define the inner function $\Theta \in H^\infty(\mathcal{L}(l^2))$ by $\Theta(\lambda)e_j = \theta_j(\lambda)e_j$, $j < \omega$. Show that the functional model $S(\Theta)$ is unitarily equivalent to the Jordan operator $\bigoplus_{j<\omega} S(\theta_j)$.

7. Show that T_Φ is unitary if and only if $\Phi(\lambda) = \Phi(0)$ is a unitary operator. Show that M_Φ is unitary if and only if Φ is two-sided inner.

8. Let $\mathscr{F}$ and $\mathscr{F}'$ be separable Hilbert spaces, and $\Theta \in H^\infty(\mathscr{L}(\mathscr{F}, \mathscr{F}'))$ a contractive function, i.e., $\|\Theta\| \le 1$. Show that the function $\Delta(\varsigma) = (I - \Theta(\varsigma)^*\Theta(\varsigma))^{1/2} \in \mathscr{L}(\mathscr{F})$, $\varsigma \in \mathbf{T}$, is strongly measurable.

9. Let Θ and Δ be as in Exercise 8. Show that for every $u \in L^2(\mathscr{F})$ the function $\varsigma \to \Delta(\varsigma)u(\varsigma)$ is measurable, and the operator $M_\Delta \in \mathscr{L}(L^2(\mathscr{F}))$ defined by $(M_\Delta u)(\varsigma) = \Delta(\varsigma)u(\varsigma)$, $\varsigma \in \mathbf{T}$, $u \in L^2(\mathscr{F})$, is a contraction commuting with $U_{\mathscr{F}}$.

10. Let Θ and Δ be as in the previous exercise. Define $\mathscr{K}_+ = H^2(\mathscr{F}') \oplus (M_\Delta L^2(\mathscr{F}))^-$, $U_+ \in \mathscr{L}(\mathscr{K}_+)$ by $U_+ = (S_{\mathscr{F}'} \oplus U_{\mathscr{F}}) \mid \mathscr{K}_+$, $\mathscr{H}(\Theta) = \mathscr{K}_+ \ominus \{T_\Theta u \oplus M_\Delta u : u \in H^2(\mathscr{F})\}$, and $S(\Theta) = P_{\mathscr{H}(\Theta)}U_+ \mid \mathscr{H}(\Theta)$. Prove that the operator $S(\Theta)$ is a completely nonunitary contraction.

11. Let T be a separably acting, completely nonunitary contraction. Prove that the function Θ_T, given by Definition 1.13, is contractive and T is unitarily equivalent to $S(\Theta_T)$.

12. Prove the analogue of Proposition 1.18 for arbitrary contractive functions Θ.

13. Assume that Θ is a two-sided inner function, and find an explicit unitary operator realizing the unitary equivalence of $S(\Theta)^*$ and $S(\Theta^\sim)$.

14. Find the characteristic function of the unilateral shift S.

15. Let $\Theta \in H^\infty(\mathscr{L}(H^2))$ be defined by $\Theta(\lambda) = S$, $\lambda \in \mathbf{D}$. Identify the operator $S(\Theta)$, i.e., represent it in terms of known operators.

16. Show that for every $\mu \in \mathbf{D}$, the operator $S_\mu = (\mu I - S)(I - \bar{\mu}S)^{-1}$ is unitarily equivalent with S. Provide an explicit unitary equivalence between S_μ and S.

17. Let Θ be an inner function, and define $\Theta_\mu(\lambda) = \Theta((\lambda + \mu)/(1 + \bar{\mu}\lambda))$, $\lambda \in \mathbf{D}$, where $\mu \in \mathbf{D}$ is fixed. Prove that $S(\Theta_\mu)$ is unitarily equivalent to $(\mu I - S(\Theta))(I - \bar{\mu}S(\Theta))^{-1}$.

18. Let T be a contraction of class $C_{\cdot 0}$, and $\mu \in \mathbf{D}$. Show that $T_\mu = (\mu I - T)(I - \bar{\mu}T)^{-1}$ is also of class $C_{\cdot 0}$, and Θ_{T_μ} is equivalent to $(\Theta_T)_\mu$, where the latter function is defined as in the preceding exercise.

19. Deduce from Exercise 18 that $\Theta_T(\mu)$ is invertible if and only if $\mu I - T$ is invertible, $\mu \in \mathbf{D}$.

20. Assume that $\|T\| < 1$. Prove that Θ_T can be extended analytically to an open neighborhood of $\mathbf{D}^-$, and the extension $\Theta_T(\varsigma)$ is unitary for $\varsigma \in \mathbf{T}$.

2. Tensor operations. The tensor operations mentioned in the title are certain *-representations of the multiplicative semigroup $\mathscr{L}(\mathscr{F})$ that arise from the consideration of tensor products of Hilbert spaces. We recall that the Hilbertian tensor product of the Hilbert spaces $\mathscr{F}_1, \mathscr{F}_2, \ldots, \mathscr{F}_n$, $n \ge 1$, is a Hilbert space denoted $\mathscr{F}_1 \otimes \mathscr{F}_2 \otimes \cdots \otimes \mathscr{F}_n$ which is the completion of the algebraic tensor product of the given spaces with respect to the unique scalar product satisfying the following relation:

$$(2.1)\qquad \begin{aligned}&(x_1 \otimes x_2 \otimes \cdots \otimes x_n, y_1 \otimes y_2 \otimes \cdots \otimes y_n)\\ &= (x_1, y_1)(x_2, y_2) \cdots (x_n, y_n), \qquad x_i, y_i \in \mathscr{F}_i,\ 1 \le i \le n.\end{aligned}$$

If $T_i \in \mathscr{L}(\mathscr{F}_i)$ then there exists a unique operator $T_1 \otimes T_2 \otimes \cdots \otimes T_n \in \mathscr{L}(\mathscr{F}_1 \otimes \mathscr{F}_2 \otimes \cdots \otimes \mathscr{F}_n)$ such that

$$(2.2)\qquad (T_1 \otimes T_2 \otimes \cdots \otimes T_n)(x_1 \otimes x_2 \otimes \cdots \otimes x_n) = T_1 x_1 \otimes T_2 x_2 \otimes \cdots \otimes T_n x_n$$

if $x_j \in \mathscr{F}_j$, $1 \le j \le n$. The equality

$$\|T_1 \otimes T_2 \otimes \cdots \otimes T_n\| = \|T_1\|\,\|T_2\| \cdots \|T_n\|$$

is easily verified. If $\mathscr{F}_1 = \mathscr{F}_2 = \cdots = \mathscr{F}_n = \mathscr{F}$, we will use the notation $\bigotimes^n \mathscr{F}$ for the tensor product $\mathscr{F}_1 \otimes \mathscr{F}_2 \otimes \cdots \otimes \mathscr{F}_n$.

Some basic representations of $\mathscr{L}(\mathscr{F})$ are obtained as follows. Fix $n \ge 1$, and consider the mapping $\Gamma_n : \mathscr{L}(\mathscr{F}) \to \mathscr{L}(\bigotimes^n \mathscr{F})$ given by

$$\Gamma_n(T) = T \otimes T \otimes \cdots \otimes T, \qquad T \in \mathscr{L}(\mathscr{F}).$$

The map Γ_n is a norm-continuous $*$-representation of the multiplicative semigroup $\mathscr{L}(\mathscr{F})$. In other words, $\Gamma_n(I) = I$, $\Gamma_n(T^*) = \Gamma_n(T)^*$, $\Gamma_n(TS) = \Gamma_n(T)\Gamma_n(S)$ for $T, S \in \mathscr{L}(\mathscr{F})$. The norm continuity of the map Γ_n follows from the fact that Γ_n is a homogeneous polynomial of degree n, induced by the continuous multilinear map $(T_1, T_2, \ldots, T_n) \to T_1 \otimes T_2 \otimes \cdots \otimes T_n$. It is also easy to verify that $\Gamma_n \mid \{T : \|T\| \le 1\}$ is continuous with respect to the corresponding weak or strong topologies. Since Γ_n is a *-representation, $\Gamma_n(T)$ is an isometry [resp., a unitary operator] provided that T is an isometry [resp., a unitary operator].

The representation Γ_n is not irreducible unless $n = 1$ or $\dim(\mathscr{F}) \le 1$. In order to determine the decomposition of Γ_n into its irreducible components it is necessary to determine the commutant $[\Gamma_n(\mathscr{L}(\mathscr{F}))]'$ and the projections in that commutant. Actually, the fact that Γ_n can be written as an orthogonal sum of irreducible representations follows from the fact that the commutant $[\Gamma_n(\mathscr{L}(\mathscr{F}))]'$ is a finite-dimensional C^*-algebra. To see this we consider the unitary representation

$$\pi_n : \mathscr{S}_n \to \mathscr{L}\left(\bigotimes^n \mathscr{F}\right),$$

where $\mathscr{S}_n$ denotes the group of permutations of $\{1, 2, \ldots, n\}$, defined by

$$(2.3)\qquad \pi_n(\sigma)(x_1 \otimes x_2 \otimes \cdots \otimes x_n) = x_{\sigma^{-1}(1)} \otimes x_{\sigma^{-1}(2)} \otimes \cdots \otimes x_{\sigma^{-1}(n)},$$

$\sigma \in \mathscr{S}_n$, $x_j \in \mathscr{F}$, $1 \le j \le n$. The operators $\{\pi_n(\sigma) : \sigma \in \mathscr{S}_n\}$ clearly generate a finite-dimensional C^*-algebra.

2.4. THEOREM. *The commutant* $[\Gamma_n(\mathscr{L}(\mathscr{F}))]'$ *coincides with the algebra generated by* $\{\pi_n(\sigma) : \sigma \in \mathscr{S}_n\}$.

PROOF. Denote by $\mathscr{B}_n$ the algebra generated by $\{\pi_n(\sigma) : \sigma \in \mathscr{S}_n\}$, and set $\mathscr{A}_n = \mathscr{B}_n'$. Since $\mathscr{B}_n$ is a finite-dimensional C^*-algebra we must also have $\mathscr{A}_n' = \mathscr{B}_n$. We also note that obviously $\Gamma_n(\mathscr{L}(\mathscr{F})) \subset \mathscr{A}_n$ and hence, to prove the theorem, it suffices to show that $\mathscr{A}_n$ is the weak*-closed algebra generated by $\Gamma_n(\mathscr{L}(\mathscr{F}))$. Suppose to the contrary that the weak*-closed algebra generated by $\Gamma_n(\mathscr{L}(\mathscr{F}))$ is strictly contained in $\mathscr{A}_n$. Then there exists a weak*-continuous functional on $\mathscr{L}(\bigotimes^n \mathscr{F})$ which is identically zero on $\Gamma_n(\mathscr{L}(\mathscr{F}))$, but not on $\mathscr{A}_n$. Equivalently, there is a trace-class operator $D \in \mathscr{L}(\bigotimes^n \mathscr{F})$ such that $\operatorname{tr}(D\Gamma_n(T)) = 0$ for all $T \in \mathscr{L}(\mathscr{F})$, but $\operatorname{tr}(DX) \neq 0$ for some X in $\mathscr{A}_n$. We set $C = (1/n!)\sum_{\sigma\in\mathscr{S}_n} \pi_n(\sigma)^{-1} D \pi_n(\sigma)$ and note that, for $X \in \mathscr{A}_n$,

$$\begin{aligned}\operatorname{tr}(CX) &= \frac{1}{n!}\sum_{\sigma\in\mathscr{S}_n} \operatorname{tr}(\pi_n(\sigma)^{-1} D \pi_n(\sigma) X) \\ &= \frac{1}{n!}\sum_{\sigma\in\mathscr{S}_n} \operatorname{tr}(\pi_n(\sigma)^{-1} D X \pi_n(\sigma)) \\ &= \operatorname{tr}(DX).\end{aligned}$$

Thus C is a trace-class operator, $C \neq 0$, $\pi_n(\sigma)C = C\pi_n(\sigma)$ for σ in $\mathscr{S}_n$, and $\operatorname{tr}(C\Gamma_n(T)) = 0$ for all T in $\mathscr{L}(\mathscr{F})$.

Now choose an orthonormal basis $\{e_i : i \in I\}$ for $\mathscr{F}$, and note that an orthonormal basis for $\bigotimes^n \mathscr{F}$ is given by $\{e_{i_1} \otimes e_{i_2} \otimes \cdots \otimes e_{i_n} : i_1, i_2, \ldots, i_n \in I\}$. The operator C has a matrix of the form $\{c(i_1, i_2, \ldots, i_n; j_1, j_2, \ldots, j_n) : i_p, j_p \in I,\ 1 \leq p \leq n\}$ with respect to this orthonormal basis. The identity $\pi_n(\sigma)C = C\pi_n(\sigma)$ for all σ in $\mathscr{S}_n$ is equivalent to

$$(2.5)\quad c(i_{\sigma(1)}, i_{\sigma(2)}, \ldots, i_{\sigma(n)}; j_{\sigma(1)}, j_{\sigma(2)}, \ldots, j_{\sigma(n)}) = c(i_1, i_2, \ldots, i_n; j_1, j_2, \ldots, j_n)$$

for all σ in $\mathscr{S}_n$ and $i_1, i_2, \ldots, i_n, j_1, j_2, \ldots, j_n \in I$. If $T \in \mathscr{L}(\mathscr{F})$ has the matrix $\{x_{ij} : i, j \in I\}$, then $\Gamma_n(T)$ has the matrix $\{x_{i_1j_1}x_{i_2j_2}\cdots x_{i_nj_n} : i_p, j_p \in I,\ 1 \leq p \leq n\}$, and the condition $\operatorname{tr}(C\Gamma_n(T)) = 0$ can be written as

$$(2.6)\qquad \sum c(i_1, i_2, \ldots, i_n; j_1, j_2, \ldots, j_n) x_{i_1j_1}x_{i_2j_2}\cdots x_{i_nj_n} = 0,$$

where the sum is extended over all choices $i_p, j_p \in I$, $1 \leq p \leq n$. If we denote the generic pair $(i,j) \in I \times I$ by α, and we set $c(\alpha_1, \alpha_2, \ldots, \alpha_n) = c(i_1, i_2, \ldots, i_n; j_1, j_2, \ldots, j_n)$ if $\alpha_p = (i_p, j_p)$, $1 \leq p \leq n$, then (2.5) and (2.6) can be rewritten as

$$f(\{x_\alpha : \alpha \in I \times I\}) = \sum c(\alpha_1, \alpha_2, \ldots, \alpha_n) x_{\alpha_1} x_{\alpha_2} \cdots x_{\alpha_n} = 0$$

and $c(\alpha_{\sigma(1)}, \alpha_{\sigma(2)}, \ldots, \alpha_{\sigma(n)}) = c(\alpha_1, \alpha_2, \ldots, \alpha_n)$, where the first identity holds at least for all finitely nonzero families of scalars $\{x_\alpha : \alpha \in I \times I\}$. But these two conditions clearly imply that $c(\alpha_1, \alpha_2, \ldots, \alpha_n) = 0$ for all $\alpha_1, \alpha_2, \ldots, \alpha_n \in I \times I$; indeed, $c(\alpha_1, \alpha_2, \ldots, \alpha_n)$ is a positive multiple of $\partial^n f/\partial x_{\alpha_1}\partial x_{\alpha_2}\cdots\partial x_{\alpha_n}$. This contradicts the fact that $C \neq 0$, and we conclude that $\mathscr{A}_n$ must coincide with the weak*-closed algebra generated by $\Gamma_n(\mathscr{L}(\mathscr{F}))$. The proof is complete.

We will need a slightly stronger version of Theorem 2.4, in which $\mathscr{L}(\mathscr{F})$ is replaced by the group $\mathscr{U}(\mathscr{F})$ of all unitary operators acting on $\mathscr{F}$.

2.7. THEOREM. *The commutant* $[\Gamma_n(\mathscr{U}(\mathscr{F}))]'$ *coincides with the algebra generated by* $\{\pi_n(\sigma) : \sigma \in \mathscr{S}_n\}$.

PROOF. Fix $X \in [\Gamma_n(\mathscr{U}(\mathscr{F}))]'$; it suffices to show that X commutes with $\Gamma_n(T)$ whenever T is a strict contraction, i.e., $\|T\| < 1$. Note that for $\|T\| < 1$, the characteristic function $\Theta_T(\lambda) = -T + \lambda D_{T^*}(I - \lambda T^*)^{-1} D_T \in \mathscr{L}(\mathscr{F})$ is defined and analytic on $\mathbf{D}^- = \{\lambda : |\lambda| \leq 1\}$, and $\Theta_T(\varsigma)$ is unitary for $\varsigma \in \mathbf{T}$. Thus we have $X\Gamma_n(\Theta_T(\varsigma)) = \Gamma_n(\Theta_T(\varsigma))X$ for $\varsigma \in \mathbf{T}$, and by analytic continuation we have this relation for ς in $\mathbf{D}$ as well. When $\varsigma = 0$ we have $\Theta_T(0) = -T$, and hence X commutes with $\Gamma_n(T)$. The theorem follows.

We see now that the reducing subspaces of $\Gamma_n(\mathscr{L}(\mathscr{F}))$ correspond to projections in a finite-dimensional C^*-algebra $\mathscr{B}_n$. The algebra $\mathscr{B}_n$ is actually the image of a homomorphism of the group C^*-algebra $C^*(\mathscr{S}_n)$. We recall that the C^*-algebra $C^*(G)$ of a finite group G consists of all formal sums $\sum_{g \in G} \alpha_g g$, $\alpha_g \in \mathbf{C}$ for $g \in G$, where multiplication is defined such that G can be regarded as a multiplicative group in $C^*(G)$, and $g^* = g^{-1}$ for g in G. The fact that $C^*(G)$ has a (unique) norm making it into a C^*-algebra is easy to verify.

The homomorphism $\pi_n : \mathscr{S}_n \to \mathscr{L}(\bigotimes^n \mathscr{F})$ can be extended to a *-homomorphism, still denoted π_n, from $C^*(\mathscr{S}_n)$ to $\mathscr{L}(\bigotimes^n \mathscr{F})$, and we have $\mathscr{B}_n = \pi_n(C^*(\mathscr{S}_n))$. Thus $\mathscr{B}_n$ is isomorphic to the quotient $C^*(\mathscr{S}_n)/\ker(\pi_n)$. In order to understand the structure of $C^*(\mathscr{S}_n)$ and its quotient algebras, we make a short digression about the structure of general finite-dimensional C^*-algebras.

Let $\mathscr{A}$ denote a finite-dimensional C^*-algebra.

2.8. LEMMA. (i) *Every element x of $\mathscr{A}$ has finite spectrum.*

(ii) *$\mathscr{A}$ has a unit.*

(iii) *$\mathscr{A}$ is spanned linearly by selfadjoint projections.*

PROOF. (i) Fix $x \in \mathscr{A}$ and note that the elements of the sequence $x, x^2, x^3, \ldots$, cannot be linearly independent. Hence there is a polynomial p, with constant term zero, such that $p(x) = 0$. By the spectral mapping theorem $\sigma(x)$ is contained in the finite set $\{z \in \mathbf{C} : p(z) = 0\}$.

(ii) We may assume that $\mathscr{A}$ is a sub C^*-algebra of $\mathscr{L}(\mathscr{H})$ for some Hilbert space $\mathscr{H}$. Let $\{x_1, x_2, \ldots, x_k\}$ be a basis of $\mathscr{A}$ consisting of positive elements. The spectrum of $x = x_1 + x_2 + \cdots + x_n$ is finite by (i), and hence the sequence $\{x^{1/n} : n \geq 0\}$ converges in norm to an element $e \in \mathscr{A}$. Clearly e is the orthogonal projection onto the range of x, and since the range of each x_i is contained in the range of x, we have $ex_i = x_i = x_i e$, $1 \leq i \leq k$. We conclude that e is the desired unit.

(iii) Let x be a positive element, and let $\sigma(x) = \{\alpha_1, \alpha_2, \dots, \alpha_r\}$. We can then write $x = \sum_{j=1}^{r} \alpha_j p_j$, where $p_j = (1/2\pi i) \int_{C_j} (\lambda e - x)^{-1} \, d\lambda$, and C_j is a small circle centered at α_j.

The lemma is proved.

From now on we will denote by 1 the unit in $\mathscr{A}$. A(selfadjoint) projection $p \in \mathscr{A}$ is said to be *minimal* if $p \neq 0$ and the only projections q such that $0 \leq q \leq p$ are 0 and p. A projection p which is minimal in the center of $\mathscr{A}$ will be called a *minimal central projection.*

2.9. LEMMA. (i) *Every selfadjoint projection p in $\mathscr{A}$ is a (finite) sum of minimal projections.*

(ii) *There are only finitely many distinct minimal central projections.*

(iii) *For every minimal projection q in $\mathscr{A}$ there is a unique minimal central projection p such that $q \leq p$.*

PROOF. (i) Note that every sequence $p_1 \geq p_2 \geq \cdots$ of projections in $\mathscr{A}$ must be stationary. Indeed, the nonzero elements in the sequence $\{p_{j+1} - p_j : j \geq 1\}$ are linearly independent. It follows at once that for every projection $p \neq 0$ there is a minimal projection $p_1 \leq p$. Analogously, if $p \neq p_1$, we can construct a minimal $p_2 \leq p - p_1$. Inductively we find $p_1, p_2, \dots$ such that $p_1 + p_2 + \cdots + p_n \leq p$ and $p_{n+1} \neq 0$ if $p \neq p_1 + p_2 + \cdots + p_n$. Again from the finite-dimensionality of $\mathscr{A}$ we deduce that $p_1 + p_2 + \cdots + p_n = p$ for some n.

(ii) By (i) we can write $1 = p_1 + p_2 + \cdots + p_k$, where $p_1, p_2, \dots, p_k$ are minimal central projections. If p is another minimal central projection we must have $p = \sum_{j=1}^{k} p_j p$, and $p_j p$ are pairwise orthogonal central projections. From the minimality of p we deduce that there is j_0 such that $p_{j_0} p = p$ and $p_j p = 0$ for $j \neq j_0$. But then $p_{j_0} p \leq p_{j_0}$ so that $p_{j_0} p = p_{j_0}$ by the minimality of p_{j_0}. Thus $p = p_{j_0}$, and $p_1, p_2, \dots, p_k$ are the only minimal central projections.

(iii) Let $p_1, p_2, \dots, p_k$ be the minimal central projections of $\mathscr{A}$, and let a be a minimal projection. As in the proof of (ii) we deduce that $q = q p_{j_0}$ for a unique j_0, as desired. The lemma is proved.

If q is a minimal projection in $\mathscr{A}$ and $p \geq q$ is a minimal central projection, then p is called the *central support* of q. Two projections p_1, p_2 in $\mathscr{A}$ are *equivalent* if there is $u \in \mathscr{A}$ such that $u^* u = p_1$ and $u u^* = p_2$.

2.10. LEMMA. (i) *Let $\mathscr{J}$ be a selfadjoint ideal in $\mathscr{A}$. Then there exists a central projection e such that $\mathscr{J} = e\mathscr{A}$.*

(ii) *Let p be a minimal central projection in $\mathscr{A}$, and let $\mathscr{J}$ be a selfadjoint ideal such that $p \notin \mathscr{J}$. Then $p\mathscr{A} \cap \mathscr{J} = \{0\}$.*

(iii) *Two minimal projections $q_1, q_2 \in \mathscr{A}$ are equivalent if and only if they have the same central support.*

PROOF. (i) Note that $\mathscr{J}$ is also a C^*-algebra, and hence it has a unit $e \in \mathscr{J}$. Thus $e^2 = e = e^*$ and $x = ex = xe$ for x in $\mathscr{J}$. For $y \in \mathscr{A}$ we have $ey, ye \in \mathscr{J}$, and hence $eye = ey$ and $eye = ye$. Thus e is central, and it is immediate that $\mathscr{J} = e\mathscr{A}$.

(ii) By (i) we have $\mathscr{J} = e\mathscr{A}$ for some central projection e. Since $p \notin \mathscr{J}$ we must have $ep = 0$, and hence $p\mathscr{A} \cap \mathscr{J} = p\mathscr{A} \cap e\mathscr{A} = pe\mathscr{A} = \{0\}$.

(iii) Let $\mathscr{J}_1$ and $\mathscr{J}_2$ be the selfadjoint ideals of $\mathscr{A}$ generated by q_1 and q_2, respectively. Clearly we have $\mathscr{J}_j = p_j\mathscr{A}$, where p_j is the central support of q_j, $j = 1, 2$. If q_1 and q_2 are equivalent, say $u^*u = q_1$, $uu^* = q_2$, then $q_2 = uu^* = uq_1u^* \in \mathscr{J}_1$ and hence $p_2q_2 = q_2$. We deduce that $p_1 = p_2$, as desired. Conversely, suppose that $p_1 = p_2$ and hence $\mathscr{J}_1 = \mathscr{J}_2$. Then $\mathscr{J}_2\mathscr{J}_1 \neq \{0\}$, and hence we must have $q_2xq_1 \neq 0$ for some $x \in \mathscr{A}$. It suffices then to choose u to be the partial isometry in the polar decomposition $q_2xq_1 = pu$, with $p = (q_1x^*q_2xq_1)^{1/2}$. Indeed, it is an easy exercise to see that this polar decomposition exists in $\mathscr{A}$. Moreover, $u^*u \leq q_1$ and $uu^* \leq q_2$, and hence $u^*u = q_1$, $uu^* = q_2$ by the minimality of q_1 and q_2.

The lemma is proved.

2.11. COROLLARY. *Let $\mathscr{B}$ be a C^*-algebra, and $\pi : \mathscr{A} \to \mathscr{B}$ a *-homomorphism. If $p \in \mathscr{A}$ is a minimal central projection such that $\pi(p) \neq 0$, then π is one-to-one on $p\mathscr{A}$.*

PROOF. Set $\mathscr{J} = \ker(\pi)$ and note that $p\mathscr{A} \cap \mathscr{J} = \{0\}$ by Lemma 2.10(ii).

Returning now to the algebra $C^*(\mathscr{S}_n)$ and the representation $\pi_n : C^*(\mathscr{S}_n) \to \mathscr{L}(\bigotimes^n \mathscr{F})$, we see that a subspace $\mathscr{M} \subset \bigotimes^n \mathscr{F}$ is reducing for $\Gamma_n(\mathscr{L}(\mathscr{F}))$ if and only if $\mathscr{M} = \pi_n(p)(\bigotimes^n \mathscr{F})$, with p a selfadjoint projection in $C^*(\mathscr{S}_n)$. If, in particular, we write $1 = q_1 + q_2 + \cdots + q_k$, with q_i minimal projections in $C^*(\mathscr{S}_n)$, $\bigotimes^n \mathscr{F}$ can be written as a direct sum

$$\bigotimes{}^n\mathscr{F} = \bigoplus_{i=1}^k \left(\pi_n(q_i)\left(\bigotimes{}^n\mathscr{F}\right)\right),$$

such that each representation $\gamma_i(T) = \Gamma_n(T) \mid \pi_n(q_i)(\bigotimes^n \mathscr{F})$ is irreducible. It follows from Lemma 2.10(iii) that two such representations γ_i and γ_j are unitarily equivalent if and only if q_i and q_j have the same central support. For the sake of uniqueness, we choose once and forever a minimal projection $q \leq p$ for each minimal central projection $p \in C^*(\mathscr{S}_n)$, and set

$$\begin{aligned} \psi_p(T) &= \Gamma_n(T) \mid \pi_n(q)\left(\bigotimes{}^n\mathscr{F}\right), \\ \phi_p(T) &= \Gamma_n(T) \mid \pi_n(p)\left(\bigotimes{}^n\mathscr{F}\right). \end{aligned} \tag{2.12}$$

It follows then that π_n is unitarily equivalent to $\phi_{p_1} \oplus \phi_{p_2} \oplus \cdots \oplus \phi_{p_m}$, where $p_1, p_2, \ldots, p_m$ are the distinct minimal central projections in $C^*(\mathscr{S}_n)$. In addition, each ϕ_p is a multiple (i.e., a finite sum of copies) of the irreducible representation ψ_p.

The structure of central projections in $C^*(\mathscr{S}_n)$ is well understood, but the general result involving Young symmetrizers will not be used here. We will, however, use the alternating projection $a_n \in C^*(\mathscr{S}_n)$ defined by

$$a_n = \frac{1}{n!}\sum_{\sigma \in \mathscr{S}_n} \varepsilon(\sigma)\sigma, \tag{2.13}$$

where $\varepsilon(\sigma)$ is the sign of σ, i.e., $\varepsilon(\sigma) = +1$ or -1 according to whether σ is an even or odd permutation.

2.14. LEMMA. *a_n is a minimal projection in $C^*(\mathscr{S}_n)$ which is also central.*

PROOF. It is easy to see that a_n is a central projection. We show that a_n is minimal. Let $p \in C^*(\mathscr{S}_n)$ be any projection such that $p \leq a_n$, so that $pa_n = p$. If $p = \sum_{\alpha\in\mathscr{S}_n} c_\alpha \alpha$, then the equation $pa_n = p$ is equivalent to

$$\begin{aligned} c_\beta = \frac{1}{n!}\sum_{\alpha\sigma=\beta} c_\alpha\varepsilon(\sigma) &= \frac{1}{n!}\sum_\alpha c_\alpha\varepsilon(\beta\alpha^{-1}) \\ &= \frac{1}{n!}\sum_\alpha \varepsilon(\beta)\varepsilon(\alpha)c_\alpha = \varepsilon(\beta)\frac{1}{n!}\sum_\alpha \varepsilon(\alpha)c_\alpha, \end{aligned}$$

and hence $p = ka_n$, where $k = (1/n!)\sum_\alpha \varepsilon(\alpha)c_\alpha$. Since $p^2 = p$, we have $k = 0$ or $k = 1$, and this concludes the proof.

2.15. LEMMA. *The representation π_n of $C^*(\mathscr{S}_n)$ is one-to-one if and only if $n \leq \dim(\mathscr{F})$.*

PROOF. If $n \leq \dim(\mathscr{F})$ and $e_1, e_2, \dots, e_n$ are orthonormal in $\mathscr{F}$, then the map $x \to \pi_n(x)(e_1 \otimes e_2 \otimes \cdots \otimes e_n)$ is one-to-one on $C^*(\mathscr{S}_n)$. Conversely, if $\dim(\mathscr{F}) < n$, it is immediate that $\pi_n(a_n) = 0$.

The previous lemma implies that the representation ϕ_p acts on a nontrivial space if $p \in C^*(\mathscr{S}_n)$ and $n \leq \dim(\mathscr{F})$. It is important to relate the representations ϕ_p and ϕ'_p with $p \in C^*(\mathscr{S}_{n+1})$ and $p' \in C^*(\mathscr{S}_n)$. In order to do this it will be convenient to regard $\mathscr{S}_n$ as a subgroup of $\mathscr{S}_{n+1}$; for example, one may regard $\mathscr{S}_n$ as the group of all permutations π of $\mathbf{Z}_+ = \{1, 2, 3, \dots, \}$ such that $\pi(k) = k$ for $k \geq n+1$. We then have $C^*(\mathscr{S}_1) \subset C^*(\mathscr{S}_2) \subset \cdots$, and expressions of the form $\pi_{n+1}(x)$ will be defined for x in $C^*(\mathscr{S}_n)$.

For every $k \in \mathscr{F}$ we define an operator $T_k : \bigotimes^n \mathscr{F} \to \bigotimes^{n+1} \mathscr{F}$ by

$$T_k(h_1 \otimes h_2 \otimes \cdots \otimes h_n) = h_1 \otimes h_2 \otimes \cdots \otimes h_n \otimes k, \qquad h_i \in \mathscr{F},\ 1 \leq i \leq n. \tag{2.16}$$

Clearly T_k is an isometry if $\|k\| = 1$, and in general

$$T_k^*(h_1 \otimes h_2 \otimes \cdots \otimes h_{n+1}) = (h_{n+1}, k)h_1 \otimes h_2 \otimes \cdots \otimes h_n. \tag{2.17}$$

The following formulas are easy to verify:

$$\pi_{n+1}(\sigma)T_k = T_k\pi_n(\sigma), \quad \sigma \in \mathscr{S}_n,\ k \in \mathscr{F}, \tag{2.18}$$

and

$$\Gamma_{n+1}(A)T_k = T_{Ak}\Gamma_n(A), \quad k \in \mathscr{F},\ A \in \mathscr{L}(\mathscr{F}). \tag{2.19}$$

2.20. LEMMA. $\bigcap_{k\in\mathscr{F}} \ker(T_k^*) = \{0\}$.

PROOF. This follows from the obvious dual equality $\bigvee_{k\in\mathscr{F}} T_k(\bigotimes^n \mathscr{F}) = \bigotimes^{n+1} \mathscr{F}$.

2.21. LEMMA. *Let $p \in C^*(\mathscr{S}_{n+1})$ be a minimal central projection such that $\pi_{n+1}(p) \neq 0$. Then we have*

$$\bigvee_{k\in\mathscr{F}} T_k^*\pi_{n+1}(p)\left(\bigotimes^{n+1}\mathscr{F}\right) = \pi_n(q)\left(\bigotimes^n\mathscr{F}\right),$$

where $q = \sum p'$, and the sum is extended over those minimal central projections $p' \in C^*(\mathscr{S}_n)$ for which $p'p \neq 0$.

PROOF. We denote $\mathscr{M} = \bigvee_{k\in\mathscr{F}} T_k^*\pi_{n+1}(p)(\bigotimes^{n+1}\mathscr{F})$. The relation

$$\begin{aligned}\pi_n(\sigma)T_k^*\pi_{n+1}(p) &= \pi_n(\sigma^{-1})^*T_k^*\pi_{n+1}(p)\\ &= T_k^*\pi_{n+1}(\sigma^{-1})^*\pi_{n+1}(p)\\ &= T_k^*\pi_{n+1}(p)\pi_{n+1}(\sigma), \quad \sigma\in\mathscr{S}_n,\end{aligned}$$

follows from (2.18) and the centrality of p. Thus $\mathscr{M}$ is invariant under $\pi_n(C^*(\mathscr{S}_n))$. Furthermore, if $U \in \mathscr{U}(\mathscr{F})$, (2.19) implies that

$$\begin{aligned}\Gamma_n(U)T_k^*\pi_{n+1}(p) &= \Gamma_n(U^*)^*T_k^*\pi_{n+1}(p)\\ &= T_{Uk}^*\Gamma_{n+1}(U)\pi_{n+1}(p)\\ &= T_{Uk}^*\pi_{n+1}(p)\Gamma_{n+1}(U),\end{aligned}$$

and this shows that $\mathscr{M}$ is also invariant under $\Gamma_n(\mathscr{U}(\mathscr{F}))$. We conclude from Theorem 2.7 that $\mathscr{M} = Q(\bigotimes^n\mathscr{F})$, where $Q \in \pi_n(C^*(\mathscr{S}_n))$ is some central projection. It is now seen from Lemma 2.10 (with $\mathscr{J} = \ker(\pi_n)$) that $Q = \pi_n(q)$ with some central projection $q \in C^*(\mathscr{S}_n)$. In order to conclude the proof we must show that a minimal central projection $p' \in C^*(\mathscr{S}_n)$ with $\pi_n(p') \neq 0$ satisfies the inequality $p' \leq q$ if and only if $p'p \neq 0$. Fix therefore a minimal central projection $p' \in C^*(\mathscr{S}_n)$ such that $\pi_n(p') \neq 0$. We have

$$\pi_n(p')T_k^*\pi_{n+1}(p) = T_k^*\pi_{n+1}(p')\pi_{n+1}(p) = T_k^*\pi_{n+1}(p'p), \quad k\in\mathscr{F},$$

and hence $p' \leq q$ if and only if $\pi_{n+1}(p'p) \neq 0$. In turn, by Corollary 2.11, $\pi_{n+1}(p'p) \neq 0$ if and only if $p'p \neq 0$. This concludes the proof of the lemma.

We will use the following immediate consequence of Lemma 2.21.

2.22. COROLLARY. *Consider two minimal central projections $p' \in C^*(\mathscr{S}_n)$ and $p \in C^*(\mathscr{S}_{n+1})$ such that $\pi_{n+1}(p) \neq 0$ and $p'p \neq 0$. We have*

$$\bigvee_{k\in\mathscr{F}} \pi_n(p')T_k^*\pi_{n+1}(p)\left(\bigotimes^{n+1}\mathscr{F}\right) = \pi_n(p')\left(\bigotimes^n\mathscr{F}\right).$$

Exercises

1. Prove the equality $\|T_1 \otimes T_2 \otimes \cdots \otimes T_n\| = \|T_1\|\,\|T_2\|\cdots\|T_n\|$, where $T_j \in \mathscr{L}(\mathscr{F}_j)$, $1 \leq j \leq n$.

2. Verify that Γ_n is indeed a *-homomorphism, and $\Gamma_n \mid \{T : \|T\| \leq 1\}$ is continuous with respect to the weak and strong topologies.

3. Verify that π_n is a unitary representation of $\mathscr{S}_n$.

4. Prove that $\Gamma_n(T)$ commutes with $\pi_n(\sigma)$ if $T \in \mathscr{L}(\mathscr{F})$ and $\sigma \in \mathscr{S}_n$.

5. Verify relation (2.5). That is, prove that (2.5) holds if and only if C commutes with $\pi_n(\sigma)$, $\sigma \in \mathscr{S}_n$.

6. Let G be a finite group, and let $L^2(G)$ denote the Hilbert space of all functions $f : G \to \mathbf{C}$ with the norm $\|f\|^2 = \sum_{g\in G} |f(g)|^2$. For $g \in G$ define the operator R_g on $L^2(G)$ by $R_g f(h) = f(hg)$, $h \in G$. Prove that $g \mapsto R_g$ is a unitary representation of G such that the operators R_g, $g \in G$, are linearly independent. Deduce the existence of the C^*-algebra $C^*(G)$.

7. Determine the commutant $\{R_g : g \in G\}'$.

8. Let $\mathscr{A}$ be a finite-dimensional subalgebra of $\mathscr{L}(\mathscr{H})$, and let $T \in \mathscr{A}$. Show that the partial isometry U in the polar decomposition $T = U(T^*T)^{1/2}$ belongs to $\mathscr{A}$.

9. Verify that a_n (defined by (2.13)) is a selfadjoint projection.

10. Prove formula (2.17).

11. Show that $s_n = (1/n!) \sum_{\sigma\in\mathscr{S}_n} \sigma$ is a minimal projection which is also central.

12. Let p and q be two minimal projections in the C^*-algebra $\mathscr{A}$. Show that the linear space $\{x : px = xq = x\}$ is at most one-dimensional.

13. Assume that the finite-dimensional C^*-algebra $\mathscr{A}$ has only one minimal central projection. Write $1 = p_1 + p_2 + \cdots + p_n$, where p_i are minimal projections. Prove that $\mathscr{A}$ is isomorphic with the C^*-algebra $\mathscr{L}(\mathbf{C}^n)$.

14. Calculate all the minimal central projections of $C^*(\mathscr{S}_3)$. How many minimal projections are there in $C^*(\mathscr{S}_3)$?

3. Scalar multiples. We finally come to the main subject of this chapter, which is the study of operators of class C_0 using their characteristic functions. Let $\mathscr{F}$ and $\mathscr{F}'$ be two separable Hilbert spaces.

3.1. DEFINITION. The function $\Theta \in H^\infty(\mathscr{L}(\mathscr{F}, \mathscr{F}'))$ is said to have a *scalar multiple* $u \in H^\infty$, $u \neq 0$, if there exists $\Omega \in H^\infty(\mathscr{L}(\mathscr{F}', \mathscr{F}))$ satisfying the relation $\Theta(\lambda)\Omega(\lambda) = u(\lambda)I_{\mathscr{F}'}$ for $\lambda \in \mathbf{D}$.

3.2. PROPOSITION. *Suppose that $\Theta \in H^\infty(\mathscr{L}(\mathscr{F}, \mathscr{F}'))$ is an inner function, and $u \in H^\infty$. Then the following assertions are equivalent:*

(i) *u is a scalar multiple of Θ;*

(ii) *$uH^2(\mathscr{F}') \subset \Theta H^2(\mathscr{F})$; and*

(iii) *$u(S(\Theta)) = 0$.*

PROOF. Let $\Omega \in H^\infty(\mathscr{L}(\mathscr{F}', \mathscr{F}))$ satisfy the relation $\Theta(\lambda)\Omega(\lambda) = u(\lambda)I_{\mathscr{F}'}$, $\lambda \in \mathbf{D}$. For every $f \in H^2(\mathscr{F}')$ we have $uf = T_\Theta(T_\Omega f) \in \Theta H^2(\mathscr{F})$. Thus we see that (i) $\Rightarrow$ (ii). Conversely, if (ii) holds, we can construct an operator

$X : H^2(\mathscr{F}') \to H^2(\mathscr{F})$ such that $uf = T_\Theta Xf$ for every $f \in H^2(\mathscr{F}')$. Since Θ is inner, we have

$$\|Xf\| = \|uf\| \le \|u\|_\infty \|f\|,$$

so that X is bounded. It is routine to verify that $X \in \mathscr{I}(S_{\mathscr{F}'}, S_{\mathscr{F}})$ so that $X = T_\Omega$ for some $\Omega \in H^\infty(\mathscr{L}(\mathscr{F}', \mathscr{F}))$. A quick look at the identity $T_\Theta T_\Omega f = uf$ applied to constant functions f in $H^2(\mathscr{F}')$ shows that $\Theta(\lambda)\Omega(\lambda) = u(\lambda)I_{\mathscr{F}'}$. We conclude that (ii) implies (i).

Now, if (ii) holds, we have

$$\begin{aligned} u(S(\Theta))\mathscr{H}(\Theta) &= P_{\mathscr{H}(\Theta)} u(S_{\mathscr{F}'})\mathscr{H}(\Theta) \subset P_{\mathscr{H}(\Theta)} uH^2(\mathscr{F}') \\ &\subset P_{\mathscr{H}(\Theta)}\Theta H^2(\mathscr{F}) = \{0\}, \end{aligned}$$

and consequently $u(S(\Theta)) = 0$. Vice versa, if $u(S(\Theta)) = 0$, a similar calculation shows that $u\mathscr{H}(\Theta) \subset \Theta H^2(\mathscr{F})$ so that

$$uH^2(\mathscr{F}') = u\mathscr{H}(\Theta) + u\Theta H^2(\mathscr{F}) \subset \Theta H^2(\mathscr{F}).$$

This shows the equivalence of (ii) and (iii), thus concluding the proof.

We have the following immediate consequence of Proposition 3.2.

3.3. COROLLARY. *Let T be a contraction of class $C_{\cdot 0}$. Then T is of class C_0 if and only if the characteristic function Θ_T has a scalar multiple $u \in H^\infty \setminus \{0\}$.*

We also record the following fact for further reference.

3.4. PROPOSITION. *Suppose that $\Theta \in H^\infty(\mathscr{L}(\mathscr{F}, \mathscr{F}'))$ is inner, $u \in H^\infty$, and $\Omega \in H^\infty(\mathscr{L}(\mathscr{F}', \mathscr{F}))$ satisfies the relation $\Theta(\lambda)\Omega(\lambda) = u(\lambda)I_{\mathscr{F}'}$, $\lambda \in \mathbf{D}$. Then we also have $\Omega(\lambda)\Theta(\lambda) = u(\lambda)I_{\mathscr{F}}$, $\lambda \in \mathbf{D}$.*

PROOF. We deduce from identities of the form $T_\Theta T_\Omega = T_{\Theta\Omega}$ that

$$\Theta(\varsigma)(\Omega(\varsigma)\Theta(\varsigma) - u(\varsigma)I_{\mathscr{F}}) = (\Theta(\varsigma)\Omega(\varsigma) - u(\varsigma)I_{\mathscr{F}'})\Theta(\varsigma) = 0$$

for almost every $\varsigma \in \mathbf{T}$. But $\Theta(\varsigma)$ is an isometry almost everywhere, hence we deduce $\Omega(\varsigma)\Theta(\varsigma) = u(\varsigma)I_{\mathscr{F}}$ for almost every $\varsigma \in \mathbf{T}$. This equality implies that $T_\Omega T_\Theta = T_{uI_{\mathscr{F}}}$ and hence the equality $\Omega(\lambda)\Theta(\lambda) = u(\lambda)I_{\mathscr{F}}$, $\lambda \in \mathbf{D}$, can be proved by looking at the constant functions in $H^2(\mathscr{F})$.

Corollary 3.3 shows that the minimal function m_T of an operator $T = S(\Theta)$ of class C_0 coincides with the least scalar multiple of Θ. It is natural to ask whether the other inner functions in the Jordan model of T can be computed in a similar way. Natural ideas from linear algebra lead us to the consideration of tensor operations—these replace the determinants and minors which are not available in the infinite-dimensional case. More precisely, let $\Phi \in H^\infty(\mathscr{L}(\mathscr{F}))$, and define a new function $\Gamma_n(\Phi) \in H^\infty(\mathscr{L}(\bigotimes^n \mathscr{F}))$ by

$$[\Gamma_n(\Phi)](\lambda) = \Gamma_n(\Phi(\lambda)), \quad \lambda \in \mathbf{D}. \tag{3.5}$$

The analyticity of $\Gamma_n(\Phi)$ follows from the fact that Γ_n is a homogeneous polynomial, and the estimate $\|\Gamma_n(\Phi)\|_\infty \le \|\Phi\|_\infty^n$ is obvious. Moreover, if

$p \in C^*(\mathscr{S}_n)$ is a minimal central projection, we can define functions $\psi_p(\Phi) \in H^\infty(\pi_n(q)(\bigotimes^n \mathscr{F}))$ and $\phi_p(\Phi) \in H^\infty(\pi_n(p)(\bigotimes^n \mathscr{F}))$ by

$$(3.6) \qquad [\psi_p(\Phi)](\lambda) = \psi_p(\Phi(\lambda)), \quad [\phi_p(\Phi)](\lambda) = \phi_p(\Phi(\lambda)), \quad \lambda \in \mathbf{D}.$$

(We recall that $q \leq p$ is a fixed minimal projection in $C^*(\mathscr{S}_n)$.)

3.7. LEMMA. *For every $\Phi \in H^\infty(\mathscr{L}(\mathscr{F}))$ we have $\Gamma_n(\Phi)(\varsigma) = \Gamma_n(\Phi(\varsigma))$ for almost every $\varsigma \in \mathbf{T}$.*

PROOF. Since $\mathscr{F}$ is separable, it suffices to show that

$$(\Gamma_n(\Phi)(\varsigma)a, b) = (\Gamma_n(\Phi(\varsigma))a, b)$$

almost everywhere for a and b in a countable generating subset of $\bigotimes^n \mathscr{F}$. Consider vectors of the form $a = a_1 \otimes a_2 \otimes \cdots \otimes a_n$ and $b = b_1 \otimes b_2 \otimes \cdots \otimes b_n$, $a_i, b_i \in \mathscr{F}$, $1 \leq i \leq n$. We have

$$(\Gamma_n(\Phi)(\lambda)a, b) = (\Phi(\lambda)a_1, b_1)(\Phi(\lambda)a_2, b_2) \cdots (\Phi(\lambda)a_n, b_n), \qquad \lambda \in \mathbf{D},$$

and, since the n functions on the right-hand side belong to H^∞,

$$\begin{aligned}(\Gamma_n(\Phi)(\varsigma)a, b) &= (\Phi(\varsigma), a_1, b_1)(\Phi(\varsigma)a_2, b_2) \cdots (\Phi(\varsigma)a_n, b_n) \\ &= (\Gamma_n(\Phi(\varsigma))a, b)\end{aligned}$$

for almost every $\varsigma \in \mathbf{T}$. The lemma obviously follows.

3.8. COROLLARY. *If $\Theta \in H^\infty(\mathscr{L}(\mathscr{F}))$ is an inner function and $p \in C^*(\mathscr{S}_n)$ is a minimal central projection, then the functions $\Gamma_n(\Theta)$, $\phi_p(\Theta)$, and $\psi_p(\Theta)$ are also inner.*

PROOF. The boundary values $\Gamma_n(\Theta)(\varsigma)$ are isometric almost everywhere by Lemmas 3.7 and 1.8. Thus a second application of Lemma 1.8 shows that $\Gamma_n(\Theta)$ is inner. The assertion about ϕ_p and ψ_p is now obvious since $\phi_p(\Theta)(\varsigma)$ and $\psi_p(\Theta)(\varsigma)$ are almost everywhere appropriate restrictions of $\Gamma_n(\Theta)(\varsigma)$.

3.9. PROPOSITION. *Assume that $\Theta \in H^\infty(\mathscr{L}(\mathscr{F}))$ is an inner function. If $S(\Theta)$ is of class C_0, and $p \in C^*(\mathscr{S}_n)$ is a minimal central projection with $\pi_n(p) \neq 0$, then $S(\psi_p(\Theta))$ is also of class C_0.*

PROOF. Assume that the function $\Omega \in H^\infty(\mathscr{L}(\mathscr{F}))$ satisfies the relation $\Theta(\lambda)\Omega(\lambda) = u(\lambda)I_{\mathscr{F}}$, $\lambda \in \mathbf{D}$, for some nonzero u in H^∞. Then we have

$$\Gamma_n(\Theta(\lambda))\Gamma_n(\Omega(\lambda)) = \Gamma_n(u(\lambda)I_{\mathscr{F}}) = u(\lambda)^n I_{\bigotimes^n \mathscr{F}}, \quad \lambda \in \mathbf{D},\ n \geq 1,$$

and this shows that $\Gamma_n(\Theta)$ has the nonzero scalar multiple u^n. Therefore $S(\Gamma_n(\Theta))$ is of class C_0 for each $n \geq 1$, and the proposition follows from the fact that the representation ψ_p is a direct summand of Γ_n.

3.10. DEFINITION. If $\Theta \in H^\infty(\mathscr{L}(\mathscr{F}))$ is such that $S(\Theta)$ is of class C_0, and $p \in C^*(\mathscr{S}_n)$ is a minimal central projection with $\pi_n(p) \neq 0$, the inner function $d^p(\Theta)$ is defined to be the minimal function of $S(\psi_p(\Theta))$. If n is a natural number, $n \leq \dim(\mathscr{F})$, we set $d_n(\Theta) = d^{a_n}(\Theta)$, where a_n is defined in (2.13).

Let us note that $d^p(\Theta)$ can be defined equivalently as the minimal function of $S(\phi_p(\Theta))$ since ϕ_p is unitarily equivalent with the orthogonal sum of finitely many copies of ψ_p.

Before proving a converse to Proposition 3.9, we need to relate the operators $S(\psi_p(\Theta))$ for various projections p. This is done in the following three results. Let us note that if $\mathscr{F}$ and $\mathscr{F}'$ are separable Hilbert spaces and $A \in \mathscr{L}(\mathscr{F}, \mathscr{F}')$, then we can construct the operator $T_A \in \mathscr{L}(H^2(\mathscr{F}), H^2(\mathscr{F}'))$, where A is regarded as a constant function in $H^\infty(\mathscr{L}(\mathscr{F}, \mathscr{F}'))$. It is convenient to denote T_A by $A \otimes I_{H^2}$. This notation is justified by the fact that $H^2(\mathscr{F})$ is naturally isomorphic with the tensor product $\mathscr{F} \otimes H^2$ via the mapping $h \otimes u \to uh$, $h \in \mathscr{F}$, $u \in H^2$.

3.11. LEMMA. *Let $\Theta \in H^\infty(\mathscr{L}(\mathscr{F}))$, and consider minimal central projections $p' \in C^*(\mathscr{S}_n)$ and $p \in C^*(\mathscr{S}_{n+1})$ with $p'p \neq 0$ and $\pi_{n+1}(p) \neq 0$. For every $k \in \mathscr{F}$ we have*

$$(\pi_n(p')T_k^*\pi_{n+1}(p) \otimes I_{H^2})\Gamma_{n+1}(\Theta)H^2\left(\bigotimes^{n+1}\mathscr{F}\right) \subset \left[(\pi_n(p') \otimes I_{H^2})\Gamma_n(\Theta)H^2\left(\bigotimes^n\mathscr{F}\right)\right]^-.$$

PROOF. The lemma follows from the stronger inclusion

$$((\pi_n(p')T_k^*) \otimes I_{H^2})\Gamma_{n+1}(\Theta)H^2\left(\bigotimes^{n+1}\mathscr{F}\right) \subset \left[(\pi_n(p') \otimes I_{H^2})\Gamma_n(\Theta)H^2\left(\bigotimes^n\mathscr{F}\right)\right]^-. \tag{3.12}$$

In order to prove (3.12) we note that the right-hand side is an invariant subspace for the shift $S_{\otimes^n\mathscr{F}}$ and hence it will suffice to show that elements of the form

$$((\pi_n(p')T_k^*) \otimes I_{H^2})\Gamma_{n+1}(\Theta)h$$

belong to that subspace if $h = h_1 \otimes h_2 \otimes \cdots \otimes h_{n+1}$ is a constant function, i.e., $h_j \in \mathscr{F}$, $1 \leq j \leq n+1$. This last assertion becomes clear from the following computation:

$$\begin{aligned}\pi_n(p')T_k^*(\Theta(\lambda)h_1 \otimes \Theta(\lambda)h_2 \otimes \cdots \otimes \Theta(\lambda)h_{n+1}) \\ = \pi_n(p')(\Theta(\lambda)h_{n+1}, k)\Gamma_n(\Theta(\lambda))(h_1 \otimes h_2 \otimes \cdots \otimes h_n) \\ = \pi_n(p')\Gamma_n(\Theta(\lambda))[(\Theta(\lambda)h_{n+1}, k)(h_1 \otimes h_2 \otimes \cdots \otimes h_n)].\end{aligned}$$

The lemma is proved.

Observe that we have

$$(\pi_{n+1}(p) \otimes I_{H^2})H^2\left(\bigotimes^{n+1}\mathscr{F}\right) = H^2\left(\pi_{n+1}(p)\left(\bigotimes^{n+1}\mathscr{F}\right)\right)$$

and, since $\pi_{n+1}(p)$ commutes with $\Gamma_{n+1}(\Theta)$, the inclusion in Lemma 3.11 can be rewritten as follows:

$$((\pi_n(p')T_k^*) \otimes I_{H^2})\phi_p(\Theta)H^2\left(\pi_{n+1}(p)\left(\bigotimes^{n+1}\mathscr{F}\right)\right) \subset \phi_{p'}(\Theta)H^2\left(\pi_n(p')\left(\bigotimes^n\mathscr{F}\right)\right)^-. \tag{3.13}$$

3.14. COROLLARY. *Under the conditions of Lemma* 3.11, *assume that* Θ *is an inner function. Then the span of all spaces of the form* $X\mathcal{H}(\phi_p(\Theta))$, *with* $X \in \mathcal{I}(S(\phi_p(\Theta)), S(\phi_{p'}(\Theta)))$, *equals* $\mathcal{H}(\phi_{p'}(\Theta))$.

PROOF. For each $k \in \mathcal{F}$ we define an operator X_k by

$$X_k = P_{\mathcal{H}(\phi_{p'}(\Theta))}((\pi_n(p')T_k^*) \otimes I_{H^2}) \mid \mathcal{H}(\phi_p(\Theta)).$$

By (3.13) X_k intertwines $S(\phi_{p'}(\Theta))$ and $S(\phi_p(\Theta))$, and hence the corollary will follow if we prove the relation

$$\bigvee_{k \in \mathcal{F}} X_k \mathcal{H}(\phi_p(\Theta)) = \mathcal{H}(\phi_{p'}(\Theta)).$$

This last relation follows from the equality

$$\begin{aligned} \bigvee_{k \in \mathcal{F}} ((\pi_n(p')T_k^*) \otimes I_{H^2}) H^2 \left(\pi_{n+1}(p) \left(\bigotimes^{n+1} \mathcal{F}\right)\right) \\ = H^2 \left(\pi_n(p') \left(\bigotimes^n \mathcal{F}\right)\right). \end{aligned} \tag{3.15}$$

Now, the left-hand side of (3.15) certainly contains all constant functions of the form $\pi_n(p')T_k^* \pi_{n+1}(p)x$ with x in $\bigotimes^{n+1} \mathcal{F}$. By Corollary 2.22, the left-hand side of (3.15) contains all constant functions in $H^2(\pi_n(p')(\bigotimes^n \mathcal{F}))$, and hence (3.15) follows from the invariance of the two spaces under the relevant shift operator $(S_{\pi_n(p')(\otimes^n \mathcal{F})})$. The proof is now complete.

3.16. LEMMA. *Consider an inner function* $\Theta \in H^\infty(\mathcal{L}(\mathcal{F}))$ *and two minimal central projections* $p' \in C^*(\mathcal{S}_n)$, $p \in C^*(\mathcal{S}_{n+1})$, *with* $p'p \neq 0$ *and* $\pi_{n+1}(p) \neq 0$. *If* $S(\psi_p(\Theta))$ *is of class* C_0 *then* $S(\psi_{p'}(\Theta))$ *is also of class* C_0 *and* $m_{S(\psi_{p'}(\Theta))}$ *divides* $m_{S(\psi_p(\Theta))}$.

PROOF. Let u denote the minimal function of $S(\psi_p(\Theta))$, and let $X \in \mathcal{I}(S(\phi_p(\Theta)), S(\phi_{p'}(\Theta)))$. By the observation following Definition 3.10, we have $u(S(\phi_p(\Theta))) = 0$, and hence

$$u(S(\phi_{p'}(\Theta)))X = Xu(S(\phi_p(\Theta))) = 0.$$

Thus $u(S(\phi_{p'}(\Theta)))$ is zero on the range of X, and Corollary 3.14 implies that $u(S(\phi_{p'}(\Theta))) = 0$ or, equivalently, $u(S(\psi_{p'}(\Theta))) = 0$. Thus $S(\psi_{p'}(\Theta))$ is of class C_0 and its minimal function divides u, as desired.

We are now able to prove the converse of Proposition 3.9.

3.17. PROPOSITION. *Consider an inner function* $\Theta \in H^\infty(\mathcal{L}(\mathcal{F}))$ *and a minimal central projection* $p \in C^*(\mathcal{S}_n)$ *such that* $\pi_n(p) \neq 0$. *If* $S(\psi_p(\Theta))$ *is of class* C_0 *then* $S(\Theta)$ *is also of class* C_0. *Moreover, if* $p' \in C^*(\mathcal{S}_m)$ *and* $p \in C^*(\mathcal{S}_n)$, $m < n$, *are minimal central projections such that* $p'p \neq 0$ *and* $\pi_n(p) \neq 0$, *then* $\pi_m(p') \neq 0$ *and* $d^{p'}(\Theta)$ *divides* $d^p(\Theta)$.

PROOF. We prove first that, if p' and p are as in the second part of the statement, and $S(\psi_p(\Theta))$ is of class C_0, then $S(\psi_{p'}(\Theta))$ is of class C_0. To do this, set $p_m = p'$ and $p_n = p$. We can define minimal central projections $p_j \in C^*(\mathcal{S}_j)$

for $m < j < n$ such that $p_i p_{i+1} \neq 0$ for $m \leq i < n$. Indeed, for $j < n-1$, it suffices to choose p_{j+1} such that $pp'p_{m+1}p_{m+2}\cdots p_j p_{j+1} \neq 0$. The fact that $\pi_n(p') \neq 0$ follows from Corollary 2.11 because $\pi_n(p'p) \neq 0$. Since

$$\pi_n(p') = \pi_m(p') \otimes \underbrace{I \otimes I \otimes \cdots \otimes I}_{n-m \text{ times}},$$

we deduce that we also have $\pi_m(p') \neq 0$. Now, $n-m$ applications of Lemma 3.16 show that $S(\psi_{p'}(\Theta))$ is of class C_0 and $m_{S(\psi_{p'}(\Theta))} \mid m_{S(\psi_p(\Theta))}$. Thus we have $d^{p'}(\Theta) \mid d^p(\Theta)$ in case $S(\Theta)$ is of class C_0.

The first assertion of the proposition is now easy to prove. We denote $p' = 1 \in C^*(\mathscr{S}_1)$ and remark that $p'p \neq 0$; moreover, $\Theta = \psi_{p'}(\Theta)$. Thus, by the proof above, $S(\Theta)$ is of class C_0 whenever $S(\psi_p(\Theta))$ is of class C_0. The proposition is proved.

We are about to prove the main result in this section, which consists in showing that the functions $d^p(\Theta)$ depend only on the quasisimilarity class of the operator $S(\Theta)$. The difficulty consists in the fact that the operators $S(\Gamma_n(\Theta))$ and $S(\Gamma_n(\Theta'))$ need not be quasisimilar, even if $S(\Theta)$ and $S(\Theta')$ are unitarily equivalent. An easy example is provided by $\Theta = \theta$ and $\Theta' = \left[\begin{smallmatrix}\theta & 0\\ 0 & 1\end{smallmatrix}\right]$, where θ is an arbitrary inner function in H^∞. We have $\Gamma_2(\Theta) = \theta^2$ while

$$\Gamma_2(\Theta') = \begin{bmatrix} \theta^2 & 0 & 0 & 0 \\ 0 & \theta & 0 & 0 \\ 0 & 0 & \theta & 0 \\ 0 & 0 & 0 & 1 \end{bmatrix}$$

with respect to a convenient basis in $\bigotimes^2 \mathbf{C}^2$. However, $S(\Gamma_2(\Theta))$ and $S(\Gamma_2(\Theta'))$ do have the same minimal function θ^2.

Let $\mathscr{F}_1$ and $\mathscr{F}_2$ be two separable Hilbert spaces, and $f_j \in H^2(\mathscr{F}_j)$, $j = 1, 2$. We will use the somewhat ambiguous notation $f_1 \otimes f_2$ for the function defined by

$$(f_1 \otimes f_2)(\lambda) = f_1(\lambda) \otimes f_2(\lambda), \qquad \lambda \in \mathbf{D}. \tag{3.18}$$

If, in addition, one of the functions f_j is in $H^\infty(\mathscr{F}_j)$, i.e., it is bounded, then $f_1 \otimes f_2$ belongs to $H^2(\mathscr{F}_1 \otimes \mathscr{F}_2)$ and

$$\|f_1 \otimes f_2\|_2 \leq \|f_1\|_\infty \|f_2\|_2 \quad \text{if } f_1 \in H^\infty(\mathscr{F}_1). \tag{3.19}$$

Assume now that $\Theta_j \in H^\infty(\mathscr{L}(\mathscr{F}_j))$ is inner, $f_j \in \Theta_j H^2(\mathscr{F}_j)$, $j = 1, 2$, and one of the functions f_j is bounded. Then we have

$$f_1 \otimes f_2 \in (\Theta_1 \otimes \Theta_2) H^2(\mathscr{F}_1 \otimes \mathscr{F}_2), \tag{3.20}$$

where $(\Theta_1 \otimes \Theta_2)(\lambda) = \Theta_1(\lambda) \otimes \Theta_2(\lambda)$. Indeed, we have $f_j = \Theta_j g_j$, $g_j \in H^2(\mathscr{F}_j)$, and one of the functions g_j is bounded because the functions Θ_j are inner. Thus we have $f_1 \otimes f_2 = (\Theta_1 \otimes \Theta_2)(g_1 \otimes g_2)$ and $g_1 \otimes g_2 \in H^2(\mathscr{F}_1 \otimes \mathscr{F}_2)$, as desired.

In the following result we need to distinguish between the representations $\pi_n : \mathscr{S}_n \to \mathscr{L}(\bigotimes^n \mathscr{F})$ and $\pi_n' : \mathscr{S}_n \to \mathscr{L}(\bigotimes^n \mathscr{F}')$.

3.21. THEOREM. *Suppose that $\Theta \in H^\infty(\mathscr{L}(\mathscr{F}))$ and $\Theta' \in H^\infty(\mathscr{L}(\mathscr{F}'))$ are two inner functions such that $S(\Theta)$ and $S(\Theta')$ are quasisimilar operators of class C_0. Then for every minimal central projection $p \in C^*(\mathscr{S}_n)$ with $\pi_n(p) \neq 0$ and $\pi'_n(p) \neq 0$ we have $d^p(\Theta) \equiv d^p(\Theta')$.*

PROOF. Set $u = d^p(\Theta)$ and note that, in order to prove that $d^p(\Theta') \mid u$, we must prove the inclusion

$$uH^2(\pi'_n(p)(\bigotimes^n \mathscr{F}')) \subset \phi_p(\Theta')H^2(\pi'_n(p)(\bigotimes^n \mathscr{F}'))$$

(cf. Proposition 3.2 above) or, equivalently, the inclusion

$$u(\pi'_n(p) \otimes I_{H^2})H^2(\bigotimes^n \mathscr{F}') \subset \Gamma_n(\Theta')(\pi'_n(p) \otimes I_{H^2})H^2(\bigotimes^n \mathscr{F}'). \tag{3.22}$$

In fact (3.22) is equivalent to the apparently weaker inclusion

$$u(\pi'_n(p) \otimes I_{H^2})H^2(\bigotimes^n \mathscr{F}') \subset \Gamma_n(\Theta')H^2(\bigotimes^n \mathscr{F}') \tag{3.23}$$

because the elements f of the left-hand side already satisfy the relation $(\pi'_n(p) \otimes I_{H^2})f = f$.

Now choose a quasiaffinity $X \in \mathscr{I}(S(\Theta), S(\Theta'))$. By Proposition 1.24 there exists a function $\Phi \in H^\infty(\mathscr{L}(\mathscr{F}, \mathscr{F}'))$ such that $Xf = P_{\mathscr{H}(\Theta')}\Phi f$, $f \in H^2(\mathscr{F})$, and $\Phi\Theta H^2(\mathscr{F}) \subset \Theta' H^2(\mathscr{F}')$. The fact that X has dense range can easily be translated into the relation

$$\Phi H^2(\mathscr{F}) \vee \Theta' H^2(\mathscr{F}') = H^2(\mathscr{F}'). \tag{3.24}$$

Indeed,

$$\begin{aligned}\Phi H^2(\mathscr{F}) \vee \Theta' H^2(\mathscr{F}') &= P_{\mathscr{H}(\Theta')}(\Phi H^2(\mathscr{F})) \vee \Theta' H^2(\mathscr{F}')\\ &= X\mathscr{H}(\Theta) \vee \Theta' H^2(\mathscr{F}')\\ &= \mathscr{H}(\Theta') \oplus \Theta' H^2(\mathscr{F}')\\ &= H^2(\mathscr{F}').\end{aligned}$$

Since $H^\infty(\mathscr{F})$ is dense in $H^2(\mathscr{F})$, it follows that functions of the form $\Phi f + \Theta' g$, $f \in H^\infty(\mathscr{F})$, $g \in H^\infty(\mathscr{F}')$, are dense in $H^2(\mathscr{F}')$. We conclude that $H^2(\bigotimes^n \mathscr{F})$ is spanned by all functions of the form

$$s = (\Phi f_1 + \Theta' g_1) \otimes (\Phi f_2 + \Theta' g_2) \otimes \cdots \otimes (\Phi f_n + \Theta' g_n), \tag{3.25}$$

where $f_j \in H^\infty(\mathscr{F})$ and $g_j \in H^\infty(\mathscr{F}')$, $1 \leq j \leq n$. Indeed, if h_1 is in $H^\infty(\mathscr{F}')$ and $\Phi f_1^{(i)} + \Theta' g_1^{(i)} \to h_1$ in $H^2(\mathscr{F}')$ as $i \to \infty$, then an appropriate application of (3.19) shows that the functions $(\Phi f_1^{(i)} + \Theta' g_1^{(i)}) \otimes (\Phi f_2 + \Theta' g_2) \otimes \cdots \otimes (\Phi f_n + \Theta' g_n)$ converge to $h_1 \otimes (\Phi f_2 + \Theta' g_2) \otimes \cdots \otimes (\Phi f_n + \Theta' g_n)$ in $H^2(\bigotimes^n \mathscr{F}')$. Several applications of this argument show that all vectors $h_1 \otimes h_2 \otimes \cdots \otimes h_n$, $h_j \in H^\infty(\mathscr{F}')$, belong to the span of vectors s of the form (3.25). In particular, all functions of the form $h(\lambda) = \lambda^k a_1 \otimes a_2 \otimes \cdots \otimes a_n$, $\lambda \in \mathbf{D}$, where $k \geq 0$ is an integer and $a_1, a_2, \ldots, a_n \in \mathscr{F}'$, belong to that span; it is clear that the functions $h(\lambda)$ generate $H^2(\bigotimes^n \mathscr{F}')$.

Returning now to (3.23), we see that if suffices to show that $u(\pi_n(p)\otimes I_{H^2})s \in \Gamma_n(\Theta')H^2(\bigotimes^n \mathscr{F}')$ whenever s has the form described in (3.25). After multiplication, the function s in (3.25) is a finite sum of functions of the form

$$(3.26)\quad r = (\pi'_n(\sigma)\otimes I_{H^2})(\Phi h_1 \otimes \Phi h_2 \otimes \cdots \otimes \Phi h_j \otimes \Theta' k_1 \otimes \Theta' k_2 \otimes \cdots \otimes \Theta' k_{n-j}),$$

where $\sigma \in \Sigma_n$, $0 \le j \le n$, $h_1, h_2, \ldots, h_j \in H^\infty(\mathscr{F})$, and $k_1, k_2, \ldots, k_{n-j} \in H^\infty(\mathscr{F}')$. Consequently (3.23) is proved if we can show that $u(\pi'_n(p)\otimes I_{H^2})r \in \Gamma_n(\Theta')H^2(\bigotimes^n \mathscr{F}')$ for all elements r given by (3.26). In fact, since $\pi'_n(p)$ and $\pi'_n(\sigma)$ commute, it suffices to consider the case in which $\sigma = 1$. We may also assume that $j \ge 1$ since for $j = 0$ we already have $r \in \Gamma_n(\Theta')H^2(\bigotimes^n \mathscr{F}')$. Under these assumptions, we use the relation $\sum p' = 1$, where the sum is extended over all minimal central projections $p' \in C^*(\mathscr{S}_j)$, in order to obtain

$$\begin{aligned}
(\pi'_n(p)\otimes I_{H^2})r
&= \sum (\pi'_n(pp')\otimes I_{H^2})r \\
&= \sum_{pp'\neq 0} (\pi'_n(pp')\otimes I_{H^2})r \\
&= \sum_{pp'\neq 0} (\pi'_n(p)\otimes I_{H^2})\{[(\pi'_j(p')\otimes I_{H^2})(\Phi h_1 \otimes \cdots \otimes \Phi h_j)] \\
&\qquad\qquad \otimes \Theta' k_1 \otimes \cdots \otimes \Theta' k_{n-j}\}.
\end{aligned}$$

It suffices then to prove that

$$u\{[(\pi'_j(p')\otimes I_{H^2})(\Phi h_1 \otimes \cdots \otimes \Phi h_j)\otimes \Theta' k_1 \otimes \cdots \otimes \Theta' k_{n-j}\} \in \Gamma_n(\Theta')H^2(\textstyle\bigotimes^n \mathscr{F}')$$

whenever $p'p \neq 0$. Since $\Gamma_n(\Theta') = \Gamma_j(\Theta')\otimes \Gamma_{n-j}(\Theta')$, (3.20) shows that it is enough to prove

$$(3.27)\qquad u(\pi'_j(p')\otimes I_{H^2})(\Phi h_1 \otimes \cdots \otimes \Phi h_j) \in \Gamma_j(\Theta')H^2\left(\textstyle\bigotimes^j \mathscr{F}'\right)$$

whenever $p'p \neq 0$. Finally, (3.27) would clearly follow from

$$(3.28)\qquad u(\pi_j(p')\otimes I_{H^2})(h_1 \otimes h_2 \otimes \cdots \otimes h_j) \in \Gamma_j(\Theta)H^2\left(\textstyle\bigotimes^j \mathscr{F}\right).$$

Now, if $p'p \neq 0$, we have $d^{p'}(\Theta) \mid d^p(\Theta) = u$ by Proposition 3.17, and (3.28) readily follows from the definition of $d^{p'}(\Theta)$. Thus the inclusion (3.22) is finally proved, and with it the relation $d^p(\Theta') \mid d^p(\Theta)$. For symmetry reasons we have $d^p(\Theta) \equiv d^p(\Theta')$. The theorem is proved.

In order to see that we gain some information from Theorem 3.21, we will analyze the action of the representation ϕ_{a_n} (with a_n as in (2.13)) on diagonal matrices. More precisely, let $\mathscr{F}'$ denote a Hilbert space with basis $\{e_j : j \ge 0\}$, and let a diagonal operator $D \in \mathscr{L}(\mathscr{F}')$ be defined such that $De_j = c_j e_j$ for all j. The operator $\Gamma_n(D)$ is also diagonal with respect to the orthonormal basis $\{e_{i_1}\otimes e_{i_2}\otimes \cdots \otimes e_{i_n} : i_1, i_2, \ldots, i_n \ge 0\}$ of $\bigotimes^n \mathscr{F}'$, and

$$\Gamma_n(D)(e_{i_1}\otimes e_{i_2}\otimes \cdots \otimes e_{i_n}) = c_{i_1}c_{i_2}\cdots c_{i_n} e_{i_1}\otimes e_{i_2}\otimes \cdots \otimes e_{i_n}.$$

Now, $\pi'_n(a_n)(\bigotimes^n \mathscr{F}')$ is a reducing subspace for $\Gamma_n(D)$, and hence $\phi_{a_n}(D)$ will also be a diagonal operator. To determine the eigenvalues of $\phi_{a_n}(D)$ we note that $\pi_n'(a_n)(e_{i_1} \otimes e_{i_2} \otimes \cdots \otimes e_{i_n}) \neq 0$ if and only if $i_p \neq i_q$ for $p \neq q$. Thus the eigenvalues of $\phi_{a_n}(D)$ are those products $c_{i_1} c_{i_2} \cdots c_{i_n}$ with the property that $i_p \neq i_q$ for $p \neq q$.

Suppose next that $\Theta' \in H^\infty(\mathscr{L}(\mathscr{F}'))$ is a diagonal inner function, i.e., $\Theta'(\lambda)e_j = \theta_j(\lambda)e_j$, $\lambda \in \mathbf{D}$, where $\theta_j \in H^\infty$ is an inner function for $j \geq 0$. Suppose in addition that θ_{j+1} divides θ_j for all j, so that $S(\Theta')$ is (unitarily equivalent to) a Jordan operator. Then we see that $\phi_{a_n}(\Theta')$ is also a diagonal inner function with diagonal entries $\theta_{i_1}\theta_{i_2}\cdots\theta_{i_n}$, where $i_p \neq i_q$ for $p \neq q$. Clearly the minimal function $d_n(\Theta')$ of $S(\phi_{a_n}(\Theta'))$ is $\theta_0\theta_1\cdots\theta_{n-1}$.

These observations prove the following consequence of Theorem 3.21.

3.29. COROLLARY. *Assume that $\Theta \in H^\infty(\mathscr{L}(\mathscr{F}))$ is an inner function such that $S(\Theta)$ is an operator of class C_0, and let $\bigoplus_{j<\omega} S(\theta_j)$ be the Jordan model of $S(\Theta)$. We have $\theta_j \equiv 1$ for $j \geq \dim(\mathscr{F})$, $\theta_0 \equiv d_1(\Theta)$, and $\theta_j \equiv d_{j+1}(\Theta)/d_j(\Theta)$ for $0 < j < \dim(\mathscr{F})$.*

Another consequence of Corollary 3.29 is the following result.

3.30. COROLLARY. *Assume that $\Theta \in H^\infty(\mathscr{L}(\mathscr{F}))$ and $\Theta' \in H^\infty(\mathscr{L}(\mathscr{F}'))$ are two inner functions such that $S(\Theta)$ and $S(\Theta')$ are quasisimilar operators of class C_0. If $\dim(\mathscr{F}) = \dim(\mathscr{F}')$ then $S(\psi_p(\Theta))$ and $S(\psi_p(\Theta'))$ are quasisimilar for every minimal central projection $p \in C^*(\mathscr{S}_n)$.*

PROOF. By Corollary 3.29, it suffices to show that $d_j(\psi_p(\Theta)) \equiv d_j(\psi_p(\Theta'))$ whenever $j \leq \dim(\pi_n(p)(\bigotimes^n \mathscr{F}))$. The representation $\phi_{a_j} \circ \psi_p$ of $\mathscr{L}(\mathscr{F})$ is a subrepresentation of Γ_{jn}, and hence it is unitarily equivalent to a sum of the form $\psi_{p_1} \oplus \psi_{p_2} \oplus \cdots \oplus \psi_{p_k}$, with $p_1, p_2, \ldots, p_k \in C^*(\mathscr{S}_{jn})$. We have therefore

$$\phi_{a_j}(\psi_p(\Theta)) \simeq \bigoplus_{i=1}^{k} \psi_{p_i}(\Theta), \qquad \phi_{a_j}(\psi_p(\Theta')) \simeq \bigoplus_{i=1}^{k} \psi_{p_i}(\Theta'),$$

so that

$$d_j(\psi_p(\Theta)) \equiv \bigvee_{i=1}^{k} d^{p_i}(\Theta), \qquad d_j(\psi_p(\Theta')) \equiv \bigvee_{i=1}^{k} d^{p_i}(\Theta').$$

The corollary now follows immediately from Theorem 3.21.

We conclude this section with a generalization of Theorem 2.4.4.

3.31. PROPOSITION. *Assume that $T \in \mathscr{L}(\mathscr{H})$ is an operator of class C_0, $\mathscr{H}'$ is an invariant subspace for T, and $T = \begin{bmatrix} T' & X \\ 0 & T'' \end{bmatrix}$ is the triangularization of T with respect to the decomposition $\mathscr{H} = \mathscr{H}' \oplus (\mathscr{H} \ominus \mathscr{H}')$. Assume also that $\mathscr{H}$ is separable, and $\bigoplus_{j<\omega} S(\theta_j)$, $\bigoplus_{j<\omega} S(\theta_j')$, and $\bigoplus_{j<\omega} S(\theta_j'')$ are the Jordan models of T, T', and T'', respectively. Then we have*

$$\theta_0\theta_1\cdots\theta_{k-1} \mid \theta_0'\theta_1'\cdots\theta_{k-1}'\theta_0''\theta_1''\cdots\theta_{k-1}''$$

for every $k \geq 1$.

PROOF. By Proposition 1.21, there exist inner functions $\Theta, \Theta', \Theta'' \in H^\infty(\mathscr{L}(\mathscr{F}))$ such that T, T', and T'' are unitarily equivalent to $S(\Theta)$, $S(\Theta')$, and $S(\Theta'')$, respectively, and $\Theta = \Theta''\Theta'$. Fix $k \geq 1$, and note that $\phi_{a_k}(\Theta) = \phi_{a_k}(\Theta'')\phi_{a_k}(\Theta')$. By Corollary 3.29 we can choose functions $\Omega', \Omega'' \in H^\infty(\mathscr{L}(\pi_k(a_k)(\bigotimes^k \mathscr{F})))$ such that $\Omega'\phi_{a_k}(\Theta') = \theta_0'\theta_1' \cdots \theta_{k-1}' I$ and $\Omega''\phi_{a_k}(\Theta'') = \theta_0''\theta_1'' \cdots \theta_{k-1}'' I$. Combining these relations we get

$$\Omega'\Omega''\phi_{a_k}(\Theta) = \theta_0'\theta_1' \cdots \theta_{k-1}'\theta_0''\theta_1'' \cdots \theta_{k-1}'' I,$$

and the proposition follows because $\theta_0\theta_1 \cdots \theta_{k-1}$ is the least scalar multiple of $\phi_{a_k}(\Theta)$.

Exercises

1. Let Θ be an inner function with a scalar multiple $u \in H^\infty \setminus \{0\}$, and let θ denote the minimal function of $S(\Theta)$. Assume that the function Ω satisfies the relation $\Theta\Omega = uI$, and show that the function Ω_1 defined by $\Omega_1(\lambda) = (\theta(\lambda)/u(\lambda))\Omega(\lambda)$, $u(\lambda) \neq 0$, can be extended to be an inner function in $\mathbf{D}$.

2. Let $\Theta \in H^\infty(\mathscr{L}(\mathbf{C}^n))$, $n \geq 2$, be an inner function defined by the matrix $[a_{ij}]_{1 \leq i,j \leq n}$ of functions in H^∞. Show that $\det(\Theta) = \det[a_{ij}]_{1 \leq i,j \leq n}$ is an inner scalar multiple of Θ, and the minimal function of $S(\Theta)$ equals $\det(\Theta)/v$, where v is the greatest common inner divisor of all minors of order $n-1$ of the matrix $[a_{ij}]_{1 \leq i,j \leq n}$.

3. Give an example showing that the relation $\Theta\Omega = uI$ does not always imply $\Omega\Theta = uI$.

4. Assume that $\Theta \in H^\infty(\mathscr{L}(\mathscr{F}, \mathscr{F}'))$, $\|\Theta\|_\infty \leq 1$, and Θ has an inner scalar multiple. Is $S(\Theta)$ an operator of class C_0?

5. Let Θ be an inner function such that $S(\Theta)$ is unitarily equivalent to the Jordan operator $\bigoplus_{j<\omega} S(\theta_j)$. Find the first three inner functions in the Jordan model of $S(\psi_{a_n}(\Theta))$.

6. Let ϕ and ψ be two inner functions in H^∞. Use Corollary 3.20 to determine the Jordan model of $S(\phi) \oplus S(\psi)$.

7. What happens with Proposition 3.22 if $\mathscr{H}$ is not a separable Hilbert space?

CHAPTER 6

Weak Contractions

The purpose of this chapter is to make the analogy between Section 5.3 and linear algebra more explicit, at least for the special case of operators of class C_0. We begin with a detailed discussion of exterior powers of Hilbert spaces and operators and a generalization of the notion of algebraic adjoint. The algebraic adjoints of higher orders are defined first in the finite-dimensional case, and the definition is then extended to trace-class perturbations of the identity operator in the infinite-dimensional case. The relationship between algebraic adjoints and minors is discussed. In Section 2, the discussion of algebraic adjoints and minors is extended to operator-valued analytic functions which are trace-class perturbations of the identity operator. It is shown that these adjoints are analytic functions. Finally, it is shown that the Jordan models of certain operators can be calculated from the minors of finite corank of analytic operator-valued functions. In Section 3 we study the class of weak contractions. In general, a contraction T is called a weak contraction if $I-T^*T$ is a trace-class operator and the spectrum of T does not cover the unit disc. We work only with those weak contractions which are of class C_{00}, and for these it suffices to assume that $I-T^*T$ is trace-class. One of the main points in this section is to show as a consequence that the spectrum of T does not cover the unit disc. Once this is done, the results of Section 2 can be applied to T. It follows that T is necessarily of class C_0, and one can calculate the Jordan model of T using the determinant and minors of finite corank of the characteristic function of T. These results are then used in Section 4 to show that an operator T of class C_0 is a weak contraction if and only if its Jordan model is a weak contraction. We also make some remarks about compact defect operators. We conclude the section with a result giving conditions under which an operator of class C_0 has a direct summand related to its Jordan model. This leads to another characterization of Jordan operators (cf. Exercise 7). In Section 5 we study a diagonalization theorem for matrices over H^∞, analogous with the diagonalization theorem for finite matrices over a principal ideal domain. We define the invariant factors of matrices over H^∞ (using, this time, minors of finite rank) and we show how to calculate the Jordan model of an operator of class C_0 using these invariant factors. The ideas in Sections 3, 4, and 5 are applied in Section 6 to the calculation of the Jordan model of a functional model

T, when we are given only a cyclic set for T^*. Interestingly, this calculation does not require knowledge of the characteristic function of T.

1. Determinants, adjoints, and minors. There are important particular cases in which the computation of the functions θ_j from Corollary 5.3.29 can be made more effective. It is clear from that corollary that the representations ϕ_{a_n} with a_n as in 5.(5.13), play a particularly important role. In this section we will study more background material connecting these representations with the notions of determinant and algebraic adjoint.

Let $\mathscr{F}$ be a Hilbert space and $n \geq 1$ a natural number. We use the notation $\bigwedge^n \mathscr{F}$ for $\pi_n(a_n)(\bigotimes^n \mathscr{F})$. It is sometimes convenient to set $\bigwedge^0 \mathscr{F} = \mathbf{C}$. The space $\bigwedge^n \mathscr{F}$ is called the nth exterior power of $\mathscr{F}$. If $n \geq 1$, $\bigwedge^n \mathscr{F}$ is generated by vectors of the form

$$k_1 \wedge k_2 \wedge \cdots \wedge k_n = (n!)^{-1/2} \sum_{\sigma \in \mathscr{S}_n} \varepsilon(\sigma) k_{\sigma(1)} \otimes k_{\sigma(2)} \otimes \cdots \otimes k_{\sigma(n)}, \tag{1.1}$$

where $k_j \in \mathscr{F}$, $1 \leq j \leq n$, and $\varepsilon(\sigma)$ denotes, as usual, the sign of the permutation σ. Note that

$$k_1 \wedge k_2 \wedge \cdots \wedge k_n = (n!)^{1/2} \pi_n(a_n)(k_1 \otimes k_2 \otimes \cdots \otimes k_n),$$

and the coefficient is chosen such that $\|e_1 \wedge e_2 \wedge \cdots \wedge e_n\| = 1$ if $\{e_1, e_2, \ldots, e_n\}$ is an orthonormal system in $\mathscr{F}$. Moreover, if $\{e_j : j \in J\}$ is an orthonormal basis for $\mathscr{F}$, and J is a totally ordered set, then the vectors

$$\{e_{i_1} \wedge e_{i_2} \wedge \cdots \wedge e_{i_n} : i_1 < i_2 < \cdots < i_n\}$$

form an orthonormal basis in $\bigwedge^n \mathscr{F}$. Of course, $\bigwedge^n \mathscr{F} = \{0\}$ if $n > \dim(\mathscr{F})$.

If $n \geq 1$ and $A \in \mathscr{L}(\mathscr{F})$, we denote by $\bigwedge^n A$ the operator $\psi_{a_n}(A) = \phi_{a_n}(A) = \Gamma_n(A) \mid \bigwedge^n \mathscr{F}$. Observe that

$$\left(\bigwedge^n A\right)(k_1 \wedge k_2 \wedge \cdots \wedge k_n) = Ak_1 \wedge Ak_2 \wedge \cdots \wedge Ak_n, \quad k_1, k_2, \ldots, k_n \in \mathscr{F}. \tag{1.2}$$

If n and m are two positive integers then there exists a continuous bilinear map $\Lambda : (\bigwedge^n \mathscr{F}) \times (\bigwedge^m \mathscr{F}) \to \bigwedge^{n+m} \mathscr{F}$ such that, using the notation $u \wedge v$ for $\Lambda(u, v)$, we have

$$(k_1 \wedge k_2 \wedge \cdots \wedge k_n) \wedge (k_{n+1} \wedge \cdots \wedge k_{n+m}) = k_1 \wedge k_2 \wedge \cdots \wedge k_{n+m}.$$

Assume now that $\mathscr{F}$ has finite dimension n. If $\{e_1, e_2, \ldots, e_n\}$ is an orthonormal basis, then $\bigwedge^n \mathscr{F}$ is the one-dimensional space generated by $e_1 \wedge e_2 \wedge \cdots \wedge e_n$, and hence we can define a bilinear form $B : (\bigwedge^k \mathscr{F}) \times (\bigwedge^{n-k} \mathscr{F}) \to \mathbf{C}$, $1 \leq k < n$, by the formula

$$B(h, g) = (h \wedge g, e_1 \wedge e_2 \wedge \cdots \wedge e_n), \qquad h \in \bigwedge^k \mathscr{F}, \quad g \in \bigwedge^{n-k} \mathscr{F}. \tag{1.3}$$

It is easy to check that $B(e_{i_1} \wedge e_{i_2} \wedge \cdots \wedge e_{i_k}, e_{i_{k+1}} \wedge \cdots \wedge e_{i_n}) = \pm 1$ if $\{i_1, \ldots, i_k\} \cup \{i_{k+1}, \ldots, i_n\} = \{1, 2, \ldots, n\}$, and therefore the linear mapping $C : \bigwedge^{n-k} \mathscr{F} \to (\bigwedge^k \mathscr{F})^d$ defined by

$$(Cg)(h) = B(h, g) \tag{1.4}$$

is an isometry onto the dual space $(\bigwedge^k \mathscr{F})^d$ of $\bigwedge^k \mathscr{F}$.

1.5. PROPOSITION. *Assume that $\mathcal{F}$ is a Hilbert space of finite dimension n. For $1 \leq k < n$ there exists a unique continuous mapping $A \to A^{\mathrm{Ad}\,k}$ from $\mathcal{L}(\mathcal{F})$ to $\mathcal{L}(\bigwedge^k \mathcal{F})$ with the following properties:*

(i) $A^{\mathrm{Ad}\,k}(\bigwedge^k A) = (\bigwedge^k A)A^{\mathrm{Ad}\,k} = \det(A)I_{\bigwedge^k \mathcal{F}}$, $A \in \mathcal{L}(\mathcal{F})$;

(ii) $(AB)^{\mathrm{Ad}\,k} = B^{\mathrm{Ad}\,k}A^{\mathrm{Ad}\,k}, A, B \in \mathcal{L}(\mathcal{F})$;

(iii) $(A^*)^{\mathrm{Ad}\,k} = (A^{\mathrm{Ad}\,k})^*, A \in \mathcal{L}(\mathcal{F})$; *and*

(iv) $\|A^{\mathrm{Ad}\,k}\| = \|\bigwedge^{n-k} A\|$, $A \in \mathcal{L}(\mathcal{F})$.

PROOF. The idea is to use the isometry C defined by (1.4). More precisely, if $A \in \mathcal{L}(\mathcal{F})$, define the operator $F \in \mathcal{L}((\bigwedge^k \mathcal{F})^d)$ by $F = C(\bigwedge^{n-k} A)C^{-1}$ and then define $A^{\mathrm{Ad}\,k} \in \mathcal{L}(\bigwedge^k \mathcal{F})$ such that $(A^{\mathrm{Ad}\,k})^d = F$. Note that this definition of $A^{\mathrm{Ad}\,k}$ depends on the mapping C, and hence on the particular orthonormal basis $\{e_1, e_2, \ldots, e_n\}$. We have

$$B(A^{\mathrm{Ad}\,k}h, g) = (Cg)(A^{\mathrm{Ad}\,k}h) = (FCg)(h) = \left(C\left(\bigwedge\nolimits^{n-k} A\right)g\right)(h)$$

for $h \in \bigwedge^k \mathcal{F}$, $g \in \bigwedge^{n-k} \mathcal{F}$, and since C is an isometry, these equalities imply (iv). Furthermore, if we substitute $\bigwedge^k Ah$ for h in the preceding identities, we obtain

$$\begin{aligned} B\left(A^{\mathrm{Ad}\,k}\left(\bigwedge\nolimits^k A\right)h, g\right) &= B\left(\left(\bigwedge\nolimits^k A\right)h, \left(\bigwedge\nolimits^{n-k} A\right)g\right) \\ &= \left(\left(\bigwedge\nolimits^n A\right)(h \wedge g), e_1 \wedge e_2 \wedge \cdots \wedge e_n\right) \\ &= \det(A)B(h, g), \end{aligned}$$

where we used the fact that $\bigwedge^n A = \det(A)I_{\bigwedge^n \mathcal{F}}$. Since B is nondegenerate we obtain the relation $A^{\mathrm{Ad}\,k}(\bigwedge^k A) = \det(A)I_{\bigwedge^k \mathcal{F}}$, $A \in \mathcal{L}(\mathcal{F})$. Note that this relation shows that $A^{\mathrm{Ad}\,k} = \det(A)(\bigwedge^k A)^{-1}$ if A is invertible, and hence $A^{\mathrm{Ad}\,k}$ is uniquely determined in this case. By continuity it follows that $A^{\mathrm{Ad}\,k}$ is uniquely determined by the relation $A^{\mathrm{Ad}\,k}(\bigwedge^k A) = \det(A)I_{\bigwedge^k \mathcal{F}}$; in particular, it does not depend on the particular orthonormal basis chosen for $\mathcal{F}$. If A is invertible we must also have $(\bigwedge^k A)A^{\mathrm{Ad}\,k} = \det(A)I_{\bigwedge^k \mathcal{F}}$, and by continuity it follows that (i) holds for all $A \in \mathcal{L}(\mathcal{F})$. Properties (ii) and (iii) are easy to verify and we leave them as an exercise.

The matrix entries of $A^{\mathrm{Ad}\,k}$ can be viewed as minors of size $n-k$ of the matrix A. More precisely, let $\{f_1, f_2, \ldots, f_k\}$ be an orthonormal system in $\mathcal{F}$, and let P denote the orthogonal projection onto the linear span of $\{f_1, f_2, \ldots, f_k\}$. Then we have

(1.6) $(A^{\mathrm{Ad}\,k}(f_1 \wedge f_2 \wedge \cdots \wedge f_k), f_1 \wedge f_2 \wedge \cdots \wedge f_k) = \det(P + (I - P)A(I - P))$.

Indeed, find $f_{k+1}, \ldots, f_n$ such that $\{f_1, f_2, \ldots, f_n\}$ is an orthonormal basis for $\mathcal{F}$, and let B' be the bilinear form defined by

$$B'(h, g) = (h \wedge g, f_1 \wedge f_2 \wedge \cdots \wedge f_n), \qquad h \in \bigwedge\nolimits^k \mathcal{F}, \quad g \in \bigwedge\nolimits^{n-k} \mathcal{F}.$$

Then we have

$$
\begin{aligned}
&(A^{\mathrm{Ad}\,k}(f_1 \wedge f_2 \wedge \cdots \wedge f_k), f_1 \wedge f_2 \wedge \cdots \wedge f_k) \\
&= (A^{\mathrm{Ad}\,k}(f_1 \wedge \cdots \wedge f_k) \wedge f_{k+1} \wedge \cdots \wedge f_n, f_1 \wedge f_2 \wedge \cdots \wedge f_n) \\
&= B'(A^{\mathrm{Ad}\,k}(f_1 \wedge f_2 \wedge \cdots \wedge f_k), f_{k+1} \wedge \cdots \wedge f_n) \\
&= B'\Big(f_1 \wedge f_2 \wedge \cdots \wedge f_k, \Big(\bigwedge\nolimits^{n-k} A\Big)(f_{k+1} \wedge \cdots \wedge f_n)\Big) \\
&= \Big(f_1 \wedge f_2 \wedge \cdots \wedge f_k \wedge \Big(\bigwedge\nolimits^{n-k} A\Big)(f_{k+1} \wedge \cdots \wedge f_n), f_1 \wedge \cdots \wedge f_n\Big) \\
&= \Big(\bigwedge\nolimits^{n}(P + A(I-P))(f_1 \wedge f_2 \wedge \cdots \wedge f_n), f_1 \wedge f_2 \wedge \cdots \wedge f_n\Big) \\
&= \det(P + A(I-P)) \\
&= \det(P + (I-P)A(I-P)),
\end{aligned}
$$

where we used the fact that the definition of $A^{\mathrm{Ad}\,k}$ does not depend on the particular orthonormal basis chosen for $\mathscr{F}$.

Proposition 1.5 can be extended to infinite-dimensional Hilbert spaces, but only for certain classes of operators A. If $\mathscr{F}$ is a separable Hilbert space, we denote by $\mathscr{C}_1(\mathscr{F})$ the ideal of nuclear (or trace-class) operators on $\mathscr{F}$, endowed with the trace norm $\|\cdot\|_1$. We recall that

$$\|X\|_1 = \operatorname{tr}(|X|), \qquad |X| = (X^*X)^{1/2}, \quad X \in \mathscr{C}_1(\mathscr{F}).$$

Assume that $A \in I + \mathscr{C}_1(\mathscr{F})$, that is, $A \in \mathscr{L}(\mathscr{F})$ and $I - A \in \mathscr{C}_1(\mathscr{F})$, and let $\{\lambda_j(A)\}$ be the eigenvalues of A, each eigenvalue being repeated according to its algebraic multiplicity. We have $\sum_j |1 - \lambda_j(A)| \le \operatorname{tr}(|I - A|) < \infty$ and therefore we can define

$$\det(A) = \prod_j \lambda_j(A), \tag{1.7}$$

where the product (if infinite) converges absolutely. The function $\det(I + X)$ is analytic as a function of $X \in \mathscr{C}_1(\mathscr{F})$. Assume that $\mathscr{F}$ is infinite-dimensional and $\{e_1, e_2, e_3, \cdots\}$ is an orthonormal basis for $\mathscr{F}$. Every operator $X \in \mathscr{C}_1(\mathscr{F})$ can be approximated in $\mathscr{C}_1(\mathscr{F})$ by the sequence $\{P_nXP_n : n \ge 1\}$, where P_n denotes the orthogonal projection onto the linear span of $\{e_1, e_2, \ldots, e_n\}$. The continuity of the function $\det(I + X)$ implies that we have

$$\det(A) = \lim_{n\to\infty} \det[(Ae_i, e_j)]_{1\le i,j\le n}, \quad A \in I + \mathscr{C}_1(\mathscr{F}). \tag{1.8}$$

The relations

$$\det(AA') = \det(A)\,\det(A'), \quad A, A' \in I + \mathscr{C}_1(\mathscr{F}) \tag{1.9}$$

and

$$\det(A^*) = \overline{\det(A)}, \quad A \in I + \mathscr{C}_1(\mathscr{F}) \tag{1.10}$$

can also be verified by using the continuity in X of $\det(I + X)$. Relations (1.8), (1.9), and (1.10) imply that

(i) $|\det(A)| = 1$ if $A \in I + \mathscr{C}_1(\mathscr{F})$ and A is unitary;
(ii) $|\det(A)| \leq 1$ if $||A|| \leq 1$, $A \in I + \mathscr{C}_1(\mathscr{F})$;
(iii) $\det(A) = 0$ if and only if A is not invertible.

The following result is an essential ingredient in the extension of Proposition 1.5 to the infinite-dimensional case.

1.11. LEMMA. *Assume that $\mathscr{F}$ is a finite-dimensional space, $A \in \mathscr{L}(\mathscr{F})$, and $1 \leq k < \dim(\mathscr{F})$. Then we have*

$$||A^{\mathrm{Ad}\,k}|| \leq \exp((1 + ||A - I||_1)^2 - 1).$$

If $A \geq 0$ we also have $A^{\mathrm{Ad}\,k} \geq 0$ and

$$||A^{\mathrm{Ad}\,k}|| \leq \exp(||A - I||_1).$$

PROOF. Assume first that $A \geq 0$ and $\lambda_1 \geq \lambda_2 \geq \cdots \geq \lambda_n \geq 0$ are the eigenvalues of A, repeated according to their multiplicities, where $n = \dim(\mathscr{F})$. Then $\bigwedge^{n-k} A$, and hence $A^{\mathrm{Ad}\,k}$, is a positive operator with eigenvalues $\{\lambda_{i_1}\lambda_{i_2}\cdots\lambda_{i_{n-k}}: 1 \leq i_1 < i_2 < \cdots < i_{n-k} \leq n\}$, and therefore

$$\begin{aligned}
||A^{\mathrm{Ad}\,k}|| &= \left\|\bigwedge\nolimits^{n-k} A\right\| \\
&= \lambda_1\lambda_2\cdots\lambda_{n-k} \\
&\leq (1 + |\lambda_1 - 1|)(1 + |\lambda_2 - 1|)\cdots(1 + |\lambda_{n-k} - 1|) \\
&\leq \exp(|\lambda_1 - 1|)\exp(|\lambda_2 - 1|)\cdots\exp(|\lambda_{n-k} - 1|) \\
&\leq \exp(\mathrm{tr}(|A - I|)) \\
&= \exp(||A - I||_1).
\end{aligned}$$

If A is not positive we can write the polar decomposition $A = UP$ of A, where $P = |A| = (A^*A)^{1/2}$, and we have

$$||A^{\mathrm{Ad}\,k}|| = ||P^{\mathrm{Ad}\,k}U^{\mathrm{Ad}\,k}|| \leq ||P^{\mathrm{Ad}\,k}|| \leq \exp(||P - I||_1). \tag{1.12}$$

A comparison of eigenvalues, using the relation $|\sqrt{t} - 1| \leq |t - 1|$ for $t \geq 0$, shows that $||P - I||_1 \leq ||A^*A - I||_1$. Thus we have

$$\begin{aligned}
||P - I||_1 &\leq ||A^*A - I||_1 \\
&\leq ||A^*(A - I)||_1 + ||A^* - I||_1 \\
&\leq ||A^*||\,||A - I||_1 + ||A - I||_1 \\
&\leq (1 + ||A - I||_1)||A - I||_1 + ||A - I||_1 \\
&= (1 + ||A - I||_1)^2 - 1,
\end{aligned}$$

and the lemma follows from these inequalities combined with (1.12).

1.13. THEOREM. *Assume that $\mathscr{F}$ is a separable infinite-dimensional Hilbert space, and $k \geq 1$. For every operator $A \in I + \mathscr{C}_1(\mathscr{F})$ there exists an operator $A^{\mathrm{Ad}\,k} \in \mathscr{L}(\bigwedge^k \mathscr{F})$ with the following properties:*

(i) $A^{\mathrm{Ad}\,k}(\bigwedge^k A) = (\bigwedge^k A)A^{\mathrm{Ad}\,k} = \det(A)I_{\bigwedge^k \mathscr{F}}$, $A \in I + \mathscr{C}_1(\mathscr{F})$;

(ii) $(AB)^{\mathrm{Ad}\,k} = B^{\mathrm{Ad}\,k}A^{\mathrm{Ad}\,k}$, $A, B \in I + \mathscr{C}_1(\mathscr{F})$;
(iii) $(A^*)^{\mathrm{Ad}\,k} = (A^{\mathrm{Ad}\,k})^*$, $A \in I + C_1(\mathscr{F})$;
(iv) $(A^{\mathrm{Ad}\,k}(f_1 \wedge f_2 \wedge \cdots \wedge f_k), f_1 \wedge f_2 \wedge \cdots \wedge f_k) = \det(P + (I - P)A(I - P))$ *if* $\{f_1, f_2, \ldots, f_k\}$ *is an orthonormal system in* $\mathscr{F}$, P *denotes the orthogonal projection onto the linear span of* $\{f_1, f_2, \ldots, f_k\}$, *and* $A \in I + \mathscr{C}_1(\mathscr{F})$;
(v) $\|A^{\mathrm{Ad}\,k}\| \leq \exp((1 + \|A - I\|_1)^2 - 1)$, $A \in I + \mathscr{C}_1(\mathscr{F})$;
(vi) $\|A^{\mathrm{Ad}\,k}\| \leq 1$ *if* $A \in I + \mathscr{C}_1(\mathscr{F})$ *and* $\|A\| \leq 1$;
(vii) $(I + X)^{\mathrm{Ad}\,k}$ *is analytic as a function of* $X \in \mathscr{C}_1(\mathscr{F})$;

PROOF. It clearly follows from Proposition 1.5 and Lemma 1.11 that operators $A^{\mathrm{Ad}\,k}$ satisfying properties (i)–(vi) can be defined for operators A such that $I - A$ has finite rank. The idea now is to extend this definition by continuity to all of $I + \mathscr{C}_1(\mathscr{F})$. Assume indeed that $A \in I + \mathscr{C}_1(\mathscr{F})$, and choose $A_n \in I + \mathscr{C}_1(\mathscr{F})$ such that $I - A_n$ has finite rank and $\lim_{n\to\infty} \|A_n - A\|_1 = 0$. If P is a finite-rank projection we must also have

$$\lim_{n\to\infty} \|(I - P)A_n(I - P) - (I - P)A(I - P)\|_1 = 0,$$

and therefore the sequence $\{\det(P + (I - P)A_n(I - P)) : n \geq 1\}$ converges to $\det(P + (I - P)A(I - P))$. Relation (iv) applied to $A_n^{\mathrm{Ad}\,k}$ shows that the sequence $\{A_n^{\mathrm{Ad}\,k} : n \geq 1\}$ converges weakly to an operator which we will denote by $A^{\mathrm{Ad}\,k}$. It is clear that the operator $A^{\mathrm{Ad}\,k}$ does not depend on the particular sequence $\{A_n\}$ used to define it (one can use the usual procedure of mixing two different sequences). It is also clear that (iv) is verified for all $A \in I + \mathscr{C}_1(\mathscr{F})$.

Assume now that $A \in I + \mathscr{C}_1(\mathscr{F})$ and set $A_n = I - P_n + P_nAP_n$, where $\{P_n : n \geq 1\}$ is a sequence of finite rank projections on $\mathscr{F}$, increasing to I. Then $\lim_{n\to\infty} \|A_n - A\|_1 = 0$ and $\|A_n\| \leq \|A\|$, $n \geq 1$. We conclude that

$$\begin{aligned}\|A^{\mathrm{Ad}\,k}\| \leq \limsup_{n\to\infty} \|A_n^{\mathrm{Ad}\,k}\| &\leq \lim_{n\to\infty} \exp((1 + \|A_n - I\|_1)^2 - 1) \\ &= \exp((1 + \|A - I\|_1)^2 - 1),\end{aligned}$$

so that (v) follows. If $\|A\| \leq 1$ we have $\|A_n\| \leq 1$ for $n \geq 1$, and hence $\|A^{\mathrm{Ad}\,k}\| \leq \limsup_{n\to\infty} \|A_n^{\mathrm{Ad}\,k}\| \leq 1$, and (vi) follows for all $A \in I + \mathscr{C}_1(\mathscr{F})$. The sequence $\{\bigwedge^k A_n : n \geq 1\}$ converges in norm to $\bigwedge^k A$, and $\{\det(A_n) : n \geq 1\}$ converges to $\det(A)$. Therefore (i) follows from the identities

$$A_n^{\mathrm{Ad}\,k}\left(\bigwedge^k A_n\right) = \left(\bigwedge^k A_n\right)A_n^{\mathrm{Ad}\,k} = \det(A_n)I.$$

Relation (iii) follows because we also have $\lim_{n\to\infty} \|A_n^* - A^*\|_1 = 0$ and the mapping $X \mapsto X^*$ is continuous in the weak operator topology. To prove (ii) assume that $B \in I + \mathscr{C}_1(\mathscr{F})$ and $I - B$ is of finite rank. Then we have $\lim_{n\to\infty} \|A_nB - AB\|_1 = 0$, and hence the relations $(A_nB)^{\mathrm{Ad}\,k} = B^{\mathrm{Ad}\,k}A_n^{\mathrm{Ad}\,k}$ imply $(AB)^{\mathrm{Ad}\,k} = B^{\mathrm{Ad}\,k}A^{\mathrm{Ad}\,k}$ via weak convergence. A repetition of this argument on the B component shows that (ii) is indeed true for all A and B in $I + \mathscr{C}_1(\mathscr{F})$.

We finally deal with the analyticity of $\det(I+X)$, $X \in \mathscr{C}_1(\mathscr{F})$. By Dunford's theorem it suffices to show that $(I+X)^{\mathrm{Ad}\,k}$ is weakly analytic, i.e., $((I+X)^{\mathrm{Ad}\,k}\xi,\eta)$ is analytic for every $\xi,\eta \in \bigwedge^k \mathscr{F}$. The inequality (v) shows that $(I+X)^{\mathrm{Ad}\,k}$ is locally bounded, and hence it suffices to consider total sets of vectors $\xi,\eta \in \bigwedge^k \mathscr{F}$. Thus, it suffices to show that the function $((I+X)^{\mathrm{Ad}\,k}(e_1\wedge e_2\wedge\cdots\wedge e_k), f_1\wedge f_2\wedge\cdots\wedge f_k)$ is analytic whenever $\{e_1,e_2,\ldots,e_k\}$ and $\{f_1,f_2,\ldots,f_k\}$ are orthonormal sequences in $\mathscr{F}$. Now, if $\{e_1,e_2,\ldots,e_k\}$ and $\{f_1,f_2,\ldots,f_k\}$ are orthonormal sequences in $\mathscr{F}$, we can choose a unitary operator $U \in I+\mathscr{C}_1(\mathscr{F})$ such that $Ue_j = f_j$, $1\le j\le k$, and hence

$$
\begin{aligned}
U^{\mathrm{Ad}\,k}(f_1\wedge f_2\wedge\cdots\wedge f_k) &= \det(U)\left(\textstyle\bigwedge^k U\right)^{-1}(f_1\wedge f_2\wedge\cdots\wedge f_k)\\
&= \det(U)(e_1\wedge e_2\wedge\cdots\wedge e_k).
\end{aligned}
$$

If P denotes the projection onto the linear span of $\{f_1,f_2,\ldots,f_k\}$, we have

$$
\begin{aligned}
&((I+X)^{\mathrm{Ad}\,k}(e_1\wedge e_2\wedge\cdots\wedge e_k), f_1\wedge f_2\wedge\cdots\wedge f_k)\\
&\quad= \det(U)^{-1}((I+X)^{\mathrm{Ad}\,k}U^{\mathrm{Ad}\,k}(f_1\wedge f_2\wedge\cdots\wedge f_k), f_1\wedge f_2\wedge\cdots\wedge f_k)\\
&\quad= \det(U)^{-1}\det(P+(I-P)U(I+X)(I-P))
\end{aligned}
$$

by (ii) and (iv). This function is analytic because $\det(I+Y)$ is analytic in $Y\in\mathscr{C}_1(\mathscr{F})$. The theorem is proved.

1.14. DEFINITION. Let $\mathscr{F}$ be a Hilbert space, $U, A \in I+\mathscr{C}_1(\mathscr{F})$, and $\mathscr{M}$, $\mathscr{N}$ two subspaces of $\mathscr{F}$; assume also that U is unitary and $U\mathscr{M}=\mathscr{N}$. The *minor* of A corresponding with the triple $(\mathscr{M},\mathscr{N},U)$ is the number $\det(UP_{\mathscr{M}}A \mid \mathscr{N})$.

Observe that the definition makes sense because $UP_{\mathscr{M}}A \mid \mathscr{N}$ belongs to $I+\mathscr{C}_1(\mathscr{N})$. In case $\mathscr{M}$ (and hence $\mathscr{N}$) has finite codimension k in $\mathscr{F}$, then $\det(UP_{\mathscr{M}}A\mid\mathscr{N})$ is called a minor of corank k of A. Assume that $\{e_1,e_2,\ldots,e_k\}$ is an orthonormal basis in $\mathscr{F}\ominus\mathscr{N}$. We then have

$$
\begin{aligned}
\det(UP_{\mathscr{M}}A\mid\mathscr{N}) &= \det(P_{\mathscr{N}}UAP_{\mathscr{N}}+(I-P_{\mathscr{N}}))\\
&= ((UA)^{\mathrm{Ad}\,k}(e_1\wedge e_2\wedge\cdots\wedge e_k), e_1\wedge e_2\wedge\cdots\wedge e_k)
\end{aligned}
$$

by Theorem 1.13(iv). Thus the minors of corank k are seen to be certain matrix entries of $(UA)^{\mathrm{Ad}\,k} = A^{\mathrm{Ad}\,k}U^{\mathrm{Ad}\,k}$.

Exercises

1. Extend the definition of $\bigwedge^k A$ and $A^{\mathrm{Ad}\,k}$ to the index $k=0$, so that all the properties in Theorem 1.13 are verified.

2. Show that $(k_1\wedge k_2\wedge\cdots\wedge k_n, h_1\wedge h_2\wedge\cdots\wedge h_n) = \det[(k_i,h_j)]_{1\le i,j\le n}$, for $k_1,\ldots,k_n,h_1,\ldots,h_n\in\mathscr{F}$.

3. Prove the existence of the operator of exterior multiplication $\Lambda_{m,n} : (\bigwedge^n \mathscr{F})\times(\bigwedge^m \mathscr{F})\to\bigwedge^{m+n}\mathscr{F}$ and compute $\|\Lambda_{m,n}\|$.

4. Show that $\bigwedge^n A = \det(A)I$ if $A\in\mathscr{L}(\mathscr{F})$ and $\dim(\mathscr{F})=n$.

5. Prove Proposition 1.5(ii) and (iii).

6. Show that all the results about determinants and adjoints can be extended to nonseparable Hilbert spaces.

7. Deduce relations (1.9) and (1.10) from the continuity of $\det(I+X)$.

8. Assume that $\mathscr{F}$ is infinite-dimensional and show that $\|A^{\mathrm{Ad}\,k}\| \geq 1$ for every $A \in I + \mathscr{C}_1(\mathscr{F})$ and $k \geq 1$.

9. Let $\{e_1, e_2, \ldots, e_k\}$ and $\{f_1, f_2, \ldots, f_k\}$ be two orthonormal sequences in $\mathscr{F}$. Show that there exists a unitary operator $U \in I + \mathscr{C}_1(\mathscr{F})$ such that $Ue_i = f_i$, $1 \leq i \leq k$. If $\mathscr{F}$ is infinite-dimensional we may assume, in addition, that $I - U$ has finite rank and $\det(U) = 1$.

10. Show that all the usual minors of a square matrix appear among the minors defined in 1.14.

2. Determinants of analytic functions. As the title suggests, in this section we study the functions $\det(\Theta(\lambda))$, where $\Theta \in H^\infty(\mathscr{L}(\mathscr{F}))$ and $I-\Theta(\lambda) \in \mathscr{C}_1(\mathscr{F})$ for $\lambda \in \mathbf{D}$. The difficult case occurs, of course, when $\mathscr{F}$ is infinite-dimensional; the results below are also true in the finite-dimensional case.

2.1. LEMMA. *Assume that $\Theta \in H^\infty(\mathscr{L}(\mathscr{F}))$ is such that $\|\Theta\|_\infty \leq 1$ and $I-\Theta(\lambda) \in \mathscr{C}_1(\mathscr{F})$ for every $\lambda \in \mathbf{D}$. Then the functions $\det(\Theta(\lambda))$ and $\Theta(\lambda)^{\mathrm{Ad}\,k}$, $\lambda \in \mathbf{D}$, $k \geq 1$, are analytic.*

PROOF. If $\{e_1, e_2, \ldots\}$ is an orthonormal basis for $\mathscr{F}$, we have

$$\det(\Theta(\lambda)) = \lim_{n\to\infty} \det[(\Theta(\lambda)e_i, e_j)]_{1\leq i,j\leq n}, \quad \lambda \in \mathbf{D},$$

by (1.8). Moreover, we have

$$|\det[(\Theta(\lambda)e_i, e_j)]_{1\leq i,j\leq n}| \leq 1, \quad \lambda \in \mathbf{D},$$

and therefore a subsequence of the form $\{\det[(\Theta(\lambda)e_i, e_j)]_{1\leq i,j\leq n_k} : k \geq 1\}$ will converge uniformly on the compact subsets of $\mathbf{D}$. It follows at once that the limit $\det(\Theta(\lambda))$ is analytic.

By Dunford's theorem, the analyticity of $\Theta(\lambda)^{\mathrm{Ad}\,k}$ will follow from the analyticity of all functions of the form

$$(\Theta(\lambda)^{\mathrm{Ad}\,k}(e_1 \wedge e_2 \wedge \cdots \wedge e_k), f_1 \wedge f_2 \wedge \cdots \wedge f_k), \quad \lambda \in \mathbf{D},$$

where $\{e_1, e_2, \ldots, e_k\}$ and $\{f_1, f_2, \ldots, f_k\}$ are arbitrary orthonormal sequences in $\mathscr{F}$. An argument similar to the one above and to the proof of Theorem 1.13(vii) shows that these functions are indeed analytic, and the lemma follows.

We will use the notation $\Theta^{\mathrm{Ad}\,k}$ for the function $\Theta^{\mathrm{Ad}\,k}(\lambda) = \Theta(\lambda)^{\mathrm{Ad}\,k}$; of course, $\Theta^{\mathrm{Ad}\,k} \in H^\infty(\mathscr{L}(\bigwedge^k \mathscr{F}))$ and $\|\Theta^{\mathrm{Ad}\,k}\|_\infty \leq 1$ by Theorem 1.13(vi). We have

$$\left(\bigwedge^k \Theta(\lambda)\right)\Theta^{\mathrm{Ad}\,k}(\lambda) = \Theta^{\mathrm{Ad}\,k}(\lambda)\left(\bigwedge^k \Theta(\lambda)\right) = \det(\Theta(\lambda)) I_{\bigwedge^k \mathscr{F}}, \tag{2.2}$$

but this does not necessarily mean that $\bigwedge^k \Theta$ has a scalar multiple since $\det(\Theta(\lambda))$ may be identically zero. We will show in Section 3 that this situation cannot occur if, in addition, Θ is required to be an inner function. In the remaining part of this section we will assume that Θ is inner and $\det(\Theta(\lambda))$ is not identically zero, as we have to wait until Section 3 to see that the latter assumption is superfluous.

2.3. THEOREM. *Assume that $\Theta \in H^\infty(\mathscr{L}(\mathscr{F}))$ is an inner function such that $I - \Theta(\lambda) \in \mathscr{C}_1(\mathscr{F})$ for $\lambda \in \mathbf{D}$ and $\det(\Theta)$ is not identically zero. Then $\det(\Theta)$ is an inner function.*

PROOF. We set $u = \det(\Theta) \in H^\infty$, and let $u = vw$ be the inner-outer decomposition of u; thus v is inner and w is outer. Since u is a scalar multiple of $\bigwedge^k \Theta$, $k \geq 1$, it follows that the minimal function of $S(\bigwedge^k \Theta)$ divides u, and hence v. Therefore v is also a scalar multiple of $\bigwedge^k \Theta$. Choose for each $k \geq 1$ a function $\Omega_k \in H^\infty(\mathscr{L}(\bigwedge^k \mathscr{F}))$ satisfying the relation $(\bigwedge^k \Theta(\lambda))\Omega_k(\lambda) = v(\lambda)I$, $\lambda \in \mathbf{D}$. Then (2.2) implies the equality

$$\left(\bigwedge^k \Theta(\lambda)\right)(\Theta(\lambda)^{\mathrm{Ad}\,k} - w(\lambda)\Omega_k(\lambda)) = 0, \quad \lambda \in \mathbf{D},$$

and, since $\bigwedge^k \Theta$ is inner, we have

$$\Theta(\lambda)^{\mathrm{Ad}\,k} = w(\lambda)\Omega_k(\lambda), \quad \lambda \in \mathbf{D}. \tag{2.4}$$

Let us note that $||w||_\infty = ||u||_\infty \leq 1$ so that $|w(\lambda)| \leq 1$, $\lambda \in \mathbf{D}$. On the other hand, if $\{e_j : 1 \leq j < \infty\}$ is an orthonormal basis of $\mathscr{F}$, we have by (2.4)

$$\begin{aligned} |w(0)| &\geq |w(0)|\,||\Omega_k(0)|| = ||\Theta(0)^{\mathrm{Ad}\,k}|| \\ &\geq |(\Theta(0)^{\mathrm{Ad}\,k}(e_1 \wedge e_2 \wedge \cdots \wedge e_k), e_1 \wedge e_2 \wedge \cdots \wedge e_k)|. \end{aligned}$$

If P_k denotes the orthogonal projection onto the orthogonal span of $\{e_1, e_2, \ldots, e_k\}$, then the preceding relation and Theorem 1.13(vii) imply that

$$|w(0)| \geq \lim_{k \to \infty} |\det(P_k + (I - P_k)\Theta(0)(I - P_k))| = \det(I) = 1.$$

By the maximum modulus principle we have $|w(\lambda)| \equiv 1$ and w is a constant. The theorem is proved.

The above line of reasoning can also be used to prove the following result.

2.5. COROLLARY. *Under the conditions of Theorem 2.3, $\det(\Theta)$ coincides with the least common inner multiple of the functions $\{d_k(\Theta): k \geq 1\}$.*

PROOF. Denote by v the least common inner multiple of the functions $\{d_k(\Theta): k \geq 1\}$, and set $u = \det(\Theta)$. Since u is a scalar multiple of $\bigwedge^k \Theta$, we have $d_k(\Theta)|u$ for $k \geq 1$, and hence $v|u$. Now write $u = vw$, $w \in H^\infty$, where $||w||_\infty = ||u||_\infty = 1$. Apply the argument in the proof of Theorem 2.3 (v is again a scalar multiple of $\bigwedge^k \Theta$, $k \geq 1$) to conclude that w is a constant function.

2.6. DEFINITION. Let $\Theta \in H^\infty(\mathscr{L}(\mathscr{F}))$ be such that $||\Theta||_\infty \leq 1$ and $I - \Theta(\lambda) \in \mathscr{C}_1(\mathscr{F})$, $\lambda \in \mathbf{D}$. A *minor of corank k* of Θ is a function of the form $\lambda \to \det(UP_{\mathscr{M}}\Theta(\lambda) \mid \mathscr{N})$, where $\mathscr{M}$ and $\mathscr{N}$ are subspaces of codimension k in $\mathscr{F}$, and U is a unitary operator in $I + \mathscr{C}_1(\mathscr{F})$ such that $U\mathscr{M} = \mathscr{N}$. We denote by $\delta_k(\Theta)$ the greatest common inner divisor of all minors of corank k of Θ, $k \geq 0$.

Note that $\delta_k(\Theta)$ is not defined if all minors of corank k of Θ are identically zero. The function $\delta_0(\Theta)$ is the inner factor of $\det(\Theta)$.

2.7. PROPOSITION. *Assume that $\Theta \in H^\infty(\mathscr{L}(\mathscr{F}))$ is an inner function such that $I - \Theta(\lambda) \in \mathscr{C}_1(\mathscr{F})$, $\lambda \in \mathbf{D}$, and $\det(\Theta)$ is not identically zero. Then we have*

$$\delta_k(\Theta)d_k(\Theta) \equiv \det(\Theta), \qquad k \geq 0,$$

where we set $d_0(\Theta) = 1$.

PROOF. By definition, $\delta_k(\Theta)$ divides all the matrix entries of $\Theta^{\mathrm{Ad}\,k}$, so that we can write $\Theta^{\mathrm{Ad}\,k} = \Omega_k\delta_k(\Theta)$ for some Ω_k in $H^\infty(\mathscr{L}(\bigwedge^k \mathscr{F}))$. We deduce from the relations

$$\left(\bigwedge^k\Theta\right)(\delta_k(\Theta)\Omega_k) = \left(\bigwedge^k\Theta\right)\Theta^{\mathrm{Ad}\,k} = \det(\Theta)I$$

that $\delta_k(\Theta) \mid \det(\Theta)$ and, following cancellation,

$$\left(\bigwedge^k\Theta\right)\Omega_k = (\det(\Theta)/\delta_k(\Theta))I.$$

We see that the minimal function of $S(\bigwedge^k \Theta)$, i.e., $d_k(\Theta)$, must divide the quotient $\det(\Theta)/\delta_k(\Theta)$, and hence

$$d_k(\Theta)\delta_k(\Theta) \mid \det(\Theta). \tag{2.8}$$

In the opposite direction, $d_k(\Theta)$ is a scalar multiple of $\bigwedge^k \Theta$, and hence we can find functions $\Phi_k \in H^\infty(\mathscr{L}(\bigwedge^k \mathscr{F}))$ satisfying the relation $(\bigwedge^k \Theta)\Phi_k = d_k(\Theta)I$. Combining this relation with (2.2) we get

$$\left(\bigwedge^k\Theta\right)(\Theta^{\mathrm{Ad}\,k} - (\det(\Theta)/d_k(\Theta))\Phi_k) = 0,$$

and hence $\Theta^{\mathrm{Ad}\,k} = (\det(\Theta)/d_k(\Theta))\Phi_k$ since $\bigwedge^k \Theta$ is inner. Therefore the function $\det(\Theta)/d_k(\Theta)$ divides all the matrix entries of $\Theta^{\mathrm{Ad}\,k}$ and hence all the minors of corank k of Θ. We conclude that $(\det(\Theta)/d_k(\Theta)) \mid \delta_k(\Theta)$ and hence $\det(\Theta) \mid \delta_k(\Theta)d_k(\Theta)$. The proposition follows from this relation and (2.8).

We can now give new formulas for the Jordan model of $S(\Theta)$, provided that Θ satisfies the hypothesis of Theorem 2.3.

2.9. PROPOSITION. *Assume that $\Theta \in H^\infty(\mathscr{L}(\mathscr{F}))$ is an inner function such that $I - \Theta(\lambda) \in \mathscr{C}_1(\mathscr{F})$, $\lambda \in \mathbf{D}$, and $\det(\Theta)$ is not identically zero. Let $\bigoplus_{j<\omega} S(\theta_j)$ be the Jordan model of $S(\Theta)$.*

(i) *$\delta_{k+1}(\Theta) \mid \delta_k(\Theta)$ and $\theta_k \equiv \delta_k(\Theta)/\delta_{k+1}(\Theta)$, $k \geq 0$.*
(ii) *The product $\prod_{j=0}^\infty |\theta_j(\lambda)|$ converges for every $\lambda \in \mathbf{D}$ and $|\delta_k(\Theta)(\lambda)| = \prod_{j=k}^\infty |\theta_j(\lambda)|$, $\lambda \in \mathbf{D}$, $k \geq 0$.*

PROOF. We know from Corollary 5.3.29 that $d_k(\Theta) \mid d_{k+1}(\Theta)$ and $\theta_k \equiv d_{k+1}(\Theta)/d_k(\Theta)$, $k \geq 0$. Then (i) follows at once from the relations $d_k(\Theta)\delta_k(\Theta) \equiv d_{k+1}(\Theta)\delta_{k+1}(\Theta) \equiv \det(\Theta)$.

Let us note that $\bigwedge_{k=0}^{\infty} \delta_k(\Theta) \equiv \det(\Theta)/\bigvee_{k=0}^{\infty} d_k(\Theta) \equiv 1$ by Proposition 2.2.4 and Corollary 2.5. This relation implies that $\lim_{k\to\infty} |\delta_k(\Theta)(\lambda)| = 1$ for all $\lambda \in \mathbf{D}$. Indeed, by (i) the sequence $\{|\delta_k(\Theta)(\lambda)| : k \geq 0\}$ is increasing, and an application of the Vitali–Montel theorem shows that there exists a function u in H^∞ such that $|u(\lambda)| = \lim_{k\to\infty} |\delta_k(\Theta)(\lambda)|$. The inequalities $|\delta_k(\Theta)(\varsigma)| \leq |u(\varsigma)| \leq 1$, $\varsigma \in \mathbf{T}$, show that u must be an inner divisor of $\delta_k(\Theta)$, $k \geq 0$, and hence $|u(\lambda)| = 1$ for $\lambda \in \mathbf{D}$.

We can now prove (ii). A repeated use of (i) yields

$$\delta_k(\Theta) \equiv \theta_k\theta_{k+1}\theta_{k+2}\cdots\theta_{k+n}\delta_{k+n+1}(\Theta),$$

and hence

$$\begin{aligned} |\delta_k(\Theta)(\lambda)| &= \lim_{n\to\infty} |\theta_k(\lambda)\theta_{k+1}(\lambda)\cdots\theta_{k+n}(\lambda)|\,|\delta_{k+n+1}(\Theta)(\lambda)| \\ &= \prod_{j=k}^{\infty} |\theta_j(\lambda)| \end{aligned} \tag{2.10}$$

by the preceding remarks. The convergence of the infinite product is immediate at points λ for which $\det(\Theta(\lambda)) \neq 0$. However, if λ is an arbitrary point in $\mathbf{D}$, we must have $\delta_k(\Theta)(\lambda) \neq 0$ for some k, and the convergence of the product follows from (2.10).

Exercises

1. Adjust the results of this section to the case in which $\mathscr{F}$ is a finite-dimensional space.
2. Assume that $\Theta \in H^\infty(\mathscr{L}(\mathscr{F}))$ is inner and $I - \Theta(\lambda) \in \mathscr{C}_1(\mathscr{F})$, $\lambda \in \mathbf{D}$. Is it always true that $I - \Theta(\varsigma) \in \mathscr{C}_1(\mathscr{F})$ for almost every $\varsigma \in \mathbf{T}$?
3. Let $\{\theta_j : 0 \leq j < \omega\}$ be a sequence of inner functions such that $\sum_{j=0}^{\infty}(1 - |\theta_j(0)|) < \infty$ and the first nonzero Taylor coefficient of θ_j at the origin is positive. Show that the formula $u(\lambda) = \prod_{j=0}^{\infty} \theta_j(\lambda)$, $\lambda \in \mathbf{D}$, defines an inner function.

3. Defect operators and weak contractions. In this section we study the relationship between the defect operators of an operator of class C_0 and the defect operators of the corresponding Jordan model. We begin with the study of weak contractions of class C_{00}, which will turn out to belong to the class C_0.

3.1. DEFINITION. Let $T \in \mathscr{L}(\mathscr{H})$ be an operator of class C_{00}. Then T is said to be a *weak contraction* if $D_T^2 = I - T^*T$ belongs to $\mathscr{C}_1(\mathscr{H})$.

An important class of weak contractions is the class $C_0(N)$, where N is a natural number. A contraction T of class C_{00} belongs to $C_0(N)$ if D_T has rank equal to N.

There are weak contractions which are not of class C_{00}, but we will not need them here.

3.2. LEMMA. *Assume that* $T \in \mathscr{L}(\mathscr{H})$ *is a contraction,* $\mathscr{H}_1 \in \operatorname{Lat}(T)$, $\mathscr{H}_2 = \mathscr{H} \ominus \mathscr{H}_1$, *and* $T = \left[\begin{smallmatrix} T_1 & X \\ 0 & T_2 \end{smallmatrix}\right]$ *is the triangularization of* T *with respect to the decomposition* $\mathscr{H} = \mathscr{H}_1 \oplus \mathscr{H}_2$.

(i) *If* T *is of class* C_{00} *then both* T_1 *and* T_2 *are of class* C_{00}.

(ii) *If* $I - T^*T \in \mathscr{C}_1(\mathscr{H})$ *then* $I - T_1^*T_1 \in \mathscr{C}_1(\mathscr{H}_1)$. *If, in addition,* $\mathscr{H}_1$ *is finite-dimensional, then we also have* $I - T_2^*T_2 \in \mathscr{C}_1(\mathscr{H}_2)$.

PROOF. Assume that T is of class C_{00}. Then we have $T_1^n = T^n \mid \mathscr{H}_1$, $T_1^{*^n} = P_{\mathscr{H}_1}T^{*^n} \mid \mathscr{H}_1$, $T_2^{*^n} = T^{*^n} \mid \mathscr{H}_2$, and $T_2^n = P_{\mathscr{H}_2}T^n \mid \mathscr{H}_2$ for $n \geq 1$; it clearly follows that T_1 and T_2 are of class C_{00}.

The second part of the lemma follows from the identities $I - T_1^*T_1 = P_{\mathscr{H}_1}(I - T^*T) \mid \mathscr{H}_1$ and $I - T_2^*T_2 = X^*X + P_{\mathscr{H}_2}(I - T^*T) \mid \mathscr{H}_2$. The first one shows that $I - T_1^*T_1 \in \mathscr{C}_1(\mathscr{H}_1)$ if $I - T^*T \in \mathscr{C}_1(\mathscr{H})$, and the second one shows that $I - T_2^*T_2 \in \mathscr{C}_1(\mathscr{H}_2)$ if $I - T^*T \in \mathscr{C}_1(\mathscr{H})$ and X has finite rank.

The relation $T(I-T^*T) = (I-TT^*)T$ suggests that there must be connections between the size of the selfadjoint operators $I - TT^*$ and $I - T^*T$. We recall the notations $D_T = (I - T^*T)^{1/2}$ and $\mathscr{D}_T = (D_T\mathscr{H})^- = ((I - T^*T)\mathscr{H})^-$.

3.3. LEMMA. *Assume that* $T \in \mathscr{L}(\mathscr{H})$ *is a contraction.*

(i) *We have* $\mathscr{D}_{T^*} = (T\mathscr{D}_T)^- \oplus \ker(T^*)$.

(ii) *Assume that* $I - T^*T$ *is compact and* $\{e_j : j \in J\}$ *is an orthonormal basis of* $\mathscr{D}_T$ *such that* $(I - T^*T)x = \sum_{j\in J} \mu_j(x, e_j)e_j$, $x \in \mathscr{H}$, *where* $\{\mu_j : j \in J\}$ *are the nonzero eigenvalues of* $I - T^*T$. *We have* $\|Te_j\| = (1 - \mu_j)^{1/2}$ *and the vectors* $f_j = (1-\mu_j)^{-1/2}Te_j$, $\mu_j \neq 1$, *form an orthonormal basis of* $(T\mathscr{D}_T)^-$ *such that* $(I - TT^*)f_j = \mu_j f_j$.

(iii) *If* $I - T^*T \in \mathscr{C}_1(\mathscr{H})$ *then*

$$\operatorname{tr}(I - T^*T) + \dim(\ker(T^*)) = \operatorname{tr}(I - TT^*) + \dim(\ker(T)),$$

so that $I - TT^* \in \mathscr{C}_1(\mathscr{H})$ *if and only if* $\ker(T^*)$ *is finite-dimensional.*

PROOF. We have $\mathscr{H} \ominus (T\mathscr{D}_T) = \mathscr{H} \ominus (T(I - T^*T)\mathscr{H})^- = \ker((I - T^*T)T^*) = \ker(T^*(I - TT^*))$. Assume that $x \in \mathscr{D}_{T^*} \ominus (T\mathscr{D}_T)^-$. The above calculations show that $(I - T^*T)T^*x = 0$ and, since $T^*x \in \mathscr{D}_T = \mathscr{H} \ominus \ker(I - T^*T)$, we conclude that $T^*x = 0$. Thus (i) follows because the inclusion $\ker(T^*) \subseteq \mathscr{D}_{T^*}$ is obvious.

Assume now that $I - T^*T$ is compact. The vectors Te_j, $j \in J$, clearly generate $(T\mathscr{D}_T)^-$, and the remaining assertions of (ii) follow from the calculations below:

$$(Te_j, Te_k) = (e_j, e_k) - ((I - T^*T)e_j, e_k) = (1 - \mu_j)\delta_{jk}, \quad j, k \in J;$$

$$\begin{aligned}(I - TT^*)f_j &= (1 - \mu_j)^{-1/2}(I - TT^*)Te_j \\ &= (1 - \mu_j)^{-1/2}T(I - T^*T)e_j = \mu_j f_j, \quad \mu_j \neq 1.\end{aligned}$$

Finally, (iii) clearly follows from the fact that the eigenvalues of $I - TT^*$ are $\{\mu_j : j \in J, \mu_j \neq 1\}$, plus the eigenvalue 1 with multiplicity $\dim\ker(T^*)$. Thus, both sides of (iii) can be written as

$$\sum_{\mu_j \neq 1} \mu_j + \dim\ker(T) + \dim\ker(T^*).$$

The following result provides the link between the functions studied in the preceding paragraph and the defect operators of contractions.

3.4. PROPOSITION. *Let $T \in \mathscr{L}(\mathscr{H})$ be an operator of class $C_{\cdot 0}$.*

(i) *Let $\Theta \in H^\infty(\mathscr{L}(\mathscr{F}))$ be an arbitrary inner function such that $S(\Theta)$ is unitarily equivalent to T. Then $I - T^*T$ is compact if and only if $I - \Theta(0)^*\Theta(0)$ is compact. Moreover, if $I - T^*T$ is compact, then $I - T^*T$ and $\Theta(0)^*\Theta(0)$ have the same nonzero eigenvalues. In particular,* $\operatorname{tr}(I - T^*T) = \operatorname{tr}(I - \Theta(0)^*\Theta(0))$.

(ii) *Assume that $I - T^*T$ is compact and* $\dim\ker(T) = \dim\ker(T^*)$. *There exists an inner function $\Theta \in H^\infty(\mathscr{L}(\mathscr{F}))$ such that $S(\Theta)$ is unitarily equivalent to T, $I - \Theta(\lambda)$ is compact for $\lambda \in \mathbf{D}$, and $\Theta(0) \geq 0$. If $I - T^*T \in \mathscr{C}_1(\mathscr{H})$ then $I - \Theta(\lambda) \in \mathscr{C}_1(\mathscr{F})$, $\lambda \in \mathbf{D}$.*

PROOF. Assume that $S(\Theta)$ is unitarily equivalent with T. With the notation of Proposition 5.1.18 we have

$$I - \Theta(0)^*\Theta(0) = U^*((I - \Theta_T(0)^*\Theta_T(0)) \oplus 0)U$$

and hence it suffices to prove (i) in case $\Theta = \Theta_T$. Now (i) clearly follows from the fact that $I - \Theta_T(0)^*\Theta_T(0) = (I - T^*T) \mid \mathscr{D}_T$ since $I - T^*T = 0$ on $\mathscr{H} \ominus \mathscr{D}_T$.

Assume that $I - T^*T$ is compact and $\dim\ker(T) = \dim\ker(T^*)$. With the notation of Lemma 3.3, we can find a basis $\{f_j : j \in J\}$ of $\mathscr{D}_{T^*}$ such that $Te_j = (1 - \mu_j)^{1/2} f_j$ for all j; indeed, it suffices to extend the basis $\{f_j : \mu_j \neq 1\}$ with a basis $\{f_j : \mu_j = 1\}$ of $\ker(T^*)$. Now define a unitary operator $V \in \mathscr{L}(\mathscr{D}_{T^*}, \mathscr{D}_T)$ by

$$Vf_j = -e_j, \quad j \in J, \tag{3.5}$$

and define $\Theta \in H^\infty(\mathscr{L}(\mathscr{D}_T))$ by

$$\Theta(\lambda) = V\Theta_T(\lambda), \quad \lambda \in \mathbf{D}. \tag{3.6}$$

Then clearly Θ is equivalent to Θ_T, so that $S(\Theta)$ is unitarily equivalent to T. Moreover, (3.6) implies

$$I_{\mathscr{D}_T} - \Theta(\lambda) = I_{\mathscr{D}_T} + VT \mid \mathscr{D}_T - \lambda V D_{T^*}(I - \lambda T^*)^{-1} D_T \mid \mathscr{D}_T, \quad \lambda \in \mathbf{D}. \tag{3.7}$$

By (3.5) we have

$$(I + VT)e_j = (1 - (1 - \mu_j)^{1/2})e_j, \quad j \in J, \tag{3.8}$$

and, since $\lim \mu_j = 0$ because $I - T^*T$ is compact, it follows that $(I + VT) \mid \mathscr{D}_T$ is also compact. Now, $D_T = (I - T^*T)^{1/2}$ is also compact and (3.7) shows that $I_{\mathscr{D}_T} - \Theta(\lambda)$ is compact for all $\lambda \in \mathbf{D}$.

If $I - T^*T$ is in $\mathscr{C}_1(\mathscr{H})$ then $I - TT^* \in \mathscr{C}_1(\mathscr{H})$ by Lemma 3.3(iii). Hence D_T and D_{T^*} are Hilbert–Schmidt operators. Moreover, because of the obvious relation $\lim_{\mu\to 0}\mu^{-1}(1-(1-\mu)^{1/2})=\frac{1}{2}$, (3.8) implies that $(I+VT)\mid\mathscr{D}_T$ is in $\mathscr{C}_1(\mathscr{D}_T)$. As before, (3.7) shows that $I-\Theta(\lambda)$ belongs to $\mathscr{C}_1(\mathscr{D}_T)$ for all $\lambda\in\mathbf{D}$. The lemma is proved.

3.9. COROLLARY. *Assume that $T\in\mathscr{L}(\mathscr{H})$ is an operator of class $C_{\cdot 0}$ with* $\dim\ker(T)=\dim\ker(T^*)$, *and $\Theta\in H^\infty(\mathscr{L}(\mathscr{F},\mathscr{F}'))$ is an inner function such that $S(\Theta)$ is unitarily equivalent to T. If $I-T^*T$ is compact [resp., trace-class] then there exists a unitary operator $W\in\mathscr{L}(\mathscr{F}',\mathscr{F})$ such that $I-W\Theta(\lambda)$ is compact [resp., trace-class] for every $\lambda\in\mathbf{D}$.*

PROOF. By Proposition 5.1.18 we can write

$$\Theta(\lambda)=U'^*(\Theta_T(\lambda)\oplus I_{\mathscr{N}})U,\quad \lambda\in\mathbf{D},$$

where $U\in\mathscr{L}(\mathscr{F},\mathscr{D}_T\oplus\mathscr{N})$ and $U'\in\mathscr{L}(\mathscr{F}',\mathscr{D}_{T^*}\oplus\mathscr{N})$ are unitary operators. Let V be the operator constructed in the proof of Proposition 3.4 (cf. (3.5)). We can then define $W=U^*(V\oplus I_{\mathscr{N}})U'$; the reader will have no difficulty in verifying that W satisfies the requirements of the corollary.

3.10. DEFINITION. Let $T\in\mathscr{L}(\mathscr{H})$ be a contraction of class $C_{\cdot 0}$ such that $I-T^*T\in\mathscr{C}_1(\mathscr{H})$, and let $\Theta\in H^\infty(\mathscr{L}(\mathscr{F}))$ be an inner function such that $S(\Theta)$ is unitarily equivalent to T and $I-\Theta(\lambda)\in\mathscr{C}_1(\mathscr{F})$, $\lambda\in\mathbf{D}$. The *determinant function* d_T of T is the function defined by $d_T(\lambda)=\det(\Theta(\lambda))$, $\lambda\in\mathbf{D}$. We also define the functions $\delta_j(T)=\delta_j(\Theta)$, $j\geq 0$.

It is easy to see that d_T (if it exists) is indeed uniquely determined by T, up to a constant multiple of absolute value one. Indeed, if $\Theta\in H^\infty(\mathscr{L}(\mathscr{F}))$ and $S(\Theta)$ is unitarily equivalent to T, then we can use Proposition 5.1.18 to write

$$\Theta(\lambda)=U'^*(V\Theta_T(\lambda)\oplus I_{\mathscr{N}})U,\qquad \lambda\in\mathbf{D},$$

with U, $U'\in\mathscr{L}(\mathscr{F},\mathscr{D}_T\oplus\mathscr{N})$. Since

$$I_{\mathscr{F}}-U'^*U=I-\Theta(0)+U'^*((V\Theta_T(\lambda)-I_{\mathscr{D}_T})\oplus 0)U\in\mathscr{C}_1(\mathscr{F}),$$

we must have

$$\begin{aligned}\det(\Theta(\lambda))&=\det[(U'^*U)U^*(V\Theta_T(\lambda)\oplus I_{\mathscr{N}})U]\\&=\det(U'^*U)\det[U^*(V\Theta_T(\lambda)\oplus I_{\mathscr{N}})U]\\&=\det(U'^*U)\det(V\Theta_T(\lambda)),\quad \lambda\in\mathbf{D},\end{aligned}$$

and $\det(U'^*U)$ is a constant of absolute value one. The same remarks apply to the functions $\delta_j(T)$, $j\geq 0$.

It is not obvious that the function d_T just defined is not identically zero, and our next task is to prove this fact. We begin with a very simple result.

3.11. LEMMA. *Assume that $T\in\mathscr{L}(\mathscr{H})$ is a contraction of class $C_{\cdot 0}$ such that $I-TT^*\in\mathscr{C}_1(\mathscr{H})$ and $\ker(T^*)=\{0\}$. Then $\ker(T)=\{0\}$ and d_T is not identically zero.*

PROOF. Consider the function $\Theta_1 \in H^\infty(\mathscr{L}(\mathscr{D}_{T^*}))$ defined by $\Theta_1(\lambda) = -\Theta_T(\lambda)T^* \mid \mathscr{D}_{T^*}$. Then we have

$$\begin{aligned} I - \Theta_1(\lambda) &= [I - TT^* + \lambda D_{T^*}(I - \lambda T^*)^{-1} D_T T^*] | \mathscr{D}_{T^*} \\ &= [I - TT^* + \lambda D_{T^*}(I - \lambda T^*)^{-1} T^* D_{T^*}] | \mathscr{D}_{T^*} \end{aligned}$$

so that $I - \Theta_1(\lambda) \in \mathscr{C}_1(\mathscr{D}_{T^*})$ because $I - TT^* \in \mathscr{C}_1(\mathscr{H})$ and D_{T^*} is a Hilbert–Schmidt operator. Moreover, $\Theta_1(0) = TT^* \mid \mathscr{D}_{T^*}$ is one-to-one and hence $\det(\Theta_1(0)) \neq 0$. Set $u(\lambda) = \det(\Theta_1(\lambda))$, $\Omega(\lambda) = \Theta_1(\lambda)^{\mathrm{Ad}\,1}$, $\lambda \in \mathbf{D}$. The relation

$$\Theta_T(\lambda)(-T^*\Omega(\lambda)) = \Theta_1(\lambda)\Omega(\lambda) = u(\lambda) I_{\mathscr{D}_{T^*}}, \quad \lambda \in \mathbf{D},$$

shows that Θ_T has a scalar multiple, and hence T is an operator of class $C_{\cdot 0}$. The equality $\ker(T) = \{0\}$ now follows from Proposition 2.4.9.

Finally, since $\ker(T) = \ker(T^*) = \{0\}$, the operator V given by (3.5) makes sense and $d_T \equiv \det(V\Theta_T)$. To conclude the proof it suffices to remark that $V\Theta_T(0)e_j = (1 - \mu_j)^{1/2} e_j$ and $\mu_j \neq 1$ for all j. Thus $V\Theta_T(0)$ is one-to-one, whence $d_T(0) \neq 0$.

The following result gives several characterizations of weak contractions of class C_{00}.

3.12. THEOREM. *Assume that $T \in \mathscr{L}(\mathscr{H})$ is a contraction of class $C_{\cdot 0}$ such that $I - T^*T \in \mathscr{C}_1(\mathscr{H})$. Then the following conditions are equivalent:*

(i) *T is of class C_0;*
(ii) *T is of class C_{00};*
(iii) $\dim \ker(T) = \dim \ker(T^*)$;
(iv) *there exists a finite-dimensional invariant subspace $\mathscr{H}_1$ for T such that $T \mid \mathscr{H}_1$ is nilpotent and $P_{\mathscr{H} \ominus \mathscr{H}_1} T \mid \mathscr{H} \ominus \mathscr{H}_1$ is invertible.*

PROOF. As we have already noted, $\ker(T)$ is finite-dimensional and $\dim \ker(T) \leq \operatorname{tr}(I - T^*T)$. The inequality

$$\dim \ker(T^k) \leq k \dim \ker(T)$$

shows that $\ker(T^k)$ is finite-dimensional for all positive integers k. We claim that $\ker(T^k) = \ker(T^{k+1})$ for some positive integer k. To prove this claim we reason by contradiction. Assume that $\ker(T^k) \neq \ker(T^{k+1})$ for all $k \geq 1$. Then the spaces

$$Y_{k,j} = T^{k-j}(\ker T^k)$$

are not zero for $1 \leq j \leq k$, and the relations

$$Y_{k+1,j} \subseteq Y_{k,j} \subseteq \ker(T^j), \quad 1 \leq j \leq k, \tag{3.13}$$

and

$$Y_{k,j} \subseteq TY_{k,j+1}, \quad 1 \leq j \leq k, \tag{3.14}$$

are easily verified. Since $\ker(T^j)$ is finite-dimensional, (3.13) implies that the sequence $\{Y_{k,j}: k \geq j\}$ is stationary, i.e., there exist nonzero subspaces $Z_j \subset$

$\ker(T^j)$, $j \geq 1$, such that $Y_{k,j} = Z_j$ for k sufficiently large. Furthermore, (3.14) shows that $Z_j \subseteq TZ_{j+1}$, $j \geq 1$. We can construct a sequence $\{x_n : 1 \leq n < \infty\}$ as follows. Choose $x_1 \in Z_1 \setminus \{0\}$ and then, inductively, $x_{n+1} \in Z_{n+1}$ such that $Tx_{n+1} = x_n$. Denote by $\mathscr{H}_0$ the space generated by $\{x_n : 1 \leq n < \infty\}$, and set $T_0 = T \mid \mathscr{H}_0$. It is clear that $\ker(T_0) \neq \{0\}$ since $x_1 \in \ker(T_0)$. Furthermore, T_0 is of class $C_{\cdot 0}$, $I - T_0^* T_0$ is of trace class, and T_0 has dense range, so that $\ker(T_0^*) = \{0\}$. This combination of properties is absurd by Lemma 3.11. We conclude that $\ker(T^k) = \ker(T^{k+1})$ for some $k \geq 1$.

Now choose an integer k such that $\ker(T^k) = \ker(T^{k+1})$, set $\mathscr{H}_1 = \ker(T^k)$, $\mathscr{H}_2 = \mathscr{H} \ominus \mathscr{H}_1$, and let

$$T = \begin{bmatrix} T_1 & X \\ 0 & T_2 \end{bmatrix}$$

be the triangularization of T with respect to the decomposition $\mathscr{H} = \mathscr{H}_1 \oplus \mathscr{H}_2$. Then clearly $\mathscr{H}_1$ is finite-dimensional and $T_1^k = \{0\}$. Moreover, we have $\ker(T_2) = \{0\}$. Indeed, $T_2 x = 0$ means $Tx \in \mathscr{H}_1 = \ker(T^k)$ so that $T^{k+1}x = T^k T x = 0$. Thus $x \in \ker(T^{k+1}) = \ker(T^k)$ so that $x \in \mathscr{H}_1 \cap \mathscr{H}_2 = \{0\}$. We also note that $T_2 \in C_{\cdot 0}$ and $I - T_2^* T_2 \in \mathscr{C}_1(\mathscr{H}_2)$ by Lemma 3.2(ii).

We now proceed to prove the equivalence of the four conditions. That (i) implies (ii) follows from Corollary 2.4.2. Assume now that (ii) holds. Then the operator T_2 is of class C_{00} and Lemma 3.11 can be applied to T_2^* to conclude that $\ker(T_2^*) = \{0\}$ and hence T_2 is invertible (the fact that T_2 has closed range follows because $I - T_2^* T_2 \in \mathscr{C}_1(\mathscr{H}_2)$). Thus we see that (ii) implies (iv). Now, if (iv) is satisfied, Lemma 3.11 applied to $P_{\mathscr{H} \ominus \mathscr{H}_1} T^* \mid \mathscr{H} \ominus \mathscr{H}_1$ shows that $P_{\mathscr{H} \ominus \mathscr{H}_1} T \mid \mathscr{H} \ominus \mathscr{H}_1$ is of class C_0 and hence T is of class C_0 by Proposition 4.4.4; thus (iv) implies (i). We finally verify the equivalence of (iii) and (iv). Assume that (iv) holds, and let T_1 and T_2 be as above so that T_2 is invertible. Then we have

$$\dim \ker(T) = \dim \ker(T_1) = \dim \ker(T_1^*)$$

because $\mathscr{H}_1$ is finite-dimensional, and

$$\ker(T^*) = \{u + [-T_2^{*^{-1}} X^* u] : u \in \ker(T_1^*)\}$$

so that $\dim \ker(T^*) = \dim \ker(T_1^*)$; (iii) follows. Assume now that (iii) holds. Then the Fredholm index of T is zero and the equality

$$T = \begin{bmatrix} I & 0 \\ 0 & T_2 \end{bmatrix} \begin{bmatrix} I & X \\ 0 & I \end{bmatrix} \begin{bmatrix} T_1 & 0 \\ 0 & I \end{bmatrix}$$

shows that

$$\operatorname{ind}(T_2) = \operatorname{ind}\left(\begin{bmatrix} I & 0 \\ 0 & T_2 \end{bmatrix} \right) - \operatorname{ind}\left(\begin{bmatrix} I & X \\ 0 & I \end{bmatrix} \right) - \operatorname{ind}\left(\begin{bmatrix} T_1 & 0 \\ 0 & I \end{bmatrix} \right) = 0.$$

Since $\ker(T_2) = \{0\}$, we conclude that T_2 is invertible, whence (iv). The theorem is proved.

3.15. COROLLARY. *Let $T \in \mathscr{L}(\mathscr{H})$ be a contraction of class C_{00}. Then T is a weak contraction if and only if T^* is a weak contraction, and* $\operatorname{tr}(I - T^*T) = \operatorname{tr}(I - TT^*)$ *if T is weak.*

PROOF. This obviously follows from Theorem 6.12 and Lemma 6.3(iii).

3.16. THEOREM. *Assume that $T \in \mathscr{L}(\mathscr{H})$ is a weak contraction of class C_{00}, $\mathscr{H}_1$ is an invariant subspace for T, and $T = \begin{bmatrix} T_1 & X \\ 0 & T_2 \end{bmatrix}$ is the triangularization of T with respect to the decomposition $\mathscr{H} = \mathscr{H}_1 \oplus (\mathscr{H} \ominus \mathscr{H}_1)$. Then*

(i) *d_T is an inner function; and*

(ii) *T_1 and T_2 are weak contractions and $d_T \equiv d_{T_1} d_{T_2}$.*

PROOF. We prove (ii) first. The fact that T_1 is a weak contraction of class C_{00} follows from Lemma 3.2. Now, T^* is also a weak contraction by the preceding corollary, so that $T_2^* = T^* \mid \mathscr{H} \ominus \mathscr{H}_1$ is a weak contraction by Lemma 3.2. Corollary 3.15 applied to T_2 shows that T_2 is a weak contraction.

By Proposition 5.1.21 there exist inner functions $\Theta \in H^\infty(\mathscr{L}(\mathscr{F},\mathscr{F}'))$, $\Theta_1 \in H^\infty(\mathscr{L}(\mathscr{F},\mathscr{K}))$, and $\Theta_2 \in H^\infty(\mathscr{L}(\mathscr{K},\mathscr{F}'))$ such that $\Theta = \Theta_2\Theta_1$, and $S(\Theta)$, $S(\Theta_1)$, and $S(\Theta_2)$ are unitarily equivalent to T, T_1, T_2, respectively. We can then apply Corollary 3.9 to find unitary operators $W \in \mathscr{L}(\mathscr{F}',\mathscr{F})$, $W_1 \in \mathscr{L}(\mathscr{K},\mathscr{F})$, $W_2 \in \mathscr{L}(\mathscr{F}',\mathscr{K})$ such that $I_\mathscr{F} - W\Theta(\lambda) \in \mathscr{C}_1(\mathscr{F})$, $I_\mathscr{F} - W_1\Theta_1(\lambda) \in \mathscr{C}_1(\mathscr{F})$, $I_\mathscr{K} - W_2\Theta_2(\lambda) \in \mathscr{C}_1(\mathscr{K})$ for $\lambda \in \mathbf{D}$. We claim that $I_\mathscr{F} - WW_2^*W_1^* \in \mathscr{C}_1(\mathscr{F})$. Indeed, we can write

$$\begin{aligned} I_\mathscr{F} - WW_2^*W_1^* &= I_\mathscr{F} - W\Theta(0) + W(\Theta(0) - W_2^*W_1^*) \\ &= I_\mathscr{F} - W\Theta(0) + W(\Theta_2(0)\Theta_1(0) - W_2^*W_1^*) \end{aligned}$$

so that we need only to verify that $\Theta_2(0)\Theta_1(0) - W_2^*W_1^*$ is a trace-class operator. But this clearly follows from the fact that $\Theta_1(0) - W_1^* = W_1^*(W_1\Theta_1(0) - I_\mathscr{F})$ and $\Theta_2(0) - W_2^* = W_2^*(W_2\Theta_2(0) - I_\mathscr{K})$. Then the obvious relation

$$W\Theta(\lambda) = (WW_2^*W_1^*)[W_1(W_2\Theta_2(\lambda))W_1^*](W_1\Theta_1(\lambda))$$

implies

$$\begin{aligned} d_T(\lambda) &= \det(W\Theta(\lambda)) \\ &= \det(WW_2^*W_1^*)\det[W_1(W_2\Theta_2(\lambda))W_1^*]\det(W_1\Theta_1(\lambda)) \\ &= \det(WW_2^*W_1^*)d_{T_2}(\lambda)d_{T_1}(\lambda). \end{aligned}$$

Part (ii) follows because $\det(WW_2^*W_1^*)$ is a constant of absolute value one.

In order to prove (i) we apply (ii) to the decomposition $T = \begin{bmatrix} T_1 & X \\ 0 & T_2 \end{bmatrix}$, where T_1 is a nilpotent operator on a finite-dimensional space and T_2 is invertible. By Theorem 2.3 we need only to show that $d_T \not\equiv 0$. We know that $d_{T_2} \not\equiv 0$ from Lemma 3.11, and it will suffice to show that $d_{T_1} \not\equiv 0$. Elementary linear algebra shows that T_1 can be written as an upper triangular matrix with zeros on the main diagonal. Then $n-1$ applications of (ii), with $n = \dim(\mathscr{H}_1)$, will show that $d_{T_1} \equiv d_{T_0}^n$, where $T_0 \in \mathscr{L}(\mathbf{C})$ is the zero operator. Since obviously $d_{T_0}(\lambda) = \lambda$, we conclude that $d_{T_1} \not\equiv 0$, thus completing the proof of our theorem.

As noted before, the Jordan operator $\bigoplus_{j<\omega} S(\theta_j)$ is unitarily equivalent to $S(\Theta)$, where $\Theta \in H^\infty(\mathscr{L}(\mathscr{F}))$ is defined by $\Theta(\lambda)e_j = \theta_j(\lambda)e_j$, $j < \omega$, and $\{e_j : j < \omega\}$ is an orthonormal basis of $\mathscr{F}$. We clearly have

$$\operatorname{tr}(I - \Theta(0)^*\Theta(0)) = \sum_{j<\omega}(1 - |\theta_j(0)|^2)$$

and, by Proposition 3.4, $\bigoplus_{j<\omega} S(\theta_j)$ is a weak contraction if and only if

$$\sum_{j<\omega}(1-|\theta_j(0)|^2) < \infty.$$

Since $1-|\theta_j(0)| \le 1-|\theta_j(0)|^2 \le 2(1-|\theta_j(0)|)$, this is equivalent to

$$\sum_{j<\omega}(1-|\theta_j(0)|) < \infty. \tag{3.17}$$

3.18. COROLLARY. *Let $T \in \mathscr{L}(\mathscr{H})$ be a weak contraction of class C_{00}.*

(i) *T is an operator of class C_0 and its Jordan model $T' = \bigoplus_{j=0}^{\infty} S(\theta_j)$ is given by the formula $\theta_j \equiv \delta_j(T)/\delta_{j+1}(T)$, $j<\omega$.*
(ii) *The Jordan model T' of T is a weak contraction and $d_T \equiv d_{T'}$. In particular, $|d_T(\lambda)| = \prod_{j<\omega} |\theta_j(\lambda)|$ for $\lambda \in \mathbf{D}$.*

PROOF. That T is of class C_0 follows from Theorem 3.12. The formulas for θ_j follow now from Definition 3.10, Corollary 3.9, and Proposition 2.9. The equality

$$|d_T(\lambda)| = \prod_{j<\omega} |\theta_j(\lambda)|, \quad \lambda \in \mathbf{D}, \tag{3.19}$$

also follows from Proposition 2.9. A consequence of this equality is the fact that $\sum_{j<\omega}(1-|\theta_j(\lambda)|) < \infty$ and hence T' is a weak contraction by the remarks preceding this corollary. The equality (3.19) also implies that $d_T \equiv d_{T'}$. Indeed, if the first nonzero coefficient of each θ_j is positive, it follows that $d_{T'}(\lambda) = \prod_{j<\omega} \theta_j(\lambda)$, and $d_T \equiv d_{T'}$ follows from (3.19) and the maximum modulus principle.

Theorem 3.16 admits a converse. For the proof we need the following technical result.

3.20. LEMMA. *Assume that*

$$T = \begin{bmatrix} T_1 & X \\ 0 & T_2 \end{bmatrix} \in \mathscr{L}(\mathscr{H}_1 \oplus \mathscr{H}_2)$$

is a contraction. Then there exists a contraction $Y \in \mathscr{L}(\mathscr{H}_2, \mathscr{H}_1)$ such that $X = D_{T_1^*} Y D_{T_2}$.

PROOF. If $A \in \mathscr{L}(\mathscr{H}, \mathscr{K})$ and $A' \in \mathscr{L}(\mathscr{H}', \mathscr{K})$ then we denote by $[A, A'] \in \mathscr{L}(\mathscr{H} \oplus \mathscr{H}', \mathscr{K})$ the operator defined by

$$[A, A'](h \oplus h') = Ah + A'h';$$

we have $[A, A']^* = \left[\begin{smallmatrix} A^* \\ A'^* \end{smallmatrix}\right]$. The fact that T is a contraction can be written as

$$\|[T_1, X](h_1 \oplus h_2)\|^2 + \|[0, T_2](h_1 \oplus h_2)\|^2 \le \|h_1 \oplus h_2\|^2$$

or, equivalently,

$$\|[T_1, X](h_1 \oplus h_2)\| \le \left\| \begin{bmatrix} I_{\mathscr{H}_1} & 0 \\ 0 & D_{T_2} \end{bmatrix} (h_1 \oplus h_2) \right\|.$$

This inequality implies the existence of a contraction $Z = [Z_1, Z_2] \in \mathscr{L}(\mathscr{H}_1 \oplus \mathscr{H}_2, \mathscr{H}_1)$ such that

$$[T_1, X] = [Z_1, Z_2] \begin{bmatrix} I_{\mathscr{H}_1} & 0 \\ 0 & D_{T_2} \end{bmatrix}$$

so that

$$T_1 = Z_1, \qquad X = Z_2 D_{T_2}. \tag{3.21}$$

Now, Z is a contraction if and only if Z^* is a contraction, so that

$$\left\| \begin{bmatrix} Z_1^* \\ Z_2^* \end{bmatrix} (h_1) \right\| \le \|h_1\|, \quad h_1 \in \mathscr{H}_1,$$

or, by (3.21),

$$\|T_1^* h_1\|^2 + \|Z_2^* h_1\|^2 \le \|h_1\|^2.$$

Equivalently, we have

$$\|Z_2^* h_1\| \le \|D_{T_1^*} h_1\|, \quad h_1 \in \mathscr{H}_1,$$

and therefore we can find a contraction Y satisfying

$$Z_2^* = Y^* D_{T_1^*}. \tag{3.22}$$

The equality $X = D_{T_1^*} Y D_{T_2}$ follows at once from (3.21) and (3.22).

3.23. PROPOSITION. *Assume that $T \in \mathscr{L}(\mathscr{H})$ is a contraction, $\mathscr{H}_1 \subseteq \mathscr{H}$ is an invariant subspace for T, and $T = \left[\begin{smallmatrix} T_1 & X \\ 0 & T_2 \end{smallmatrix}\right]$ is the triangularization of T with respect to the decomposition $\mathscr{H} = \mathscr{H}_1 \oplus (\mathscr{H} \ominus \mathscr{H}_1)$. If T_1 and T_2 are weak contractions of class C_{00} then T is a weak contraction of class C_{00}.*

PROOF. Theorem 3.12 implies that T_1 and T_2 are operators of class C_0, and hence T is of class C_0 by Proposition 2.4.4. Thus it suffices to show that $I - T^*T \in \mathscr{C}_1(\mathscr{H})$. We have

$$I - T^*T = \begin{bmatrix} I - T_1^*T_1 & -T_1^*X \\ -X^*T_1 & I - T_2^*T_2 - X^*X \end{bmatrix}.$$

Since $I - T_j^*T_j \in \mathscr{C}_1(\mathscr{H}_j)$, $j = 1, 2$, the proposition will be proved if we show that X is a trace-class operator. But this clearly follows from Lemma 3.20: $X = D_{T_1^*} Y D_{T_2}$, where $D_{T_1^*}$ and D_{T_2} are Hilbert–Schmidt operators.

Exercises

1. Let $T_j \in \mathscr{L}(\mathscr{H}_j)$, $j = 1, 2$, and $Y \in \mathscr{L}(\mathscr{H}_2, \mathscr{H}_1)$ be contractions. Show that

$$\begin{bmatrix} T_1 & D_{T_1^*} Y D_{T_2} \\ 0 & T_2 \end{bmatrix} \in \mathscr{L}(\mathscr{H}_1 \oplus \mathscr{H}_2)$$

is a contraction.

2. Let T be a contraction such that $\dim \ker(T) = \dim \ker(T^*)$. Prove that D_T is unitarily equivalent to D_{T^*}. Is the converse true?

3. Let $\Theta \in H^\infty(\mathcal{L}(\mathcal{F}))$ be an inner function such that $I - \Theta(\lambda) \in \mathcal{C}_1(\mathcal{F})$ for all $\lambda \in \mathbf{D}$. Show that the function $\Psi: \mathbf{D} \to \mathcal{C}_1(\mathcal{F})$ given by $\Psi(\lambda) = I - \Theta(\lambda)$ is analytic.

4. Assume that $\{e_n: 0 \le n < \infty\}$ is an orthonormal basis of $\mathcal{H}$ and let $\{\alpha_n: 0 \le n < \infty\}$ be a bounded sequence in $\mathbf{C}\setminus\{0\}$. Define the operator $T \in \mathcal{L}(\mathcal{H})$ by $Te_n = \alpha_n e_{n+1}$, $0 \le n < \infty$. Show that the sequence $\{\alpha_n: 0 \le n < \infty\}$ can be chosen such that T is of class C_{00} and $I - T^*T$ is compact. Obviously $\ker(T) = \{0\} \neq \ker(T^*)$.

5. Show that there exist contractions $T \in \mathcal{L}(\mathcal{H})$ of class $C_{\cdot 0}$ such that $I - T^*T \in \mathcal{C}_1(\mathcal{H})$ and $\ker(T) = \{0\} \neq \ker(T^*)$.

6. Show that there exist contractions T of class C_{00} such that $I - T^*T$ is compact but $I - TT^*$ is not compact.

7. Let T be an operator of class C_0. Prove that D_T is compact if and only if D_{T^*} is compact.

8. Assume that $T = \begin{bmatrix} T_1 & X \\ 0 & T_2 \end{bmatrix}$ is an operator of class C_0. Show that D_T is compact if and only if D_{T_1} and D_{T_2} are compact.

9. Let T be a completely nonunitary contraction such that $I - T^*T$ is compact. Prove that T acts on a separable Hilbert space.

10. Let T be an algebraic contraction such that $I - T^*T$ is compact. Show that T acts on a finite-dimensional space.

11. Let V be an accretive operator such that $\sigma(V) = \{0\}$, $\ker(V) = \{0\}$, and $V + V^*$ has finite trace. Show that the operator $T = (I - V)(I + V)^{-1}$ is of class C_0.

12. Let T, T_1, and T_2 be as in Exercise 8. Show that D_T has finite rank if and only if D_{T_1} and D_{T_2} have finite rank. If T, T_1, and T_2 are of class $C_0(N)$, $C_0(N_1)$, and $C_0(N_2)$, respectively, then

$$\max\{N_1, N_2\} \le N \le N_1 + N_2.$$

4. Defect operators and splitting. The material in this section is a logical continuation of Section 3, and it is included in a separate section in order to reduce the length of Section 3. We have shown in Corollary 3.18 that the Jordan model of a weak contraction of class C_0 is also a weak contraction. We will now prove the converse of this result, along with a few related facts.

Let $A \in \mathcal{L}(\mathcal{K})$, and let $\mathcal{M}$ be a closed subspace of $\mathcal{K}$. We set

$$\gamma(A, \mathcal{M}) = \inf\{\|Ak\|: k \in \mathcal{M}, \|k\| = 1\},$$

and

$$\gamma_j(A) = \sup\{\gamma(A, \mathcal{M}): \dim(\mathcal{K} \ominus \mathcal{M}) = j\}, \quad 0 \le j < \dim(\mathcal{K}). \tag{4.1}$$

Since $\|Ak\| = \| |A|k\|$, $|A| = (A^*A)^{1/2}$, for all $k \in \mathcal{K}$, it follows that

$$\gamma_j(A) = \gamma_j(|A|), \quad 0 \le j < \dim(\mathcal{K}). \tag{4.2}$$

The numbers $\gamma_j(A)$ can be described in terms of the spectral characteristics of $|A|$. If, for example, $\mathscr{K}$ is finite-dimensional, then $\{\gamma_j(A): 0 \leq j < \dim(\mathscr{K})\}$ are the eigenvalues of $|A|$ in increasing order and repeated according to their multiplicities. If $\mathscr{K}$ is infinite-dimensional, then not all eigenvalues of $|A|$ will usually figure in the sequence $\{\gamma_j(A): j \geq 0\}$. More precisely, let us set

$$\alpha = \inf\{\lambda: \lambda \in \sigma(|A|)\}$$

and

$$\beta = \inf\{\lambda: \lambda \in \sigma_e(|A|)\},$$

where σ_e denotes the essential spectrum. The set

$$\sigma_0 = \{\lambda: \lambda \in \sigma(|A|), \alpha \leq \lambda < \beta\}$$

consists of eigenvalues of $|A|$ with finite multiplicities. If the sum of the multiplicities of the eigenvalues in σ_0 is infinite, then $\{\gamma_j(A): j \geq 0\}$ is an increasing list of these eigenvalues, each repeated according to its multiplicity. If the sum of the multiplicities of the eigenvalues in σ_0 is $n < \infty$, then $\{\gamma_j(A): 0 \leq j < n\}$ is a list of these eigenvalues, and $\gamma_j(A) = \beta$ for $j \geq n$. We will need these facts only in the case in which $|A|$ has finite spectrum, and in this case they are obvious.

4.3. LEMMA. *For every $A \in \mathscr{L}(\mathscr{K})$ and every $j < \dim(\mathscr{K})$ we have*

$$\gamma_0\left(\bigwedge^{j+1} A\right) = \gamma_0(A)\gamma_1(A)\cdots\gamma_j(A).$$

PROOF. It is clear that $|\bigwedge^{j+1} A| = \bigwedge^{j+1}|A|$ and hence we may assume that A is a positive operator. Moreover, the definition (4.1) of γ_j clearly implies that

$$|\gamma_j(A) - \gamma_j(B)| \leq \|A - B\|, \quad A, B \in \mathscr{L}(\mathscr{K}),$$

and hence it suffices to prove Lemma 4.3 for a dense set of positive operators A. Such a dense set consists of all positive operators with finite spectrum. Assume then that A is positive and has finite spectrum. There is an orthonormal basis $\{e_\alpha\}$ of $\mathscr{K}$ consisting of eigenvalues of A, say $Ae_\alpha = \lambda_\alpha e_\alpha$. Then $\bigwedge^{j+1}\mathscr{K}$ also has a basis of eigenvectors $e_{\alpha_1} \wedge e_{\alpha_2} \wedge \cdots \wedge e_{\alpha_{j+1}}$ of $\bigwedge^{j+1} A$, and the eigenvalues of $\bigwedge^{j+1} A$ are all the products of $j+1$ distinct eigenvalues of A. It is clear from the discussion above that the least eigenvalue of $\bigwedge^{j+1} A$, that is, $\gamma_0(\bigwedge^{j+1} A)$, coincides with $\gamma_0(A)\gamma_1(A)\cdots\gamma_j(A)$. The lemma follows.

4.4. COROLLARY. *Assume that T is an operator of class C_0 with Jordan model $\bigoplus_{j<\omega} S(\theta_j)$, and $\Theta \in H^\infty(\mathscr{L}(\mathscr{F}))$ is an inner function such that T is unitarily equivalent to $S(\Theta)$. For every $j < \dim(\mathscr{F})$ and every $\lambda \in \mathbf{D}$ we have*

$$\gamma_0(\Theta(\lambda))\gamma_1(\Theta(\lambda))\cdots\gamma_j(\Theta(\lambda)) \geq |\theta_0(\lambda)\theta_1(\lambda)\cdots\theta_j(\lambda)|.$$

PROOF. By Corollary 5.3.29 we have $d_{j+1}(\Theta) \equiv \theta_0\theta_1\cdots\theta_j$, $j < \dim(\mathscr{F})$, and hence $\theta_0\theta_1\cdots\theta_j$ is a scalar multiple of $\bigwedge^{j+1}\Theta$. Choose a function $\Omega_j \in H^\infty(\mathscr{L}(\bigwedge^{j+1}\mathscr{F}))$ such that $\Omega_j(\lambda)(\bigwedge^{j+1}\Theta(\lambda)) = \theta_0(\lambda)\theta_1(\lambda)\cdots\theta_j(\lambda)I_{\bigwedge^{j+1}\mathscr{F}}$,

$\lambda \in \mathbf{D}$. Note that $\|\Omega_j\|_\infty \leq 1$ and, in fact, Ω_j is inner since $\bigwedge^{j+1}\Theta$ and $\theta_0\theta_1\cdots\theta_j$ are inner functions. Therefore

$$\begin{aligned}|\theta_0(\lambda)\theta_1(\lambda)\cdots\theta_j(\lambda)| &= \gamma_0(\theta_0(\lambda)\theta_1(\lambda)\cdots\theta_j(\lambda)I_{\bigwedge^{j+1}\mathscr{F}}) \\ &= \gamma_0(\Omega_j(\lambda)\left(\bigwedge^{j+1}\Theta(\lambda))\right) \\ &\leq \gamma_0\left(\bigwedge^{j+1}\Theta(\lambda)\right)\end{aligned}$$

and the corollary follows from Lemma 4.3 applied to $A = \Theta(\lambda)$.

We need one more technical device before the proof of our next result. Assume that $T \in \mathscr{L}(\mathscr{H})$ and $\mu \in \mathbf{D}$, and define $T_\mu \in \mathscr{L}(\mathscr{H})$ by

$$T_\mu = (T - \mu I)(I - \overline{\mu}T)^{-1}. \tag{4.5}$$

The operator T_μ, called a *Möbius transform* of T, inherits many of the properties of T. If T is unitary [resp., completely nonunitary] then T_μ is also unitary [resp., completely nonunitary]. More important for us is the obvious fact that T_μ is of class C_0 whenever T is of class C_0 and m_{T_μ} can be computed as follows:

$$m_{T_\mu}(\lambda) = m_T\left(\frac{\lambda+\mu}{1+\overline{\mu}\lambda}\right), \quad \lambda \in \mathbf{D}. \tag{4.6}$$

Since $m_{T_\mu}(0) = m_T(\mu)$, the operator T_μ is invertible if $m_T(\mu) \neq 0$. An easy calculation shows that

$$I - T_\mu^*T_\mu = (1-|\mu|^2)(I-\mu T^*)^{-1}(I-T^*T)(I-\overline{\mu}T)^{-1},$$

and hence $I - T_\mu^*T_\mu$ is compact [resp., of trace-class] if and only if $I - T^*T$ is compact [resp., of trace-class]. Moreover, $\dim(\mathscr{D}_T) = 1$ if and only if $\dim(\mathscr{D}_{T_\mu}) = 1$, so that T is unitarily equivalent to a Jordan block if and only if T_μ is unitarily equivalent to a Jordan block.

4.7. THEOREM. *Let $T \in \mathscr{L}(\mathscr{H})$ be an operator of class C_0 with Jordan model $T' = \bigoplus_{j<\omega} S(\theta_j)$. Then T is a weak contraction if and only if T' is a weak contraction.*

PROOF. The Jordan model of T_μ is unitarily equivalent to T'_μ, $\mu \in \mathbf{D}$, and hence by the remarks preceding this theorem there is no loss of generality in assuming that T is invertible. We already know from Corollary 3.18 that T' is a weak contraction whenever T is a weak contraction. Assume therefore that T' is a weak contraction, and choose an inner function $\Theta \in H^\infty(\mathscr{L}(\mathscr{F}))$ such that T is unitarily equivalent to $S(\Theta)$. We may of course assume that $\mathscr{F}$ is infinite-dimensional since otherwise $S(\Theta)$ is clearly a weak contraction. Now, the fact that T' is a weak contraction means that $\sum_{j<\omega}(1-|\theta_j(0)|) < \infty$ (cf. (3.17)) and, since $\theta_j(0) \neq 0$ for all j, $\prod_{j<\omega}|\theta_j(0)|$ is different from zero. We deduce from Corollary 4.4 that $\prod_{j<\omega}\gamma_j(\Theta(0)) > 0$ and therefore

$$\sum_{j<\omega}(1-\gamma_j(\Theta(0))^2) \leq 2\sum_{j<\omega}(1-\gamma_j(\Theta(0))) < \infty.$$

This last inequality is easily seen to be equivalent to the inequality $\operatorname{tr}(I - \Theta(0)^*\Theta(0)) < \infty$ and thus T is a weak contraction by Proposition 3.4(i). The theorem is proved.

A similar, but weaker, result can be proved in the case of compact defect operators.

4.8. PROPOSITION. *Let $T \in \mathscr{L}(\mathscr{H})$ be an operator of class C_0 with Jordan model $T' = \bigoplus_{j<\omega} S(\theta_j)$. If $I - T'^*T'$ is compact then $I - T^*T$ is also compact.*

PROOF. As in the proof of the preceding theorem, there is no loss of generality in assuming that T and T' are invertible. Proposition 3.4(i) applied to T' shows that $\lim_{j\to\infty} |\theta_j(0)| = 1$, and this implies that

$$\lim_{j\to\infty} |\theta_0(0)\theta_1(0)\cdots\theta_j(0)|^{1/(j+1)} = 1.$$

Now, if Θ is chosen as in the proof of Theorem 4.7, Corollary 4.4 implies that

$$\lim_{j\to\infty} (\gamma_0(\Theta(0))\gamma_1(\Theta(0))\cdots\gamma_j(\Theta(0)))^{1/(j+1)} = 1$$

and, since the sequence $\{\gamma_j(\Theta(0)) : j \geq 0\}$ is increasing, we deduce that $\lim_{j\to\infty}\gamma_j(\Theta(0)) = 1$. This relation implies that $I - \Theta(0)^*\Theta(0)$ is compact. Indeed, if $\varepsilon > 0$ and $\gamma_j(\Theta(0)) > 1 - \varepsilon$, we can find a subspace $\mathscr{M} \subseteq \mathscr{F}$, $\dim(\mathscr{F} \ominus \mathscr{M}) = j$, such that $\gamma(\Theta(0), \mathscr{M}) > 1 - \varepsilon$. It is clear then that $I - \Theta(0)^*\Theta(0)$ can be approximated within ε by the finite rank operator $(I - \Theta(0)^*\Theta(0))(I - P_{\mathscr{M}})$. Finally, our result follows from Proposition 3.4(i).

We turn now to the splitting alluded to in the title of this section. It is interesting to know when an operator T of class C_0 is unitarily equivalent to its Jordan model. It is thus relevant to study the reducing subspaces of T. The constants $\gamma_j(\Theta(0))$ used in the proof of the two preceding results can be used in certain instances to produce reducing subspaces. The basic idea is the following result from linear algebra.

4.9. LEMMA. *Assume that $\mathscr{F}$ is a Hilbert space, $T \in \mathscr{L}(\mathscr{F})$ is an invertible operator, and $\{e_1, e_2, \ldots, e_n\}$ is an orthonormal system in $\mathscr{F}$ such that $e_1 \wedge e_2 \wedge \cdots \wedge e_n$ is an eigenvalue for $\bigwedge^n T$. Then the space $\mathscr{M}$ generated by $\{e_1, e_2, \ldots, e_n\}$ is invariant for T.*

PROOF. We note first that, if $h_1, h_2, \ldots, h_n, k_1, k_2, \ldots, k_n \in \mathscr{F}$, we have

$$(h_1 \wedge h_2 \wedge \cdots \wedge h_n, k_1 \wedge k_2 \wedge \cdots \wedge k_n) = \det[(h_i, k_j)]_{1\leq i,j\leq n}. \tag{4.10}$$

Indeed, this is an easy consequence of (1.1).

If $\mathscr{M} = \mathscr{F}$ there is nothing to prove. Assume therefore that e_{n+1} is a unit vector in $\mathscr{F}$, orthogonal onto $\mathscr{M}$, and denote by $x_j \in \mathbf{C}^n$, $1 \leq j \leq n+1$, the vector with components

$$(x_j)_i = (Te_i, e_j), \quad 1 \leq i \leq n.$$

The eigenvalue of $\bigwedge^n T$ corresponding with $e_1 \wedge e_2 \wedge \cdots \wedge e_n$ must be different from zero because T is invertible. Thus (4.10) with $k_i = e_i$ and $h_i = Te_i$, $1 \leq i \leq n$, shows that the vectors $x_1, x_2, \dots x_n$ are linearly independent. Similar arguments show that the vectors $x_1, x_2, \dots x_{i-1}, x_{i+1}, \dots x_{n+1}$ are linearly dependent for $1 \le i \le n$, and hence we can write $x_{n+1} = \sum_{j \ne i} \alpha_j^i x_j$, $\alpha_j^i \in \mathbf{C}$. Now, the coefficients α_j^i must not depend on i because $x_1, x_2, \dots, x_n$ are independent. Hence we conclude that $x_{n+1} = 0$ or, equivalently, $(Te_i, e_{n+1}) = 0$ for $1 \le i \le n$. Since e_{n+1} is arbitrary in $\mathscr{F} \ominus \mathscr{M}$, we conclude that $Te_i \in \mathscr{M}$, $1 \le i \le n$, and hence $\mathscr{M}$ is invariant for T.

4.11. THEOREM. *Assume that $T \in \mathscr{L}(\mathscr{H})$ is an operator of class C_0 with Jordan model $\bigoplus_{j<\omega} S(\theta_j)$, $\Theta \in H^\infty(\mathscr{L}(\mathscr{F}))$ is an inner function such that $S(\Theta)$ is unitarily equivalent to T, and $0 \le j < \dim(\mathscr{D}_T)$. Suppose also that $\theta_0(0) \ne 0$ and $|\theta_j(0)| < \lim_{n\to\infty} |\theta_n(0)|$. Then the following conditions are equivalent:*

(i) $|\theta_0(0)\theta_1(0)\cdots\theta_j(0)| = \gamma_0(\Theta(0))\gamma_1(\Theta(0))\cdots\gamma_j(\Theta(0))$;

(ii) *$\mathscr{H}$ can be written as $\mathscr{H}_1 \oplus \mathscr{H}_2$, where $\mathscr{H}_1$ is reducing for T, the Jordan model of $T_1 = T \mid \mathscr{H}_1$ is $\bigoplus_{n=0}^{j} S(\theta_n)$, the Jordan model of $T_2 = T \mid \mathscr{H}_2$ is $\bigoplus_{n<\omega} S(\theta_{n+j})$, and $\dim(\mathscr{D}_{T_1}) = j+1$.*

PROOF. Assume first that (ii) holds. Then the function Θ can be factored as $\Theta = \Theta_2\Theta_1$, where $\Theta_j \in H^\infty(\mathscr{L}(\mathscr{F}))$ is an inner function equivalent to Θ_{T_j}, $j = 1, 2$. Since $\dim(\mathscr{D}_{T_1}) = j+1$, we have $\bigwedge^{j+1} \Theta_{T_1} = \theta_0\theta_1 \dots \theta_j V$, where V is a unitary operator from $\bigwedge^{j+1} \mathscr{D}_{T_1}$ to $\bigwedge^{j+1} \mathscr{D}_{T_1^*}$ (these two spaces are one-dimensional). The coincidence of Θ_1 with Θ_{T_1} shows that

$$\gamma_0(\Theta_1(0))\cdots\gamma_j(\Theta_1(0)) = \gamma_0\left(\bigwedge^{j+1}\Theta_1(0)\right) = |\theta_0(0)\theta_1(0)\cdots\theta_j(0)|;$$

here we used, of course, Theorem 5.3.21 and Lemma 4.3. The equality $\Theta(0) = \Theta_2(0)\Theta_1(0)$ combined with $\|\Theta_2(0)\| \le 1$ shows that $\gamma_i(\Theta(0)) \le \gamma_1(\Theta_1(0))$ for all i, and we deduce

$$\gamma_0(\Theta(0))\gamma_1(\Theta(0))\cdots\gamma_j(\Theta(0)) \le |\theta_0(0)\theta_1(0)\cdots\theta_j(0)|.$$

Thus (i) follows now from Corollary 4.4. Observe that in this part of the proof we have not used the entire hypothesis of the theorem.

Conversely, assume that (i) holds. As we saw before, there is no loss of generality in assuming that $\Theta(0)$ is a positive operator on $\mathscr{F}$. Indeed, the condition $\theta_0(0) \ne 0$ implies that T, and hence $\Theta(0)$, is invertible. If $\Theta(0) = UP$ is the polar decomposition of $\Theta(0)$, with $P = (\Theta(0)^*\Theta(0))^{1/2}$, then U is unitary, and we may replace Θ by the equivalent function $U^*\Theta$.

With the additional assumption that $\Theta(0) \ge 0$, we claim that $\gamma_j(\Theta(0))$ is less than the first essential eigenvalue of $\Theta(0)$ $(= (\Theta(0)^*\Theta(0))^{1/2})$. Indeed, if this were not true, we would have

$$\gamma_j(\Theta(0)) = \gamma_{j+1}(\Theta(0)) = \cdots$$

and, as in the proof of Proposition 4.8, we would have

$$\begin{aligned}\gamma_j(\Theta(0)) &= \lim_{n\to\infty} \gamma_n(\Theta(0)) \\ &= \lim_{n\to\infty} (\gamma_0(\Theta(0))\gamma_1(\Theta(0))\cdots\gamma_{n-1}(\Theta(0)))^{1/n} \\ &\geq \lim_{n\to\infty} |\theta_n(0)| \\ &> |\theta_j(0)|.\end{aligned}$$

This inequality and (i) would then yield

$$|\theta_0(0)\theta_1(0)\cdots\theta_{j-1}(0)| > \gamma_0(\Theta(0))\gamma_1(\Theta(0))\cdots\gamma_{j-1}(\Theta(0)),$$

and this would contradict Corollary 4.4. We conclude therefore that $\gamma_0(\Theta(0)), \ldots, \gamma_j(\Theta(0))$ are eigenvalues of $\Theta(0)$, so that we can choose an orthonormal system $\{e_0, e_1, \ldots, e_j\}$ satisfying

$$\Theta(0)e_k = \gamma_k(\Theta(0))e_k, \quad k = 0, 1, \ldots, j.$$

The vector $f = e_0 \wedge e_1 \wedge \cdots \wedge e_j \in \bigwedge^{j+1} \mathscr{F}$ is not zero, and the preceding equalities imply that

$$\left(\bigwedge\nolimits^{j+1} \Theta(0)\right) f = \gamma_0(\Theta(0))\gamma_1(\Theta(0))\cdots\gamma_j(\Theta(0))f. \tag{4.12}$$

Corollary 5.3.20 now implies that $\theta_0\theta_1\cdots\theta_j$ is a scalar multiple of $\bigwedge^{j+1}\Theta$ and hence there exists an inner function $\Omega \in H^\infty(\mathscr{L}(\bigwedge^{j+1}\mathscr{F}))$ such that

$$\Omega(\lambda)\left(\bigwedge\nolimits^{j+1}\Theta(\lambda)\right) = \theta_0(\lambda)\theta_1(\lambda)\cdots\theta_j(\lambda)I_{\bigwedge^{j+1}\mathscr{F}}. \tag{4.13}$$

Relations (4.11), (4.12), and (i) show that we must have $\|\Omega(0)f\| = \|f\|$. The maximum modulus principle and the Schwarz inequality applied to the scalar analytic function $(\Omega(\lambda)f, f)$, $\lambda \in \mathbf{D}$, show the existence of a scalar μ of absolute value one satisfying

$$\Omega(\lambda)f = \mu f, \quad \lambda \in \mathbf{D}.$$

This last equality and (4.13) imply that

$$\left(\bigwedge\nolimits^{j+1}\Theta(\lambda)\right) f = \mu^{-1}\theta_0(\lambda)\theta_1(\lambda)\cdots\theta_{j-1}(\lambda)f, \quad \lambda \in \mathbf{D}.$$

Lemma 4.9 now implies that the space $\mathscr{F}_1$ generated by $\{e_0, e_1, \ldots e_j\}$ is invariant for $\Theta(\lambda)$ whenever $\theta_0(\lambda) \neq 0$. By continuity, $\mathscr{F}_1$ is invariant under $\Theta(\lambda)$ for all $\lambda \in \mathbf{D}$, and hence $\mathscr{F}_1$ is invariant for $\Theta(\varsigma)$ for almost every ς in $\mathbf{T}$. We conclude that in fact $\mathscr{F}_1$ is reducing for $\Theta(\varsigma)$ and for $\Theta(\lambda)$, $\lambda \in \mathbf{D}$. Upon setting $\mathscr{F}_2 = \mathscr{F} \ominus \mathscr{F}_1$, $\Theta_i(\lambda) = \Theta(\lambda) \mid \mathscr{F}_i$, $i = 1, 2$, we have $\Theta = \Theta_1 \oplus \Theta_2$, and hence T is unitarily equivalent to $S(\Theta_1) \oplus S(\Theta_2)$. To conclude the proof we will show that $T_1 = S(\Theta_1)$ is quasisimilar to $\bigoplus_{i=0}^{j} S(\theta_i)$, $T_2 = S(\Theta_2)$ is quasisimilar to $\bigoplus_{i=j+1}^{\infty} S(\theta_i)$, and $\dim(\mathscr{D}_{T_1}) = j+1$. The facts that $\Theta_{T_1}(0) = -T_1 \mid \mathscr{D}_{T_1}$ and Θ_1 is equivalent to Θ_{T_1} show that $\dim(\mathscr{D}_{T_1}) = \operatorname{rank}(I - \Theta_1(0)^*\Theta_1(0))$, and it is clear that $I - \Theta_1(0)^*\Theta_1(0)$ is a diagonal operator with eigenvalues $1 - \gamma_i(\Theta(0))^2$, $0 \leq i \leq j$. The equality $\dim(\mathscr{D}_{T_1}) = j+1$ follows at once.

Assume now that $S(\theta_0') \oplus S(\theta_1') \oplus \cdots \oplus S(\theta_j')$ is the Jordan model of T_1 and $\bigoplus_{i<\omega} S(\theta_i'')$ is the Jordan model of T_2. We know from Corollary 5.3.20 that $\theta_0\theta_1 \cdots \theta_j$ is the least scalar multiple of the function $\bigwedge^{j+1}(\Theta_1 \oplus \Theta_2)$. Furthermore, it follows from Theorem 5.3.12 that $\theta_0\theta_1 \cdots \theta_j$ is the least scalar multiple of the function $\bigwedge^{j+1} \Theta_3$, where Θ_3 is the diagonal matrix with diagonal entries $\theta_0', \theta_1', \ldots, \theta_j', \theta_0'', \theta_1'', \ldots$. A moment's thought now shows that

$$\theta_0\theta_1 \cdots \theta_j \equiv \left(\bigvee_{k=0}^{j-1} \theta_0'\theta_1' \cdots \theta_k'\theta_0''\theta_1'' \cdots \theta_{j-k-1}'' \right) \vee (\theta_0''\theta_1'' \cdots \theta_j''). \tag{4.14}$$

On the other hand, assumption (ii) and the fact that $\gamma_i(\Theta_1(0)) = \gamma_i(\Theta(0))$, $0 \le i \le j$, imply that

$$\begin{aligned} |\theta_0'(0)\theta_1'(0) \cdots \theta_j'(0)| &= \gamma_0(\Theta_1(0))\gamma_1(\Theta_1(0)) \cdots \gamma_j(\Theta_1(0)) \\ &= |\theta_0(0)\theta_1(0) \cdots \theta_j(0)|, \end{aligned}$$

and by the maximum modulus principle we have $\theta_0\theta_1 \cdots \theta_j \equiv \theta_0'\theta_1' \cdots \theta_j'$. Relation (4.14) now implies that $\theta_0'\theta_1' \cdots \theta_{j-1}'\theta_0'' \mid \theta_0'\theta_1' \cdots \theta_j'$, and hence $\theta_0'' \mid \theta_j'$. We conclude that $(\bigoplus_{i=0}^{j} S(\theta_i')) \oplus (\bigoplus_{i<\omega} S(\theta_i''))$ is a Jordan operator quasisimilar to $T_1 \oplus T_2$. The uniqueness of the Jordan model implies that $\theta_i' \equiv \theta_i$, $0 \le i \le j$, and $\theta_i'' \equiv \theta_{j+i+1}$, $i < \omega$, thus concluding the proof.

Exercises

1. Verify the description of the sequence $\{\gamma_j(A): j \ge 0\}$ given at the beginning of Section 4.

2. Show that the set of positive operators with finite spectrum is (norm-) dense in the set of all positive operators in $\mathscr{L}(\mathscr{H})$.

3. Under the assumptions of Corollary 4.4, show that the inequality $\gamma_1(\Theta(\lambda)) \ge |\theta_1(\lambda)|$ is not always true.

4. Verify the properties of T_μ described after (4.5). Show that T_μ is a unilateral shift if and only if T is a unilateral shift.

5. Let μ be a finite positive measure on $[0, 2\pi]$. Assume that μ is continuous (i.e., without atoms) but singular with respect to Lebesgue measure. Define the functions

$$\theta_{j,n}(\lambda) = \exp\left[-\int_{2\pi(j-1)/n}^{2\pi j/n} \frac{e^{it} + \lambda}{e^{it} - \lambda} \, d\mu(t) \right], \quad 1 \le j \le n, \quad n \ge 1, \quad \lambda \in \mathbf{D},$$

and the operator $T = \bigoplus_{n=1}^{\infty} (\bigoplus_{j=1}^{n} S(\theta_{j,n}))$. Prove that T is an operator of class C_0; $I - T^*T$ is compact but this property is not inherited by the Jordan model of T. Thus the converse of Proposition 4.8 is false.

6. Show that Theorem 4.10 may fail if $|\theta_j(0)| = \lim_{n\to\infty} |\theta_n(0)|$. (A counterexample can be obtained as follows. Choose $\alpha \in (0, 1)$ and $\varepsilon_n \in (0, 1 - \alpha^2)$ such

that $\lim_{n\to\infty}\varepsilon_n = 1-\alpha^2$. Define $T = \bigoplus_{n=0}^{\infty}\left[\begin{smallmatrix}\alpha & -\varepsilon_n\\ 0 & \alpha\end{smallmatrix}\right]$, and let Θ be equivalent to Θ_T. Then we have $\gamma_j(\Theta(0)) = |\theta_j(0)| = \alpha^2$ for all $j<\omega$.)

7. Let $T_0 = \bigoplus_{j<\omega} S(\theta_j)$ be a Jordan operator such that $\theta_0(0)\neq 0$ and $\lim_{j\to\infty}|\theta_j(0)| = 1$. Define the classes $\mathscr{T}_j$, $-1\le j<\omega$, as follows. The class $\mathscr{T}_{-1}$ is the class of all contractions $T\in\mathscr{L}(\mathscr{H})$ that are quasisimilar to T_0. Then, inductively, $\mathscr{T}_j = \{T\in\mathscr{T}_{j-1}\colon \gamma_j(T) = \inf\{\gamma_j(T')\colon T'\in\mathscr{T}_{j-1}\}\}$. Prove that every element of $\bigcap_{j<\omega}\mathscr{T}_j$ is unitarily equivalent to T_0.

5. Invariant factors and quasiequivalence. It is known that the Jordan model of an operator T acting on a finite-dimensional Hilbert space can be calculated from the invariant factors of the polynomial characteristic matrix $\lambda I - T$, $\lambda\in\mathbf{C}$. We will show shortly that an appropriate generalization of this fact holds for all operators of class C_0.

5.1. DEFINITION. Assume that $\Theta\in H^\infty(\mathscr{L}(\mathscr{F}))$, and $\phi\in H^\infty$ is an inner function. We say that ϕ *divides* Θ if there exists $\Theta'\in H^\infty(\mathscr{L}(\mathscr{F}))$ satisfying $\Theta(\lambda) = \phi(\lambda)\Theta'(\lambda)$, $\lambda\in\mathbf{D}$. The greatest inner divisor of Θ will be denoted (if it exists) by $\mathscr{D}(\Theta)$.

It is clear that $\mathscr{D}(\Theta)$ fails to exist if and only if $\Theta\equiv 0$. Thus $\mathscr{D}(\Theta)$ exists if Θ is inner and $\mathscr{F}\neq\{0\}$. It is sometimes convenient to write $\mathscr{D}(\Theta) = 0$ if $\Theta\equiv 0$. Note that ϕ divides Θ if and only if $\phi H^2(\mathscr{F})\supset\Theta H^2(\mathscr{F})$ and, by invariance under $S_{\mathscr{F}}$, if and only if $\Theta f\in\phi H^2(\mathscr{F})$ for all $f\in\mathscr{F}$.

5.2. DEFINITION. Assume that $\Theta\in H^\infty(\mathscr{L}(\mathscr{F}))$, and k is an integer such that $1\le k\le\dim(\mathscr{F})$. Then the kth *elementary divisor* $\mathscr{D}_k(\Theta)$ is defined by $\mathscr{D}_k(\Theta) = \mathscr{D}(\bigwedge^k\Theta)$.

If Θ is viewed as a matrix with respect to some fixed orthonormal basis, then $\mathscr{D}_k(\Theta)$ coincides with the greatest common inner divisors of all minors of rank k of Θ (provided that at least one such minor is not identically zero). This observation makes the following result quite obvious.

5.3. LEMMA. *Assume that $\Theta\in H^\infty(\mathscr{L}(\mathscr{F}))$ and $1\le k\le\dim(\mathscr{F})$. Then $\mathscr{D}_k(\Theta)$ divides $\mathscr{D}_{k+1}(\Theta)$.*

PROOF. Denote $\phi = \mathscr{D}_k(\Theta)$. By the remark preceding Definition 5.2 it suffices to show that $(\bigwedge^{k+1}\Theta)h\in\phi H^2(\bigwedge^{k+1}\mathscr{F})$ for all $h\in\bigwedge^{k+1}\mathscr{F}$. We may, of course, restrict ourselves to a generating set of vectors h, for example, to vectors of the form

$$h = f_1\wedge f_2\wedge\cdots\wedge f_{k+1},\quad f_j\in\mathscr{F},\quad 1\le j\le k+1.$$

Then we have $(\bigwedge^k\Theta)(f_1\wedge f_2\wedge\cdots\wedge f_k) = \phi g$ for some $g\in H^2(\bigwedge^k\mathscr{F})$ by the definition of ϕ, and therefore

$$\left(\bigwedge\nolimits^{k+1}\Theta\right)(f_1\wedge f_2\wedge\cdots\wedge f_k\wedge f_{k+1}) = (\phi g)\wedge(\Theta f_{k+1}) = \phi(g\wedge(\Theta f_{k+1}))$$

belongs to $\phi H^2(\bigwedge^{k+1}\mathscr{F})$, as desired. The lemma follows.

5.4. DEFINITION. The *invariant factors* of a function $\Theta \in H^\infty(\mathscr{L}(\mathscr{F}))$ are defined by the following formulas:

$$\mathscr{E}_1(\Theta) = \mathscr{D}_1(\Theta),$$
$$\mathscr{E}_k(\Theta) = \mathscr{D}_k(\Theta)/\mathscr{D}_{k-1}(\Theta), \quad 2 \le k \le \dim(\mathscr{F}), \quad \mathscr{D}_k(\Theta) \neq 0,$$

and

$$\mathscr{E}_k(\Theta) = 0, \qquad\qquad 2 \le k \le \dim(\mathscr{F}), \quad \mathscr{D}_k(\Theta) = 0.$$

Note that for an inner function Θ all elementary divisors and all invariant factors are inner functions (i.e., they are not identically zero).

5.5. LEMMA. *Assume that Θ, $\Omega \in H^\infty(\mathscr{L}(\mathscr{F}))$, $\phi \in H^\infty$, Θ and ϕ are inner, and $\Omega(\lambda)\Theta(\lambda) = \phi(\lambda)I_{\mathscr{F}}$, $\lambda \in \mathbf{D}$. Then Θ and Ω are two-sided inner, $S(\Theta)$ and $S(\Omega)$ are operators of class C_0, and*

$$\mathscr{D}(\Omega)m_{S(\Theta)} \equiv \phi. \tag{5.6}$$

PROOF. Parts of this result have been proved and used before. Thus, we know from Proposition 5.3.4 that the hypothesis implies $\Theta(\lambda)\Omega(\lambda) = \phi(\lambda)I_{\mathscr{F}}$, so that Ω has the scalar multiple ϕ. These relations also show that

$$\Theta(\varsigma)\Omega(\varsigma) = \Omega(\varsigma)\Theta(\varsigma) = \phi(\varsigma)I_{\mathscr{F}}$$

for almost every $\varsigma \in \mathbf{T}$, and this clearly implies that $\Theta(\varsigma)$ and $\Omega(\varsigma)$ are unitary operators for almost every $\varsigma \in \mathbf{T}$. We conclude that Θ and Ω are two-sided inner, and $S(\Theta)$ and $S(\Omega)$ are of class C_0 by Proposition 5.3.2.

Now denote by ψ the minimal function of $S(\Theta)$; thus $\psi \mid \phi$. Proposition 5.3.2 implies the existence of a function $\Omega_1 \in H^\infty(\mathscr{L}(\mathscr{F}))$ such that $\Omega_1(\lambda)\Theta(\lambda) = \psi(\lambda)I_{\mathscr{F}}$, $\lambda \in \mathbf{D}$. Consequently we have $\Theta(\lambda)((\psi/\phi)(\lambda)\Omega_1(\lambda) - \Omega(\lambda)) = 0$, $\lambda \in \mathbf{D}$, and, since T_Θ is an isometry, $\Omega = (\psi/\phi)\Omega_1$. We conclude that

$$(\psi/\phi) \mid \mathscr{D}(\Omega). \tag{5.7}$$

On the other hand we have $\Omega = \mathscr{D}(\Omega)\Omega'$ for some $\Omega' \in H^\infty(\mathscr{L}(\mathscr{F}))$, and an easy computation shows that $\Omega'\Theta = (\phi/\mathscr{D}(\Omega))I_{\mathscr{F}}$. We conclude that $\psi \mid (\phi/\mathscr{D}(\Omega))$ and this relation, together with (5.7), implies (5.6). The lemma is proved.

We can now convert the formulas in Corollary 5.3.29 into formulas involving the invariant factors.

5.8. PROPOSITION. *Assume that Ω, $\Theta \in H^\infty(\mathscr{L}(\mathscr{F}))$, and $\phi \in H^\infty$ are inner functions such that $\Omega(\lambda)\Theta(\lambda) = \phi(\lambda)I_{\mathscr{F}}$, $\lambda \in \mathbf{D}$; let $\bigoplus_{j<\omega} S(\theta_j)$ be the Jordan model of $S(\Theta)$. Then we have*

$$\theta_j \equiv \phi/\mathscr{E}_{j+1}(\Omega), \quad 0 \le j < \dim(\mathscr{F})$$

and

$$\theta_j \equiv 1 \quad \textit{for } j \ge \dim(\mathscr{F}).$$

PROOF. It suffices to consider the case in which $j < \dim(\mathscr{F})$. We have

$$\left(\bigwedge^k \Omega(\lambda)\right)\left(\bigwedge^k \Theta(\lambda)\right) = \bigwedge^k(\phi(\lambda)I_{\mathscr{F}}) = \phi(\lambda)^k I_{\bigwedge^k \mathscr{F}}, \qquad \lambda \in \mathbf{D},$$

for $1 \leq k \leq \dim(\mathscr{F})$. By Corollary 5.3.20 the minimal function of $S(\bigwedge^k \Theta)$ is $\theta_0\theta_1 \cdots \theta_{k-1}$ so that (5.6) implies

$$\mathscr{D}_k(\Omega)\theta_0\theta_1 \cdots \theta_{k-1} \equiv \phi^k, \quad 1 \leq k \leq \dim(\mathscr{F}). \tag{5.9}$$

Relations (5.9) for $k = j$ and $k = j+1$ clearly imply

$$\begin{aligned} \phi^{j+1} &\equiv \mathscr{D}_{j+1}(\Omega)\theta_0\theta_1 \cdots \theta_j \\ &\equiv \mathscr{E}_{j+1}(\Omega)\mathscr{D}_j(\Omega)\theta_0\theta_1 \cdots \theta_j \\ &\equiv \mathscr{E}_{j+1}(\Omega)\phi^j\theta_j, \end{aligned}$$

and the proposition follows at once.

5.10. COROLLARY. *Assume that $\Theta \in H^\infty(\mathscr{L}(\mathscr{F}))$ is an inner function such that $S(\Theta)$ is an operator of class C_0. Then $\mathscr{E}_j(\Theta)$ divides $\mathscr{E}_{j+1}(\Theta)$ for $1 \leq j < \dim(\mathscr{F})$.*

PROOF. Proposition 5.8 implies that $\mathscr{E}_j(\Omega) \equiv \phi/\theta_{j-1}$ divides $\phi/\theta_j \equiv \mathscr{E}_{j+1}(\Omega)$, $1 \leq j < \dim(\mathscr{F})$. The corollary follows because the roles of Θ and Ω are interchangeable by Lemma 5.5.

Remark that Proposition 5.8 is not a perfect analogue of the classical linear algebra theorem because it requires the calculation of the elementary factors of a function (Ω) which is different from the characteristic function of the operator in question (Θ). The situation is better if $\mathscr{F}$ is a finite-dimensional space.

5.11. PROPOSITION. *Assume that $\Theta \in H^\infty(\mathscr{L}(\mathscr{F}))$ is an inner function, and $\dim(\mathscr{F}) = n < \infty$. If the Jordan model of $S(\Theta)$ is $\bigoplus_{j<\omega} S(\theta_j)$ then*

$$\theta_j \equiv \mathscr{E}_{n-j}(\Theta), \quad 0 \leq j < n,$$

and

$$\theta_j \equiv 1, \quad n \leq j < \omega.$$

PROOF. It suffices to consider the case in which $j < n$. Note that

$$\Theta^{\mathrm{Ad}\,k}(\lambda)\left(\bigwedge^k \Theta\right)(\lambda) = \det(\Theta(\lambda))I_{\bigwedge^k \mathscr{F}}, \quad \lambda \in \mathbf{D}, \quad 1 \leq k \leq n,$$

so that (5.6) and Corollary 5.3.20 imply

$$\mathscr{D}(\Theta^{\mathrm{Ad}\,k})\theta_0\theta_1 \cdots \theta_{k-1} \equiv \det(\Theta). \tag{5.12}$$

As is easily seen from the proof of Proposition 1.5, the matrix of $\Theta^{\mathrm{Ad}\,k}$ coincides (in a suitable orthonormal basis) with the transpose of the matrix of $\bigwedge^{n-k} \Theta$. Consequently $\mathscr{D}(\Theta^{\mathrm{Ad}\,k}) \equiv \mathscr{D}(\bigwedge^{n-k} \Theta) \equiv \mathscr{D}_{n-k}(\Theta)$ and (5.12) implies

$$\det(\Theta) \equiv \mathscr{D}_{n-j-1}(\Theta)\theta_0\theta_1 \cdots \theta_j \equiv \mathscr{D}_{n-j}(\Theta)\theta_0\theta_1 \cdots \theta_{j-1}$$

and therefore $\theta_j \equiv \mathscr{D}_{n-j}(\Theta)/\mathscr{D}_{n-j-1}(\Theta) = \mathscr{E}_{n-j}(\Theta)$.

Proposition 5.11 cannot be extended in a meaningful way to infinite-dimensional spaces $\mathscr{F}$. For example, assume that $\Theta = \operatorname{diag}(\phi, 1, 1, \cdots)$, where ϕ is an inner function in H^∞. We clearly have $\mathscr{E}_j(\Theta) \equiv 1$ for all $j \geq 1$, and hence no information can be extracted from these invariant factors. Still, $S(\Theta)$ is a nice operator—it is unitarily equivalent to $S(\phi)$.

A natural question is whether Proposition 5.11 can be deduced on a different path, more similar to the classical equivalence theory of matrices over a principal ideal domain. One basic problem is the fact that H^∞ is not a principal ideal domain, and the greatest common inner divisor of two functions in H^∞ does not coincide with their greatest common divisor (when it exists) in the ring H^∞. In particular, the equivalence relation for matrices over H^∞ will not lead to a good diagonalization theorem. It turns out that the weaker relation of quasiequivalence, to be introduced shortly, leads to a certain diagonalization theorem.

In the sequel we will say that a function $\Phi \in H^\infty(\mathcal{L}(\mathcal{F}))$ has a (two-sided) scalar multiple $\phi \in H^\infty \backslash \{0\}$ if there exists $\Phi' \in H^\infty(\mathcal{L}(\mathcal{F}))$ satisfying the relations $\Phi'(\lambda)\Phi(\lambda) = \Phi(\lambda)\Phi'(\lambda) = \phi(\lambda)I_{\mathcal{F}}$, $\lambda \in \mathbf{D}$. We will generally omit the words "two-sided."

5.13. DEFINITION. Assume that Θ_1, $\Theta_2 \in H^\infty(\mathcal{L}(\mathcal{F}))$, and $\omega \in H^\infty$ is an inner function. Then Θ_1 and Θ_2 are said to be *ω-equivalent* if there exist functions Φ, $\Psi \in H^\infty(\mathcal{L}(\mathcal{F}))$ with two-sided multiples ϕ and ψ, respectively, such that

$$\phi \wedge \omega \equiv 1, \qquad \psi \wedge \omega \equiv 1, \tag{5.14}$$

and

$$\Phi\Theta_1 = \Theta_2\Psi. \tag{5.15}$$

The functions Θ_1 and Θ_2 are *quasiequivalent* if they are ω-equivalent for every inner function ω in H^∞.

The fact that ω-equivalence (and hence quasiequivalence) is reflexive and transitive is obvious. Symmetry is proved as follows: assume that Φ, Ψ, ϕ, ψ satisfy (5.14) and (5.15), and Φ', $\Psi' \in H^\infty(\mathcal{L}(\mathcal{F}))$ are such that $\Phi\Phi' = \Phi'\Phi = \phi I_{\mathcal{F}}$, and $\Psi\Psi' = \Psi'\Psi = \psi I_{\mathcal{F}}$. Then the functions $\phi\Psi'$ and $\psi\Phi'$ both have the scalar multiple $\phi\psi$, $(\phi\psi) \wedge \omega \equiv 1$, and, by (5.15),

$$(\psi\Phi')\Theta_2 = \Phi'\Theta_2\psi = \Phi'\Theta_2\Psi\Psi' = \Phi'\Phi\Theta_1\Psi' = \Theta_1(\phi\Psi'), \tag{5.16}$$

which shows that Θ_2 is ω-equivalent to Θ_1. Thus ω-equivalence and quasiequivalence are indeed equivalence relations.

It is obvious that quasiequivalence could be defined on $H^\infty(\mathcal{L}(\mathcal{F}, \mathcal{F}'))$, but this would complicate notation. The reader will have no difficulty in extending the relevant definitions and results to the case in which $\mathcal{F} \neq \mathcal{F}'$.

5.17. PROPOSITION. *Assume that Θ_1 and Θ_2 are two quasiequivalent functions in $H^\infty(\mathcal{L}(\mathcal{F}))$.*

(i) *The functions $\bigwedge^k \Theta_1$ and $\bigwedge^k \Theta_2$ are quasiequivalent, $1 \le k \le \dim(\mathcal{F})$.*
(ii) *$\mathcal{D}_k(\Theta_1) \equiv \mathcal{D}_k(\Theta_2)$, $1 \le k \le \dim(\mathcal{F})$.*
(iii) *If Θ_1 and Θ_2 are inner then $S(\Theta_1)$ is of class C_0 if and only if $S(\Theta_2)$ is of class C_0.*
(iv) *If $S(\Theta_1)$ and $S(\Theta_2)$ are of class C_0 then they are quasisimilar.*

PROOF. Assume that Φ, Ψ, ϕ, ψ satisfy (5.14) and (5.15), and Φ', Ψ' are such that $\Phi\Phi' = \Phi'\Phi = \phi I_{\mathscr{F}}$ and $\Psi\Psi' = \Psi'\Psi = \psi I_{\mathscr{F}}$. We clearly have $(\bigwedge^k \Phi)(\bigwedge^k \Theta_1) = (\bigwedge^k \Theta_1)(\bigwedge^k \Psi)$ and $(\bigwedge^k \Phi)(\bigwedge^k \Phi') = (\bigwedge^k \Phi')(\bigwedge^k \Phi) = \bigwedge^k (\phi I_{\mathscr{F}}) = \phi^k I_{\bigwedge^k \mathscr{F}}$ for $1 \leq k \leq \dim(\mathscr{F})$. Moreover, $\phi^k \wedge \omega \equiv \psi^k \wedge \omega \equiv 1$ (cf. Proposition 2.2.12) and we conclude that $\bigwedge^k \Theta_1$ and $\bigwedge^k \Theta_2$ are ω-equivalent. Since ω is an arbitrary inner function, (i) is proved.

In view of (i), (ii) will follow if we can prove that $\mathscr{D}(\Theta_1) \equiv \mathscr{D}(\Theta_2)$. Let Φ, Φ', Ψ, Ψ', ϕ, ψ be as above. Relation (5.15) shows that $\Theta_1(\lambda) = 0$ if and only if $\Theta_2(\lambda) = 0$, provided that $\phi(\lambda) \neq 0$ and $\psi(\lambda) \neq 0$; indeed, for such values of λ the operators $\Phi(\lambda)$ and $\Psi(\lambda)$ are invertible. Since the zeros of $\phi\psi$ form a discrete set we conclude that $\Theta_1 = 0$ if and only if $\Theta_2 = 0$, hence $\mathscr{D}(\Theta_1) = 0$ if and only if $\mathscr{D}(\Theta_2) = 0$. We may then assume that $\mathscr{D}(\Theta_1) \neq 0$, $\mathscr{D}(\Theta_2) \neq 0$, and we choose $\omega = \mathscr{D}(\Theta_1)$. We have $\Theta_1 = \mathscr{D}(\Theta_1)\Theta_1'$ for some $\Theta_1' \in H^\infty(\mathscr{L}(\mathscr{F}))$ by the definition of $\mathscr{D}(\Theta_1)$. By (5.15) we have $\mathscr{D}(\Theta_1)\Phi\Theta_1'\Psi' = \Phi\Theta_1\Psi' = \Theta_2\Psi\Psi' = \psi\Theta_2$ and, since $\psi \wedge \mathscr{D}(\Theta_1) \equiv 1$, $\mathscr{D}(\Theta_1)$ must divide Θ_2. We conclude that $\mathscr{D}(\Theta_1) \mid \mathscr{D}(\Theta_2)$ and, by symmetry, $\mathscr{D}(\Theta_1) \equiv \mathscr{D}(\Theta_2)$.

Assume next that Θ_1 and Θ_2 are inner, and $S(\Theta_2)$ is of class C_0. By Proposition 5.3.2, Θ_2 has a scalar multiple $\theta \in H^\infty$, so that $\Omega_2\Theta_2 = \theta I_{\mathscr{F}}$ for some $\Omega_2 \in H^\infty(\mathscr{L}(\mathscr{F}))$. Relation (5.15) now implies

$$\Psi'\Omega_2\Phi\Theta_1 = \Psi'\Omega_2\Theta_2\Psi = \theta\Psi'\Psi = \theta\psi I_{\mathscr{F}}$$

so that Θ_1 has the scalar multiple $\theta\psi \neq 0$. By Proposition 5.3.2, $S(\Theta_1)$ is of class C_0, and (iii) follows for reasons of symmetry.

Assume finally that $S(\Theta_1)$ and $S(\Theta_2)$ are of class C_0 with minimal functions θ_1 and θ_2, respectively. Let Φ, Φ',Ψ,Ψ',ϕ, ψ be as above, with $\omega = \theta_1\theta_2$. Relations (5.15) and (5.16) and Proposition 5.1.24 show that we can define operators $Y \in \mathscr{I}(S(\Theta_2), S(\Theta_1))$ and $X \in \mathscr{I}(S(\Theta_1), S(\Theta_2))$ by the formulas $X = P_{\mathscr{H}(\Theta_2)}T_\Phi \mid \mathscr{H}(\Theta_1)$ and $Y = P_{\mathscr{H}(\Theta_1)}T_{\psi\Phi'} \mid \mathscr{H}(\Theta_2)$. Moreover, we have

$$\begin{aligned} XY &= P_{\mathscr{H}(\Theta_2)}T_\Phi T_{\psi\Phi'} \mid \mathscr{H}(\Theta_2) \\ &= P_{\mathscr{H}(\Theta_2)}T_{\psi\Phi\Phi'} \mid \mathscr{H}(\Theta_2) \\ &= P_{\mathscr{H}(\Theta_2)}T_{\phi\psi I} \mid \mathscr{H}(\Theta_2) \\ &= (\phi\psi)(S(\Theta_2)) \end{aligned} \tag{5.18}$$

and, in an analogous manner,

$$YX = (\phi\psi)(S(\Theta_1)). \tag{5.19}$$

We have $(\phi\psi) \wedge \theta_1 \equiv (\phi\psi) \wedge \theta_2 \equiv 1$ so that the operators $(\phi\psi)(S(\Theta_1))$ and $(\phi\psi)(S(\Theta_2))$ are quasiaffinities by Proposition 2.4.9. Relations (5.18) and (5.19) show that X and Y are also quasiaffinities, hence $S(\Theta_1) \sim S(\Theta_2)$.

The proof of the proposition is now complete.

5.20. REMARK. The above proof of (ii) shows that, in fact, $\mathscr{D}(\Theta_1) \equiv \mathscr{D}(\Theta_2)$ whenever Θ_1 and Θ_2 are ω-equivalent, $\mathscr{D}(\Theta_1) \mid \omega$ and $\mathscr{D}(\Theta_2) \mid \omega$. Analogously, if Θ_1 and Θ_2 are ω-equivalent, $\mathscr{D}_k(\Theta_1) \mid \omega$, and $\mathscr{D}_k(\Theta_2) \mid \omega$, then $\mathscr{D}_k(\Theta_1) \equiv \mathscr{D}_k(\Theta_2)$. More generally, if Θ_1, Θ_2, Φ, Ψ, ϕ, ψ satisfy (5.14) and (5.15), then

$$\mathscr{D}_k(\Theta_1) \mid \psi_0^k \mathscr{D}_k(\Theta_2) \quad \text{and} \quad \mathscr{D}_k(\Theta_2) \mid \phi_0^k \mathscr{D}_k(\Theta_1), \tag{5.21}$$

where ϕ_0 and ψ_0 denote the inner factors of ϕ and ψ, respectively. Indeed, if we write $\bigwedge^k \Theta_1 = \mathscr{D}_k(\Theta_1)\Omega_1$, then $\mathscr{D}_k(\Theta_1)(\bigwedge^k \Phi)\Omega_1(\bigwedge^k \Psi') = \psi^k(\bigwedge^k \Theta_2)$, and the first relation in (5.21) follows at once. The second relation follows by symmetry.

From this point on we will fix an orthonormal basis $\{e_j: 0 \le j < \dim(\mathscr{F})\}$ of the separable Hilbert space $\mathscr{F}$, so that every function $\Theta \in H^\infty(\mathscr{L}(\mathscr{F}))$ can be described via its matrix $[\theta_{ij}]_{0 \le i,j < \dim(\mathscr{F})}$ over H^∞, where $\theta_{ij}(\lambda) = (\Theta(\lambda)e_i, e_j)$, $\lambda \in \mathbf{D}$, $0 \le i, j < \dim(\mathscr{F})$. For an integer $k \le \dim(\mathscr{F})$ we will denote by $\mathscr{F}_k$ the space generated by $\{e_j: j \ge k\}$; thus $\dim(\mathscr{F} \ominus \mathscr{F}_k) = k$.

If $\theta_1, \theta_2, \ldots, \theta_k \in H^\infty$ and $\theta_0 \in H^\infty(\mathscr{L}(\mathscr{F}_k))$, then we can build a matrix $\Theta = \operatorname{diag}(\theta_1, \theta_2, \ldots, \theta_k, \Theta_0) \in H^\infty(\mathscr{L}(\mathscr{F}))$ by setting

$$\begin{aligned} \Theta(\lambda)e_j &= \theta_{j+1}(\lambda)e_j \quad && \text{if } 0 \le j \le k-1, \\ &= \Theta_0(\lambda)e_j && \text{if } j \ge k. \end{aligned}$$

We recall that the elementary divisor $\mathscr{D}_k(\Theta)$ of a function $\Theta = [\theta_{ij}]_{0 \le i,j < \dim(\mathscr{F})}$ can be defined as the greatest common inner divisor of all minors of rank k of $[\theta_{ij}]_{0 \le i,j < \dim(\mathscr{F})}$, i.e., the greatest common inner divisor of all functions of the form $\det[\theta_{i_p j_q}]_{1 \le p,q \le k}$, where $0 \le i_1 < i_2 < \cdots < i_k < \dim(\mathscr{F})$ and $0 \le j_1 < j_2 < \cdots < j_k < \dim(\mathscr{F})$. In particular, $\mathscr{D}_1(\Theta) = \bigwedge\{\theta_{ij}: 0 \le i, j < \dim(\mathscr{F})\}$.

5.22. LEMMA. *Assume that $\Theta \in H^\infty(\mathscr{L}(\mathscr{F}))$, and let $\omega \in H^\infty$ be an inner function. There exists a function $\Theta_1 \in H^\infty(\mathscr{L}(\mathscr{F}_1))$ such that Θ is ω-equivalent to* $\operatorname{diag}(\mathscr{D}(\Theta), \Theta_1)$ *and $\mathscr{D}(\Theta)$ divides $\mathscr{D}(\Theta_1)$. We have $\Theta_1 = 0$ if and only if $\mathscr{D}_2(\Theta) = 0$.*

PROOF. The lemma is trivial if $\Theta = 0$, so we assume that $\Theta \ne 0$. It is clear that ω'-equivalence implies ω-equivalence if $\omega \mid \omega'$. There is therefore no loss of generality in assuming that $\mathscr{D}(\Theta)^2 \mid \omega$. Let $[\theta_{ij}]_{0 \le i,j < \dim(\mathscr{F})}$ be the matrix of Θ. An application of Theorem 3.1.14 provides a sequence $\{x_j: 0 \le j < \dim(\mathscr{F})\}$ of scalars such that

$$x_0 \ne 0, \qquad \sum_{0 \le j < \dim(\mathscr{F})} |x_j| < \infty, \tag{5.23}$$

and

$$\Big(\sum_{0 \le j < \dim(\mathscr{F})} x_j \theta_{ij}\Big) \wedge \omega \equiv \Big(\bigwedge\nolimits_{0 \le j < \dim(\mathscr{F})} \theta_{ij}\Big) \wedge \omega, \quad 0 \le i < \dim(\mathscr{F}). \tag{5.24}$$

In order to realize simultaneously the conditions in (5.24) we must use the known fact that a finite or countable intersection of dense G_δ sets is also a dense G_δ

set. If we set $\theta_i = \sum_{0\le j<\dim(\mathscr{F})} x_j\theta_{ij}$, then we have

$$\begin{aligned}\left(\bigwedge\nolimits_{0\le i<\dim(\mathscr{F})}\theta_i\right)\wedge\omega &\equiv \bigwedge\nolimits_{0\le i<\dim(\mathscr{F})}(\theta_i\wedge\omega)\\ &\equiv \bigwedge\nolimits_{0\le i<\dim(\mathscr{F})}\left[\left(\bigwedge\nolimits_{0\le j<\dim(\mathscr{F})}\theta_{ij}\right)\wedge\omega\right]\\ &\equiv \left(\bigwedge\{\theta_{ij}\colon 0\le i,j<\dim(\mathscr{F})\}\right)\wedge\omega\\ &\equiv \mathscr{D}(\Theta)\wedge\omega\\ &\equiv \mathscr{D}(\Theta)\end{aligned}$$

because of the assumption that $\mathscr{D}(\Theta)^2 \mid \omega$. One further application of Theorem 3.1.14 provides a sequence $\{y_i\colon 0\le i<\dim(\mathscr{F})\}$ such that

$$y_0\ne 0,\qquad \sum_{0\le i<\dim(\mathscr{F})}|y_i|<\infty, \tag{5.25}$$

and

$$\left(\sum_{0\le i<\dim(\mathscr{F})} y_i\theta_i\right)\wedge\omega\equiv\mathscr{D}(\Theta). \tag{5.26}$$

We now construct invertible constant matrices $X=[x_{ij}]_{0\le i,j<\dim(\mathscr{F})}$ and $Y=[y_{ij}]_{0\le i,j<\dim(\mathscr{F})}$ as follows: $x_{ij}=x_j$ if $i=0$, $x_{ij}=\delta_{ij}$ if $i\ge 1$, $y_{ij}=y_i$ if $j=0$, and $y_{ij}=\delta_{ij}$ if $j\ge 1$ (where $\delta_{ij}=0$ if $i\ne j$ and $\delta_{ii}=1$); X and Y are bounded and invertible by (5.23) and (5.25). The matrix Θ is clearly equivalent to $\Theta'=Y\Theta X=[\theta'_{ij}]$, and we have $\theta'_{00}=\sum_{0\le i<\dim(\mathscr{F})} y_i\theta_i$ so that $\theta'_{00}\wedge\omega\equiv\mathscr{D}(\Theta)\equiv\mathscr{D}(\Theta')$ by (5.26). Now, Proposition 2.2.4 shows that $(\theta'_{00}/\mathscr{D}(\Theta))\wedge(\omega/\mathscr{D}(\Theta))\equiv 1$ and, since $\mathscr{D}(\Theta)\mid(\omega/\mathscr{D}(\Theta))$, we also have $(\theta'_{00}/\mathscr{D}(\Theta))\wedge\mathscr{D}(\Theta)\equiv 1$. The last two equalities imply (via Proposition 2.2.12(iv)) that $(\theta'_{00}/\mathscr{D}(\Theta))\wedge\omega\equiv 1$ so that we can write

$$\theta'_{00}=\phi\mathscr{D}(\Theta),\qquad \phi\wedge\omega\equiv 1. \tag{5.27}$$

We perform next an ω-equivalence in order to get rid of the factor ϕ. Let us set $\Theta''=[\theta''_{ij}]_{0\le i,j<\dim(\mathscr{F})}$, where

$$\begin{aligned}\theta''_{ij} &= \mathscr{D}(\Theta) && \text{if } i=j=0,\\ &= \theta'_{ij} && \text{if } i=0 \text{ or } j=0 \text{ but } i\ne j,\\ &= \phi\theta'_{ij} && \text{if } i\ne 0 \text{ and } j\ne 0.\end{aligned}$$

We also set $\Phi=\operatorname{diag}(1,\phi I_{\mathscr{F}_1})$ and $\Psi=\operatorname{diag}(\phi,I_{\mathscr{F}_1})$. Both Φ and Ψ have the scalar multiple ϕ, and clearly $\Phi\Theta'=\Theta''\Psi$ so that Θ and Θ'' are ω-equivalent. Note that $\mathscr{D}(\Theta)$ continues to divide all the entries of Θ''. The final step in our proof eliminates the entries θ_{i0} and θ_{0i} for $i\ne 0$. We define two invertible matrices $\Phi_1=[\phi_{ij}]_{0\le i,j<\dim(\mathscr{F})}$ and $\Psi_1=[\psi_{ij}]_{0\le i,j<\dim(\mathscr{F})}$ as follows:

$$\begin{aligned}\phi_{ij} &= -\theta''_{ij}/\mathscr{D}(\Theta) && \text{if } j=0 \text{ and } i\ge 1,\\ &= \delta_{ij} && \text{if } j>0 \text{ or } i=j=0,\end{aligned}$$

and

$$\psi_{ij} = -\theta''_{ij}/\mathscr{D}(\Theta) \quad \text{if } i = 0 \text{ and } j \geq 1,$$
$$= \delta_{ij} \quad \text{if } i > 0 \text{ or } i = j = 0.$$

The boundedness of Ψ is verified as follows. If $f = \sum_{0 \leq j < \dim(\mathscr{F})} f_j e_j \in \mathscr{F}$, then we have

$$\left| \sum_{1 \leq j < \dim(\mathscr{F})} \psi_{0j}(\varsigma) f_j \right| = \left| \sum_{1 \leq j < \dim(\mathscr{F})} \theta''_{0j}(\varsigma) f_j \right| \leq \|\Theta''(\varsigma) f\| \leq \|\Theta''\|_\infty \|f\|$$

for almost every $\varsigma \in \mathbf{T}$, and hence $\|\Psi(\varsigma)\| \leq 1 + \|\Theta''\|_\infty$ for almost every $\varsigma \in \mathbf{T}$. The boundedness of Φ is treated analogously. It is also clear that Φ and Ψ are invertible (for example, $\Phi^{-1} = 2I - \Phi$, $\Psi^{-1} = 2I - \Psi \in H^\infty(\mathscr{L}(\mathscr{F}))$). Thus Θ'' is quasiequivalent to $\Theta''' = \Phi\Theta''\Psi$ so that $\mathscr{D}(\Theta''') \equiv \mathscr{D}(\Theta'') \equiv \mathscr{D}(\Theta)$. It is clear now that Θ''' has the form $\operatorname{diag}(\mathscr{D}(\Theta), \Theta_1)$, and $\mathscr{D}(\Theta)$ divides all the entries of Θ_1. The lemma is proved.

5.28. THEOREM. *Assume that $\Theta \in H^\infty(\mathscr{L}(\mathscr{F}))$, and $\omega \in H^\infty$ is an inner function.*

(i) *We have $\mathscr{E}_j(\Theta) \mid \mathscr{E}_{j+1}(\Theta)$ for all integers j, $1 \leq j < \dim(\mathscr{F})$.*
(ii) *If k is an integer, $1 \leq k \leq \dim(\mathscr{F})$, there exists a function $\Theta_k \in H^\infty(\mathscr{L}(\mathscr{F}_k))$ such that $\mathscr{E}_k(\Theta) \mid \mathscr{D}(\Theta_k)$ and Θ is ω-equivalent to $\operatorname{diag}(\mathscr{E}_1(\Theta), \mathscr{E}_2(\Theta), \ldots, \mathscr{E}_k(\Theta), \Theta_k)$. We have $\Theta_k = 0$ if and only if $\mathscr{D}_{k+1}(\Theta) = 0$.*

PROOF. We prove (ii) first. As in the proof of Lemma 5.22, there is no loss of generality in assuming that $\mathscr{D}_j(\Theta) \mid \omega$ whenever $j \leq k$ and $\mathscr{D}_j(\Theta) \neq 0$. Let us now set $\Theta_0 = \Theta$. An inductive application of Lemma 5.22 shows the existence of functions $\Theta_j \in H^\infty(\mathscr{L}(\mathscr{F}_j))$, $1 \leq j \leq k$, with the following properties: Θ_j is ω-equivalent to $\operatorname{diag}(\mathscr{D}(\Theta_j), \Theta_{j+1})$ and $\mathscr{D}(\Theta_j) \mid \mathscr{D}(\Theta_{j+1})$, $0 \leq j < k$. Define functions $\delta_1, \delta_2, \ldots, \delta_k \in H^\infty$ by $\delta_j = \mathscr{D}(\Theta_{j-1})$, $1 \leq j \leq k$. Then Θ is ω-equivalent to $\Theta' = \operatorname{diag}(\delta_1, \delta_2, \ldots, \delta_k, \Theta_k)$, $\delta_j \mid \delta_{j+1}$, $1 \leq j < k$, and $\delta_k \mid \mathscr{D}(\Theta_k)$. We now want to relate the functions δ_j with $\mathscr{D}_j(\Theta)$ and $\mathscr{E}_j(\Theta)$. To do this we note that by Remark 5.20 (cf. particularly (5.21)), there exist inner functions $\phi_0, \psi_0 \in H^\infty$ satisfying the relations

$$\phi_0 \wedge \omega \equiv \psi_0 \wedge \omega \equiv 1 \tag{5.29}$$

and

$$\mathscr{D}_j(\Theta) \mid \psi_0^j \mathscr{D}_j(\Theta'), \qquad \mathscr{D}_j(\Theta') \mid \phi_0^j \mathscr{D}_j(\Theta) \tag{5.30}$$

for all integers j, $1 \leq j \leq \dim(\mathscr{F})$. Now, it is clear that $\mathscr{D}_j(\Theta') \equiv \delta_1\delta_2\cdots\delta_j$ for $1 \leq j \leq k$, and (5.30) can be translated into

$$\mathscr{D}_j(\Theta) \mid \psi_0^j \delta_1\delta_2\cdots\delta_j, \qquad \delta_1\delta_2\cdots\delta_j \mid \phi_0^j \mathscr{D}_j(\Theta), \quad 1 \leq j \leq k. \tag{5.31}$$

Relations (5.31) show, in particular, that the first j for which $\delta_j = 0$ coincides with the first j for which $\mathscr{D}_j(\Theta) = 0$. Assume that j is such that $\delta_j \neq 0$, hence

$\mathscr{D}_j(\Theta) \neq 0$. Then we have $\mathscr{D}_j(\Theta) \wedge \psi_0^j \equiv 1$ (by (5.29), because $\mathscr{D}_j(\Theta) \mid \omega$) and hence (5.31) implies that $\mathscr{D}_j(\Theta) \mid \delta_1\delta_2\cdots\delta_j$. Thus we have $\mathscr{E}_j(\Theta)\mathscr{D}_{j-1}(\Theta) \mid \delta_1\delta_2\cdots\delta_j$ and an application of the second relation in (5.31) with j replaced by $j-1$ shows that $\mathscr{E}_j(\Theta)\mathscr{D}_{j-1}(\Theta) \mid \phi_0^{j-1}\mathscr{D}_{j-1}(\Theta)\delta_j$ or, equivalently, $\mathscr{E}_j(\Theta) \mid \phi_0^{j-1}\delta_j$. Now, $\mathscr{E}_j(\Theta) \mid \mathscr{D}_j(\Theta) \mid \omega$ so that $\mathscr{E}_j(\Theta) \wedge \phi_0^{j-1} \equiv 1$. We conclude that $\mathscr{E}_j(\Theta) \mid \delta_j$. The second relation in (5.29) now shows that we can write

$$\delta_j = \mathscr{E}_j(\Theta)\eta_j, \quad \eta_j \in H^\infty, \quad \eta_j \wedge \omega \equiv 1. \tag{5.32}$$

(In fact (5.31) shows that $\eta_1\eta_2\cdots\eta_j \mid \phi_0^j$ because $\mathscr{D}_j(\Theta) = \mathscr{E}_1(\Theta)\mathscr{E}_2(\Theta)\cdots\mathscr{E}_j(\Theta)$.) Relation (5.32) can be extended to all $j \leq k$, by setting $\eta_j = 1$ if $\mathscr{D}_j(\Theta) = \delta_j = 0$. We now apply one further ω-equivalence to eliminate the factors η_j. Define $\Psi = \operatorname{diag}(\eta_1, \eta_2, \ldots, \eta_k, I_{\mathscr{F}_k})$ and note that

$$\Theta' = \operatorname{diag}(\mathscr{E}_1(\Theta), \mathscr{E}_2(\Theta), \ldots, \mathscr{E}_k(\Theta), \Theta_k)\Psi.$$

Moreover, Ψ has the scalar multiple $\eta_1\eta_2\cdots\eta_k$ with $(\eta_1\eta_2\cdots\eta_k) \wedge \omega \equiv 1$. Thus the ω-equivalence in (ii) is proved. The relation $\mathscr{E}_k(\Theta) \mid \mathscr{D}(\Theta_k)$ follows from $\delta_k \mid \mathscr{D}(\Theta_k)$ and (5.32). Finally, $\mathscr{D}_{k+1}(\Theta) = 0$ if and only if $\mathscr{D}_{k+1}(\operatorname{diag}(\mathscr{E}_1(\Theta), \mathscr{E}_2(\Theta), \ldots, \mathscr{E}_k(\Theta), \Theta_k)) = 0$, and this happens if and only if $\Theta_k = 0$.

In order to prove (i) for some integer $j < \dim(\mathscr{F})$, choose an integer k such that $j < k \leq \dim(\mathscr{F})$, and perform the construction above. In relation (5.32) we have $\mathscr{E}_j(\Theta) \wedge \eta_{j+1} \equiv 1$ because $\mathscr{E}_j(\Theta) \mid \omega$ if $\mathscr{E}_j(\Theta) \neq 0$. Thus the relations $\mathscr{E}_j(\Theta) \mid \delta_j \mid \delta_{j+1} = \mathscr{E}_{j+1}(\Theta)\eta_{j+1}$ imply that $\mathscr{E}_j(\Theta) \mid \mathscr{E}_{j+1}(\Theta)$, as desired. The proof is now complete.

It is clear from the construction in Lemma 5.22 that the function Θ_k constructed above usually depends on ω. Thus the preceding theorem does not normally yield quasiequivalence, unless $\Theta_k = 0$ or $k = \dim(\mathscr{F})$. In the latter case $\mathscr{F}$ is finite-dimensional and Θ_k is an empty matrix (i.e., it acts on the space $\{0\}$). We also see from the example $\Theta = \operatorname{diag}(\phi, 1, 1, \ldots)$ that we cannot hope to achieve the ω-equivalence of Θ with the matrix $\operatorname{diag}(\mathscr{E}_1(\Theta), \mathscr{E}_2(\Theta), \ldots)$.

Exercises

1. Give a proof of Lemma 5.3 based on the representation of $\mathscr{D}_k(\Theta)$ as the greatest common inner divisor of a family of minors.

2. Use Theorem 5.28 to prove the existence of the Jordan model of $S(\Theta)$ if Θ is an inner function on a finite-dimensional space.

3. Find a proof of Proposition 5.11 based on Exercise 2.

4. Let Θ_1 and Θ_2 be two quasiequivalent inner functions. Assume that $S(\Theta_1)$ is of class C_0 and give a direct proof of the fact that $S(\Theta_1)$ and $S(\Theta_2)$ have the same minimal function.

5. Find a proof of Proposition 5.17(iv) based on Exercise 4.

6. Formulate analogues of Lemma 5.22 and Theorem 5.28 for the case in which $\Theta \in H^\infty(\mathscr{L}(\mathscr{F}, \mathscr{F}'))$.

7. Assume that ϕ_1, ϕ_2, $\phi_3 \in H^\infty$ are inner functions. Show that the Jordan model of $S(\phi_1) \oplus S(\phi_2) \oplus S(\phi_3)$ is $S(\theta_0) \oplus S(\theta_1) \oplus S(\theta_2)$, where $\theta_0 \equiv \phi_1 \vee \phi_2 \vee \phi_3$, $\theta_2 \equiv \phi_1 \wedge \phi_2 \wedge \phi_3$, and $\theta_1 \equiv \phi_1\phi_2\phi_3/\theta_0\theta_2$.

8. Assume that Θ_1 and Θ_2 are two quasiequivalent inner functions. Are $S(\Theta_1)$ and $S(\Theta_2)$ necessarily quasisimilar?

6. The calculation of Jordan models from $*$-cyclic sets. Given a subset $\mathscr{M} \subset H^2(\mathscr{F})$, form the subspace

$$\mathscr{H} = \bigvee_{n=0}^{\infty} S_{\mathscr{F}}^{*^n} \mathscr{M},$$

invariant for $S_{\mathscr{F}}^*$. Of course, there is an inner function $\Theta \in H^\infty(\mathscr{L}(\mathscr{G}, \mathscr{F}))$ such that $\mathscr{H} = \mathscr{H}(\Theta)$. Suppose that the operator $S(\Theta)$ is of class C_0. We are interested in calculating the Jordan model of $S(\Theta)$ without first figuring out what Θ looks like. The fact that this is possible will follow from facts proved in the preceding section.

6.1. LEMMA. *Let $\mathscr{M} \subset H^2(\mathscr{F})$ be an arbitrary subset, and let $\Theta \in H^\infty(\mathscr{L}(\mathscr{G}, \mathscr{F}))$ be an inner function such that $\mathscr{H}(\Theta) = \bigvee_{n=0}^{\infty} S_{\mathscr{F}}^{*^n} \mathscr{M}$. The operator $S(\Theta)$ is of class C_0 if and only if there exists an inner function $\phi \in H^\infty$ such that*

$$\overline{\phi}\mathscr{M} \subset L^2(\mathscr{F}) \ominus H^2(\mathscr{F}). \tag{6.2}$$

Moreover, $\phi(S(\Theta)) = 0$ if and only if (6.2) holds.

PROOF. By definition, the set $\mathscr{M}$ is cyclic for $S(\Theta)^*$. Therefore we have $\phi(S(\Theta)) = 0$ if and only if $\phi(S(\Theta))^*\mathscr{M} = \{0\}$. This happens if and only if $\phi(S_{\mathscr{F}})^*\mathscr{M} = \{0\}$ or, equivalently, if and only if

$$P_{H^2(\mathscr{F})}\phi(U_{\mathscr{F}})^*\mathscr{M} = P_{H^2(\mathscr{F})}(\overline{\phi}\mathscr{M}) = \{0\}.$$

This equivalence proves the lemma.

6.3. LEMMA. *Let $\mathscr{M}$ and Θ be as in the preceding lemma, and let ϕ be an inner function. Then $L^2(\mathscr{F}) \ominus \overline{\phi}M_\Theta H^2(\mathscr{G})$ is the invariant subspace for U^* generated by $\overline{\phi}\mathscr{M}$ and $L^2(\mathscr{F}) \ominus \overline{\phi}H^2(\mathscr{F})$.*

PROOF. Since multiplication by $\overline{\phi}$ is a unitary operator on $L^2(\mathscr{F})$ commuting with $U_{\mathscr{F}}^*$, it suffices to show that $L^2(\mathscr{F}) \ominus M_\Theta H^2(\mathscr{F})$ is the invariant subspace for $U_{\mathscr{F}}^*$ generated by $\mathscr{M}$ and $L^2(\mathscr{F}) \ominus H^2(\mathscr{F})$. Thus, we have to prove that

$$L^2(\mathscr{F}) \ominus M_\Theta H^2(\mathscr{G}) = (L^2(\mathscr{F}) \ominus H^2(\mathscr{F})) \vee \left(\bigvee_{n=0}^{\infty} U_{\mathscr{F}}^{*^n} \mathscr{M} \right).$$

Since both sides contain $L^2(\mathscr{F}) \ominus H^2(\mathscr{F})$, this equality is equivalent to

$$H^2(\mathscr{F}) \ominus M_\Theta H^2(\mathscr{G}) = \bigvee_{n=0}^{\infty} P_{H^2(\mathscr{F})} U_{\mathscr{F}}^{*^n} \mathscr{M}.$$

Since $H^2(\mathscr{F}) \ominus M_\Theta H^2(\mathscr{G}) = \mathscr{H}(\Theta)$, and $P_{H^2(\mathscr{F})} U_{\mathscr{F}}^{*^n} \mid H^2(\mathscr{F}) = S_{\mathscr{F}}^{*^n}$, this last equality is true by the definition of Θ. The lemma follows.

We recall that, if $S(\Theta)$ is of class C_0 then Θ is two-sided inner, and hence $\mathscr{F}$ and $\mathscr{G}$ must have the same dimension. In order to simplify notation, Θ will be chosen from now on in $H^\infty(\mathscr{L}(\mathscr{F}))$.

6.4. LEMMA. *Let $\mathscr{M}$ and $\Theta \in H^\infty(\mathscr{L}(\mathscr{F}))$ be as in the preceding results. Assume that $\phi \in H^\infty$ is inner, $\phi(S(\Theta)) = 0$, and $\Omega \in H^\infty(\mathscr{L}(\mathscr{F}))$ is such that $\Omega(\lambda)\Theta(\lambda) = \Theta(\lambda)\Omega(\lambda) = \phi(\lambda)I$, $\lambda \in \mathbf{D}$. Then $\overline{\phi} M_\Theta H^2(\mathscr{F}) = M_\Omega^* H^2(\mathscr{F})$.*

PROOF. This is an immediate consequence of the equalities

$$\overline{\phi}(\varsigma)\Theta(\varsigma) = (\phi(\varsigma)I)^*\Theta(\varsigma) = \Omega(\varsigma)^*\Theta(\varsigma)^*\Theta(\varsigma) = \Omega(\varsigma)^*, \quad \varsigma \in \mathbf{T},$$

which follow from the fact that ϕ and Θ are inner.

In the following result we will use the unitary operator $R: L^2(\mathscr{F}) \to L^2(\mathscr{F})$ defined by

$$(Rf)(\varsigma) = \overline{\varsigma} f(\overline{\varsigma}), \quad \varsigma \in \mathbf{T}, f \in L^2(\mathscr{F}).$$

We note three important, easily verifiable properties of R:

$$R = R^* = R^{-1}, \qquad U_{\mathscr{F}} R = R U_{\mathscr{F}}^*, \qquad R(L^2(\mathscr{F}) \ominus H^2(\mathscr{F})) = H^2(\mathscr{F}).$$

6.5. PROPOSITION. *Let $\mathscr{M} \subset H^2(\mathscr{F})$ and $\Theta \in H^\infty(\mathscr{L}(\mathscr{F}))$ be such that $\mathscr{H}(\Theta) = \bigvee_{n=0}^\infty S_{\mathscr{F}}^{*^n} \mathscr{M}$. Assume that ϕ is inner, $\phi(S(\Theta)) = 0$, and $\Omega \in H^\infty(\mathscr{L}(\mathscr{F}))$ satisfies $\Omega(\lambda)\Theta(\lambda) = \Theta(\lambda)\Omega(\lambda) = \phi(\lambda)I$, $\lambda \in \mathbf{D}$. Then $M_{\Omega^\sim} H^2(\mathscr{F})$ is the invariant subspace for $U_{\mathscr{F}}$ (or $S_{\mathscr{F}}$) generated by $R(\overline{\phi}\mathscr{M})$ and $\phi^\sim H^2(\mathscr{F})$.*

PROOF. The identity $RM_\Omega^* = M_{\Omega^\sim} R$ and Lemma 6.4 imply that

$$\begin{aligned} R(L^2(\mathscr{F}) \ominus \overline{\phi} M_\Theta H^2(\mathscr{F})) &= R(L^2(\mathscr{F}) \ominus M_\Omega^* H^2(\mathscr{F})) \\ &= R M_\Omega^* (L^2(\mathscr{F}) \ominus H^2(\mathscr{F})) \\ &= M_{\Omega^\sim} R (L^2(\mathscr{F}) \ominus H^2(\mathscr{F})) \\ &= M_{\Omega^\sim} H^2(\mathscr{F}). \end{aligned}$$

We conclude that $M_{\Omega^\sim} H^2(\mathscr{F})$ is the cyclic space for $U_{\mathscr{F}}$ generated by $R(\overline{\phi}\mathscr{M})$ and $R(L^2(\mathscr{F}) \ominus \overline{\phi} H^2(\mathscr{F}))$. The proposition now follows immediately because $R(L^2(\mathscr{F}) \ominus \overline{\phi} H^2(\mathscr{F})) = R\overline{\phi}(L^2(\mathscr{F}) \ominus H^2(\mathscr{F})) = \phi^\sim R(L^2(\mathscr{F}) \ominus H^2(\mathscr{F})) = \phi^\sim H^2(\mathscr{F})$.

Proposition 5.8 shows that we can calculate the inner functions in the Jordan model of $S(\Theta)$ if we can calculate $\mathscr{E}_j(\Omega)$ or, equivalently, the functions $\mathscr{E}_j(\Omega^\sim) = \mathscr{E}_j(\Omega)^\sim$. Proposition 6.5 above raises the natural question of whether one can calculate these functions using just a cyclic set for $S_{\mathscr{F}} \mid M_{\Omega^\sim} H^2(\mathscr{F})$. In order to show that this is indeed the case, we extend the definition of the functions $\mathscr{D}_j$ and $\mathscr{E}_j$ as follows.

6.6. DEFINITION. Let $\mathscr{X} \subset H^2(\mathscr{F})$ be an arbitrary set, and $1 \leq j \leq \dim(\mathscr{F})$ an integer. Let $\{e_1, e_2, \ldots, e_j\}$ be an orthonormal system in $\mathscr{F}$, and let $x_1, x_2, \ldots, x_j \in \mathscr{X}$. The function

$$\det[(x_i(\lambda), e_k)]_{1 \leq i,k \leq j}, \quad \lambda \in \mathbf{D},$$

is called a *minor of order* j of $\mathscr{X}$. The function $\mathscr{D}_j(\mathscr{X})$ is the greatest common inner divisor of all minors of order j of $\mathscr{X}$. The *invariant factors* $\mathscr{E}_j(\mathscr{X})$ are defined as follows:

$$\begin{aligned} &\mathscr{E}_1(\mathscr{X}) = \mathscr{D}_1(\mathscr{X}), \\ &\mathscr{E}_k(\mathscr{X}) = \mathscr{D}_k(\mathscr{X})/\mathscr{D}_{k-1}(\mathscr{X}), \quad 2 \leq k \leq \dim(\mathscr{F}), \mathscr{D}_k(\mathscr{X}) \neq 0, \end{aligned}$$

and

$$\mathscr{E}_k(\mathscr{X}) = 0, \qquad 2 \leq k \leq \dim(\mathscr{F}), \mathscr{D}_k(\mathscr{X}) = 0.$$

We should emphasize here that a minor of order j of $\mathscr{X}$ is not generally an element of H^p for some fixed $p > 0$. These determinants belong, however, to the Nevanlinna class N^+, so they can be factored as an inner factor times an outer factor. Thus the definition of $\mathscr{D}_j(\mathscr{X})$ makes sense since the outer factors are irrelevant. We did not prove it, but it follows from our next result that $\mathscr{D}_{j-1}(\mathscr{X}) \mid \mathscr{D}_j(\mathscr{X})$ for $j \geq 2$; this is also easy to prove directly.

6.7. THEOREM. *Let $\mathscr{X} \subset H^2(\mathscr{F})$ be an arbitrary set and let $\Phi \in H^\infty(\mathscr{L}(\mathscr{G}, \mathscr{F}))$ be an inner function such that $\bigvee_{n=0}^\infty S_{\mathscr{F}}^n = M_\Phi H^2(\mathscr{G})$. Then we have $\mathscr{D}_j(\mathscr{X}) \equiv \mathscr{D}_j(\Phi)$ and $\mathscr{E}_j(\mathscr{X}) \equiv \mathscr{E}_j(\Phi)$ for all j, $1 \leq j \leq \dim(\mathscr{F})$.*

PROOF. Denote by $\mathscr{Y}$ the linear manifold generated by $\bigcup_{n=0}^\infty S_{\mathscr{F}}^n \mathscr{M}$, and set $\mathscr{Z} = \mathscr{Y}^-$, so that $\mathscr{Z} = M_\Phi H^2(\mathscr{G})$. We claim that it suffices to prove that

$$\mathscr{D}_j(\mathscr{X}) \equiv \mathscr{D}_j(\mathscr{Z}) \tag{6.8}$$

for all j. Indeed, assume this equality has been proved for all subsets $\mathscr{X} \subset H^2(\mathscr{F})$. Fix an orthonormal basis $\{e_i\}_i$ of $\mathscr{G}$, and set $\mathscr{X}' = \{M_\Phi e_i\}_i$. It is clear that

$$\bigvee_{n=0}^\infty S_{\mathscr{F}}^n \mathscr{X}' = \bigvee_{n=0}^\infty \bigvee_i M_\Phi S_{\mathscr{G}}^n e_i = M_\Phi H^2(\mathscr{G}) = \mathscr{Z},$$

so that we would have from (6.8) that $\mathscr{D}_j(\mathscr{X}) \equiv \mathscr{D}_j(\mathscr{X}') \equiv \mathscr{D}_j(\mathscr{Z})$ for all j. But it is clear that $\mathscr{D}_j(\mathscr{X}') \equiv \mathscr{D}_j(\Phi)$ because the vectors $M_\Phi e_i$ are exactly the "columns" of the matrix Φ. We set out therefore to prove (6.8) for a fixed integer j. We first note that the inclusions $\mathscr{X} \subset \mathscr{Y} \subset \mathscr{Z}$ imply that

$$\mathscr{D}_j(\mathscr{Z}) \mid \mathscr{D}_j(\mathscr{Y}) \mid \mathscr{D}_j(\mathscr{X}).$$

Assume on the other hand that $y_1, y_2, \ldots, y_j \in \mathscr{Y}$ so that each y_k is a linear combination of vectors of the form $S_{\mathscr{F}}^n x$ with $n \geq 0$ and $x \in \mathscr{X}$. Assume further that $\{e_1, e_2, \ldots, e_j\}$ is an orthonormal system in $\mathscr{F}$. The multilinearity of the determinant implies that the minor $\det[(y_i(\lambda), e_k)]_{1 \le i,k \le j}$ is a linear combination of functions of the form

$$\det[(\lambda^{n_i} x_i(\lambda), e_k)]_{1 \le i,k \le j} = \lambda^n \det[(x_i(\lambda), e_k)]_{1 \le i,k \le j},$$

where $n = n_1 + n_2 + \cdots + n_j$. It follows therefore that $\mathscr{D}_j(\mathscr{X})$ divides every minor of order j of $\mathscr{Y}$, and hence $\mathscr{D}_j(\mathscr{X}) \mid \mathscr{D}_j(\mathscr{Y})$. Thus we proved that $\mathscr{D}_j(\mathscr{X}) \equiv \mathscr{D}_j(\mathscr{Y})$, and it remains to show that $\mathscr{D}_j(\mathscr{Y}) \equiv \mathscr{D}_j(\mathscr{Z})$. Assume therefore that $z_1, z_2, \ldots, z_j \in \mathscr{Z}$, and $\{e_1, e_2, \ldots, e_j\}$ is an orthonormal system in $\mathscr{F}$. For each i, $1 \le i \le j$, choose a sequence $y_{i,n} \in \mathscr{Y}$ such that $\lim_{n \to \infty} \|z_i - y_{i,n}\| = 0$. Next, choose an outer function $\psi \in H^\infty$ such that all the functions $(\psi(\lambda) y_{i,n}(\lambda), e_k)$ and $(\psi(\lambda) z_i(\lambda), e_k)$ are bounded (see the exercises below). Clearly $\mathscr{D}_j(\mathscr{Y})$ divides the functions

$$\psi(\lambda)^j \det[(y_{i,n_i}(\lambda), e_k)]_{1 \le i,k \le j} = \det[\psi(\lambda)(y_{i,n_i}(\lambda), e_k)]_{1 \le i,k \le j}$$

for all choices of natural numbers $n_1, n_2, \ldots, n_j$. In order to simplify notation for the remainder of this proof, a determinant of the form $\det[(x_i(\lambda), e_k)]_{1 \le i,k \le j}$ will be denoted $D(x_1, x_2, \ldots, x_j)$. Since all the entries in $D(\psi y_{1,n_1}, \psi y_{2,n_2}, \ldots, \psi y_{j,n_j})$ are bounded functions, we deduce that

$$\lim_{n_1 \to \infty} D(\psi y_{1,n_1}, \psi y_{2,n_2}, \ldots, \psi y_{j,n_j}) = D(\psi z_1, \psi y_{2,n_2}, \ldots, \psi y_{j,n_j})$$

in the space H^2. Thus $\mathscr{D}_j(\mathscr{Y})$ divides the limit function. We proceed analogously, letting $n_2, n_3, \ldots, n_j$ tend to infinity one at a time, and conclude that $\mathscr{D}_j(\mathscr{Y})$ divides

$$D(\psi z_1, \psi z_2, \ldots, \psi z_j) = \psi^j \det[(z_i(\lambda), e_k)]_{1 \le i,k \le j}.$$

Since ψ is outer, we conclude that $\mathscr{D}_j(\mathscr{Y})$ divides every minor of order j of $\mathscr{Z}$, whence $\mathscr{D}_j(\mathscr{Y}) \mid \mathscr{D}_j(\mathscr{Z})$. The equality $\mathscr{D}_j(\mathscr{Y}) \equiv \mathscr{D}_j(\mathscr{Z})$ follows, and the theorem is proved.

6.9. COROLLARY. *With the notation of Proposition 6.5, we have* $\mathscr{E}_j(\Omega)^\sim = \mathscr{E}_j(\mathscr{X})$, *where* $\mathscr{X} = R(\overline{\phi}\mathscr{M}) \cup \phi^\sim H^2(\mathscr{F})$.

PROOF. This follows immediately from Proposition 6.5 and Theorem 6.7.

We want to state the formulas for the Jordan model of $S(\Theta)$ in a more elegant form.

6.10. THEOREM. *Let* $\mathscr{M} \subset H^2(\mathscr{F})$ *and the inner function* $\phi \in H^\infty$ *be such that* $\overline{\phi}\mathscr{M} \subset L^2(\mathscr{F}) \ominus H^2(\mathscr{F})$. *There exists an inner function* $\Theta \in H^\infty(\mathscr{L}(\mathscr{F}))$ *such that* $\mathscr{H}(\Theta) = \bigvee_{n=0}^\infty S_{\mathscr{F}}^{*n} \mathscr{M}$ *and* $\phi(S(\Theta)) = 0$. *Moreover, the inner functions in the Jordan model* $\bigoplus_{j=0}^\infty S(\theta_j)$ *of* $S(\Theta)$ *can be calculated as follows:*

$$\theta_j \equiv \phi / (\mathscr{E}_{j+1}(R(\overline{\phi}\mathscr{M}))^\sim \wedge \phi) \quad \text{if } j < \dim(\mathscr{F}),$$
$$\equiv 1 \quad \text{otherwise.}$$

PROOF. We need only to prove the formula for θ_j if $j < \dim(\mathscr{F})$, and this formula follows immediately from Corollary 6.9 if we prove that

$$\mathscr{E}_j(\mathscr{X}) = \mathscr{E}_j(R(\overline{\phi}\mathscr{M})) \wedge \phi^\sim, \quad j \leq \dim(\mathscr{F}), \tag{6.11}$$

where $\mathscr{X}$ is as in that corollary. Let us set $\mathscr{D}_j = \mathscr{D}_j(R(\overline{\phi}\mathscr{M}))$, $\mathscr{E}_j = \mathscr{E}_j(R(\overline{\phi}\mathscr{M}))$, and $\Delta_j = \mathscr{D}_j(\mathscr{X})$. Note that a minor of order j of $\mathscr{X}$ is formed by p vectors from $\phi^\sim H^2(\mathscr{F})$, and $j-p$ vectors from $R(\overline{\phi}\mathscr{M})$, where $0 \leq p \leq j$. Every such minor is a sum of factors of the form AB, where A is a minor of order p of $\phi^\sim H^2(\mathscr{F})$ and B a minor of order $j-p$ of $R(\overline{\phi}\mathscr{M})$. Now, $(\phi^\sim)^p$ divides A, and we conclude that the function

$$\bigwedge\nolimits_{p=0}^{j} \phi^{\sim p} \mathscr{D}_{j-p}$$

divides Δ_j. Conversely, fix p, an orthonormal system $\{e_1, e_2, \dots, e_{j-p}\}$ in $\mathscr{F}$, and vectors $x_1, x_2, \dots, x_{j-p}$ in $R(\overline{\phi}\mathscr{M})$. Choose $e_{j-p+1}, \dots, e_j$ such that $\{e_1, e_2, \dots, e_j\}$ is an orthonormal system, and set $x_i = \phi^\sim e_i$ for $j-p+1 \leq i \leq j$. Then we have $\det[(x_i(\lambda), e_k)]_{1\leq i,k\leq j} = \phi^{\sim p} \det[(x_i(\lambda), e_k]_{1 \leq i,k \leq j-p}$ and we conclude that Δ_j must divide $\phi^{\sim p}\mathscr{D}_{j-p}$. Thus we have the formula

$$\Delta_j \equiv \bigwedge\nolimits_{p=0}^{j} \phi^{\sim p} \mathscr{D}_{j-p} \equiv \bigwedge\nolimits_{p=0}^{j} \phi^{\sim p} \mathscr{E}_1 \mathscr{E}_2 \cdots \mathscr{E}_{j-p}. \tag{6.12}$$

Relation (6.11) is equivalent to

$$\Delta_j \equiv (\mathscr{E}_1 \wedge \phi^\sim)(\mathscr{E}_2 \wedge \phi^\sim) \cdots (\mathscr{E}_j \wedge \phi^\sim), \qquad j \leq \dim(\mathscr{F}). \tag{6.13}$$

We prove this relation by induction. For $j=1$ this relation follows immediately from (6.12). Assume that $j \geq 2$ and (6.13) has already been proved with j replaced by $j-1$. Then (6.12) implies that

$$\begin{aligned}
\Delta_j &\equiv \mathscr{D}_j \wedge \left(\bigwedge\nolimits_{p=1}^{j} \phi^{\sim p} \mathscr{D}_{j-p} \right) \\
&\equiv \mathscr{D}_j \wedge \phi^\sim \left(\bigwedge\nolimits_{q=0}^{j-1} \phi^{\sim q} \mathscr{D}_{(j-1)-q} \right) \\
&\equiv \mathscr{D}_j \wedge \phi^\sim \Delta_{j-1}.
\end{aligned}$$

We next apply the inductive hypothesis to get

$$\begin{aligned}
\Delta_j &\equiv \mathscr{D}_j \wedge [\phi^\sim (\mathscr{E}_1 \wedge \phi^\sim)(\mathscr{E}_2 \wedge \phi^\sim) \cdots (\mathscr{E}_{j-1} \wedge \phi^\sim)] \\
&\equiv \mathscr{E}_1 \mathscr{E}_2 \cdots \mathscr{E}_j [\phi^\sim (\mathscr{E}_1 \wedge \phi^\sim)(\mathscr{E}_2 \wedge \phi^\sim) \cdots (\mathscr{E}_{j-1} \wedge \phi^\sim)] \\
&\equiv (\mathscr{E}_1 \wedge \phi^\sim)(\mathscr{E}_2 \wedge \phi^\sim) \cdots (\mathscr{E}_{j-1} \wedge \phi^\sim) \\
&\qquad \left[\left(\frac{\mathscr{E}_1}{\mathscr{E}_1 \wedge \phi^\sim} \frac{\mathscr{E}_2}{\mathscr{E}_2 \wedge \phi^\sim} \cdots \frac{\mathscr{E}_{j-1}}{\mathscr{E}_{j-1} \wedge \phi^\sim} \mathscr{E}_j \right) \wedge \phi^\sim \right] \\
&\equiv (\mathscr{E}_1 \wedge \phi^\sim)(\mathscr{E}_2 \wedge \phi^\sim) \cdots (\mathscr{E}_j \wedge \phi^\sim) \\
&\qquad \left[\left(\frac{\mathscr{E}_1}{\mathscr{E}_1 \wedge \phi^\sim} \frac{\mathscr{E}_2}{\mathscr{E}_2 \wedge \phi^\sim} \cdots \frac{\mathscr{E}_j}{\mathscr{E}_j \wedge \phi^\sim} \right) \wedge \left(\frac{\phi^\sim}{\mathscr{E}_j \wedge \phi^\sim} \right) \right].
\end{aligned}$$

To conclude the proof of (6.13) it suffices to show that $\phi^\sim/(\mathscr{E}_j \wedge \phi^\sim)$ is relatively prime to $\mathscr{E}_i/(\mathscr{E}_i \wedge \phi^\sim)$ for $i \leq j$. Now, we know that $\mathscr{E}_i \mid \mathscr{E}_j$ for $i \leq j$, and hence $\phi^\sim/(\mathscr{E}_j \wedge \phi^\sim) | \phi^\sim/(\mathscr{E}_i \wedge \phi^\sim)$ for $i \leq j$. It suffices then to prove that $\mathscr{E}_j/(\mathscr{E}_j \wedge \phi^\sim)$ and $\phi^\sim/(\mathscr{E}_j \wedge \phi^\sim)$ are relatively prime, and this is obvious. This concludes our proof.

Exercises

1. Prove that R has the properties stated before Proposition 6.5.

2. Show that $RM_\Omega^* = M_{\Omega^\sim} R$ for every $\Omega \in H^\infty(\mathscr{L}(\mathscr{F}))$.

3. Let $\{\phi_i : i \geq 0\}$ be a sequence of functions in H^2. Show that there exists an outer function $\psi \in H^\infty$ such that $\psi\phi_i$ is bounded for every i.

4. Let $\mathscr{X}$ and $\mathscr{Y}$ be two subsets of $H^2(\mathscr{F})$. Is the following formula always true?

$$\mathscr{D}_j(\mathscr{X} \cup \mathscr{Y}) = \bigwedge_{p=0}^{j} \mathscr{D}_p(\mathscr{X})\mathscr{D}_{j-p}(\mathscr{Y})$$

5. Find a characterization of the cyclic vectors of S^*, where S denotes the unilateral shift of multiplicity one.

CHAPTER 7

Fredholm Theory

In this chapter we give a generalization, related to the class C_0, of the classical Fredholm theory. In this generalization the algebra of all operators on a Hilbert space is replaced by the commutant of a given operator. We start in Section 1 with the observation that weak contractions T have the following finiteness property. If X commutes with T then $\ker X$ and $\ker X^*$ have the same "dimension" in the sense that the compressions $T \mid \ker X$ and $(T^* \mid \ker X^*)^*$ have the same determinant function. (It follows that $\ker X$ and $\ker X^*$ have the same linear dimension as well.) In particular, $\ker X = \{0\}$ if and only if $\ker X^* = \{0\}$. This leads to the introduction of property (P) for an operator T. Next, we characterize the operators of class C_0 with the finiteness property (P) and give some applications. Section 2 provides a characterization of those operators of class C_0 that have a stronger finiteness property than (P). This provides a nice application of the techniques developed in Chapter 6. In Section 3 we consider the inner functions in the Jordan model of $T \mid \mathscr{K}$, where T is of class C_0 and $\mathscr{K}$ is invariant for T. It is shown that these functions have a continuity property relative to increasing sequences of invariant subspaces. This property is important for the reduction of certain proofs from the case of general operators T to the case of operators with finite multiplicity (and therefore weak). In order to introduce a notion of a "dimension" for operators with property (P) we require an extension of the notion of an inner function. This extension is presented in Section 4, along with some basic properties of the newly defined "dimension." In Section 5 we show that the notion of dimension is well suited for our purposes. The main result states that two operators T_1 and T_2, with property (P), have the same dimension if and only if they can be realized (up to quasisimilarity) as $T \mid \ker X$ and $(T^* \mid \ker X^*)^*$, where T has property (P) and X commutes with T. The techniques of Chapter 6 for calculating Jordan models are very important here. After these preliminaries we are ready to study Fredholm theory in Sections 6 and 7. We introduce the class of C_0-Fredholm operators in Section 6. We also show that the product of Fredholm operators is Fredholm, and the index behaves as expected. In Section 7 we prove the stability of C_0-Fredholm operators under "finite-rank" perturbations. We conclude with a characterization of the C_0-Fredholm operators in the double commutant of an operator, and with an example showing that C_0-Fredholm operators need not have closed range.

1. Finiteness properties. It is well known that a Hilbert space $\mathscr{H}$ is finite-dimensional if and only if every operator $X \in \mathscr{L}(\mathscr{H})$, with the property $\ker(X) = \{0\}$, also satisfies $\ker(X^*) = \{0\}$. Recalling that the commutant $\{0\}'$ of the zero operator in $\mathscr{L}(\mathscr{H})$ coincides with $\mathscr{L}(\mathscr{H})$, the following definition is a natural extension of finite dimensionality.

1.1. DEFINITION. An operator $T \in \mathscr{L}(\mathscr{H})$ is said to have *property* (P) if every operator $X \in \{T\}'$ with the property that $\ker(X) = \{0\}$ is a quasiaffinity, i.e., $\ker(X^*) = \ker(X) = \{0\}$.

We see therefore that $\mathscr{H}$ is finite-dimensional if and only if the zero operator on $\mathscr{H}$ has property (P). Now, if $\mathscr{H}$ is finite-dimensional and $X \in \mathscr{L}(\mathscr{H})$, the fact that $\ker(X) = \{0\}$ implies $\ker(X^*) = \{0\}$ is a consequence of the equality

$$\dim \ker(X) = \dim \ker(X^*). \tag{1.2}$$

In the context of Definition 1.1, one may attempt to associate with every $X \in \{T\}'$ a certain measure or "dimension" of $\ker(X)$ and $\ker(X^*)$, and try to prove that T has property (P) by showing that $\ker(X)$ and $\ker(X^*)$ have the same "dimension." The natural thing to do is to compare the operators $T \mid \ker(X)$ and $T^* \mid \ker(X^*)$. One extension of (1.2) (which is not the most suitable for the class C_0) is given as follows.

1.3. DEFINITION. An operator $T \in \mathscr{L}(\mathscr{H})$ is said to have *property* (Q) if the relation $T \mid \ker(X) \sim (T^* \mid \ker(X^*))^*$ holds for every X in $\{T\}'$.

We recall that the sign "$\sim$" indicates quasisimilarity. It is quite obvious that property (Q) implies property (P), and it is easy to see that the converse is not true (cf. Exercise 1 below).

Before we give examples of operators with property (P) on an infinite-dimensional space, we need to review some of the properties of the function d_T from Definition 6.3.10. This function was defined for all weak contractions T of class C_0. It turned out that d_T is an inner function which is a quasisimilarity invariant, i.e., $d_T \equiv d_{T'}$ if T and T' are quasisimilar (cf. Corollary 6.3.18). It will be convenient to define $d_T = 0$ if T is a contraction of class C_0 which is not a weak contraction. Then Theorem 6.3.16 and Proposition 6.3.23 can be given a unified statement, as follows.

1.4. THEOREM. *Assume that $T \in \mathscr{L}(\mathscr{H})$ is an operator of class C_0, and $\mathscr{H}_1 \subseteq \mathscr{H}$ is an invariant subspace for T. If*

$$T = \begin{bmatrix} T_1 & X \\ 0 & T_2 \end{bmatrix}$$

is the triangularization of T with respect to the decomposition $\mathscr{H} = \mathscr{H}_1 \oplus (\mathscr{H} \ominus \mathscr{H}_1)$, then we have $d_T \equiv d_{T_1} d_{T_2}$.

PROOF. If T is a weak contraction this follows from Theorem 6.3.16(ii). If T is not a weak contraction then at most one of the contractions T_1 and T_2 can be weak by Proposition 6.3.23. In this case we have $d_T = d_{T_1} d_{T_2} = 0$.

We introduce the notation $\text{Lat}_{1/2}(T)$ for the collection of all semi-invariant subspaces of an operator $T \in \mathcal{L}(\mathcal{H})$. Observe that

$$\text{Lat}_{1/2}(T) = \text{Lat}_{1/2}(T^*) \tag{1.5}$$

because, if $\mathcal{M}, \mathcal{N} \in \text{Lat}(T)$ and $\mathcal{M} \supseteq \mathcal{N}$, then $\mathcal{H} \ominus \mathcal{M}, \mathcal{H} \ominus \mathcal{N} \in \text{Lat}(T^*)$, and $\mathcal{M} \ominus \mathcal{N} = (\mathcal{H} \ominus \mathcal{N}) \ominus (\mathcal{H} \ominus \mathcal{M})$. If T is of class C_0, $\mathcal{M} \in \text{Lat}_{1/2}(T)$, and $T_{\mathcal{M}}$ denotes the compression of T to $\mathcal{M}$, then we define

$$d_T(\mathcal{M}) = d_{T_{\mathcal{M}}}. \tag{1.6}$$

Since we have $(T_{\mathcal{M}})^* = (T^*)_{\mathcal{M}}$, we deduce the equality

$$d_T(\mathcal{M}) = d_{T^*}(\mathcal{M})^{\sim}, \quad \mathcal{M} \in \text{Lat}_{1/2}(T).$$

We can now give a first extension of (1.2) to operators acting on infinite-dimensional Hilbert spaces.

1.7. PROPOSITION. *Assume that T is a weak contraction of class C_0. Then the relation $d_T(\ker(X)) \equiv d_T(\ker(X^*))$ holds for every X in $\{T\}'$. In particular, T has property (P).*

PROOF. Assume that T acts on $\mathcal{H}$. Fix $X \in \{T\}'$ and note that the relations

$$d_T \equiv d_T(\ker(X))d_T(\mathcal{H} \ominus \ker(X)) \equiv d_T(\ker(X^*))d_T((X\mathcal{H})^-) \tag{1.8}$$

are consequences of Theorem 1.4 and of the equality $(X\mathcal{H})^- = \mathcal{H} \ominus \ker(X^*)$. The operator $Y \in \mathcal{L}(\mathcal{H} \ominus \ker(X), (X\mathcal{H})^-)$ defined by $Yh = Xh$, $h \in \mathcal{H} \ominus \ker(X)$, is clearly a quasiaffinity and

$$(T \mid (X\mathcal{H})^-)Y = YT_{\mathcal{H} \ominus \ker(X)}.$$

By Proposition 3.5.32, the operators $T \mid (X\mathcal{H})^-$ and $T_{\mathcal{H} \ominus \ker(X)}$ are quasisimilar. Since d_T is a quasisimilarity invariant we deduce that $d_T((X\mathcal{H})^-) \equiv d_T(\mathcal{H} \ominus \ker(X))$, so that (1.8) can be rewritten as

$$d_T \equiv d_T(\ker(X))d_T(\mathcal{H} \ominus \ker(X)) \equiv d_T(\ker(X^*))d_T(\mathcal{H} \ominus \ker(X)).$$

The assumption that $d_T \neq 0$ now implies that we may cancel the factor $d_T(\mathcal{H} \ominus \ker(X))$ and obtain the desired conclusion that $d_T(\ker(X)) \equiv d_T(\ker(X^*))$. The fact that T has property (P) now follows from the fact that $d_T(\mathcal{M}) \equiv 1$ if and only if $\mathcal{M} = \{0\}$. The proposition is proved.

The preceding result provides many operators of class C_0 with property (P), but not all of them.

1.9. THEOREM. *Assume that $T \in \mathcal{L}(\mathcal{H})$ is an operator of class C_0 with Jordan model $\bigoplus_\alpha S(\theta_\alpha)$. Then T has property (P) if and only if*

$$\bigwedge_{j<\omega} \theta_j \equiv 1. \tag{1.10}$$

Thus, if T has property (P), $\mathcal{H}$ is separable and T^* also has property (P).

PROOF. Assume first that (1.10) holds and $X \in \{T\}'$ is such that $\ker(X) = \{0\}$. The subspace $\mathcal{H}_j = (\theta_j(T)\mathcal{H})^-$ is hyperinvariant for T, and $T \mid \mathcal{H}_j$ has finite multiplicity ($\leq j$) for $j < \omega$ (cf. Definition 3.5.24 and Corollary 3.5.25). By Theorem 6.5.7, $T \mid \mathscr{H}_j$ is a weak contraction and hence it has property (P) by Proposition 1.7 above. Since obviously $X \mid \mathscr{H}_j \in \{T \mid \mathscr{H}_j\}'$ and $\ker(X \mid \mathscr{H}_j) = \{0\}$, we conclude that $(X\mathscr{H}_j)^- = \mathscr{H}_j$, and hence $(X\mathscr{H})^- \supseteq \mathscr{H}_j$. Thus, in order to show that $\ker(X^*) = \{0\}$ or, equivalently, that X has dense range, it suffices to show that $\bigvee_{j<\omega} \mathscr{H}_j = \mathscr{H}$. But Corollary 2.4.8 implies that

$$\bigvee_{j<\omega} \mathscr{H}_j = \bigvee_{j<\omega} (\theta_j(T)\mathscr{H})^- = ((\bigwedge_{j<\omega} \theta_j)(T)\mathscr{H})^- = \mathscr{H}$$

by virtue of (1.10). Since $X \in \{T\}'$ was arbitrary with $\ker(X) = \{0\}$, it follows that T has property (P).

Conversely, let us assume that $\theta \equiv \bigwedge_{j<\omega} \theta_j$ is not a constant. By Corollary 3.5.10 we have

$$\left(\bigoplus_{j<\omega} S(\theta_j)\right) \oplus S(\theta) \sim \bigoplus_{j<\omega} S(\theta_j),$$

and it follows by comparison of Jordan models that $T \oplus S(\theta) \sim T$. Choose a quasiaffinity $A \in \mathscr{L}(\mathscr{H}, \mathscr{H} \oplus \mathscr{H}(\theta))$ such that $(T \oplus S(\theta))A = AT$, and define

$$\begin{aligned} \mathscr{M} &= (A^*(\{0\} \oplus \mathscr{H}(\theta)))^- \in \operatorname{Lat}(T^*), \\ \mathscr{N} &= \mathscr{H} \ominus \mathscr{M} = A^{-1}(\mathscr{H} \oplus \{0\}) \in \operatorname{Lat}(T). \end{aligned} \tag{1.11}$$

The idea now is to show that $T \mid \mathscr{N}$ is quasisimilar to T. Indeed, this would imply the existence of a quasiaffinity $X \in \mathscr{L}(\mathscr{H}, \mathscr{N})$ such that $(T \mid \mathscr{N})X = XT$. Of course X can be regarded as an operator acting on $\mathscr{H}$, and as such it commutes with T, $\ker(X) = \{0\}$, and $\ker(X^*) = \mathscr{M} \neq \{0\}$. This shows that T does not have property (P).

Now, by Proposition 3.5.32, it suffices to show that $T \mid \mathscr{N} \prec T$, and this would be achieved if we can show that

$$(A\mathscr{N})^- = \mathscr{H} \oplus \{0\}. \tag{1.12}$$

Indeed, if (1.12) holds and P_1 denotes the projection of $\mathscr{H} \oplus \mathscr{H}(\theta)$ onto $\mathscr{H}$, then $B = P_1 A \mid \mathscr{N}$ is a quasiaffinity satisfying $TB = B(T \mid \mathscr{N})$. To prove (1.12), denote $\mathscr{K} = (\mathscr{H} \oplus \mathscr{H}(\theta)) \ominus (A\mathscr{N})^- \in \operatorname{Lat}(T \oplus S(\theta))^*$ and observe that $A^*\mathscr{K} \subseteq \mathscr{H} \ominus \mathscr{N} = \mathscr{M}$. Since we clearly have $\{0\} \oplus \mathscr{H}(\theta) \subseteq \mathscr{K}$, we deduce from (1.11) that

$$\mathscr{M} = (A^*(\{0\} \oplus \mathscr{H}(\theta)))^- = (A^*\mathscr{K})^-.$$

These relations show that we have

$$(T \oplus S(\theta))^* \mid \{0\} \oplus \mathscr{H}(\theta) \prec T^* \mid \mathscr{M} \quad \text{and} \quad (T \oplus S(\theta))^* \mid \mathscr{K} \prec T^* \mid \mathscr{M}, \tag{1.13}$$

the quasiaffinities being $A^* \mid \{0\} \oplus \mathscr{H}(\theta)$ and $A^* \mid \mathscr{M}$, respectively. By Proposition 3.5.32, the three operators in (1.13) are pairwise quasisimilar and, in particular,

$$(T \oplus S(\theta))^* \mid \mathscr{K} \sim (T \oplus S(\theta))^* \mid \{0\} \oplus \mathscr{H}(\theta) \simeq S(\theta).$$

Thus $(T \oplus S(\theta))^* \mid \mathscr{K}$ is a multiplicity-free operator, and Theorem 3.2.13(iii) (with $\mathscr{K}' = \{0\} \oplus \mathscr{H}(\theta))$ shows that $\mathscr{K} = \{0\} \oplus \mathscr{H}(\theta)$. Hence $(A\mathscr{N})^- = (\mathscr{H} \oplus \mathscr{H}(\theta)) \ominus \mathscr{K} = \mathscr{H} \oplus \{0\}$, thus concluding the proof of (1.12).

Assume now that T has property (P). Then (1.10) implies $\theta_\omega \equiv 1$ and hence $\bigoplus_\alpha S(\theta_\alpha) = \bigoplus_{j<\omega} S(\theta_j)$, so that T acts on a separable space. The fact that T^* also has property (P) follows from the first part of the theorem, combined with the fact that $T^* \sim \bigoplus_{j<\omega} S(\theta_j^\sim)$. The theorem is proved.

1.14. REMARK. Condition (1.10) is easily seen to be equivalent to the equality $\lim_{j\to\infty} |\theta_j(0)| = 1$. We know that T is a weak contraction if and only if $\sum_{j<\omega}(1 - |\theta_j(0)|) < \infty$. Thus there are many contractions that are not weak but do have property (P). We note for further use that $\lim_{j\to\infty} |\theta_j(0)| = 1$ if and only if $\lim_{j\to\infty} |\theta_j(\lambda)| = 1$ for some $\lambda \in \mathbf{D}$. This is an easy consequence of the equality $\lim_{j\to\infty} |\theta_j(\lambda)| = |\theta(\lambda)|$, where $\theta \equiv \bigwedge_{j<\omega} \theta_j$.

1.15. REMARK. It is clear from the argument in the proof of Theorem 1.9 that an operator T of class C_0 fails to have property (P) if, and only if, T is quasisimilar to $T \mid \mathscr{N}$, where $\mathscr{N}$ is a proper invariant subspace for T.

1.16. COROLLARY. *Let T and T' be two quasisimilar operators of class C_0. Then T has property (P) if and only if T' has property (P).*

PROOF. The operators T and T' have the same Jordan model, so the corollary is obvious by Theorem 1.9.

1.17. COROLLARY. *Assume that $T \in \mathscr{L}(\mathscr{H})$ is an operator of class C_0, $\mathscr{H}'$ is an invariant subspace for T, and*

$$T = \begin{bmatrix} T' & X \\ 0 & T'' \end{bmatrix}$$

is the matrix of T with respect to the decomposition $\mathscr{H} = \mathscr{H}' \oplus (\mathscr{H} \ominus \mathscr{H}')$. Then T has property (P) if and only if T' and T'' have property (P).

PROOF. Let $\bigoplus_\alpha S(\theta_\alpha)$, $\bigoplus_\alpha S(\theta'_\alpha)$, and $\bigoplus_\alpha S(\theta''_\alpha)$ be the Jordan models of T, T', and T'', respectively. Assume first that T has property (P), so that (1.10) holds. We clearly have $T' \prec^i T$ and $T''^* \prec^i T^*$ so that Proposition 3.5.31 implies the relations $\theta'_\alpha \mid \theta_\alpha$ and $\theta''_\alpha \mid \theta_\alpha$ for all ordinals α. We deduce that $(\bigwedge_{j<\omega} \theta'_j) \mid (\bigwedge_{j<\omega} \theta_j)$, $(\bigwedge_{j<\omega} \theta''_j) \mid (\bigwedge_{j<\omega} \theta_j)$, hence $\bigwedge_{j<\omega} \theta'_j \equiv \bigwedge_{j<\omega} \theta''_j \equiv 1$, and T' and T'' have property (P) by Theorem 1.9.

Conversely, assume that T' and T'' have property (P). Then $\mathscr{H}'$ and $\mathscr{H}''$ are separable, and an application of Proposition 5.3.22 shows that $\theta_0\theta_1 \cdots \theta_j \mid \theta'_0\theta'_1 \cdots \theta'_j\theta''_0\theta''_1 \cdots \theta''_j$ for all $j < \omega$. We have therefore

$$|\theta_0(\lambda)\theta_1(\lambda) \cdots \theta_j(\lambda)| \geq |\theta'_0(\lambda)\theta'_1(\lambda) \cdots \theta'_j(\lambda)|\, |\theta''_0(\lambda)\theta''_1(\lambda) \cdots \theta''_j(\lambda)|$$

for $\lambda \in \mathbf{D}$ and $j < \omega$. If we now choose a point $\lambda \in \mathbf{D}$ such that $\theta_0(\lambda) \neq 0$, we get

$$\begin{aligned}\lim_{j\to\infty} |\theta_j(\lambda)| &= \lim_{j\to\infty} |\theta_0(\lambda)\theta_1(\lambda)\cdots\theta_j(\lambda)|^{1/(j+1)} \\ &\geq \lim_{j\to\infty} |\theta_0'(\lambda)\theta_1'(\lambda)\cdots\theta_j'(\lambda)|^{1/(j+1)}|\theta_0''(\lambda)\theta_1''(\lambda)\cdots\theta_j''(\lambda)|^{1/(j+1)} \\ &= \lim_{j\to\infty} |\theta_j'(\lambda)| \lim_{j\to\infty} |\theta_j''(\lambda)|.\end{aligned}$$

Since T' and T'' have property (P), we deduce that $\lim_{j\to\infty} |\theta_j(0)| = 1$ (cf. Remark 1.14 above), and hence T has property (P) by Theorem 1.9. The corollary is proved.

1.18. COROLLARY. *If T is an operator of class C_0 having property (P), and $\mathscr{M} \in \mathrm{Lat}_{1/2}(T)$, then $T_{\mathscr{M}}$ also has property (P).*

PROOF. Assume that $\mathscr{M} = \mathscr{U} \ominus \mathscr{V}$, where $\mathscr{U}, \mathscr{V} \in \mathrm{Lat}(T)$ and $\mathscr{U} \supseteq \mathscr{V}$. Then we have $T^*_{\mathscr{M}} = (T \mid \mathscr{U})^* \mid \mathscr{V}$, and the conclusion that $T_{\mathscr{M}}$ has property (P) follows from two applications of Corollary 1.17.

Returning now to the analogy with finite-dimensional spaces, we note that an operator $X \in \mathscr{L}(\mathscr{H})$, where $\dim(\mathscr{H}) < \infty$, is actually invertible if $\ker(X) = \{0\}$. Now, if T has property (P) and $X \in \{T\}'$ is a quasiaffinity, it is not usually true that X is invertible. Counterexamples can be found for operators like $T = S(\theta)$. Invertibility is replaced by a weaker property, to be defined below.

1.19. DEFINITION. Assume that T and T' are two operators and $X \in \mathscr{I}(T,T')$ (i.e., $T'X = XT$). Define a mapping $X_*\colon \mathrm{Lat}(T) \to \mathrm{Lat}(T')$ by $X_*(\mathscr{M}) = (X\mathscr{M})^-$. The operator X is said to be a *(T,T')-lattice-isomorphism* if X_* is a bijection of $\mathrm{Lat}(T)$ onto $\mathrm{Lat}(T')$.

We will use the name lattice-isomorphism instead of (T,T')-lattice-isomorphism when no confusion may arise. The following result facilitates the use of lattice-isomorphisms.

1.20. LEMMA. *Assume that T and T' are two operators, and $X \in \mathscr{I}(T,T')$. Then the mapping X_* is onto $\mathrm{Lat}(T')$ if and only if $(X^*)_*$ is one-to-one on $\mathrm{Lat}(T'^*)$.*

PROOF. Let $T \in \mathscr{L}(\mathscr{H})$, $T' \in \mathscr{L}(\mathscr{H}')$, and assume first that $(X^*)_*$ is one-to-one. Take $\mathscr{M}' \in \mathrm{Lat}(T')$ and define $\mathscr{M} = X^{-1}(\mathscr{M}')$ and $\mathscr{M}_1' = (X\mathscr{M})^-$. We want to show that $\mathscr{M}_1' = \mathscr{M}'$, so that $\mathscr{M}'$ belongs to the range of X_*. To do this we will show that $(X^*)_*(\mathscr{H}' \ominus \mathscr{M}') = (X^*)_*(\mathscr{H}' \ominus \mathscr{M}_1')$ and then use the injectivity of $(X^*)_*$. Indeed,

$$\begin{aligned}(X^*)_*(\mathscr{H}' \ominus \mathscr{M}_1') &= (\mathrm{ran}(X^* P_{\mathscr{H}'\ominus\mathscr{M}_1'}))^- \\ &= \mathscr{H} \ominus \ker(P_{\mathscr{H}'\ominus\mathscr{M}_1'} X) \\ &= \mathscr{H} \ominus (X^{-1}(\mathscr{M}_1')),\end{aligned}$$

and, analogously,

$$(X^*)_*(\mathscr{H}' \ominus \mathscr{M}') = \mathscr{H} \ominus (X^{-1}(\mathscr{M}')).$$

The equality $(X^*)_*(\mathcal{H}' \ominus \mathcal{M}') = (X^*)_*(\mathcal{H}' \ominus \mathcal{M}_1')$ follows from the obvious equality $X^{-1}(\mathcal{M}') = X^{-1}(\mathcal{M}_1')$. We conclude that $\mathcal{M}' = \mathcal{M}_1'$, so that $\mathcal{M}' = X_*(\mathcal{M})$ belongs to the range of X_*. The surjectivity of X_* is proved.

Conversely, assume that X_* is onto and $\mathcal{M}' \in \operatorname{Lat}(T'^*)$. If we set $\mathcal{M} = X^{-1}(\mathcal{H}' \ominus \mathcal{M}')$ we have therefore $\mathcal{H}' \ominus \mathcal{M}' = (X\mathcal{M})^-$. Thus

$$\begin{aligned}\mathcal{M}' &= \mathcal{H}' \ominus (\mathcal{H}' \ominus \mathcal{M}') = \mathcal{H}' \ominus (X\mathcal{M})^- \\ &= \mathcal{H}' \ominus (\operatorname{ran}(XP_{\mathcal{M}}))^- = \ker(P_{\mathcal{M}}X^*) \\ &= X^{*^{-1}}(\mathcal{H} \ominus \mathcal{M}) = X^{*^{-1}}(\mathcal{H} \ominus \ker(P_{\mathcal{M}'}X)) \\ &= X^{*^{-1}}((\operatorname{ran}(X^*P_{\mathcal{M}'}))^-) = X^{*^{-1}}((X^*)_*(\mathcal{M}'))\end{aligned}$$

and hence $\mathcal{M}'$ is uniquely determined by $(X^*)_*(\mathcal{M}')$. It follows at once that $(X^*)_*$ is one-to-one, thus concluding the proof.

1.21. PROPOSITION. *Assume that $T \in \mathcal{L}(\mathcal{H})$ and $T' \in \mathcal{L}(\mathcal{H}')$ are two quasisimilar operators of class C_0, and $X \in \mathcal{I}(T, T')$ is an injection. If T has property (P) then X is a lattice-isomorphism.*

PROOF. Assume that T has property (P). Then T', T^*, and T'^* also have property (P). By Lemma 1.20, we have to show that X_* and $(X^*)_*$ are one-to-one. By the above remarks, it will suffice to show that X has dense range and X_* is one-to-one; indeed, in this case $X^* \in \mathcal{I}(T'^*, T^*)$ is another injective operator. The fact that X has dense range follows easily from the fact that T' and $T' \mid (X\mathcal{H})^-$ are quasisimilar (cf. Remark 1.15 above). Assume therefore that $\mathcal{M}_1, \mathcal{M}_2 \in \operatorname{Lat}(T)$ and $X_*(\mathcal{M}_1) = X_*(\mathcal{M}_2) = \mathcal{M}'$. If we set $\mathcal{M} = \mathcal{M}_1 \vee \mathcal{M}_2$ we clearly have $X\mathcal{M}_1 \subseteq X\mathcal{M} \subseteq \mathcal{M}'$ and hence $X_*(\mathcal{M}) = \mathcal{M}'$. Now, the operators $T \mid \mathcal{M}$, $T \mid \mathcal{M}_1$, and $T \mid \mathcal{M}_2$ are quasisimilar to $T' \mid \mathcal{M}'$ because X realizes the relations $T \mid \mathcal{M} \prec T' \mid \mathcal{M}'$, $T \mid \mathcal{M}_1 \prec T' \mid \mathcal{M}'$, and $T \mid \mathcal{M}_2 \prec T' \mid \mathcal{M}'$. Thus $T \mid \mathcal{M}$, which has property (P) by Corollary 1.17, is quasisimilar to its restrictions to $\mathcal{M}_1$ and $\mathcal{M}_2$. Thus (see Remark 1.15) we must have $\mathcal{M} = \mathcal{M}_1 = \mathcal{M}_2$, whence the injectivity of X_*. The proposition follows.

The preceding result enables us to improve Theorem 3.6.20 and Theorem 4.2.8 for operators with property (P).

1.22. COROLLARY. *Assume that $T \in \mathcal{L}(\mathcal{H})$ is an operator of class C_0 with property (P), and let $\bigoplus_{j<\omega} S(\theta_j)$ be the Jordan model of T. There exist invariant subspaces $\mathcal{H}_j \in \operatorname{Lat}(T)$ such that*

(i) $\mathcal{H} = \bigvee_{j<\omega} \mathcal{H}_j$ *and this decomposition is quasidirect; and*
(ii) $T \mid \mathcal{H}_j \sim S(\theta_j)$.

PROOF. Choose an arbitrary quasiaffinity X satisfying the relation $TX = X(\bigoplus_{j<\omega} S(\theta_j))$, and set $\mathcal{H}_j = X_*(\mathcal{H}(\theta_j)) = (X\mathcal{H}(\theta_j))^-$; here $\mathcal{H}(\theta_j)$ is regarded as the j-component space in $\bigoplus_{j<\omega} \mathcal{H}(\theta_j)$. Let $\{K_\alpha : \alpha \in A\}$ be a family of sets of natural numbers, and define $K = \bigcap_{\alpha \in A} K_\alpha$. Since X_* is a lattice-isomorphism, we have

$$\begin{aligned}\bigcap_{\alpha\in A}\left(\bigvee_{j\in K_\alpha}\mathscr{H}_j\right) &= X_*\left(\bigcap_{\alpha\in A}\left(\bigoplus_{j\in K}\mathscr{H}(\theta_j)\right)\right)\\ &= X_*\left(\bigoplus_{j\in K}\mathscr{H}(\theta_j)\right)\\ &= \bigvee_{j\in K}\mathscr{H}_j.\end{aligned}$$

The corollary follows (cf. Definition 3.6.1).

Recall that the preceding (as well as the following) corollary applies, in particular, to weak contractions.

1.23. COROLLARY. *Assume that T and T' are two quasisimilar operators of class C_0, and define η: $\operatorname{Lat}(\{T\}') \to \operatorname{Lat}(\{T'\}')$, ξ: $\operatorname{Lat}(\{T'\}') \to \operatorname{Lat}(\{T\}')$ by the familiar formulas*

$$\eta(\mathscr{M}) = \bigvee\{X\mathscr{M} : X \in \mathscr{I}(T,T')\}, \quad \mathscr{M} \in \operatorname{Lat}(\{T\}'),$$

and

$$\xi(\mathscr{M}') = \bigvee\{Y\mathscr{M}' : Y \in \mathscr{I}(T',T)\}, \quad \mathscr{M}' \in \operatorname{Lat}(\{T'\}').$$

Assume, in addition, that T has property (P) and $B \in \mathscr{I}(T',T)$, $A \in \mathscr{I}(T,T')$ are quasiaffinities. Then

(i) *η is a bijection and $\xi = \eta^{-1}$; and*
(ii) *$\eta(\mathscr{M}) = (A\mathscr{M})^- = B^{-1}(\mathscr{M})$, $\mathscr{M} \in \operatorname{Lat}(\{T\}')$.*

PROOF. We know from Proposition 1.21 that A and B are lattice-isomorphisms. Fix $\mathscr{M} \in \operatorname{Lat}(\{T\}')$ and observe that $BA\mathscr{M} \subseteq \mathscr{M}$ because $BA \in \{T\}'$. Thus $BA \mid \mathscr{M} \in \{T \mid \mathscr{M}\}'$, $\ker(BA \mid \mathscr{M}) = \{0\}$, and $T \mid \mathscr{M}$ has property (P) by Corollary 1.17. We conclude that $BA \mid \mathscr{M}$ has dense range, i.e., $(BA\mathscr{M})^- = \mathscr{M}$. Since B is a lattice-isomorphism, we deduce

$$B^{-1}(\mathscr{M}) = (A\mathscr{M})^-. \tag{1.24}$$

If $X \in \mathscr{I}(T,T')$ is arbitrary, we have $BX \in \{T\}'$ and hence $BX\mathscr{M} \subseteq \mathscr{M}$. By (1.24) we then have $X\mathscr{M} \subseteq B^{-1}(\mathscr{M}) = (A\mathscr{M})^-$. We deduce that $\eta(\mathscr{M}) \subseteq (A\mathscr{M})^-$ and (ii) follows because the inclusion $(A\mathscr{M})^- \subseteq \eta(\mathscr{M})$ is obvious. A symmetric argument will show that $\xi(\mathscr{M}') = (B\mathscr{M}')^-$ for $\mathscr{M}' \in \operatorname{Lat}(\{T'\}')$. Thus we have

$$\xi(\eta(\mathscr{M})) = \xi((A\mathscr{M})^-) = (B(A\mathscr{M})^-)^- = (BA\mathscr{M})^- = \mathscr{M}, \qquad \mathscr{M} \in \operatorname{Lat}(\{T\}'),$$

that is, $\xi \circ \eta = \mathrm{id}_{\operatorname{Lat}(\{T\}')}$. Now (i) follows by symmetry and the corollary is proved.

Exercises

1. Let $\mathscr{H}$ be a Hilbert space with the orthonormal basis $\{e_1, e_2, e_3\}$. Define $T, X \in \mathscr{L}(\mathscr{H})$ by $T(\alpha_1 e_1 + \alpha_2 e_2 + \alpha_3 e_3) = \alpha_2 e_1$, and $X(\alpha_1 e_1 + \alpha_2 e_2 + \alpha_3 e_3) = \alpha_3 e_1$. Prove that $X \in \{T\}'$ but $T \mid \ker(X)$ and $(T^* \mid \ker(X^*))^*$ are not quasisimilar.

2. Assume that T is an algebraic operator. Show that T has property (P) if and only if it acts on a finite-dimensional Hilbert space.

3. Show that there exist nonseparably acting operators with property (P).

4. Give a characterization of all separably acting normal operators that have property (P).

5. Show that a normal operator with property (P) also has property (Q). More precisely, if T is normal, has property (P), and $X \in \{T\}'$, then $T \mid \ker(X)$ is unitarily equivalent to $(T^* \mid \ker(X^*))^*$.

6. Let T and T' be operators of class C_0 and $X \in \mathscr{I}(T, T')$. Show that $d_T d_T(\ker(X^*)) \equiv d_{T'} d_T(\ker(X))$.

7. Let T and T' be operators of class C_0 and $X \in \mathscr{I}(T, T')$. Assume that $d_T \equiv d_{T'} \not\equiv 0$ and $\ker(X) = \{0\}$. Show that X is a quasiaffinity.

8. Let T be an operator of class C_0. Show that T fails to have property (P) if and only if there exists an operator T' of class C_0, acting on a nontrivial space, such that $T \sim T \oplus T'$.

9. Let $T = \bigoplus_{j<\omega} S(\theta_j)$ be a Jordan operator. Show that the following conditions are equivalent:

(i) D_T is a compact operator;

(ii) $\bigwedge_{j<\omega} \theta_j \equiv 1$;

(iii) $\lim_{j\to\infty} |\theta_j(\lambda)| = 1$ for some $\lambda \in \mathbf{D}$; and

(iv) $\lim_{j\to\infty} |\theta_j(\lambda)| = 1$ for every $\lambda \in \mathbf{D}$.

10. Let T be an operator of class C_0. Show that D_T is compact if T has property (P), but not conversely.

11. Let T be a multiplicity-free operator of class C_0. Show that T has property (Q).

12. Assume that $T = \left[\begin{smallmatrix} T' & X \\ 0 & T'' \end{smallmatrix}\right]$ is an operator of class C_0, and $\bigoplus_\alpha S(\theta_\alpha)$, $\bigoplus_\alpha S(\theta'_\alpha)$, $\bigoplus_\alpha S(\theta''_\alpha)$ are the Jordan models of T, T', T'', respectively. Show that $\nu_T(\theta'_\alpha \theta''_\alpha) \leq 2\,\mathrm{card}(\alpha)$ for every ordinal α, and deduce that $\theta_{2\alpha} \mid \theta'_\alpha \theta''_\alpha$ (cf. §3.5).

13. Give a new proof of Corollary 1.17, based on Exercise 12.

14. Let θ be an inner function which is not a finite Blaschke product. Show that there exist noninvertible quasiaffinities $X \in \{S(\theta)\}'$.

15. Assume that V is an accretive operator such that $\ker(V) = \{0\}$, $\sigma(V) = \{0\}$, and $V + V^*$ is a trace-class operator. Show that V has property (P).

16. Take $T \in \mathscr{L}(\mathscr{H})$, $T' \in \mathscr{L}(\mathscr{H}')$, $X \in \mathscr{I}(T, T')$, and define a mapping $X_-\colon \operatorname{Lat}(T') \to \operatorname{Lat}(T)$ by $X_-(\mathscr{M}') = X^{-1}(\mathscr{M}')$, $\mathscr{M}' \in \operatorname{Lat}(T')$. Show that $X_*(X_-(\mathscr{H}' \ominus \mathscr{M}')) = \mathscr{H}' \ominus X_-^*((X^*)_*(\mathscr{M}'))$ for every $\mathscr{M}' \in \operatorname{Lat}(T'^*)$.

17. Assume that $T = T' = 0 \in \mathscr{L}(\mathscr{H})$. Show that an operator $X \in \mathscr{L}(\mathscr{H}) = \mathscr{I}(T', T)$ is a (T', T)-lattice-isomorphism if and only if it is invertible.

18. Let S denote, as usual, the unilateral shift on H^2. Show that S^* has property (P) but S does not have property (P).

19. Let U denote the bilateral shift on L^2. Show that U has property (P). (Note that $S = U \mid H^2$, so that not all restrictions of U inherit property (P).)

20. Assume that $T = A \oplus B$ has property (P). Show that both A and B have property (P).

21. Let $V \in \mathscr{L}(\mathscr{H})$ be an accretive operator such that $\sigma(V) = \{0\}$, $\ker(V) = \{0\}$, and $V + V^*$ is a trace-class operator. Show that $\mathscr{H}$ can be written as a quasidirect sum $\mathscr{H} = \bigvee_{j<\omega} \mathscr{H}_j$ of invariant subspaces for V such that

(i) $V \mid \mathscr{H}_j$ is quasisimilar to some Volterra integral operator V_{t_j}, $j < \omega$, $t_j \geq 0$ (for convenience, we define V_0 as an operator acting on the trivial space $\{0\}$); and

(ii) $t_0 \geq t_1 \geq \cdots$ and $\sum_{j<\omega} t_j = \operatorname{tr}(V + V^*)$.

2. Operators with property (Q). In this section we will give a complete characterization of operators of class C_0 with property (Q). This result is somewhat marginal to our development of Fredholm theory, but it is a nice application of previously developed techniques.

2.1. LEMMA. *Assume that T and T' are two quasisimilar operators of class C_0. Then T has property (Q) if and only if T' has property (Q).*

PROOF. Assume that T has property (Q). Then T has property (P), and it follows from Corollary 1.16 that T' also has property (P). Choose quasiaffinities $Y \in \mathscr{I}(T, T')$ and $X \in \mathscr{I}(T', T)$, and recall that X and Y are actually lattice-isomorphisms by Proposition 1.21. Finally, take $A \in \{T'\}'$ and denote $B = XAY \in \{T\}'$. Since X is one-to-one, the equality $\ker(B) = Y^{-1}(\ker(A))$ is obvious. Furthermore, because Y is a lattice-isomorphism, we deduce that $(Y(\ker(B)))^- = \ker(A)$, and therefore $Y \mid \ker(B)$ is a quasiaffinity from $\ker(B)$ to $\ker(A)$. Since $Y \mid \ker(B) \in \mathscr{I}(T \mid \ker(B), T' \mid \ker(A))$, it follows from Proposition 3.5.32 that $T' \mid \ker(A)$ and $T \mid \ker(B)$ are quasisimilar. In an analogous fashion we can see that $T^* \mid \ker(B^*) \sim T'^* \mid \ker(A^*)$ and hence

$$T' \mid \ker(A) \sim T \mid \ker(B) \sim (T^* \mid \ker(B^*))^* \sim (T'^* \mid \ker(A^*))^*,$$

where the second quasisimilarity follows because T has property (Q). Since A is arbitrary in $\{T'\}'$, we deduce that T' has property (Q). The lemma follows for reasons of symmetry.

The next result produces the basic building block of operators with property (Q).

2.2. PROPOSITION. *Assume that $\theta \in H^\infty$ is an inner function and $k \geq 1$ is a natural number. Then the operator*

$$T = S(\theta)^{(k)} = \underbrace{S(\theta) \oplus S(\theta) \oplus \cdots \oplus S(\theta)}_{k \text{ times}}$$

has property (Q).

PROOF. Fix $X \in \{T\}'$. An application of Proposition 5.1.26 (or, alternatively, k^2 applications of Corollary 3.1.20) provides a matrix $A = [a_{ij}]_{1 \leq i,j \leq k}$ over H^∞ such that

$$Xh = P_{\mathscr{H}} Ah, \qquad h = h_1 \oplus h_2 \oplus \cdots \oplus h_k \in \mathscr{H} = \mathscr{H}(\theta)^{(k)}. \tag{2.3}$$

We can now apply a θ-equivalence to the matrix A and obtain a diagonal matrix (cf. Theorem 6.5.28). That is, there are $k \times k$ matrices $B = [b_{ij}]_{1 \leq i,j \leq k}$, U and V over H^∞ such that U and V have scalar multiples ϕ and ψ, respectively, with $\phi \wedge \theta \equiv \psi \wedge \theta \equiv 1$, $b_{ij} = 0$ for $i \neq j$, and

$$AU = VB. \tag{2.4}$$

Let U' and V' be $k \times k$ matrices satisfying the relations

$$UU' = U'U = \phi I_{\mathbf{C}^k} \quad \text{and} \quad VV' = V'V = \psi I_{\mathbf{C}^k}. \tag{2.5}$$

The matrices B, U, V, U', and V' determine, via formulas analogous to (2.3), operators Y, K, L, K', and L' in $\{T'\}$, respectively. By (2.5) we have $KK' = K'K = \phi(T)$, $LL' = L'L = \psi(T)$. The operators $\phi(T)$ and $\psi(T)$ are quasiaffinities because $\phi \wedge \theta \equiv \psi \wedge \theta \equiv 1$ (recall that $\theta \equiv m_T$) so that the above relations prove that K and L are quasiaffinities. We deduce from Proposition 1.21 that K and L are (T,T)-lattice-isomorphisms.

Next we note that $XK = LY$ by (2.4), and hence we have $K(\ker(Y)) \subseteq \ker(X)$ and $K^{-1}(\ker(X)) \subseteq \ker(Y)$. Since K is a lattice-isomorphism, we deduce that $(K(\ker(Y)))^- = \ker(X)$ and this clearly implies that $T \mid \ker(X)$ and $T \mid \ker(Y)$ are quasisimilar. In an analogous manner it follows that $T^* \mid \ker(X^*) \sim T^* \mid \ker(Y^*)$. The operators $T \mid \ker(Y)$ and $(T^* \mid \ker(Y^*))^*$ are relatively easy to understand. Namely, we have $Y = \bigoplus_{j=1}^k b_{jj}(S(\theta))$ so that $\ker(Y) = \bigoplus_{j=1}^k \ker(b_{jj}(S(\theta)))$ and $\ker(Y^*) = \bigoplus_{j=1}^k \ker(b_{jj}(S(\theta))^*)$. A quick look at Corollary 3.1.12 shows that both $T \mid \ker(Y)$ and $(T^* \mid \ker(Y^*))^*$ are unitarily equivalent to $\bigoplus_{j=1}^k S(\theta_j)$, where $\theta_j \equiv b_{jj} \wedge \theta$, $1 \leq j \leq k$. Thus we have

$$T \mid \ker(X) \sim T \mid \ker(Y) \simeq (T^* \mid \ker(Y^*))^* \sim (T^* \mid \ker(X^*))^*.$$

Since X is arbitrary in $\{T'\}$, we reach the desired conclusion that T has property (Q).

2.6. LEMMA. *If $T \oplus S$ has property (Q) then T and S have property (Q).*

PROOF. If $X \in \{T\}'$ then $X \oplus I \in \{T \oplus S\}'$ and $T \mid \ker(X) \simeq (T \oplus S) \mid \ker(X \oplus I)$, $T^* \mid \ker(X^*) \simeq (T \oplus S)^* \mid \ker((X \oplus I)^*)$. The lemma obviously follows from this observation.

We need two more observations about Jordan models. These can now be easily proved in view of the results of Chapter 6.

2.7. LEMMA. *Let ϕ and ψ be two inner functions in H^∞. The Jordan model of $S(\phi) \oplus S(\psi)$ is $S(\phi \vee \psi) \oplus S(\phi \wedge \psi)$.*

PROOF. Denote $T = S(\phi) \oplus S(\psi)$, and let $T' = S(\theta_0) \oplus S(\theta_1)$ be the Jordan model of T. It is clear that $\theta_0 \equiv m_T \equiv \phi \vee \psi$. Furthermore, since $d_T \equiv \phi\psi$ and $d_{T'} \equiv \theta_0\theta_1$, Corollary 2.2.5 and the invariance of d_T under quasisimilarities imply that

$$\theta_1 \equiv d_{T'}/\theta_0 \equiv d_T/\theta_0 \equiv \phi\psi/\phi \vee \psi \equiv \phi \wedge \psi.$$

The lemma follows.

2.8. LEMMA. *Assume that $\{\phi_j : j \in J\}$ is a family of pairwise relatively prime inner functions in H^∞ such that $\phi = \bigvee\{\phi_j : j \in J\}$ exists. Then the Jordan model of $\bigoplus_{j\in J} S(\phi_j)$ is $S(\phi)$.*

PROOF. There is no loss of generality in assuming that $\theta_j \not\equiv 1$ for all j, and hence J is at most countable by Proposition 2.2.14. Denote by $\mathscr{F}$ a Hilbert space with basis $\{e_j : j \in J\}$, and define the functions Θ, $\Omega \in H^\infty(\mathscr{L}(\mathscr{F}))$ by $\Theta(\lambda)e_j = \phi_j(\lambda)e_j$ and $\Omega(\lambda)e_j = (\phi/\phi_j)(\lambda)e_j$, $j \in J$, $\lambda \in \mathbf{D}$. We now use Proposition 6.5.5 in order to determine the Jordan model $\bigoplus_{j<\omega} S(\theta_j)$ of $\bigoplus_{j\in J} S(\phi_j) \simeq S(\Theta)$. To do this we note that $\Theta\Omega = \Omega\Theta = \phi I_{\mathscr{F}}$,

$$\mathscr{D}_1(\Omega) \equiv \bigwedge_{j\in J} (\phi/\phi_j) \equiv \phi/ \bigvee_{j\in J} \phi_j \equiv \phi/\phi = 1,$$

and

$$\begin{aligned}
\mathscr{D}_k(\Omega) &\equiv \bigwedge\{\phi^k/\phi_{j_1}\phi_{j_2}\cdots\phi_{j_k} : \operatorname{card}\{j_1, j_2, \ldots, j_k\} = k\} \\
&\equiv \phi^k/\bigvee\{\phi_{j_1}\phi_{j_2}\cdots\phi_{j_k} : \operatorname{card}\{j_1, j_2, \ldots, j_k\} = k\} \\
&\equiv \phi^k/\phi \\
&= \phi^{k-1}, \quad k \geq 2.
\end{aligned}$$

This leads immediately to the conclusion that $\theta_0 \equiv \phi/\mathscr{D}_1(\Omega) \equiv \phi$, $\theta_1 \equiv \phi/\mathscr{E}_2(\Omega) \equiv \phi/\phi = 1$. The lemma follows.

2.9. THEOREM. *Assume that $T \in \mathscr{L}(\mathscr{H})$ is an operator of class C_0 with Jordan model $\bigoplus_\alpha S(\theta_\alpha)$. Then T has property (Q) if and only if the following two conditions are fulfilled:*

$$\bigwedge_{j<\omega} \theta_j \equiv 1; \tag{2.10}$$

(2.11) $$(\theta_j/\theta_{j+1}) \wedge (\theta_k/\theta_{k+1}) \equiv 1 \quad \text{for } j \neq k, \quad j,k < \omega.$$

Thus, if T has property (Q), $\mathscr{H}$ is separable and T^* also has property (Q).

PROOF. Assume first that T has property (Q). Then T has property (P), and hence (2.10) must be satisfied by Theorem 1.9. Furthermore, the Jordan model $\bigoplus_\alpha S(\theta_\alpha)$ has property (Q) by Lemma 2.1, and hence $S(\theta_j) \oplus S(\theta_{j+1})$ has property (Q) for $j < \omega$ by Lemma 2.6. Let us set $\phi_j = \theta_j/\theta_{j+1}$, $j < \omega$, and define an operator $X \in \{S(\theta_j) \oplus S(\theta_{j+1})\}'$ by $X(h_1 \oplus h_2) = \phi_j(S(\theta_j))h_1 \oplus 0$, $h_1 \oplus h_2 \in \mathscr{H}(\theta_j) \oplus \mathscr{H}(\theta_{j+1})$. Clearly X is a partial isometry, $\ker(X) = \mathscr{H}(\theta_j) \oplus \{0\}$, and $\operatorname{ran}(X) = (\phi_j H^2 \ominus \theta_j H^2) \oplus \{0\}$. We deduce that $\ker(X^*) = \mathscr{H}(\phi_j) \oplus \mathscr{H}(\theta_{j+1})$ and $((S(\theta_j) \oplus S(\theta_{j+1}))^* \mid \ker(X^*))^* = S(\phi_j) \oplus S(\theta_{j+1})$ by Proposition 3.1.10. Now, $(S(\theta_j) \oplus S(\theta_{j+1})) \mid \ker(X)$ is unitarily equivalent to $S(\theta_j)$, and we deduce from property (Q) that $S(\phi_j) \oplus S(\theta_{j+1}) \sim S(\theta_j)$. Since $\theta_j \equiv \phi_j\theta_{j+1}$, an application of Lemma 2.7 shows that $(\theta_j/\theta_{j+1}) \wedge \theta_{j+1} = \phi_j \wedge \theta_{j+1} \equiv 1$. If $k > j$ we have $(\theta_k/\theta_{k+1}) \mid \theta_{j+1}$, and therefore conditions (2.11) must be satisfied.

Conversely, assume that (2.10) and (2.11) are satisfied. By Lemma 2.1 it will suffice to show that $\bigoplus_{j<\omega} S(\theta_j)$ has property (Q). Let us again use the notation $\phi_j = \theta_j/\theta_{j+1}$ for $j < \omega$. Then we clearly have

$$\theta_j = \phi_j\theta_{j+1} = \phi_j\phi_{j+1}\theta_{j+2} = \cdots = \phi_j\phi_{j+1}\cdots\phi_{j+n}\theta_{j+n+1}, \quad j,n < \omega,$$

from which we infer, via (2.10) and (2.11), that

$$\bigvee\{\phi_n : n \geq j\} = \bigvee\{\phi_j\phi_{j+1}\cdots\phi_{j+n} : n < \omega\}$$
$$\equiv \theta_j / \bigwedge\{\theta_{j+n+1} : n < \omega\} \equiv \theta_j, \quad j < \omega.$$

Lemma 2.8 now implies that $S(\theta_j)$ is quasisimilar to $\bigoplus_{n \geq j} S(\phi_n)$, and hence T is quasisimilar to $T' = \bigoplus_{j<\omega} S(\phi_j)^{(j+1)}$, where, as usual, $S(\phi_j)^{(j+1)}$ denotes the orthogonal sum of $j+1$ copies of $S(\phi_j)$. By Lemma 2.1, it suffices to show that T' has property (Q). Before doing so, we note that $\theta_0/\phi_j \equiv (\bigvee_{n<\omega} \phi_n)/\phi_j \equiv \bigvee_{n \neq j} \phi_n$, $j < \omega$, and hence $(\theta_0/\phi_j)(S(\phi_j))$ is a quasiaffinity by Proposition 2.4.9, while $(\theta_0/\phi_j)(S(\phi_k)) = 0$ for $k \neq j$. We deduce that

$$(\operatorname{ran}((\theta_0/\phi_j)(T')))^- = \mathscr{H}(\phi_j)^{(j+1)}, \quad j < \omega,$$

where $\mathscr{H}(\phi_j)^{(j+1)}$ is regarded as the j-component space in the direct sum $\bigoplus_{i<\omega}(\mathscr{H}(\phi_i)^{(i+1)})$. Therefore $\mathscr{H}(\phi_j)^{(j+1)}$ is in $\operatorname{Lat}(\{T'\}')$ for $j < \omega$. Assume now that $X \in \{T'\}'$. By the preceding remarks, we have $X(\mathscr{H}(\phi_j)^{(j+1)}) \subseteq \mathscr{H}(\phi_j)^{(j+1)}$ for all j, and hence X can be written as $X = \bigoplus_{j<\omega} X_j$ with $X_j \in \{S(\phi_j)^{(j+1)}\}'$. We know from Proposition 2.2 that $S(\phi_j)^{(j+1)} \mid \ker(X_j) \sim ((S(\phi_j)^{(j+1)})^* \mid \ker(X_j^*))^*$, and this easily implies that $T' \mid \ker(X) = \bigoplus_{j<\omega}(S(\phi_j)^{(j+1)} \mid \ker(X_j)) \sim (T'^* \mid \ker(X^*))^*$. Since X is arbitrary in $\{T'\}'$, T' has property (Q). The theorem is proved.

Operators with property (Q) can also be distinguished by the relatively simple structure of their lattice of hyperinvariant subspaces.

2.12. PROPOSITION. *Assume that T is an operator of class C_0 and has property (P). Then T has property (Q) if and only if*

$$\operatorname{Lat}(\{T\}') = \{(\operatorname{ran}(\phi(T)))^- : \phi \ \textit{inner}, \phi \mid m_T\}. \tag{2.13}$$

PROOF. Let $T' = \bigoplus_{j<\omega} S(\theta_j)$ be the Jordan model of T, and choose a quasiaffinity $X \in \mathscr{I}(T',T)$. Corollary 1.23 shows that X_* realizes a bijection of $\operatorname{Lat}(\{T'\}')$ onto $\operatorname{Lat}(\{T\}')$. Since we clearly have $X_*((\operatorname{ran}(u(T')))^-) = (\operatorname{ran}(u(T)))^-$, $u \in H^\infty$, and since T' also has property (P), it will suffice to prove the proposition under the additional assumption that $T = T'$. Assume first that (2.13) holds, and fix $j < \omega$. Then the hyperinvariant subspace

$$\ker(\theta_{j+1}(T)) = \left[\bigoplus_{i\le j}((\theta_i/\theta_{j+1})H^2 \ominus \theta_i H^2)\right] \oplus \left[\bigoplus_{j+1\le i<\omega} \mathscr{H}(\theta_i)\right]$$

must have the form $(\operatorname{ran}(u(T)))^-$ for some inner divisor u of θ_0. A look at the first component in this decomposition shows that $\operatorname{ran}(u(S(\theta_0))) = (\theta_0/\theta_{j+1})H^2 \ominus \theta_0 H^2$, and hence $u \equiv \theta_0/\theta_{j+1}$. Next we look at the $(j+1)$st component. We must have $(\operatorname{ran}(u(S(\theta_{j+1}))))^- = \mathscr{H}(\theta_{j+1})$, and hence $u \wedge \theta_{j+1} \equiv 1$ (cf. Proposition 2.4.9). Thus we have $(\theta_0/\theta_{j+1}) \wedge \theta_{j+1} \equiv 1$ and, since $(\theta_j/\theta_{j+1}) \mid (\theta_0/\theta_{j+1})$ and $(\theta_k/\theta_{k+1}) \mid \theta_{j+1}$, $k > j$, we see that conditions (2.11) are satisfied. Condition (2.10) is satisfied because of the hypothesis that T has property (P). We conclude that T has property (Q).

Conversely, let us assume that T has property (Q). We set, as before, $\phi_j = \theta_j/\theta_{j+1}$, $j < \omega$, and note that T is quasisimilar to $T' = \bigoplus_{j<\omega} S(\phi_j)^{(j+1)}$. A second use of Corollary 1.23 shows that it suffices to prove that

$$\operatorname{Lat}(\{T'\}') = \{(\operatorname{ran}(\phi(T)))^- : \phi \,\text{inner}, \phi \mid \theta_0\}. \tag{2.14}$$

Assume now that $\mathscr{M} \in \operatorname{Lat}(T')$, and set $\mathscr{M}_j = ((\theta_0/\phi_j)(T')\mathscr{M})^-$. We claim that

$$\mathscr{M} = \bigoplus_{j<\omega} \mathscr{M}_j \quad \text{and} \quad \mathscr{M}_j = \mathscr{M} \cap \mathscr{H}(\phi_j)^{(j+1)}, \quad j < \omega. \tag{2.15}$$

The inclusion $\mathscr{M} \supseteq \bigoplus_{j<\omega} \mathscr{M}_j$ is obvious, and the opposite inclusion follows from the following calculation:

$$\begin{aligned}
\mathscr{M} \ominus \left(\bigoplus_{j<\omega} \mathscr{M}_j\right) &= \bigcap_{j<\omega} (\mathscr{M} \ominus (\operatorname{ran}((\theta_0/\phi_j)(T' \mid \mathscr{M})))^-) \\
&= \bigcap_{j<\omega} \ker((\theta_0/\phi_j)^\sim (T' \mid \mathscr{M})^*) \\
&= \ker((\bigwedge_{j<\omega} \theta_0/\phi_j)^\sim (T' \mid \mathscr{M})^*) \\
&= \{0\},
\end{aligned}$$

where we used Theorem 2.4.6. The equality $\mathscr{M}_j = \mathscr{M} \cap \mathscr{H}(\phi_j)^{(j+1)}$ follows from the fact that $\mathscr{M}_j \subseteq \mathscr{H}(\phi_j)^{(j+1)}$ and $\mathscr{M} = \bigoplus_{j<\omega} \mathscr{M}_j$. Assume now that, in addition, $\mathscr{M}$ is hyperinvariant for T'. Then (2.15) shows that $\mathscr{M}_j$

is hyperinvariant for $S(\phi_j)^{(j+1)}$, and it follows from Proposition 4.2.1 that $\mathscr{M}_j = \psi_j(S(\phi_j)^{(j+1)})\mathscr{H}(\phi_j)^{(j+1)}$ for some inner factor ψ_j of ϕ_j. If we now set $\psi \equiv \bigvee_{i<\omega}\psi_i$, we have $(\psi/\psi_j) \wedge \phi_j \equiv 1$, $j < \omega$, so that $(\psi/\psi_j)(S(\phi_j))$ is a quasiaffinity and, consequently, $\mathscr{M}_j = (\psi(S(\phi_j)^{(j+1)})\mathscr{H}(\phi_j)^{(j+1)})^-$. It is clear now that $\mathscr{M} = (\operatorname{ran}(\psi(T')))^-$, thus concluding the proof of (2.14). The proposition follows.

Exercises

1. Show that an operator T has property (Q) if and only if T^* has property (Q).

2. Let V be an accretive operator such that $\sigma(V) = \{0\}$, $\ker(V) = \{0\}$, and $V + V^*$ is a trace-class operator. Show that V has property (Q) if and only if it has a cyclic vector.

3. Assume that θ is an inner function and $k \geq 1$ is a natural number. Show that $S(\theta)^{(k)}$ has property $(*)$ (cf. Definition 4.1.10).

4. Show that $T \oplus S$ may fail to have property (Q) if T and S have property (Q).

5. Give a characterization of all Jordan operators T satisfying relation (2.13).

6. Show that property (2.13) is not invariant under quasisimilarity.

7. Denote by $\mathscr{L}_\theta^k$ the lattice $\operatorname{Lat}(S(\theta)^{(k)})$. Assume that $T = \bigoplus_{j<\omega} S(\theta_j)$ is a Jordan operator and has property (Q). Show that $\operatorname{Lat}(T)$ is isomorphic to $\prod_{j<\omega} \mathscr{L}_{\phi_j}^{j+1}$, where $\phi_j = \theta_j/\theta_{j+1}$, $j < \omega$.

8. Assume that T is an operator of class C_0 with property (P). Show that T has property (Q) if and only if

$$\operatorname{Lat}(\{T\}') = \{\ker(\phi(T)) \colon \phi \text{ inner}, \phi \mid m_T\}.$$

3. A continuity property of the Jordan model. Let $T \in \mathscr{L}(\mathscr{H})$ be an operator of class C_0, and let $\{\mathscr{H}_j \colon j \geq 0\}$ be a sequence of invariant subspaces for T. If $\mathscr{H} = \bigvee_{j=0}^\infty \mathscr{H}_j$ then we certainly have $m_T \equiv \bigvee\{m_{T|\mathscr{H}_j} \colon j \geq 0\}$. We will now show that an analogous property holds for all the functions in the Jordan model of T. We recall from §3.5 that the Jordan model of T is the operator $\bigoplus_\alpha S(M_T(\alpha))$, where

$$M_T(\alpha) = \bigwedge\{\phi \colon \phi \text{ inner}, \nu_T(\phi) \leq \operatorname{card}(\alpha)\} \tag{3.1}$$

for every ordinal number α.

3.2. THEOREM. *Assume that $T \in \mathscr{L}(\mathscr{H})$ is an operator of class C_0, $\{\mathscr{H}_j \colon j \geq 0\}$ is an increasing sequence of invariant subspaces for T, and $\mathscr{H} = \bigvee_{j=0}^\infty \mathscr{H}_j$. Then the relation*

$$M_T(\alpha) \equiv \bigvee\{M_{T|\mathscr{H}_j}(\alpha) \colon j \geq 0\} \tag{3.3}$$

holds for every ordinal number α.

PROOF. We obviously have $T \mid \mathscr{H}_j \prec^i T \mid \mathscr{H}_{j+1} \prec^i T$ so that Proposition 3.5.31 implies the relations

$$M_{T|\mathscr{H}_j}(\alpha) \mid M_{T|\mathscr{H}_{j+1}}(\alpha) \mid M_T(\alpha) \tag{3.4}$$

for all integers j and all ordinals α. Now fix an ordinal $\alpha \geq \omega$ and set $\theta \equiv \bigvee\{M_{T|\mathscr{H}_j}(\alpha) : j \geq 0\}$. By (3.4), θ divides $M_T(\alpha)$. Because $M_{T|\mathscr{H}_j}(\alpha)$ divides θ we have

$$\mu_{T|(\theta(T)\mathscr{H}_j)^-} = \nu_{T|\mathscr{H}_j}(\theta) \leq \operatorname{card}(\alpha), \qquad j \geq 0,$$

and the obvious equality $(\theta(T)\mathscr{H})^- = \bigvee_{j=0}^{\infty} \theta(T)\mathscr{H}_j$ implies

$$\nu_T(\theta) = \mu_{T|(\theta(T)\mathscr{H})^-} \leq \aleph_0 \operatorname{card}(\alpha) = \operatorname{card}(\alpha).$$

We conclude that $M_T(\alpha)$ divides θ, and hence (3.3) is proved for $\alpha \geq \omega$.

For the case in which $\alpha < \omega$, we want to reduce the problem to separably acting operators. To do this, we recall that Theorem 3.6.4 associates with each limit ordinal β a reducing subspace $\mathscr{M}_\beta$ for T such that $T \mid \mathscr{M}_\beta \sim \bigoplus_{j<\omega} S(\theta_{\beta+j})$, and $\mathscr{H} = \bigoplus_\beta \mathscr{M}_\beta$. We next define $\mathscr{K}_j = (P_{\mathscr{M}_0}\mathscr{H}_j)^-$, $j \geq 0$, and observe that $\mathscr{M}_0 = \bigvee_{j=0}^{\infty} \mathscr{K}_j$. Furthermore, we have $(T \mid \mathscr{K}_j)^* \prec^i (T \mid \mathscr{H}_j)^*$, the injection being realized by $P_{\mathscr{K}_j} \mid \mathscr{K}_j$. We deduce from Proposition 3.5.31 that $M_{T|\mathscr{K}_j}(\alpha) \mid M_{T|\mathscr{H}_j}(\alpha)$ and, since $M_T(\alpha) = M_{T|\mathscr{M}_0}(\alpha)$ for $\alpha < \omega$, it suffices to prove (3.3) for $\alpha < \omega$ under the additional assumption that $\mathscr{H} = \mathscr{M}_0$ and $\mathscr{H}_j = \mathscr{K}_j$, $j \geq 0$. That is, we may assume that $\mathscr{H}$ is a separable space. We may, of course, assume that $T = S(\Theta)$, where $\Theta \in H^\infty(\mathscr{L}(\mathscr{F}))$ is an inner function. Proposition 5.1.21 now implies the existence of inner functions $\Theta_j^{(1)}, \Theta_j^{(2)} \in H^\infty(\mathscr{L}(\mathscr{F}))$ such that $\Theta = \Theta_j^{(2)}\Theta_j^{(1)}$, $\mathscr{H}_j = \Theta_j^{(2)} H^2(\mathscr{F}) \ominus H^2(\mathscr{F})$, and $T \mid \mathscr{H}_j$ is unitarily equivalent to $S(\Theta_j^{(1)})$, $j \geq 0$. Next, the condition $\mathscr{H} = \bigvee_{j=0}^{\infty} \mathscr{H}_j$ is equivalent to $H^2(\mathscr{F}) = \bigvee_{j=0}^{\infty} \Theta_j^{(2)} H^2(\mathscr{F})$. In particular, if $f \in \mathscr{F} \subseteq H^2(\mathscr{F})$, we must have $\lim_{j\to\infty} \|f - P_{\Theta_j^{(2)} H^2(\mathscr{F})} f\| = 0$ (here we need the fact that $\mathscr{H}_j \subseteq \mathscr{H}_{j+1}$ for all j). It is easy to see that

$$P_{\Theta_j^{(2)} H^2(\mathscr{F})} f = T_{\Theta_j^{(2)}} \Theta_j^{(2)}(0)^* f, \quad f \in \mathscr{F}, j \geq 0.$$

Indeed, an easy calculation shows that

$$(f - T_{\Theta_j^{(2)}} \Theta_j^{(2)}(0)^* f, S_{\mathscr{F}}^n T_{\Theta_j^{(2)}} g) = 0, \quad j \geq 0, \quad n \geq 0, \quad f, g \in \mathscr{F}.$$

We conclude that $f = \lim_{j\to\infty} T_{\Theta_j^{(2)}} \Theta_j^{(2)}(0)^* f$, $f \in \mathscr{F}$. The functions $(T_{\Theta_j^{(2)}} \Theta_j^{(2)}(0)^* f)(\lambda) = \Theta_j^{(2)}(\lambda)\Theta_j^{(2)}(0)^* f$ are uniformly bounded by $\|f\|$ for $\lambda \in \mathbf{D}$, and we deduce that

$$f_0 \wedge f_1 \wedge \cdots \wedge f_k = \lim_{j\to\infty} T_{\wedge^{k+1}\Theta_j^{(2)}} \left(\bigwedge\nolimits^{k+1} \Theta_j^{(2)}(0)\right)^* (f_0 \wedge f_1 \wedge \cdots \wedge f_k)$$

for all $k \geq 0$ and $f_0, f_1, \ldots, f_k \in \mathscr{F}$. Therefore the span

$$\bigvee_{j=0}^{\infty} (\bigwedge\nolimits^{k+1} \Theta_j^{(2)}) H^2(\bigwedge\nolimits^{k+1} \mathscr{F})$$

contains all constant functions. Since that span is invariant under $S_{\bigwedge^{k+1}\mathscr{F}}$, we necessarily have

$$H^2\left(\bigwedge^{k+1}\mathscr{F}\right)=\bigvee_{j=0}^{\infty}\left(\bigwedge^{k+1}\Theta_j^{(2)}\right)H^2\left(\bigwedge^{k+1}\mathscr{F}\right),\qquad k\geq 0. \tag{3.5}$$

The subspaces

$$\mathscr{H}_j^k=\left[\left(\bigwedge^{k+1}\Theta_j^{(2)}\right)H^2\left(\bigwedge^{k+1}\mathscr{F}\right)\right]\ominus\left[\left(\bigwedge^{k+1}\Theta\right)H^2\left(\bigwedge^{k+1}\mathscr{F}\right)\right], j,k\geq 0,$$

are invariant under $S(\bigwedge^{k+1}\Theta)$. Moreover, the factorization $\bigwedge^{k+1}\Theta = (\bigwedge^{k+1}\Theta_j^{(2)})(\bigwedge^{k+1}\Theta_j^{(1)})$ shows that $S(\bigwedge^{k+1}\Theta)\mid\mathscr{H}_j^k$ is unitarily equivalent to $S(\bigwedge^{k+1}\Theta_j^{(1)})$. Finally, (3.5) easily implies that $\mathscr{H}(\bigwedge^{k+1}\Theta)=\bigvee_{j=0}^{\infty}\mathscr{H}_j^k$, $k\geq 0$, so that

$$m_{S(\bigwedge^{k+1}\Theta)}\equiv\bigvee_{j=0}^{\infty}m_{S(\bigwedge^{k+1}\Theta)|\mathscr{H}_j^k}\equiv\bigvee_{j=0}^{\infty}m_{S\left(\bigwedge^{k+1}\Theta_j^{(1)}\right)}. \tag{3.6}$$

By Corollary 5.3.29, we are almost done. Indeed, we have $m_{S(\bigwedge^{k+1}\Theta)}\equiv M_T(0)M_T(1)\cdots M_T(k)$, and $m_{S(\bigwedge^{k+1}\Theta_j^{(1)})}\equiv M_{T|H_j}(0)M_{T|\mathscr{H}_j}(1)\cdots M_{T|\mathscr{H}_j}(k)$, $j,k\geq 0$, so that (3.6) is equivalent to

$$M_T(0)M_T(1)\cdots M_T(k)\equiv\bigvee_{j=0}^{\infty}M_{T|\mathscr{H}_j}(0)M_{T|\mathscr{H}_j}(1)\cdots M_{T|\mathscr{H}_j}(k), \tag{3.7}$$

for $k\geq 0$. It is easy now to deduce from (3.4) that

$$\bigvee_{j=0}^{\infty}M_{T|\mathscr{H}_j}(0)M_{T|\mathscr{H}_j}(1)\cdots M_{T|\mathscr{H}_j}(k)\equiv\theta_0\theta_1\cdots\theta_k,$$

where $\theta_i\equiv\bigvee_{j=0}^{\infty}M_{T|\mathscr{H}_j}(i)$, $i\geq 0$. Thus a comparison of relations (3.7) for $k=\alpha-1$ and $k=\alpha<\omega$ shows that (3.3) is true for $\alpha<\omega$. The proof is now complete.

3.8. COROLLARY. *Assume that $T\in\mathscr{L}(\mathscr{H})$ is a weak contraction of class C_0, and $\mathscr{H}_j\in\operatorname{Lat}(T)$ for $j\geq 0$.*

(i) *If $\mathscr{H}_j\subseteq\mathscr{H}_{j+1}$ for all j and $\mathscr{H}=\bigvee_{j=0}^{\infty}\mathscr{H}_j$, then $d_T\equiv\bigvee_{j=0}^{\infty}d_T(\mathscr{H}_j)$.*

(ii) *If $\mathscr{H}_j\subseteq\mathscr{H}_{j+1}$ for all j and $\bigcap_{j=0}^{\infty}\mathscr{H}_j=\{0\}$, then $\bigwedge_{j=0}^{\infty}d_T(\mathscr{H}_j)\equiv 1$.*

PROOF. We prove (i) first. It is clear that $\bigvee_{j=0}^{\infty}d_T(\mathscr{H}_j)$ divides d_T. On the other hand, $M_{T|\mathscr{H}_j}(0)M_{T|\mathscr{H}_j}(1)\cdots M_{T|\mathscr{H}_j}(k)$ divides $d_T(\mathscr{H}_j)$ for all $j,k\geq 0$, and Theorem 3.2 implies that $M_T(0)M_T(1)\cdots M_T(k)$ divides $\bigvee_{j=0}^{\infty}d_T(\mathscr{H}_j)$. Now, the equality in (i) follows because $d_T\equiv\bigvee_{j=0}^{\infty}M_T(0)M_T(1)\cdots M_T(k)$.

To prove (ii) we recall that T^* is also a weak contraction (cf. Corollary 6.3.15) and (i) implies that $d_T\equiv\bigvee_{j=0}^{\infty}d_T(\mathscr{H}\ominus\mathscr{H}_j)$. By Theorem 1.4 we have $d_T\equiv d_T(\mathscr{H}_j)d_T(\mathscr{H}\ominus\mathscr{H}_j)$, and we deduce from Proposition 2.2.4 that

$$d_T\equiv\left(\bigwedge_{j=0}^{\infty}d_T(\mathscr{H}_j)\right)\left(\bigvee_{j=0}^{\infty}d_T(\mathscr{H}\ominus\mathscr{H}_j)\right)\equiv\left(\bigwedge_{j=0}^{\infty}d_T(\mathscr{H}_j)\right)d_T.$$

The corollary follows.

The following result gives a new characterization of operators of class C_0 that have property (P).

3.9. PROPOSITION. *Let $T \in \mathcal{L}(\mathcal{H})$ be an operator of class C_0. Then T has property (P) if and only if the following holds: for every decreasing sequence $\{\mathcal{H}_j : j \geq 0\} \subseteq \operatorname{Lat}(T)$ such that $\bigcap_{j=0}^{\infty} \mathcal{H}_j = \{0\}$, we have $\bigwedge_{j=0}^{\infty} m_{T|\mathcal{H}_j} \equiv 1$.*

PROOF. By Theorem 3.6.10, $\mathcal{H}$ can be written as an almost direct sum

$$\mathcal{H} = \bigvee_{\alpha} \mathcal{M}_\alpha, \tag{3.10}$$

where $\mathcal{M}_\alpha \in \operatorname{Lat}(T)$ and $T \mid \mathcal{M}_\alpha \sim S(\theta_\alpha)$ for each ordinal α, where $\bigoplus_\alpha S(\theta_\alpha)$ is the Jordan model of T. For each $j < \omega$ set $\mathcal{H}_j = \bigvee_{j \leq k < \omega} \mathcal{M}_k$. Since the decomposition (3.10) is almost direct, we have $\bigcap_{j=0}^{\infty} \mathcal{H}_j = \{0\}$. Moreover, it is clear that $m_{T|\mathcal{H}_j} \equiv \theta_j$, $j < \omega$. Thus, if T has the property mentioned in the statement of the proposition, we must have $\bigwedge_{j<\omega} \theta_j \equiv 1$, so that T has property (P) by Theorem 1.9.

Conversely, assume that T has property (P), and let $\{\mathcal{H}_j : j \geq 0\}$ be an arbitrary decreasing sequence in $\operatorname{Lat}(T)$ such that $\bigcap_{j=0}^{\infty} \mathcal{H}_j = \{0\}$. Denote by $\bigoplus_{j<\omega} S(\theta_j)$ the Jordan model of T, and write $\mathcal{H}_j = \mathcal{M}_j^k \oplus \mathcal{N}_j^k$, where $\mathcal{M}_j^k = (\theta_k(T)\mathcal{H}_j)^-$ and $\mathcal{N}_j^k = \mathcal{H}_j \ominus \mathcal{M}_j^k$ for $j, k \geq 0$. It is quite clear that $m_{T_{\mathcal{N}_j^k}}$ divides θ_k so that we have

$$m_{T|\mathcal{H}_j} \mid m_{T|\mathcal{M}_j^k} \theta_k, \quad j, k \geq 0, \tag{3.11}$$

by Proposition 2.4.3. The operator $T|(\theta_k(T)\mathcal{H})^-$ has finite multiplicity and is therefore a weak contraction (cf. Theorem 6.4.7). Observe then that $\mathcal{M}_j^k \subseteq (\theta_k(T)\mathcal{H})^-$ and $\bigcap_{j=0}^{\infty} \mathcal{M}_j^k \subseteq \bigcap_{j=0}^{\infty} \mathcal{H}_j = \{0\}$. The preceding corollary (applied to $T \mid (\theta_k(T)\mathcal{H})^-$) implies that $\bigwedge_{j=0}^{\infty} d_T(\mathcal{M}_j^k) \equiv 1$, hence, a fortiori, $\bigwedge_{j=0}^{\infty} m_{T|\mathcal{M}_j^k} \equiv 1$ for $k \geq 0$. Relation (3.11) now shows that $\bigwedge_{j=0}^{\infty} m_{T|\mathcal{H}_j}$ divides θ_k, and the conclusion that $\bigwedge_{j=0}^{\infty} m_{T|\mathcal{H}_j} \equiv 1$ follows from the assumption that T has property (P), i.e., $\bigwedge_{k=0}^{\infty} \theta_k \equiv 1$. The proposition is proved.

We can now prove two important criteria for an operator T of class C_0 to have property (P).

3.12. PROPOSITION. *Assume that $T \in \mathcal{L}(\mathcal{H})$ is an operator of class C_0, $\{\mathcal{H}_j : j \geq 0\} \subseteq \operatorname{Lat}(T)$ is an increasing sequence, $\mathcal{H}_0 = \{0\}$, $\mathcal{H} = \bigvee_{j=0}^{\infty} \mathcal{H}_j$, and $\mathcal{K}_j = \mathcal{H}_{j+1} \ominus \mathcal{H}_j \in \operatorname{Lat}_{1/2}(T)$, $j \geq 0$. Then T has property (P) if and only if the following two conditions are satisfied:*

(i) *$T_{\mathcal{K}_j}$ has property (P) for all $j \geq 0$;*

(ii) *$\bigwedge_{j=0}^{\infty} m_{T_{\mathcal{H} \ominus \mathcal{H}_j}} \equiv 1$.*

PROOF. Assume first that T has property (P). Then (i) follows from Corollary 1.18, and (ii) follows from Proposition 3.9.

Conversely, assume that (i) and (ii) are satisfied. Define $\phi_j = m_{T_{\mathscr{H}\ominus\mathscr{K}_j}}$, and set $\mathscr{L}_j = (\phi_j(T)\mathscr{H})^- \in \operatorname{Lat}(\{T\}')$. Corollary 2.4.8 shows, via (ii), that $\bigvee_{j=0}^{\infty} \mathscr{L}_j = \mathscr{H}$. We see now, as in the proof of Theorem 1.9, that it suffices to prove that $T \mid \mathscr{L}_j$ has property (P) for every j. Since clearly $\mathscr{L}_j \subseteq \mathscr{H}_j$, Corollary 1.17 shows that it suffices to show that $T \mid \mathscr{H}_j$ has property (P). This last statement follows by an easy induction, based on Corollary 1.17 and on the decomposition $T \mid \mathscr{H}_{j+1} = \left[\begin{smallmatrix} T|\mathscr{H}_j & X_j \\ 0 & T_{\mathscr{K}_j} \end{smallmatrix}\right]$, $j \geq 0$. The proposition follows.

3.13. COROLLARY. *Assume that $T \in \mathscr{L}(\mathscr{H})$ is an operator of class C_0, $\{\mathscr{H}_j : j \geq 0\} \subseteq \operatorname{Lat}(T)$ is a decreasing sequence, $\mathscr{H}_0 = \mathscr{H}$, $\bigcap_{j=0}^{\infty} \mathscr{H}_j = \{0\}$, and $\mathscr{K}_j = \mathscr{H}_j \ominus \mathscr{H}_{j+1} \in \operatorname{Lat}_{1/2}(T)$, $j \geq 0$. Then T has property (P) if and only if the following two conditions are satisfied:*

(i) *$T_{\mathscr{K}_j}$ has property (P) for all $j \geq 0$;*

(ii) $\bigwedge_{j=0}^{\infty} m_{T|\mathscr{H}_j} \equiv 1$.

PROOF. We know that T has property (P) if and only if T^* has property (P). It suffices then to apply Proposition 3.12 to T^* and the increasing sequence $\{\mathscr{H} \ominus \mathscr{H}_j : j \geq 0\} \subseteq \operatorname{Lat}(T^*)$. Note that $\mathscr{K}_j = (\mathscr{H} \ominus \mathscr{H}_{j+1}) \ominus (\mathscr{H} \ominus \mathscr{H}_j)$, and $T^*_{\mathscr{K}_j} = (T_{\mathscr{K}_j})^*$.

Exercises

1. Show that Theorem 3.2 is not true if one replaces the sequence $\{\mathscr{H}_j : j \geq 0\}$ by an arbitrary totally ordered family of invariant subspaces for T.

2. Show that Theorem 3.2 remains true if one replaces the sequence $\{\mathscr{H}_j : j \geq 0\}$ by a totally ordered family $\{\mathscr{H}_i : i \in I\} \subseteq \operatorname{Lat}(T)$, under the additional assumption that T acts on a separable space.

3. Assume that $T \in \mathscr{L}(\mathscr{H})$ is an operator of class C_0, $\{\mathscr{H}_j : -\infty < j < \infty\} \subseteq \operatorname{Lat}(T)$ is an increasing family, $\bigcap_{j=-\infty}^{\infty} \mathscr{H}_j = \{0\}$, $\bigvee_{j=-\infty}^{\infty} \mathscr{H}_j = \mathscr{H}$, and $\mathscr{K}_j = \mathscr{H}_{j+1} \ominus \mathscr{H}_j$. Show that T has property (P) if and only if the following two conditions are satisfied:

(i) $T_{\mathscr{K}_j}$ has property (P) for all j;

(ii) $\bigwedge_{j=-\infty}^{0} m_{T|\mathscr{H}_j} \equiv \bigwedge_{j=0}^{\infty} m_{T_{\mathscr{H}\ominus\mathscr{H}_j}} \equiv 1$.

4. Let $T \in \mathscr{L}(\mathscr{H})$ be an operator of class C_0. An invariant subspace $\mathscr{K} \in \operatorname{Lat}(T)$ is said to be maximal cyclic for T if (i) $T \mid \mathscr{K}$ has a cyclic vector; and (ii) if $\mathscr{K}' \in \operatorname{Lat}(T)$, $\mathscr{K} \subseteq \mathscr{K}'$, and $T \mid \mathscr{K}'$ has a cyclic vector, then $\mathscr{K}' = \mathscr{K}$. Prove that each vector $h \in \mathscr{H}$ is contained in a maximal cyclic subspace for T.

4. Generalized inner functions. As we saw in Proposition 1.7, the fact that weak contractions T of class C_0 have property (P) is a consequence of the relation $d_T(\ker(X)) \equiv d_T(\ker(X^*))$, $X \in \{T\}'$. In attempting to extend Proposition 1.7 to arbitrary operators with property (P) one sees easily that d_T no longer is an appropriate "dimension." Indeed, there are operators T of class

C_0 that have property (P) but are not weak contractions. For such T we have $d_T = 0$.

Let us recall from §2.2 that an inner function is uniquely determined (up to a constant factor with absolute value one) by its multiplicity function μ and by a finite Borel measure ν on $\mathbf{T}$, singular with respect to Lebesgue measure. More precisely, given a function $\mu\colon \mathbf{D} \to \{0, 1, 2, \dots\}$ such that

$$\sum_{\lambda \in \mathbf{D}} \mu(\lambda)(1 - |\lambda|) < \infty, \tag{4.1}$$

and a singular measure ν on $\mathbf{T}$, one defines an inner function θ by $\theta = b_\mu s_\nu$, where b_μ is the Blaschke product with multiplicity function μ (cf. (2.2.10)) and s_ν is the singular inner function determined by ν. Moreover, every inner function θ has the form $\alpha b_\mu s_\nu$, with $\alpha \in \mathbf{C}$, $|\alpha| = 1$.

We will denote by Γ the set of all pairs (μ, ν), where $\mu\colon \mathbf{D} \to \{0, 1, 2, \dots\}$ satisfies (4.1) and ν is a finite Borel measure on $\mathbf{T}$, singular with respect to Lebesgue measure. If $\theta = \alpha b_\mu s_\nu$ is an inner function with $\alpha \in \mathbf{C}$, $|\alpha| = 1$, we set

$$\gamma(\theta) = (\mu, \nu) \in \Gamma. \tag{4.2}$$

One can define the operations $\vee$, $\wedge$, and $+$ componentwise on Γ:

$$(\mu, \nu) + (\mu', \nu') = (\mu + \mu', \nu + \nu'),$$
$$(\mu, \nu) \vee (\mu', \nu') = (\mu \vee \mu', \nu \vee \nu'),$$
$$(\mu, \nu) \wedge (\mu', \nu') = (\mu \wedge \mu', \nu \wedge \nu').$$

If $\gamma, \gamma' \in \Gamma$, we will write $\gamma \le \gamma'$ if $\gamma \wedge \gamma' = \gamma$. If $\{\gamma_i : i \in I\}$ is a nonempty family of elements of Γ then the (possibly infinite) operation $\bigwedge\{\gamma_i : i \in I\}$ makes sense. The supremum $\bigvee\{\gamma_i : i \in I\}$ might not exist. Assume that $\{\gamma_n : n \ge 0\} \subseteq \Gamma$ is a sequence, where $\gamma_n = (\mu_n, \nu_n)$, $n \ge 0$. We define the sum $\sum_{n=0}^\infty \gamma_n = (\mu, \nu)$ by $\mu(\lambda) = \sum_{n=0}^\infty \mu_n(\lambda)$, $\lambda \in \mathbf{D}$, and $\nu(\omega) = \sum_{n=0}^\infty \nu_n(\omega)$, $\omega \in \mathbf{T}$, ω a Borel set. These sums may, of course, be infinite, so that (μ, ν) might not be an element of Γ.

We define one more operation on the set Γ. Let $j\colon \mathbf{D} \cup \mathbf{T} \to \mathbf{D} \cup \mathbf{T}$ be defined by $j(\lambda) = \overline{\lambda}$. If $\gamma = (\mu, \nu) \in \Gamma$, we set

$$\gamma^\sim = (\mu^\sim, \nu^\sim), \quad \text{where } \mu^\sim = \mu \circ j \quad \text{and } \nu^\sim = \nu \circ j. \tag{4.3}$$

The formulas in (4.3) mean $\mu^\sim(\lambda) = \mu(\overline{\lambda})$, $\lambda \in \mathbf{D}$, and $\mu^\sim(\omega) = \mu(j(\omega))$, for a Borel set $\omega \subseteq \mathbf{T}$.

4.4. LEMMA. *Assume that $\theta_0, \theta_1, \theta_2, \dots \in H^\infty$ are inner functions.*

(i) $\gamma(\theta_0\theta_1) = \gamma(\theta_0) + \gamma(\theta_1)$.

(ii) $\theta_0 \mid \theta_1$ *if and only if* $\gamma(\theta_0) \le \gamma(\theta_1)$.

(iii) $\theta_0 \equiv \theta_1$ *if and only if* $\gamma(\theta_0) = \gamma(\theta_1)$.

(iv) $\gamma(\theta_0^\sim) = \gamma(\theta_0)^\sim$.

(v) *The family* $\{\theta_0\theta_1 \cdots \theta_j : j \ge 0\}$ *has a least inner multiple* θ *if and only if* $\sum_{j=0}^\infty \gamma(\theta_j) \in \Gamma$. *If* θ *exists, we have* $\gamma(\theta) = \sum_{j=0}^\infty \gamma(\theta_j)$.

PROOF. The verification of (i)-(iv) is straightforward. To prove (v), assume first that θ exists. Since $\theta_0\theta_1\cdots\theta_j \mid \theta$ for all j, we must have $\sum_{i=0}^{j}\gamma(\theta_i) \leq \gamma(\theta)$ by (i) and (ii). We deduce that $\sum_{i=0}^{\infty}\gamma(\theta_i) \leq \gamma(\theta)$ so that $\sum_{i=0}^{\infty}\gamma(\theta_i) \in \Gamma$. Conversely, assume that $\sum_{i=0}^{\infty}\gamma(\theta_i) = (\mu,\nu) \in \Gamma$, and define an inner function $\theta' = b_\mu s_\nu$. A reverse application of (i) and (ii) shows that $\theta_0\theta_1\cdots\theta_j \mid \theta'$ for all j, and hence θ' is a common inner multiple of the family $\{\theta_0\theta_1\cdots\theta_j : j \geq 0\}$. Finally, assume that θ exists. Then $\theta \mid \theta'$ so that $\gamma(\theta) \leq \gamma(\theta') = \sum_{i=0}^{\infty}\gamma(\theta_i)$, which, together with the opposite inequality proved above, concludes the proof of (v). The lemma is proved.

4.5. COROLLARY. *Assume that T is a weak contraction of class C_0, and let $\bigoplus_{j<\omega} S(\theta_j)$ be the Jordan model of T. Then we have $\gamma(d_T) = \sum_{j=0}^{\infty}\gamma(\theta_j)$.*

PROOF. By Lemma 4.4(v), it suffices to verify that d_T is the least common inner multiple of the family $\{\theta_0\theta_1\cdots\theta_j : j \geq 0\}$. This last property of d_T is easily deduced from Corollary 6.3.18.

The idea in extending the notion of dimension to operators with property (P) is to take the sum $\sum_{j=0}^{\infty}\gamma(\theta_j)$ as the dimension of the Jordan operator $\bigoplus_{j<\omega} S(\theta_j)$. We have, however, to find an appropriate extension of the set Γ. We begin by extending the class of singular measures we consider. Denote by $\mathscr{B}$ the class of all Borel subsets of $\mathbf{T}$.

4.6. DEFINITION. Let $\nu, \nu' : \mathscr{B} \to [0,+\infty]$ be two countably additive measures. We say that ν is *absolutely continuous* with respect to ν', and we write $\nu \ll \nu'$, if there exists a Borel function $f : \mathbf{T} \to [0,+\infty]$ such that $\nu(\omega) = \int_\omega f\,d\nu'$, $\omega \in \mathscr{B}$.

We will denote by $\mathscr{N}$ the class of those measures $\nu : \mathscr{B} \to [0,+\infty]$ with the property that $\nu \ll \nu'$ for some finite measure ν', singular with respect to Lebesgue measure on $\mathbf{T}$. Furthermore, denote by $\mathscr{N}_0$ the class of σ-finite measures in $\mathscr{N}$, and by $\mathscr{N}_\infty$ the class of measures $\nu \in \mathscr{N}$ with the property that $\nu(\mathscr{B}) \subseteq \{0,+\infty\}$.

4.7. DEFINITION. A *generalized inner function* is a pair $\gamma = (\mu,\nu)$, where $\nu \in \mathscr{N}$, and $\mu : \mathbf{D} \to \{0,1,2,\ldots\}$ is a function satisfying the relation $\sum_{\mu(\lambda)\neq 0}(1-|\lambda|) < \infty$. We denote by $\tilde{\Gamma}$ the set of all generalized inner functions, by Γ_0 the set of those $\gamma = (\mu,\nu) \in \tilde{\Gamma}$ such that $\nu \in \mathscr{N}_0$, and by Γ_∞ the set of those $\gamma = (0,\nu) \in \tilde{\Gamma}$ such that $\nu \in \mathscr{N}_\infty$.

Let us note that every element $\gamma \in \tilde{\Gamma}$ has a unique decomposition of the form $\gamma = \gamma_0 + \gamma_\infty$, with $\gamma_0 \in \Gamma_0$ and $\gamma_\infty \in \Gamma_\infty$, and $\gamma_0 \wedge \gamma_\infty = 0$. Indeed, if $\gamma = (\mu,\nu)$ it suffices to decompose ν into a sum $\nu_0 + \nu_\infty$, with $\nu_0 \in \mathscr{N}_0$, $\nu_\infty \in \mathscr{N}_\infty$, and $\nu_0 \wedge \nu_\infty = 0$. To do this, we write $\nu(\omega) = \int_\omega f\,d\nu'$ with a finite measure ν'. Denote $A = \{\varsigma \in \mathbf{T} : f(\varsigma) = +\infty\}$, and define

$$\nu_0(\omega) = \nu(\omega \setminus A), \qquad \nu_\infty(\omega) = \nu(\omega \cap A), \quad \omega \in \mathscr{B}. \tag{4.8}$$

The uniqueness of ν_0 and ν_∞ is easily verified.

4.9. PROPOSITION. *Assume that* $\{\gamma_j : j \geq 0\} \subseteq \Gamma$ *is a sequence with the following properties*:

$$\gamma_j \geq \gamma_{j+1}, j \geq 0, \quad \text{and} \quad \bigwedge_{j=0}^{\infty} \gamma_j = 0. \tag{4.10}$$

Then the element γ *defined by*

$$\gamma = \sum_{j=0}^{\infty} \gamma_j \tag{4.11}$$

belongs to $\tilde{\Gamma}$. *Conversely, every element* $\gamma \in \tilde{\Gamma}$ *has a representation of the form* (4.11), *where the elements* $\gamma_j \in \Gamma$ *satisfy* (4.10).

PROOF. Assume first that the sequence $\{\gamma_j : j \geq 0\}$ satisfies (4.10). Write $\gamma_j = (\mu_j, \nu_j)$, and note that $\mu_j \geq \mu_{j+1}$, $\nu_j \geq \nu_{j+1}$ for $j \geq 0$, $\bigwedge_{j=0}^{\infty} \mu_j = 0$, and $\bigwedge_{j=0}^{\infty} \nu_j = 0$. The equality $(\bigwedge_{j=0}^{\infty} \mu_j)(\lambda) = \inf\{\mu_j(\lambda) : j \geq 0\} = 0$ means that for each $\lambda \in \mathbf{D}$ there exists an integer n_λ such that $\mu_{n_\lambda}(\lambda) = 0$. Consequently $\mu(\lambda) = \sum_{j=0}^{\infty} \mu_j(\lambda) = \sum_{j=0}^{n_\lambda} \mu_j(\lambda)$ is a finite integer. We clearly have $\sum_{\mu(\lambda)\neq 0}(1 - |\lambda|) \leq \sum_{\lambda \in \mathbf{D}} \mu_0(\lambda)(1 - |\lambda|) < \infty$. Now, the inequality $\nu_j \leq \nu_0$ implies by the Radon–Nikodym theorem the existence of Borel functions $f_j : \mathbf{T} \to [0, 1]$ such that $\nu_j(\omega) = \int_\omega f_j \, d\nu_0$, $j \geq 0$, $\omega \in \mathscr{B}$. Thus, if ν is defined by $\nu(\omega) = \sum_{j=0}^{\infty} \nu_j(\omega)$, $\omega \in \mathscr{B}$, we have $\nu(\omega) = \int_\omega f \, d\nu_0$, $\omega \in \mathscr{B}$, where $f = \sum_{j=0}^{\infty} f_j$. We conclude that $\gamma = (\mu, \nu)$ is indeed a generalized inner function.

Conversely, let $\gamma = (\mu, \nu)$ be a generalized inner function. We want to define $\gamma_j = (\mu_j, \nu_j) \in \Gamma$, $j \geq 0$, satisfying (4.10) and (4.11). We define μ_j first:

$$\begin{aligned} \mu_j(\lambda) &= 1 \quad \text{if } j < \mu(\lambda); \\ &= 0 \quad \text{if } j \geq \mu(\lambda), \quad \lambda \in \mathbf{D}. \end{aligned}$$

The inequalities $\mu_j \geq \mu_{j+1}$ and the relations $\mu = \sum_{j=0}^{\infty} \mu_j$ and $\bigwedge_{j=0}^{\infty} \mu_j = 0$ are obviously satisfied. Furthermore,

$$\sum_{\lambda \in \mathbf{D}} \mu_j(\lambda)(1 - |\lambda|) = \sum_{\mu_j(\lambda) \neq 0} (1 - |\lambda|) \leq \sum_{\mu(\lambda) \neq 0} (1 - |\lambda|) < \infty.$$

Assume next that $\nu(\omega) = \int_\omega f \, d\nu'$, $\omega \in \mathscr{B}$, for some finite singular measure ν' and some Borel function $f : \mathbf{T} \to [0, +\infty]$. To conclude the proof it will suffice to define $\nu_j(\omega) = \int_\omega f_j \, d\nu'$, $j \geq 0$, $\omega \in \mathscr{B}$, where $f_j : \mathbf{T} \to [0, 1/(j + 1)]$ are Borel functions such that $f_j \geq f_{j+1}$ and $f = \sum_{j=0}^{\infty} f_j$. This is easily done as follows. Define the sets $\omega_j = \{\varsigma \in \mathbf{T} : f(\varsigma) > 1 + \frac{1}{2} + \cdots + 1/(j + 1)\}$, $j \geq 0$, $\omega_{-1} = \mathbf{T}$, and set

$$\begin{aligned} f_j(\varsigma) &= 1/(j + 1) && \text{if } \varsigma \in \omega_j; \\ &= f(\varsigma) - (1 + \tfrac{1}{2} + \cdots + 1/j) && \text{if } \varsigma \in \omega_{j-1} \setminus \omega_j; \\ &= 0 && \text{if } \varsigma \in \mathbf{T} \setminus \omega_{j-1}, j \geq 0. \end{aligned}$$

The relation $\bigwedge_{j=0}^{\infty} f_j = 0$, and hence $\bigwedge_{j=0}^{\infty} \nu_j = 0$, is satisfied because $f_j \leq 1/(j + 1)$ for all j. The inequalities $f_j \geq f_{j+1}$ and the identity $f = \sum_{j=0}^{\infty} f_j$ are easily verified and the proposition follows.

Assume now that T is an operator of class C_0 with Jordan model $\bigoplus_{j<\omega} S(\theta_j)$. If T has property (P) then the sequence $\{\gamma_j : j \geq 0\}$ defined by $\gamma_j = \gamma(\theta_j)$ satisfies conditions (4.10) by Theorem 1.9. This enables us to give the following definition.

4.12. DEFINITION. Assume that T is an operator of class C_0 that has property (P), and let $\bigoplus_{j<\omega} S(\theta_j)$ be the Jordan model of T. Then the *dimension* $\gamma_T \in \tilde{\Gamma}$ of T is given by $\gamma_T = \sum_{j=0}^{\infty} \gamma(\theta_j)$. Furthermore, assume that T is an arbitrary operator of class C_0 and $\mathscr{M} \in \mathrm{Lat}_{1/2}(T)$ is such that T has property (P). Then the *T-dimension* $\gamma_T(\mathscr{M})$ of $\mathscr{M}$ is given by $\gamma_T(\mathscr{M}) = \gamma_{T_{\mathscr{M}}}$.

We will use the terminology C_0-dimension for T-dimension when no confusion may arise. We note in the following remark some more or less obvious properties of C_0-dimension.

4.13. REMARK. (i) If T has property (P) then $\gamma_{T^*} = (\gamma_T)^{\sim}$. If $T_{\mathscr{M}}$ has property (P) then $\gamma_{T^*}(\mathscr{M}) = \gamma_T(\mathscr{M})^{\sim}$.

(ii) $\gamma_T = 0$ if and only if T acts on the space $\{0\}$.

(iii) If $T \sim T'$ then $\gamma_T = \gamma_{T'}$.

(iv) Assume that T has property (P). Then T is a weak contraction if and only if $\gamma_T \in \Gamma$. If $\gamma_T \in \Gamma$, then $\gamma_T = \gamma(d_T)$.

(v) Every generalized inner function can be written under the form γ_T, where T is an operator of class C_0 with property (P).

(vi) If $T \prec^i T'$ or $T^* \prec^i T'^*$ then $\gamma_T \leq \gamma_{T'}$.

Many properties of γ_T can be deduced from the better understood case in which T is a weak contraction. The basic tool is given by the following result.

4.14. LEMMA. *Assume that $T \in \mathscr{L}(\mathscr{H})$ is an operator of class C_0 and $\{\mathscr{H}_j : j \geq 0\} \subseteq \mathrm{Lat}(T)$ is an increasing sequence. If $\mathscr{H} = \bigvee_{j=0}^{\infty} \mathscr{H}_j$, and if T has property (P), then $\gamma_T = \bigvee_{j=0}^{\infty} \gamma_T(\mathscr{H}_j)$.*

PROOF. The inequalities $\gamma_T(\mathscr{H}_j) \leq \gamma_T$ follow because $T \mid \mathscr{H}_j \prec^i T$. Thus we have $\bigvee_{j=0}^{\infty} \gamma_T(\mathscr{H}_j) \leq \gamma_T$. In the opposite direction, for fixed $k < \omega$ we have $\gamma_T(\mathscr{H}_j) \geq \sum_{i=0}^{k} \gamma(M_{T|\mathscr{H}_j}(i))$ so that

$$\bigvee_{j=0}^{\infty} \gamma_T(\mathscr{H}_j) \geq \bigvee_{j=0}^{\infty} \left(\sum_{i=0}^{k} \gamma(M_{T|\mathscr{H}_j}(i)) \right) = \sum_{i=0}^{k} \left(\bigvee_{j=0}^{\infty} \gamma(M_{T|\mathscr{H}_j}(i)) \right).$$

Theorem 3.2 implies that $\bigvee_{j=0}^{\infty} \gamma(M_{T|\mathscr{H}_j}(i)) = \gamma(M_T(i))$, $i < \omega$, so that the last inequality above can be rewritten as $\bigvee_{j=0}^{\infty} \gamma_T(\mathscr{H}_j) \geq \sum_{i=0}^{k} \gamma(M_T(i))$. Since k is arbitrary, we deduce that $\gamma_T \leq \bigvee_{j=0}^{\infty} \gamma_T(\mathscr{H}_j)$, and this completes the proof.

4.15. REMARK. The conclusion of Lemma 4.14 holds under the modified assumption that $\mathscr{H}_j \in \mathrm{Lat}(T^*)$, $j \geq 0$. This follows from Remark 4.13(i).

We are now ready to extend Proposition 1.7 to all operators of class C_0 with property (P).

4.16. PROPOSITION. *Assume that T is an operator of class C_0 with property (P). Then the equality $\gamma_T(\ker(X)) = \gamma_T(\ker(X^*))$ holds for every $X \in \{T\}'$.*

PROOF. Assume that $T \in \mathcal{L}(\mathcal{H})$, and let $\bigoplus_{j<\omega} S(\theta_j)$ be the Jordan model of T. As we noted before (cf. the proof of Theorem 1.9) the spaces $\mathcal{H}_j = (\theta_j(T)\mathcal{H})^-$ are hyperinvariant for T, $T \mid \mathcal{H}_j$ has finite multiplicity, $\mathcal{H}_j \subseteq \mathcal{H}_{j+1}$, and $\mathcal{H} = \bigvee_{j=0}^{\infty} \mathcal{H}_j$. We have therefore $\mathcal{H}_j \in \operatorname{Lat}(X)$, so we may define the operators $X_j = X \mid \mathcal{H}_j \in \{T \mid \mathcal{H}_j\}'$, $j \geq 0$. Because $T_j = T \mid \mathcal{H}_j$ has finite multiplicity, and hence it is a weak contraction, we deduce from Proposition 1.7 that

$$\gamma_T(\ker(X_j)) = \gamma_T(\ker(X_j^*)), \quad j \geq 0. \tag{4.17}$$

We clearly have $X\theta_j(T) \mid \ker(X) = 0$, so that $\ker(X_j) \supseteq (\theta_j(T)\ker(X))^-$. We then have $\ker(X) \supseteq \bigvee_{j=0}^{\infty} \ker(X_j) \supseteq \bigvee_{j=0}^{\infty} (\theta_j(T)\ker(X))^- = \ker(X)$ by Theorem 2.4.6 (applied to $T \mid \ker(X)$). Lemma 4.14 now implies

$$\gamma_T(\ker(X)) = \bigvee_{j=0}^{\infty} \gamma_T(\ker(X_j)). \tag{4.18}$$

On the other hand, since $X_j = X \mid \mathcal{H}_j$, we have $X_j^* P_{\mathcal{H}_j} = P_{\mathcal{H}_j} X^*$, and this intertwining relation shows that $P_{\mathcal{H}_j}(\ker(X^*)) \subseteq \ker(X_j^*)$. We also have $T_j^* P_{\mathcal{H}_j} = P_{\mathcal{H}_j} T^*$ and therefore we deduce

$$(T_j^* \mid \ker(X_j^*))(P_{\mathcal{H}_j} \mid \ker(X^*)) = (P_{\mathcal{H}_j} \mid \ker(X^*))(T^* \mid \ker(X^*)). \tag{4.19}$$

The operator $P_{\mathcal{H}_j} \mid \ker(X^*)$ is not one-to-one, but it becomes one-to-one when restricted to the space

$$\mathcal{K}_j = \ker(X^*) \ominus \ker(P_{\mathcal{H}_j} \mid \ker(X^*)) = \ker(X^*) \ominus (\ker(X^*) \cap (\mathcal{H} \ominus \mathcal{H}_j)).$$

If $Y_j = P_{\mathcal{H}_j} \mid \mathcal{K}_j$, then Y_j is one-to-one and (4.19) implies $(T_j^* \mid \ker(X_j^*))Y_j = Y_j T_{\mathcal{K}_j}^*$. We deduce from Remark 4.13(vi) that

$$\gamma_T(\mathcal{K}_j) \leq \gamma_T(\ker(X_j^*)), \quad j \geq 0. \tag{4.20}$$

Note that

$$\bigvee_{j=0}^{\infty} \mathcal{K}_j = \ker(X^*) \ominus \left[\ker(X^*) \cap \left(\bigcap_{j=0}^{\infty} (\mathcal{H} \ominus \mathcal{H}_j)\right)\right] = \ker(X^*)$$

so that Lemma 4.15 (applied to $T^* \mid \ker(X^*)$) shows that $\gamma_T(\ker(X^*)) = \bigvee_{j=0}^{\infty} \gamma_T(\mathcal{K}_j)$. Thus (4.20) and (4.18) imply

$$\begin{aligned}
\gamma_T(\ker(X^*)) &= \bigvee_{j=0}^{\infty} \gamma_T(\mathcal{K}_j) \\
&\leq \bigvee_{j=0}^{\infty} \gamma_T(\ker(X_j^*)) \\
&= \bigvee_{j=0}^{\infty} \gamma_T(\ker(X_j)) \\
&= \gamma_T(\ker(X)).
\end{aligned}$$

The same argument applied to T^* and X^* yields the opposite inequality $\gamma_T(\ker(X)) \leq \gamma_T(\ker(X^*))$. The proposition follows.

We conclude this section with the following consequences of Proposition 4.16.

4.21. COROLLARY. *Assume that $T \in \mathscr{L}(\mathscr{H})$ is an operator of class C_0 and $\mathscr{M} \in \mathrm{Lat}(T)$. If T has property (P) then $\gamma_T = \gamma_T(\mathscr{M}) + \gamma_T(\mathscr{H} \ominus \mathscr{M})$.*

PROOF. Consider the operator $T' = T \oplus (T \mid \mathscr{M})$, and the operator $X \in \{T'\}'$ defined by $X(u \oplus v) = v \oplus 0$, $u \in \mathscr{H}$, $v \in \mathscr{M}$. It is quite obvious that $\ker(X) = \mathscr{H} \oplus \{0\}$ and $\ker(X^*) = (\mathscr{H} \ominus \mathscr{M}) \oplus \mathscr{M}$, so that $T' \mid \ker(X)$ and $T'_{\ker(X^*)}$ are unitarily equivalent to T and $T_{\mathscr{H}\ominus\mathscr{M}} \oplus (T \mid \mathscr{M})$, respectively. If T has property (P) then T' has property (P) by Corollary 1.17 and, in this case, the equality $\gamma_T = \gamma_{T_{\mathscr{H}\ominus\mathscr{M}} \oplus (T|\mathscr{M})}$ follows from Proposition 4.16. It suffices then to show that $\gamma_{T_{\mathscr{H}\ominus\mathscr{M}} \oplus (T|\mathscr{M})} = \gamma_{T_{\mathscr{H}\ominus\mathscr{M}}} + \gamma_{T|\mathscr{M}}$ or, more generally, that $\gamma_{R\oplus R'} = \gamma_R + \gamma_{R'}$ whenever R and R' are operators of class C_0 with property (P). By Remark 4.13(iii) we may assume that $R = \bigoplus_{j<\omega} S(\theta_j)$ and $R' = \bigoplus_{j<\omega} S(\theta'_j)$ are Jordan operators with $\bigwedge_{j=0}^{\infty} \theta_j \equiv \bigwedge_{j=0}^{\infty} \theta'_j \equiv 1$. Let us consider the invariant subspaces

$$\mathscr{H}_j = \mathscr{H}(\theta_0) \oplus \mathscr{H}(\theta_1) \oplus \cdots \oplus \mathscr{H}(\theta_j) \in \mathrm{Lat}(R),$$
$$\mathscr{H}'_j = \mathscr{H}(\theta'_0) \oplus \mathscr{H}(\theta'_1) \oplus \cdots \oplus \mathscr{H}(\theta'_j) \in \mathrm{Lat}(R'),$$

and $\mathscr{K}_j = \mathscr{H}_j \oplus \mathscr{H}'_j \in \mathrm{Lat}(R \oplus R')$, $j \geq 0$. By Lemma 4.14 we have $\gamma_R = \bigvee_{j=0}^{\infty} \gamma_R(\mathscr{H}_j)$, $\gamma_{R'} = \bigvee_{j=0}^{\infty} \gamma_{R'}(\mathscr{H}'_j)$, and $\gamma_{R\oplus R'} = \bigvee_{j=0}^{\infty} \gamma_{R\oplus R'}(\mathscr{K}_j)$. It suffices therefore to prove that $\gamma_{R\oplus R'}(\mathscr{K}_j) = \gamma_R(\mathscr{H}_j) + \gamma_{R'}(\mathscr{H}'_j)$, and this clearly follows from Theorem 1.4. The corollary is proved.

4.22. COROLLARY. *Assume that $T \in \mathscr{L}(\mathscr{H})$ is an operator of class C_0, $\{\mathscr{H}_j : j \geq 0\} \subseteq \mathrm{Lat}(T)$ is a decreasing sequence, $\mathscr{H}_0 = \mathscr{H}$, and $\bigcap_{j=0}^{\infty} \mathscr{H}_j = \{0\}$. If T has property (P) then $\gamma_T = \sum_{j=0}^{\infty} \gamma_T(\mathscr{H}_j \ominus \mathscr{H}_{j+1})$.*

PROOF. As noted in Remark 4.15, we have $\gamma_T = \bigvee_{j=0}^{\infty} \gamma_T(\mathscr{H} \ominus \mathscr{H}_j)$. Since we have

$$T_{\mathscr{H}\ominus\mathscr{H}_{j+1}} = \begin{bmatrix} T_{\mathscr{H}_j\ominus\mathscr{H}_{j+1}} & X_j \\ 0 & T_{\mathscr{H}\ominus\mathscr{H}_j} \end{bmatrix}$$

with respect to the decomposition $\mathscr{H} \ominus \mathscr{H}_{j+1} = (\mathscr{H}_j \ominus \mathscr{H}_{j+1}) \oplus (\mathscr{H} \ominus \mathscr{H}_j)$, an inductive application of Corollary 4.21 shows that $\gamma_T(\mathscr{H} \ominus \mathscr{H}_{j+1}) = \sum_{i=0}^{j} \gamma_T(\mathscr{H}_i \ominus \mathscr{H}_{i+1})$. The corollary now becomes obvious.

Exercises

1. Verify assertions (i)–(iv) in Lemma 4.4.

2. Check the uniqueness of the components $\nu_0 \in \mathscr{N}_0$ and $\nu_\infty \in \mathscr{N}_\infty$ in the decomposition $\nu = \nu_0 + \nu_\infty$ of a measure in $\mathscr{N}$, $\nu_0 \wedge \nu_\infty = 0$.

3. Assume that the sequence $\{\gamma_j : j \geq 0\} \subseteq \Gamma$ satisfies (4.10). Show that if, in addition, $(\gamma_j - \gamma_{j+1}) \wedge (\gamma_k - \gamma_{k+1}) = 0$ for $j \neq k$, then $\gamma = \sum_{j=0}^{\infty} \gamma_j \in \Gamma_0$.

Conversely, every $\gamma \in \Gamma_0$ can be written as $\gamma = \sum_{j=0}^{\infty} \gamma_j$, where $\{\gamma_j : j \geq 0\} \subseteq \Gamma$ satisfies (4.10) and the conditions $(\gamma_j - \gamma_{j+1}) \wedge (\gamma_k - \gamma_{k+1}) = 0$ for $j \neq k$.

4. Show that $\Gamma_0 = \{\gamma_T : T \text{ is of class } C_0,\ T \text{ has property (Q)}\}$.

5. Verify the inequalities $f_j \geq f_{j+1}$ and the relation $f = \sum_{j=0}^{\infty} f_j$ in the proof of Proposition 4.9.

6. Verify the statements in Remark 4.13.

7. The proof of Proposition 1.7 was based on the analogue of Corollary 4.21 for weak contractions. Can we deduce Proposition 4.16 as a consequence of Corollary 4.21? (Note that Corollary 4.21 can be given a proof independent of Proposition 4.16.)

8. Assume that T is an operator of class C_0 having property (P). We say that T has property (P_0) [respectively, (P_∞)] if $\gamma_T \in \Gamma_0$ [respectively, $\gamma_T \in \Gamma_\infty$]. Show that T has property (P_0) if and only if the following holds: for every decreasing sequence $\{\mathscr{H}_j : j \geq 0\} \subseteq \mathrm{Lat}(T)$ such that $\bigcap_{j=0}^{\infty} \mathscr{H}_j = \{0\}$ we have $\bigwedge_{j=0}^{\infty} \gamma_T(\mathscr{H}_j) = 0$.

9. Let $T \in \mathscr{L}(\mathscr{H})$ be an operator of class C_0 with property (P). Show that there exist subspaces $\mathscr{H}', \mathscr{H}'' \in \mathrm{Lat}(\{T\}')$ such that $\mathscr{H}' \cap \mathscr{H}'' = \{0\}$, $\mathscr{H}' \vee \mathscr{H}'' = \mathscr{H}$, $T \mid \mathscr{H}'$ has property (P_0), and $T \mid \mathscr{H}''$ has property (P_∞).

10. Let $T \in \mathscr{L}(\mathscr{H})$ be an operator of class C_0 with property (P). Assume that $\{\mathscr{H}_j : -\infty < j < +\infty\} \subseteq \mathrm{Lat}(T)$ is an increasing family of subspaces such that $\bigvee_{j=0}^{\infty} \mathscr{H}_j = \mathscr{H}$ and $\bigcap_{j=-\infty}^{0} \mathscr{H}_j = \{0\}$. Show that $\gamma_T = \sum_{j=-\infty}^{\infty} \gamma_T(\mathscr{H}_{j+1} \ominus \mathscr{H}_j)$.

11. Let $T = 0 \in \mathscr{L}(\mathscr{H})$, and $\mathscr{M} \in \mathrm{Lat}_{1/2}(T)$. Assume that $T_{\mathscr{M}}$ has property (P), and show that $\gamma_T(\mathscr{M}) = (\mu, 0)$, where $\mu(\lambda) = 0$ for $\lambda \neq 0$ and $\mu(0) = \dim(\mathscr{M})$.

12. Let $\{\gamma_j : j \geq 0\} \subseteq \tilde{\Gamma}$ be a sequence of generalized inner functions. Is the sum $\sum_{j=0}^{\infty} \gamma_j$ an element of $\tilde{\Gamma}$?

13. Let $\{\gamma_j : j \geq 0\}$ and $\{\gamma_j' : j \geq 0\}$ be two sequences of generalized inner functions. Assume that $\sum_{j=0}^{\infty} \gamma_j = \sum_{j=0}^{\infty} \gamma_j' \in \tilde{\Gamma}$, and prove the existence of an array $\{\gamma_{ij} : i, j \geq 0\} \subseteq \tilde{\Gamma}$ such that $\sum_{j=0}^{\infty} \gamma_{ij} = \gamma_i$, $i \geq 0$, and $\sum_{i=0}^{\infty} \gamma_{ij} = \gamma_j'$, $j \geq 0$.

5. Operators with equal dimensions. We have seen in Proposition 4.16 that $\gamma_T(\ker(X)) = \gamma_T(\ker(X^*))$ if $X \in \{T\}'$ and T has property (P). It is natural to ask whether this result is best possible, that is, whether there are no other relations between $T \mid \ker(X)$ and $T_{\ker(X^*)}$ that must always hold. In order to answer this question, we introduce a natural relation on the class of operators of class C_0 with property (P).

5.1. DEFINITION. Assume that T_1 and T_2 are two operators of class C_0 with property (P). We say that T_1 and T_2 are in relation ρ, and we write $T_1 \rho T_2$, if there exist an operator T of class C_0 and $X \in \{T\}'$ such that T has property (P), $T_1 \sim T \mid \ker(X)$, and $T_2 \sim T_{\ker(X^*)}$.

Proposition 4.16 can be reformulated as follows: $T_1 \rho T_2$ implies $\gamma_{T_1} = \gamma_{T_2}$. We will prove in this section that the converse implication is also true. We begin with a particular case that will illustrate the difficulties involved.

5.2. PROPOSITION. *Assume that T and T' are weak contractions of class C_0. If $\gamma_T = \gamma_{T'}$ then $T\rho T'$.*

PROOF. Assume that $\gamma_T = \gamma_{T'}$ or, equivalently, $d_T \equiv d_{T'}$. We may assume that $T = \bigoplus_{j<\omega} S(\theta_j)$ and $T' = \bigoplus_{j<\omega} S(\theta'_j)$ are Jordan operators, so that

$$d_T \equiv \bigvee_{j=0}^{\infty} \theta_0\theta_1 \cdots \theta_j \equiv \bigvee_{j=0}^{\infty} \theta'_0\theta'_1 \cdots \theta'_j.$$

Denote $\phi_0 = d_T$ and for an integer $j \geq 1$ set $\phi_j = \phi_0/(\theta_0\theta_1 \cdots \theta_{j-1})$, $\phi_{-j} = \phi_0/(\theta'_0\theta'_1 \cdots \theta'_{j-1})$. We certainly have

$$\bigwedge_{j=0}^{\infty} \phi_j \equiv \bigwedge_{j=-\infty}^{0} \phi_j \equiv 1$$

from which we deduce, via Proposition 3.12, that the operator $R = \bigoplus_{j=-\infty}^{\infty} S(\phi_j)$ has property (P). In fact Theorem 1.9 applies to each of the two operators $\bigoplus_{j=1}^{\infty} S(\phi_j)$ and $\bigoplus_{j=-\infty}^{0} S(\phi_j)$, which are actually (unitarily equivalent to) Jordan operators. Next we define an operator $X \in \{R\}'$ by $X(\bigoplus_{j=-\infty}^{\infty} h_j) = \bigoplus_{j=-\infty}^{\infty} k_j$, where

$$\begin{aligned} k_j &= P_{\mathscr{H}(\phi_j)} h_{j-1} && \text{if } j \geq 1, \\ &= (\phi_j/\phi_{j-1}) h_{j-1} && \text{if } j < 0. \end{aligned}$$

The fact that $X \in \{R\}'$ can be verified by looking at Theorem 3.1.16. It is also clear that the mapping $h_{j-1} \to k_j$ defined by the above formulas is one-to-one or onto according to whether $j \leq 0$ or $j > 0$. Hence $\ker(X) = \bigoplus_{j=0}^{\infty} \ker(X \mid \mathscr{H}(\phi_j))$ and $\ker(X^*) = \bigoplus_{j=-\infty}^{0} \ker(X^* \mid \mathscr{H}(\phi_j))$. For $j \geq 0$ we have $\ker(X \mid \mathscr{H}(\phi_j)) = \phi_{j+1}H^2 \ominus \phi_j H^2$, and $S(\phi_j) \mid \ker(X \mid \mathscr{H}(\phi_j))$ is unitarily equivalent to $S(\phi_j/\phi_{j+1}) = S(\theta_j)$ (cf. Proposition 3.1.10). We conclude that $R \mid \ker(X)$ is unitarily equivalent to $\bigoplus_{j=0}^{\infty} S(\theta_j) = T$. In an analogous manner it can be shown that $R^* \mid \ker(X^*)$ is unitarily equivalent to T'^*. Therefore $T\rho T'$ and the proposition is proved.

Note how the fact that T and T' are weak contractions was used in the preceding proof. We could afford to construct an operator X which is a certain kind of operator-weighted shift, whose weights consist of an infinite string of injective operators followed by an infinite string of surjective operators. In the general case of operators T and T' with property (P) this construction is impossible because the central piece ($S(d_T)$) is inexistent. The construction will have to be more subtle, using a weighted shift whose injective and surjective weights blend in a different way.

5.3. LEMMA. *Let ϕ and ψ be two inner functions in H^∞. There exist inner functions ϕ', ϕ'', ψ', and ψ'' with the following properties:*

$$\phi = \phi'\phi'', \qquad \psi = \psi'\psi'', \qquad \phi' \wedge \phi'' \equiv \phi' \wedge \psi'' \equiv \psi' \wedge \psi'' \equiv \psi' \wedge \phi'' \equiv 1,$$
$$\text{and} \quad \phi' \mid \psi', \psi'' \mid \phi''.$$

These functions are not generally unique, but the quotients ψ'/ϕ' and ϕ''/ψ'' are uniquely determined by ϕ and ψ (up to a constant factor of absolute value one).

PROOF. Denote $\gamma(\phi) = (\mu_1, \nu_1)$, $\gamma(\psi) = (\mu_2, \nu_2)$, and choose a measure ν (e.g., $\nu = \nu_1 + \nu_2$) such that $\nu_1 \leq \nu$ and $\nu_2 \leq \nu$. We can then write $d\nu_1 = f_1\, d\nu$ and $d\nu_2 = f_2\, d\nu$, where f_1 and f_2 are two Borel functions on $\mathbf{T}$, $0 \leq f_j(\varsigma) \leq 1$, $\varsigma \in \mathbf{T}$, $j = 1, 2$. Define the set $A = \{\varsigma \in \mathbf{T}\colon f_1(\varsigma) \leq f_2(\varsigma)\}$, and the measures ν_1', ν_2', ν_1'', ν_2'' as follows:

$$\nu_1'(\omega) = \int_{\omega \cap A} f_1\, d\nu, \qquad \nu_2'(\omega) = \int_{\omega \cap A} f_2\, d\nu,$$
$$\nu_1''(\omega) = \int_{\omega \setminus A} f_1\, d\nu, \qquad \nu_2''(\omega) = \int_{\omega \setminus A} f_2\, d\nu, \quad \omega \in \mathscr{B}.$$

Analogously, define functions $\mu_1', \mu_1'', \mu_2', \mu_2''\colon \mathbf{D} \to \{0, 1, 2, \dots\}$ by

$$\begin{aligned} \mu_1'(\lambda) &= \mu_1(\lambda) && \text{if } \mu_1(\lambda) \leq \mu_2(\lambda), \\ &= 0 && \text{if } \mu_1(\lambda) > \mu_2(\lambda), \lambda \in \mathbf{D}, \\ \mu_1'' &= \mu_1 - \mu_1', \\ \mu_2'(\lambda) &= \mu_2(\lambda) && \text{if } \mu_1(\lambda) \leq \mu_2(\lambda), \\ &= 0 && \text{if } \mu_1(\lambda) > \mu_2(\lambda), \lambda \in \mathbf{D}, \end{aligned}$$

and

$$\mu_2'' = \mu_2 - \mu_2'.$$

It is clear that

$$(\mu_1', \nu_1') + (\mu_1'', \nu_1'') = (\mu_1, \nu_1), \qquad (\mu_2', \nu_2') + (\mu_2'', \nu_2'') = (\mu_2, \nu_2), \tag{5.4}$$

$$\begin{aligned} (\mu_1', \nu_1') \wedge (\mu_1'', \nu_1'') &= (\mu_1', \nu_1') \wedge (\mu_2'', \nu_2'') = (\mu_2', \nu_2') \wedge (\mu_2'', \nu_2'') \\ &= (\mu_2', \nu_2') \wedge (\mu_1'', \nu_1'') = 0, \end{aligned} \tag{5.5}$$

and

$$(\mu_1', \nu_1') \leq (\mu_2', \nu_2'), \qquad (\mu_1'', \nu_1'') \geq (\mu_2'', \nu_2''). \tag{5.6}$$

It suffices therefore to choose the functions ϕ', ϕ'', ψ', and ψ'' such that $\gamma(\phi') = (\mu_1', \nu_1')$, $\gamma(\phi'') = (\mu_1'', \nu_1'')$, $\gamma(\psi') = (\mu_2', \nu_2')$, and $\gamma(\psi'') = (\mu_2'', \nu_2'')$.

The fact that these functions are not unique can be seen from the case in which $\phi = \psi$. In this case we can take $\phi' = \psi'$, $\phi'' = \psi''$, where $\phi' \wedge \phi'' \equiv 1$ and $\phi = \phi'\phi''$. In general an inner function has many such decompositions into relatively prime factors.

Now let $\phi = \phi'\phi''$ and $\psi = \psi'\psi''$ be two decompositions with the properties listed in the statement of the lemma. Denote $\gamma(\phi) = (\mu_1, \nu_1)$, $\gamma(\psi) = (\mu_2, \nu_2)$, $\gamma(\phi') = (\mu_1', \nu_1')$, $\gamma(\phi'') = (\mu_1'', \nu_1'')$, $\gamma(\psi') = (\mu_2', \nu_2')$, and $\gamma(\psi'') = (\mu_2'', \nu_2'')$. Relations (5.4), (5.5), and (5.6) must then hold. We see that necessarily

$$\mu_2'(\lambda) - \mu_1'(\lambda) = \max\{\mu_2(\lambda) - \mu_1(\lambda), 0\}$$

and

$$\mu_1''(\lambda) - \mu_2''(\lambda) = \max\{\mu_1(\lambda) - \mu_2(\lambda), 0\}.$$

In an analogous manner, the measures $\nu_2' - \nu_1'$ and $\nu_1'' - \nu_2''$ are the two components of the (unique) Hahn decomposition of the real measure $\nu_2 - \nu_1$. The lemma clearly follows because $\gamma(\psi'/\phi') = (\mu_2' - \mu_1', \nu_2' - \nu_1')$ and $\gamma(\phi''/\psi'') = (\mu_1'' - \mu_2'', \nu_1'' - \nu_2'')$.

5.7. DEFINITION. The inner functions ϕ''/ψ'' and ψ'/ϕ' defined in Lemma 5.3 are denoted $(\phi/\psi)_+$ and $(\psi/\phi)_+$, respectively.

5.8. LEMMA. *Assume that ϕ, $\psi \in H^\infty$ are two inner functions. Then there exists an operator $X \in \mathscr{I}(S(\phi), S(\psi))$ such that $S(\phi) \mid \ker(X)$ and $S(\psi)_{\ker(X^*)}$ are unitarily equivalent to $S((\phi/\psi)_+)$ and $S((\psi/\phi)_+)$, respectively.*

PROOF. Find decompositions $\phi = \phi'\phi''$ and $\psi = \psi'\psi''$ satisfying the conditions of Lemma 5.3. Then the function $(\psi'/\phi')\phi$ is a multiple of ψ and hence we can define X by $Xh = P_{\mathscr{H}(\psi)}(\psi'/\phi')h$, $h \in \mathscr{H}(\phi)$. Note now that, if $h \in \mathscr{H}(\phi)$, then $h \in \ker(X)$ if and only if $h\psi'/\phi' \in \psi H^2$ or, equivalently, $h \in \phi'\psi'' H^2$. Thus $\ker(X) = \phi'\psi'' H^2 \ominus \phi'\phi'' H^2$ and $S(\phi) \mid \ker(X) \simeq S(\phi'\phi''/\phi'\psi'') = S(\phi''/\psi'')$ by Proposition 3.1.10.

On the other hand, $X\mathscr{H}(\phi) = (\psi'/\phi')H^2 \ominus \psi H^2$ so that $S(\psi)_{\ker(X^*)} = S(\psi'/\phi')$ by Proposition 3.1.10. The lemma is proved.

The following result will give the basic method for producing operators in relation ρ.

5.9. COROLLARY. *Assume that $\{\phi_j\colon -\infty < j < \infty\} \subset H^\infty$ is a sequence of inner functions, and set $T = \bigoplus_{j=-\infty}^{\infty} S(\phi_j)$. There exists an operator $X \in \{T\}'$ such that $T \mid \ker(X)$ and $T_{\ker(X^*)}$ are unitarily equivalent to $\bigoplus_{j=-\infty}^{\infty} S((\phi_j/\phi_{j+1})_+)$ and $\bigoplus_{j=-\infty}^{\infty} S((\phi_{j+1}/\phi_j)_+)$, respectively.*

PROOF. Choose for every j an operator $X_j \in \mathscr{I}(S(\phi_j), S(\phi_{j+1}))$ such that $S(\phi_j) \mid \ker(X_j) \simeq S((\phi_j/\phi_{j+1})_+)$, $S(\phi_{j+1})_{\ker(X_j^*)} \simeq S((\phi_{j+1}/\phi_j)_+)$, and $\|X_j\| \leq 1$. We can now define X as follows: $X(\bigoplus_{j=-\infty}^{\infty} h_j) = \bigoplus_{j=-\infty}^{\infty} k_j$, where $k_{j+1} = X_j h_j$. It is easy to verify that X is bounded, $\|X\| \leq 1$, $X \in \{T\}'$, and $T \mid \ker(X)$, $T_{\ker(X^*)}$ have the required properties.

In order to give an effective use of Corollary 5.9, we must answer the following two questions: (i) When is the operator $T = \bigoplus_{j=-\infty}^{\infty} S(\phi_j)$ an operator of class C_0 with property (P)? (ii) What is the Jordan model of an operator of the form $\bigoplus_{j=-\infty}^{\infty} S(\psi_j)$? We begin with a partial answer to the first problem.

5.10. PROPOSITION. *Let ϕ and $\{\phi_j\colon -\infty < j < \infty\}$ be inner functions, and denote $\gamma(\phi) = (\mu, \nu)$, $\gamma(\phi_j) = (\mu_j, \nu_j)$, $-\infty < j < \infty$. Assume that*

(i) *$\mu_j(\lambda) \leq \mu(\lambda)$, $-\infty < j < \infty$, and $\lim_{|j|\to\infty} \mu_j(\lambda) = 0$, $\lambda \in \mathbf{D}$; and*
(ii) *there exist Borel functions $f_j\colon \mathbf{T} \to [0,1]$ such that $d\nu_j = f_j\, d\nu$, $-\infty < j < \infty$, and $\lim_{|j|\to\infty} f_j(\varsigma) = 0$, $\varsigma \in \mathbf{T}$.*

Then the operator $T = \bigoplus_{j=-\infty}^{\infty} S(\phi_j)$ is of class C_0 and has property (P).

PROOF. Conditions (i) and (ii) imply, in particular, that $\phi_j \mid \phi$ for all j, hence $\phi(T) = 0$. To prove property (P) we use Proposition 3.9. The operator $\bigoplus_{j=-n}^{n} S(\phi_j)$, $n \geq 1$, has finite multiplicity, and hence it has property (P). It remains to show that $\bigwedge_{n=1}^{\infty} \psi_n \equiv 1$, where ψ_n is the minimal function of $\bigoplus_{|j|>n} S(\phi_j)$. Now, ψ_n is easy to compute. We must have $\gamma(\psi_n) = \sup\{(\mu_j, \nu_j)\colon |j| > n\}$ and we easily deduce that $\gamma(\psi_n) = (\mu_n', \nu_n')$, where

$$\mu_n'(\lambda) = \max\{\mu_j(\lambda)\colon |j| > n\}, \quad n \geq 1, \lambda \in \mathbf{D},$$

and

$$d\nu_n' = g_n\, d\nu, \qquad g_n(\varsigma) = \sup\{f_j(\varsigma)\colon |j| > n\}, \quad n \geq 1, \varsigma \in \mathbf{T}.$$

We clearly have

$$\left(\bigwedge_{n=1}^{\infty} \mu_n'\right)(\lambda) = \inf\{\mu_n'(\lambda)\colon n \geq 1\} = \limsup_{|j|\to\infty} \mu_j(\lambda) = 0, \quad \lambda \in \mathbf{D}.$$

Similarly,

$$\inf\{g_n(\varsigma)\colon n \geq 1\} = \limsup_{|j|\to\infty} f_j(\varsigma) = 0, \quad \varsigma \in \mathbf{T},$$

and the monotone convergence theorem shows that $\bigwedge_{n=1}^{\infty} \nu_n' = 0$. Thus $\bigwedge_{n=1}^{\infty} \gamma(\psi_n) = 0$ or, equivalently, $\bigwedge_{n=1}^{\infty} \psi_n \equiv 1$. The proposition is proved.

To answer the second question stated before Proposition 5.10, we need to define an operation on sequences of real numbers. We will denote by $\mathscr{S}$ the collection of all bounded sequences $\{x_j\colon -\infty < j < \infty\}$ of nonnegative real numbers. We denote by $\mathscr{S}_+$ the collection of all nonincreasing sequences $\{y_j\colon j \geq 0\}$ of nonnegative real numbers. We now want a sorting mapping $\operatorname{sort}\colon \mathscr{S} \to \mathscr{S}_+$ which takes a sequence in $\mathscr{S}$ and rearranges it in decreasing order. Simple examples show that this cannot always be done, so we introduce the following definition: if $x = \{x_j\colon -\infty < j < \infty\} \in \mathscr{S}$, then $\operatorname{sort}(x) = y = \{y_j\colon j \geq 0\}$ is given by

$$y_0 = \sup\{x_j\colon -\infty < j < \infty\}, \tag{5.11}$$

and

$$y_j = \inf\{\sup\{x_i\colon i \notin \{k_1, k_2, \ldots, k_j\}\}\colon -\infty < k_1 < k_2 < \cdots < k_j < \infty\} \tag{5.12}$$

for $j \geq 1$. This definition gives a decreasing rearrangement of x if, for example, $x_j \neq 0$ for all j and $\lim_{|j|\to\infty} x_j = 0$.

5.13. LEMMA. *Assume that* $x = \{x_j\colon -\infty < j < \infty\} \in \mathcal{S}$, *and* $\{y_j : j \geq 0\} = \operatorname{sort}(x)$. *Then we have*

$$y_0 + y_1 + \cdots + y_j = \sup\{x_{i_0} + x_{i_1} + \cdots + x_{i_j} : -\infty < i_0 < i_1 < \cdots < i_j < \infty\}$$

for all $j \geq 0$.

PROOF. It is more convenient to write the equality to be proved under the equivalent form

$$\begin{aligned} y_0 + y_1 + \cdots + y_j & \\ &= \sup\{x_{i_0} + x_{i_1} + \cdots + x_{i_j} : i_s \neq i_t \quad \text{for } s \neq t, 0 \leq s, t \leq j\}. \end{aligned} \tag{5.14}$$

Fix an integer $j \geq 0$ and a positive number ε. We first show that

$$p = \operatorname{card}\{i\colon x_i > y_j\} \leq j. \tag{5.15}$$

Indeed, assume that the elements $x_{i_0}, x_{i_1}, \ldots, x_{i_j}$, $i_0 < i_1 < \cdots < i_j$, satisfy the inequalities $x_{i_t} > y_j$, $0 \leq t \leq j$. Then no set of the form $\{k_1, k_2, \ldots, k_j\}$ can contain $\{i_0, i_1, \ldots, i_j\}$, and hence (5.12) implies the absurd inequality

$$y_j \geq \min\{x_{i_0}, x_{i_1}, \ldots, x_{i_j}\} > y_j.$$

On the other hand, we claim that

$$\operatorname{card}\{i\colon x_i > y_j - \varepsilon\} > j. \tag{5.16}$$

Indeed, suppose to the contrary that $\{i\colon x_i > y_j - \varepsilon\} = \{x_{i_1}, x_{i_2}, \ldots, x_{i_n}\}$, where $i_1 < i_2 < \cdots < i_n$ and $n \leq j$. Then we deduce the absurd inequality

$$y_j \leq y_n \leq \sup\{x_i : i \notin \{i_1, i_2, \ldots, i_n\}\} \leq y_j - \varepsilon.$$

Relations (5.15) and (5.16) imply the existence of a set of $j + 1$ integers $\{k_0, k_1, \ldots, k_j\}$ such that

$$x_{k_0} \geq x_{k_1} \geq \cdots \geq x_{k_{p-1}} > y_j \geq x_{k_p} \geq \cdots \geq x_{k_j} > y_j - \varepsilon,$$

and such that $x_i \leq y_j$ whenever $i \notin \{k_0, k_1, \ldots, k_j\}$. The numbers $y_0, y_1, \ldots, y_{j-1}$ are now easy to determine. First, we clearly have $y_0 = x_{k_0}$, $y_1 = x_{k_1}$, $\ldots, y_{p-1} = x_{k_{p-1}}$, and the inequalities

$$y_j \leq y_{j-1} \leq \cdots \leq y_p \leq \sup\{x_i : i \notin \{k_0, k_1, \ldots, k_{p-1}\}\} \leq y_j$$

show that $y_p = y_{p+1} = \cdots = y_j$. Relation (5.14) now follows from the following obvious calculation:

$$\begin{aligned} y_0 + y_1 + \cdots + y_j &= y_0 + y_1 + \cdots + y_{p-1} + y_p + \cdots + y_j \\ &= x_{k_0} + x_{k_1} + \cdots + x_{k_{p-1}} + (j - p + 1)y_j \\ &\leq x_{k_0} + x_{k_1} + \cdots + x_{k_{p-1}} + x_{k_p} + \cdots + x_{k_j} + (j - p + 1)\varepsilon \\ &\leq \sup\{x_{i_0} + x_{i_1} + \cdots + x_{i_j} : i_s \neq i_t \quad \text{for } s \neq t\} + (j - p + 1)\varepsilon \\ &\leq x_{k_0} + x_{k_1} + \cdots + x_{k_j} + (j + 1)\varepsilon + (j - p + 1)\varepsilon \\ &\leq y_0 + y_1 + \cdots + y_j + (j + 1)\varepsilon + (j - p + 1)\varepsilon. \end{aligned}$$

The next to last inequality follows from the fact that we have $x_{i_s} \leq x_{k_s} + \varepsilon$, $0 \leq j$, if $x_{i_0} \geq x_{i_1} \geq x_{i_j}$. The lemma follows because ε is an arbitrary positive number.

5.17. PROPOSITION. *Let ψ and $\{\psi_j\colon -\infty < j < \infty\}$ be inner functions, and denote $\gamma(\psi) = (\mu, \nu)$, $\gamma(\psi_j) = (\mu_j, \nu_j)$, $-\infty < j < \infty$. Assume that*

(i) *$\mu_j(\lambda) \le \mu(\lambda)$, $\lambda \in \mathbf{D}$, $-\infty < j < \infty$, and*

(ii) *$d\nu_j = f_j\, d\nu$, where $f_j\colon \mathbf{T} \to [0,1]$ are Borel functions for $-\infty < j < \infty$.*

Define functions $\mu_n'\colon \mathbf{D} \to \{0,1,2,\dots\}$ and $f_n'\colon \mathbf{T} \to [0,1]$, $n \ge 0$, such that

$$\{\mu_n'(\lambda)\colon n \ge 0\} = \operatorname{sort}(\{\mu_j(\lambda)\colon -\infty < j < \infty\}), \quad \lambda \in \mathbf{D},$$

and

$$\{f_n'(\varsigma)\colon n \ge 0\} = \operatorname{sort}(\{f_j(\varsigma)\colon -\infty < j < \infty\}), \quad \varsigma \in \mathbf{T}.$$

Finally, define inner functions θ_n, $n \ge 0$, such that $\gamma(\theta_n) = (\mu_n', \nu_n')$, $d\nu_n' = f_n'\, d\nu$. Then $T = \bigoplus_{j=-\infty}^{\infty} S(\psi_j)$ is an operator of class C_0, and $\bigoplus_{n<\omega} S(\theta_n)$ is the Jordan model of T.

PROOF. As in Proposition 5.10, we have $\psi(T) = 0$, so that T is indeed of class C_0. Note next that the functions f_n' are Borel, as one can easily see from relations (5.11) and (5.12). Thus the functions θ_n exist.

Denote by $\mathscr{F}$ a Hilbert space with an orthonormal basis $\{e_j\colon -\infty < j < \infty\}$, and define $\Theta = \operatorname{diag}\{\psi_j\colon -\infty < j < \infty\} \in H^\infty(\mathscr{L}(\mathscr{F}))$, that is, $\Theta(\lambda)e_j = \psi_j(\lambda)e_j$, $\lambda \in \mathbf{D}$, $-\infty < j < \infty$. We clearly have $T = S(\Theta)$, and a moment's thought shows that $S(\bigwedge^j \Theta) \simeq \bigoplus_{i_0<i_1<\cdots<i_{j-1}} S(\psi_{i_0}\psi_{i_1}\cdots\psi_{i_{j-1}})$. Corollary 5.3.29 shows that we can conclude the proof by showing that $\theta_0\theta_1\cdots\theta_{j-1}$ is the minimal function of $S(\bigwedge^j \Theta)$, $j \ge 1$, or, equivalently, that $\theta_0\theta_1\cdots\theta_{j-1} = \bigvee\{\psi_{i_0}\psi_{i_1}\cdots\psi_{i_{j-1}}\colon i_0 < i_1 < \cdots < i_{j-1}\}$. This relation can be rewritten as

$$\sum_{s=0}^{j-1}(\mu_s', \nu_s') = \bigvee\left\{\sum_{s=0}^{j-1}(\mu_{i_s}, \nu_{i_s})\colon i_0 < i_1 < \cdots < i_{j-1}\right\}, \quad j \ge 1.$$

This is equivalent to verifying that

$$\begin{aligned}&\mu_0'(\lambda) + \mu_1'(\lambda) + \cdots + \mu_{j-1}'(\lambda)\\ &\quad = \sup\{\mu_{i_0}(\lambda) + \mu_{i_1}(\lambda) + \cdots + \mu_{i_{j-1}}(\lambda)\colon i_0 < i_1 < \cdots < i_{j-1}\}, \quad \lambda \in \mathbf{D},\end{aligned}$$

and

$$\begin{aligned}&f_0'(\varsigma) + f_1'(\varsigma) + \cdots + f_{j-1}'(\varsigma)\\ &\qquad = \sup\{f_{i_0}(\varsigma) + f_{i_1}(\varsigma) + \cdots + f_{i_{j-1}}(\varsigma)\colon i_0 < i_1 < \cdots < i_{j-1}\}\end{aligned}$$

for $d\nu$-almost every $\varsigma \in \mathbf{T}$. Since these relations follow from the definitions of μ_n', f_n' and from Lemma 5.13, the proposition is proved.

We can now extend Proposition 5.2 to all pairs of operators with equal C_0-dimensions.

5.18. THEOREM. *Assume that T and T' are contractions of class C_0, both having property (P). Then $T\rho T'$ if and only if $\gamma_T = \gamma_{T'}$.*

PROOF. We already know that $T\rho T'$ implies that $\gamma_T = \gamma_{T'}$, so we need only to prove the converse. Assume therefore that $\gamma_T = \gamma_{T'}$, and let $\bigoplus_{j<\omega} S(\theta_j)$ and

$\bigoplus_{j<\omega} S(\theta'_j)$ be the Jordan models of T and T', respectively. By Corollary 5.9 it suffices to produce a sequence $\{\phi_j\colon -\infty < j < \infty\}$ of inner functions such that $\bigoplus_{j=-\infty}^{\infty} S(\phi_j)$ is of class C_0, has property (P), $\bigoplus_{j=-\infty}^{\infty} S((\phi_j/\phi_{j+1})_+)$ is quasisimilar to $\bigoplus_{j<\infty} S(\theta_j)$, and $\bigoplus_{j=-\infty}^{\infty} S((\phi_{j+1}/\phi_j)_+)$ is quasisimilar to $\bigoplus_{j<\omega} S(\theta'_j)$. Let us reformulate these conditions in terms of $\gamma(\theta_j)$, $\gamma(\theta'_j)$, and $\gamma(\phi_j)$. Denote first $\theta = \theta_0^2 \theta_0'^2$, $\gamma(\theta) = (\mu, \nu)$, and let $\gamma(\theta_j) = (\mu_j, \nu_j)$, $\gamma(\theta'_j) = (\mu'_j, \nu'_j)$, $j \geq 0$. It is clear that we can find Borel functions f_j, $f'_j\colon \mathbf{T} \to [0, \frac{1}{2}]$ such that $d\nu_j = f_j\, d\nu$, $d\nu'_j = f'_j\, d\nu$, $j \geq 0$. Also note the obvious inequality $2\max\{\mu_j(\lambda), \mu'_j(\lambda)\} \leq \mu(\lambda)$, $j \geq 0$. The facts that T and T' have property (P) and $\gamma_T = \gamma_{T'}$ can be translated into the equalities

$$\sum_{j=0}^{\infty} \mu_j(\lambda) = \sum_{j=0}^{\infty} \mu'_j(\lambda) < \infty, \quad \lambda \in \mathbf{D}, \tag{5.19}$$

and

$$\sum_{j=0}^{\infty} f_j(\varsigma) = \sum_{j=0}^{\infty} f'_j(\varsigma), \qquad \lim_{j\to\infty} f_j(\varsigma) = \lim_{j\to\infty} f'_j(\varsigma) = 0, \tag{5.20}$$

for $d\nu$-almost every $\varsigma \in \mathbf{T}$. We may, of course, assume that (5.20) holds for every $\varsigma \in \mathbf{T}$; this is achieved by redefining f_j and f'_j to be zero on the ν-negligible Borel set on which the relations in (5.20) fail.

We want to find inner functions $\{\phi_j\colon -\infty < j < \infty\}$ such that $\gamma(\phi_j) = (\mu''_j, \nu''_j)$, where $d\nu''_j = g_j\, d\nu$, and the following conditions are satisfied:

$$\mu''_j \leq \mu, \quad -\infty < j < \infty, \qquad \lim_{|j|\to\infty} \mu''_j(\lambda) = 0, \quad \lambda \in \mathbf{D}; \tag{5.21}$$

$$g_j\colon \mathbf{T} \to [0,1] \text{ are Borel and } \lim_{|j|\to\infty} g_j(\varsigma) = 0, \quad \varsigma \in \mathbf{T}; \tag{5.22}$$

$$\operatorname{sort}(\{(\mu''_j(\lambda) - \mu''_{j+1}(\lambda))_+\colon -\infty < j < \infty\}) = \{\mu_j(\lambda)\colon j \geq 0\}, \ \lambda \in \mathbf{D}; \tag{5.23}$$

$$\operatorname{sort}(\{(g_j(\varsigma) - g_{j+1}(\varsigma))_+\colon -\infty < j < \infty\}) = \{f_j(\varsigma)\colon j \geq 0\}, \ \varsigma \in \mathbf{T}; \tag{5.24}$$

$$\operatorname{sort}(\{(\mu''_{j+1}(\lambda) - \mu''_j(\lambda))_+\colon -\infty < j < \infty\}) = \{\mu'_j(\lambda)\colon j \geq 0\}, \ \lambda \in \mathbf{D}; \tag{5.25}$$

and

$$\operatorname{sort}(\{(g_{j+1}(\varsigma) - g_j(\varsigma))_+\colon -\infty < j < \infty\}) = \{f'_j(\varsigma)\colon j \geq 0\}, \quad \varsigma \in \mathbf{T}. \tag{5.26}$$

Conditions (5.21) and (5.22) imply, via Proposition 5.10, that $\bigoplus_{j=-\infty}^{\infty} S(\phi_j)$ is of class C_0 and has property (P). Conditions (5.23) and (5.24) imply, via Proposition 5.17, that $\bigoplus_{j=-\infty}^{\infty} S((\phi_j/\phi_{j+1})_+)$ is quasisimilar to $\bigoplus_{j<\omega} S(\theta_j)$. Analogously, (5.25) and (5.26) imply that $\bigoplus_{j<\omega} S(\theta'_j)$ is the Jordan model of $\bigoplus_{j=-\infty}^{\infty} S((\phi_{j+1}/\phi_j)_+)$. It will suffice therefore to construct functions μ''_j and g_j satisfying conditions (5.21)–(5.26). We begin with the μ''_j. Fix a point $\lambda \in \mathbf{D}$. We claim that there exist nondecreasing functions

$$u_\lambda\colon \{0,1,2,\dots\} \to \{0,1,2,\dots\} \quad \text{and} \quad v_\lambda\colon \{0,1,2,\dots\} \to \{-1,0,1,2,\dots\}$$

with the following properties:

$$u_\lambda(0) = 0, \qquad v_\lambda(0) = -1, \qquad u_\lambda(n) + v_\lambda(n) = n-1, \quad n \geq 1; \tag{5.27}$$

and

$$(5.28)\qquad 0 \leq \sum_{i=0}^{u_\lambda(n)} \mu_i'(\lambda) - \sum_{j=0}^{v_\lambda(n)} \mu_j(\lambda) \leq \mu'_{u_\lambda(n)}(\lambda) + \mu_{v_\lambda(n)+1}(\lambda).$$

The second sum in (5.28) is taken to be zero if $v_\lambda(n) = -1$. Assume that the $u_\lambda(k)$ and $v_\lambda(k)$ have been defined for all $k \leq n$. If $\sum_{i=0}^{u_\lambda(n)} \mu_i'(\lambda) - \sum_{j=0}^{v_\lambda(n)} \mu_j(\lambda) \geq \mu_{v_\lambda(n)+1}(\lambda) > 0$ then we set $u_\lambda(n+1) = u_\lambda(n)$ and $v_\lambda(n+1) = v_\lambda(n) + 1$. If $\sum_{i=0}^{u_\lambda(n)} \mu_i'(\lambda) - \sum_{j=0}^{v_\lambda(n)} \mu_j(\lambda) < \mu_{v_\lambda(n)+1}(\lambda)$ then we set $u_\lambda(n+1) = u_\lambda(n) + 1$ and $v_\lambda(n+1) = v_\lambda(n)$. Finally, if $\mu_{v_\lambda(n)+1}(\lambda) = 0$, we set $u_\lambda(n+1) = u_\lambda(n)$, $v_\lambda(n+1) = v_\lambda(n) + 1$ or $v_\lambda(n+1) = v_\lambda(n)$, $u_\lambda(n+1) = u_\lambda(n) + 1$ according to whether n is odd or even. The fact that u and v are nondecreasing and surjective and property (5.27) are obvious. It remains to prove that (5.28), which is trivially satisfied for $n = 0$, is preserved by this inductive process. We have three cases to analyze. If $\sum_{i=0}^{u_\lambda(n)} \mu_i'(\lambda) - \sum_{j=0}^{v_\lambda(n)} \mu_j(\lambda) \geq \mu_{v_\lambda(n)+1}(\lambda) > 0$ then

$$\begin{aligned}\sum_{i=0}^{u_\lambda(n+1)} \mu_i'(\lambda) - \sum_{j=0}^{v_\lambda(n+1)} \mu_j(\lambda) &= \sum_{i=0}^{u_\lambda(n)} \mu'(\lambda) - \sum_{j=0}^{v_\lambda(n)} \mu_j(\lambda) - \mu_{v_\lambda(n)+1}(\lambda)\\ &\leq \mu'_{u_\lambda(n)}(\lambda) + \mu_{v_\lambda(n)+1}(\lambda) - \mu_{v_\lambda(n)+1}(\lambda)\\ &\leq \mu'_{u_\lambda(n+1)}(\lambda) + \mu_{v_\lambda(n+1)+1}(\lambda)\end{aligned}$$

for the simple reason that $\mu_{v_\lambda(n+1)+1}(\lambda) \geq 0$. If $\sum_{i=0}^{u_\lambda(n)} \mu_i'(\lambda) - \sum_{j=0}^{v_\lambda(n)} \mu_j(\lambda) < \mu_{v_\lambda(n)+1}(\lambda)$ then obviously

$$\begin{aligned}\sum_{i=0}^{u_\lambda(n+1)} \mu_i'(\lambda) - \sum_{j=0}^{v_\lambda(n+1)} \mu_j(\lambda) &= \mu'_{u_\lambda(n+1)}(\lambda) + \left(\sum_{i=0}^{u_\lambda(n)} \mu_i'(\lambda) - \sum_{j=0}^{v_\lambda(n)} \mu_j(\lambda)\right)\\ &< \mu'_{u_\lambda(n+1)}(\lambda) + \mu_{v_\lambda(n+1)+1}(\lambda).\end{aligned}$$

Finally, if $\mu_{v_\lambda(n)+1} = 0$, then $\sum_{j=0}^{v_\lambda(n)} \mu_j(\lambda) = \sum_{j=0}^{\infty} \mu_j(\lambda)$ so that

$$\begin{aligned}\sum_{i=0}^{u_\lambda(n+1)} \mu_i'(\lambda) - \sum_{j=0}^{v_\lambda(n+1)} \mu_j(\lambda) &= \sum_{i=0}^{u_\lambda(n+1)} \mu_i'(\lambda) - \sum_{j=0}^{\infty} \mu_j(\lambda)\\ &\leq \sum_{i=0}^{\infty} \mu_i'(\lambda) - \sum_{j=0}^{\infty} \mu_j(\lambda) = 0\end{aligned}$$

by (5.19). Let us also note that the above construction is made possible by the fact that $\mu_i'(\lambda) = 0$ for i sufficiently large, so that the inequality $\sum_{i=0}^{u} \mu_i'(\lambda) > \mu_0(\lambda)$ is satisfied for sufficiently large u, and hence v_λ is surjective. Thus the existence of the functions u_λ and v_λ is proved by induction, and we can now set

$$\begin{aligned}\mu_n''(\lambda) &= 0 && \text{if } n < 0,\\ &= \sum_{i=0}^{u_\lambda(n)} \mu_i'(\lambda) - \sum_{j=0}^{v_\lambda(n)} \mu_j(\lambda) && \text{if } n \geq 0, \lambda \in \mathbf{D}.\end{aligned}$$

Conditions (5.21) are satisfied by virtue of (5.28) (remember that $\mu_i'(\lambda) \le \frac{1}{2}\mu(\lambda)$ and $\mu_j(\lambda) \le \frac{1}{2}\mu(\lambda)$, $\lambda \in \mathbf{D}$, $i, j \ge 0$). Conditions (5.23) and (5.25) are also easy to verify. Indeed, we have

$$\begin{aligned}(\mu_n''(\lambda) - \mu_{n+1}''(\lambda))_+ &= 0 && \text{if } n < 0,\\ &= \mu_{v_\lambda(n+1)}(\lambda) && \text{if } n \ge 0 \text{ and } v_\lambda(n+1) = v_\lambda(n),\end{aligned}$$

while

$$\begin{aligned}(\mu_{n+1}''(\lambda) - \mu_n''(\lambda))_+ &= 0 && \text{if } n < -1\\ &= \mu_0'(\lambda) && \text{if } n = -1\\ &= \mu_{u_\lambda(n+1)}'(\lambda) && \text{if } n \ge 0 \text{ and } u_\lambda(n+1) = u_\lambda(n) + 1.\end{aligned}$$

The idea in the construction of the functions g_n is similar. We need, however, to distinguish between two cases. Define the Borel set $\omega \subseteq \mathbf{T}$ consisting of those points ς such that $f_j(\varsigma) \ne 0$ for all $j \ge 0$. If $\varsigma \in \omega$ we can construct nondecreasing surjective functions $u_\varsigma\colon \{0,1,2,\dots\} \to \{0,1,2,\dots\}$ and $v_\varsigma\colon \{0,1,2,\dots\} \to \{-1,0,1,2,\dots\}$ satisfying the following conditions:

$$(5.29)\quad u_\varsigma(0) = 0, \qquad v_\varsigma(0) = -1, \qquad u_\varsigma(n) + v_\varsigma(n) = n-1, \quad n \ge 1, \varsigma \in \omega;$$

and

$$(5.30)\qquad 0 \le \sum_{i=0}^{u_\varsigma(n)} f_i'(\varsigma) - \sum_{j=0}^{v_\varsigma(n)} f_j(\varsigma) \le f_{u_\varsigma(n)}'(\varsigma) + f_{v_\varsigma(n)+1}(\varsigma), \quad \varsigma \in \omega.$$

The construction works because for every integer $v > 0$ there exists an integer u such that $\sum_{i=0}^{u} f_i'(\varsigma) - \sum_{j=0}^{v} f_j(\varsigma) > 0$; this assertion follows from the first equality in (5.20), and it ensures that v_ς is onto. Also note that the functions $\varsigma \to u_\varsigma(n)$ and $\varsigma \to v_\varsigma(n)$ are measurable, and we can define, for $\varsigma \in \omega$,

$$\begin{aligned}g_n(\varsigma) &= 0 && \text{if } n < 0,\\ &= \sum_{i=0}^{u_\varsigma(n)} f_i'(\varsigma) - \sum_{j=0}^{v_\varsigma(n)} f_j(\varsigma) && \text{if } n \ge 0.\end{aligned}$$

Properties (5.22), (5.24), and (5.26) are readily verified for $\varsigma \in \omega$. On the complement $\mathbf{T} \setminus \omega$ we can follow a "symmetrical" strategy. Namely, for $\varsigma \in \mathbf{T} \setminus \omega$ we can define nondecreasing surjective functions $u_\varsigma\colon \{0,1,2,\dots\} \to \{0,1,2,\dots\}$ and $v_\varsigma\colon \{0,1,2,\dots\} \to \{-1,0,1,2,\dots\}$ satisfying the conditions

$$(5.31)\qquad u_\varsigma(0) = 0, \qquad v_\varsigma(0) = -1, \qquad u_\varsigma(n) + v_\varsigma(n) = n-1, \quad n \ge 1,$$

and

$$(5.32)\qquad 0 \le \sum_{i=0}^{u_\varsigma(n)} f_i(\varsigma) - \sum_{j=0}^{v_\varsigma(n)} f_j'(\varsigma) \le f_{u_\varsigma(n)}(\varsigma) + f_{v_\varsigma(n)+1}'(\varsigma), \quad n \ge 0.$$

For $\varsigma \in \mathbf{T} \setminus \omega$ we now define

$$\begin{aligned}g_n(\varsigma) &= 0 && \text{if } n > 0,\\ &= \sum_{i=0}^{u_\varsigma(-n)} f_i(\varsigma) - \sum_{j=0}^{v_\varsigma(-n)} f_j'(\varsigma) && \text{if } n \le 0.\end{aligned}$$

Properties (5.22), (5.24), and (5.26) are easy to verify on $\mathbf{T}\setminus\omega$. The theorem is proved.

Exercises

1. Assume that T and T' are contractions of class C_0, $\mu_T = m < \infty$, $\mu'_T = n < \infty$, and $\gamma_T = \gamma_{T'}$. Show that the construction in the proof of Proposition 5.2 yields an operator R such that $\mu_R \leq m + n - 1$.

2. Let T and T' be as in Exercise 1. Show that there exists an operator R of class C_0 and $X \in \{T\}'$ such that $\mu_R \leq \max\{m,n\}$, $R|\ker(X) \sim T$, and $R_{\ker(X^*)} \sim T'$.

3. Formulate a converse to Proposition 5.10. Is this converse true?

4. Assume that $x = \{x_j\colon -\infty < j < \infty\}$, and $\operatorname{sort}(x) = \{y_j\colon j \geq 0\}$. Give necessary conditions on x for the existence of a bijection $\pi\colon \{\dots,-2,-1,0,1,2,\dots\} \to \{0,1,2,\dots\}$ such that $x_j = y_{\pi(j)}$ for all j.

5. Let x and y be as in Exercise 4. Show that a real number t satisfies the inequality $t \geq y_j$ if and only if $\operatorname{card}\{i\colon t < x_i\} \leq j$.

6. Check the details in the proofs of (5.23)–(5.26).

7. Find an operator T of class C_0 with property (P) such that $T\rho(T\oplus T)$.

8. Show that T has property (P_0) if and only if it has property (P) and $T\rho(T\oplus T')$ implies that T' acts on a trivial space.

6. C_0-Fredholm operators.

6.1. DEFINITION. Assume that $T \in \mathscr{L}(\mathscr{H})$ and $T' \in \mathscr{L}(\mathscr{H}')$ are operators of class C_0 and $X \in \mathscr{I}(T,T')$. Then X is said to be *(T,T')-semi-Fredholm* if the following two conditions are satisfied:

(i) $X \mid (\mathscr{H} \ominus \ker(X))$ is a $(T_{\mathscr{H}\ominus\ker(X)}, T' \mid (X\mathscr{H})^-)$-lattice-isomorphism;

(ii) either $T \mid \ker(X)$ or $T'^* \mid \ker(X^*)$ has property (P).

The operator X is said to be *(T,T')-Fredholm* if it satisfies (i) and

(iii) both $T \mid \ker(X)$ and $T'^* \mid \ker(X^*)$ have property (P).

The set of all (T,T')-semi-Fredholm [respectively, (T,T')-Fredholm] operators will be denoted $s\mathscr{F}(T,T')$ [respectively, $\mathscr{F}(T,T')$]. We write $s\mathscr{F}(T)$ [respectively, $\mathscr{F}(T)$] for $s\mathscr{F}(T,T)$ [respectively, $\mathscr{F}(T,T)$].

We will use the name C_0-Fredholm operators instead of (T',T)-Fredholm operators whenever no confusion may arise.

It is readily seen that $\mathscr{F}(T)$ coincides with the set of classical Fredholm operators on $\mathscr{H}$ if $T = 0 \in \mathscr{L}(\mathscr{H})$. In this section, and in the following section, we will extend to the context of C_0-Fredholm operators some of the results of classical Fredholm theory. The first problem is to introduce an appropriate notion of index for C_0-Fredholm operators. The natural attempt is to embed

$\tilde{\Gamma}$ into a group $\mathscr{G}$, and define $\text{ind}(X) = \gamma_T(\ker(X)) - \gamma_{T'}(\ker(X^*))$ if X is a (T,T')-Fredholm operator. The problem is that the semigroup $\tilde{\Gamma}$ contains "infinite" elements and it cannot be embedded in any group. More precisely, we have the following result.

6.2. LEMMA. *Assume that $\gamma, \gamma', \gamma'' \in \tilde{\Gamma}$, and $\gamma = \gamma_0 + \gamma_\infty$ is that decomposition of γ such that $\gamma_0 \in \Gamma_0$, $\gamma_\infty \in \Gamma_\infty$, and $\gamma_0 \wedge \gamma_\infty = 0$.*

(i) *The equality $\gamma' + \gamma = \gamma'' + \gamma$ always implies $\gamma' = \gamma''$ if and only if $\gamma_\infty = 0$.*

(ii) *The equality $\gamma' + \gamma = \gamma'' + \gamma$ implies $\gamma' = \gamma''$ whenever $\gamma_\infty \le \gamma' \wedge \gamma''$.*

PROOF. That $\gamma' + \gamma = \gamma'' + \gamma$ implies $\gamma' = \gamma''$ is obvious if $\gamma \in \Gamma_0$. Vice versa, if $\gamma_\infty \ne 0$, we have $\gamma' = 0 \ne \gamma'' = \gamma_\infty$, and $\gamma' + \gamma = \gamma'' + \gamma$. Now assume that $\gamma' + \gamma = \gamma'' + \gamma$ and $\gamma_\infty \le \gamma' \wedge \gamma''$. Then we have $\gamma' + \gamma_\infty = \gamma'' + \gamma_\infty$ because $\gamma_0 \in \Gamma_0$, and, since $\gamma_\infty \le \gamma' \wedge \gamma''$, $\gamma' = \gamma' + \gamma_\infty = \gamma'' + \gamma_\infty = \gamma''$. The lemma is proved.

The Cartesian product $\tilde{\Gamma} \times \tilde{\Gamma}$ is a semigroup (under componentwise addition), and we introduce a relation " $\sim$" on $\tilde{\Gamma} \times \tilde{\Gamma}$ by writing $(\gamma, \gamma') \sim (\gamma_1, \gamma_1')$ if and only if $\gamma + \gamma_1' = \gamma' + \gamma_1$. Lemma 6.2 shows that this is not an equivalence relation.

6.3. DEFINITION. Let T and T' be two operators of class C_0, and let $X \in \mathscr{F}(T,T')$. The *index* of X is defined by

$$\text{ind}(X) = (\gamma_T(\ker(X)), \gamma_{T'}(\ker(X^*))) \in \tilde{\Gamma} \times \tilde{\Gamma}.$$

Furthermore, if $X \in s\mathscr{F}(T,T')$ but $X \notin \mathscr{F}(T,T')$, we set

$$\begin{aligned} \text{ind}(X) &= +\infty \qquad \text{if } T'_{\ker(X^*)} \text{ has property (P)}, \\ &= -\infty \qquad \text{if } T \mid \ker(X) \text{ has property (P)}. \end{aligned}$$

It will be useful to extend the addition on $\tilde{\Gamma} \times \tilde{\Gamma}$ by setting $(\gamma, \gamma') + \infty = +\infty$, $(\gamma, \gamma') + (-\infty) = -\infty$, $(+\infty) + (+\infty) = +\infty$, $(-\infty) + (-\infty) = -\infty$. The symbols $(+\infty) + (-\infty)$ and $(-\infty) + (+\infty)$ are left undefined. We define two more operations on $\tilde{\Gamma} \times \tilde{\Gamma}$ as follows:

$$(\gamma, \gamma')^\sim = (\gamma^\sim, \gamma'^\sim), \quad (\gamma, \gamma') \in \tilde{\Gamma} \times \tilde{\Gamma}, \tag{6.4}$$

and

$$-(\gamma, \gamma') = (\gamma', \gamma), \quad (\gamma, \gamma') \in \tilde{\Gamma} \times \tilde{\Gamma}. \tag{6.5}$$

Quite clearly we have $(\gamma, \gamma') + (-(\gamma, \gamma')) \sim (0,0)$. We also set $-(\pm\infty) = \mp\infty$, $(\pm\infty)^\sim = \pm\infty$.

6.6. LEMMA. *Let $T \in \mathscr{L}(\mathscr{H})$ and $T' \in \mathscr{L}(\mathscr{H}')$ be two operators of class C_0. Then $s\mathscr{F}(T,T')^* = \{X^* : X \in s\mathscr{F}(T,T')\} = s\mathscr{F}(T'^*, T^*)$, $\mathscr{F}(T,T')^* = \mathscr{F}(T'^*, T^*)$, and $\text{ind}(X^*) = -\text{ind}(X)^\sim$ for $X \in s\mathscr{F}(T,T')$.*

PROOF. By Lemma 1.20, the operator $X \mid (\mathscr{H} \ominus \ker(X)) \in \mathscr{L}(\mathscr{H} \ominus \ker(X), (X\mathscr{H})^-)$ is a lattice-isomorphism if and only if its adjoint

$$\begin{aligned} (X \mid (\mathscr{H} \ominus \ker(X)))^* = X^* \mid (\mathscr{H}' \ominus \ker(X^*)) &\in \mathscr{L}((X\mathscr{H})^-, \mathscr{H} \ominus \ker(X)) \\ &= \mathscr{L}(\mathscr{H}' \ominus \ker(X^*), (X^*\mathscr{H}')^-) \end{aligned}$$

is a lattice-isomorphism. Furthermore, it is clear that $\gamma_T(\ker(X)) = \gamma_{T^*}(\ker(X))^\sim$, $\gamma_{T'}(\ker(X^*)) = \gamma_{T'^*}(\ker(X^*))^\sim$. A quick look at definitions 6.1, 6.3, (6.4), and (6.5) convinces us that the lemma is true.

We now give some simple examples of C_0-Fredholm operators. These correspond in the classical case (i.e., the case in which $T = 0$ and $T' = 0$) to operators acting between finite-dimensional spaces.

6.7. PROPOSITION. *Assume that $T \in \mathscr{L}(\mathscr{H})$ and $T' \in \mathscr{L}(\mathscr{H}')$ are two operators of class C_0.*

(i) *If both T and T' have property (P), then $\mathscr{F}(T,T') = \mathscr{I}(T,T')$, and $\operatorname{ind}(X) \sim (\gamma_T, \gamma_{T'})$ for $X \in \mathscr{I}(T,T')$.*

(ii) *If exactly one of the operators T and T' has property (P), then $\mathscr{F}(T,T') = \varnothing$, $s\mathscr{F}(T,T') = \mathscr{I}(T,T')$, and*

$$\begin{aligned} \operatorname{ind}(X) &= +\infty \quad \text{if } T' \text{ has property } (P), X \in \mathscr{I}(T,T'), \\ &= -\infty \quad \text{if } T \text{ has property } (P), X \in \mathscr{I}(T,T'). \end{aligned}$$

PROOF. Assume that $X \in \mathscr{I}(T',T)$ and note that $X \mid \mathscr{H} \ominus \ker(X)$ is a quasiaffinity in $\mathscr{I}(T_{\mathscr{H}\ominus\ker(X)}, T' \mid (X\mathscr{H})^-)$. If either T or T' has property (P) then one, and hence both, of the operators $T' \mid (X\mathscr{H})^-$ and $T_{\mathscr{H}\ominus\ker(X)}$ must have property (P) by Corollary 1.17. We conclude that $X \mid \mathscr{H} \ominus \ker(X)$ is a lattice-isomorphism by virtue of Proposition 1.21. Assume now that both T and T' have property (P). The fact that $X \in \mathscr{F}(T,T')$ follows from Corollary 1.17. Furthermore, Corollary 4.21 yields the relations $\gamma_T = \gamma_T(\ker(X)) + \gamma_T(\mathscr{H} \ominus \ker(X))$ and $\gamma_{T'} = \gamma_{T'}(\ker(X^*)) + \gamma_{T'}((X\mathscr{H})^-)$. The fact that $T' \mid (X\mathscr{H})^- \sim T_{\mathscr{H}\ominus\ker(X)}$ implies that $\gamma_{T'}((X\mathscr{H})^-) = \gamma_T(\mathscr{H} \ominus \ker(X))$. We set $\gamma = \gamma_T(\mathscr{H} \ominus \ker(X))$ so that the above equalities can be written as $\gamma_T = \gamma_T(\ker(X)) + \gamma$ and $\gamma_{T'} = \gamma_{T'}(\ker(X^*)) + \gamma$. These easily imply that

$$\gamma_T + \gamma_{T'}(\ker(X^*)) + \gamma = \gamma_{T'} + \gamma_T(\ker(X)) + \gamma$$

and, since $\gamma \leq \gamma_T \wedge \gamma_{T'}$, Lemma 6.2 implies that

$$\gamma_T + \gamma_{T'}(\ker(X^*)) = \gamma_{T'} + \gamma_T(\ker(X)).$$

This last equality is equivalent to $\operatorname{ind}(X) \sim (\gamma_T, \gamma_{T'})$, and (i) is proved.

Assume now that T' has property (P) but T does not have property (P). Corollary 1.17 implies that $T \mid \ker(X)$ does not have property (P), hence $\operatorname{ind}(X) = +\infty$. Analogously, if T has property (P) and T' does not, then $T'_{\ker(X^*)}$ cannot have property (P) by Corollary 1.17, and hence $\operatorname{ind}(X) = -\infty$. The proposition follows.

6.8. REMARK. The preceding proposition shows, in particular, that given $(\gamma, \gamma') \in \tilde{\Gamma} \times \tilde{\Gamma}$, there exist operators T, T' and $X \in \mathscr{I}(T',T)$ such that $\operatorname{ind}(X) = (\gamma, \gamma')$. Indeed, choose T and T' to have property (P), $\gamma_T = \gamma$, $\gamma_{T'} = \gamma'$, and set $X = 0 \in \mathscr{I}(T',T)$. We can do better than that. Namely, we can produce an operator R of class C_0, and $Y \in \mathscr{F}(R)$ such that $\operatorname{ind}(Y) = (\gamma, \gamma')$. Indeed, let T and T' be as above, and define $R = (T \oplus T \oplus T \oplus \cdots) \oplus (T' \oplus T' \oplus \cdots)$. Next define $Y \in \{R\}'$ by

$$Y((h_0 \oplus h_1 \oplus \cdots) \oplus (k_0 \oplus k_1 \oplus \cdots)) = (h_1 \oplus h_2 \oplus \cdots) \oplus (0 \oplus k_0 \oplus k_1 \oplus \cdots).$$

It is easy to check that Y has the required properties.

The following result will help us compensate for the fact that lattice-isomorphisms are not, generally, invertible.

6.9. PROPOSITION. *Assume that* $T \in \mathcal{L}(\mathcal{H})$, $T' \in \mathcal{L}(\mathcal{H}')$, $T'' \in \mathcal{L}(\mathcal{H}'')$ *are operators of class* C_0, $A \in \mathcal{I}(T', T)$, $B \in \mathcal{I}(T'', T)$, *and* $A\mathcal{H}' \subseteq (B\mathcal{H}'')^-$. *If, in addition,* $T \mid (B\mathcal{H}'')^-$ *has property (P) then*

(i) $(A^{-1}(B\mathcal{H}''))^- = \mathcal{H}'$; *and*

(ii) $(A\mathcal{H}' \cap B\mathcal{H}'')^- \supseteq A\mathcal{H}'$.

PROOF. Note that $A\mathcal{H}' \cap B\mathcal{H}'' = A(A^{-1}(B\mathcal{H}''))$ and therefore (ii) is an immediate consequence of (i). Thus we need only to prove (i). There is no loss of generality in assuming that A is one-to-one, B is a quasiaffinity, and T has property (P). Indeed, we may replace T, T', T'', A, and B by $T \mid (B\mathcal{H}'')^-$, $T'_{\mathcal{H}' \ominus \ker(A)}$, $T''_{\mathcal{H}'' \ominus \ker(B)}$, $A \mid (\mathcal{H}' \ominus \ker(A))$, and $B \mid (\mathcal{H}'' \ominus \ker(B))$, respectively. (Note that $A^{-1}(B\mathcal{H}'') = [A \mid (\mathcal{H}' \ominus \ker(A))]^{-1}(B\mathcal{H}'') \oplus \ker(A)$.) Under these new assumptions we have $T'' \prec T$ and $T' \prec^i T$ so that T' and T'' have property (P) by Theorem 1.9 and Proposition 3.5.31. Now, $T'' \prec T$ implies that $T'' \sim T$, so that the operators $T' \oplus T''$ and $T' \oplus T$ are quasisimilar and have property (P) by Corollary 1.17. Define an operator $X \in \mathcal{I}(T' \oplus T'', T' \oplus T)$ by

$$X(h' \oplus h'') = h' \oplus (Ah' - Bh''), \quad h' \oplus h'' \in \mathcal{H}' \oplus \mathcal{H}''. \tag{6.10}$$

Observe that X is one-to-one. Indeed, $X(h' \oplus h'') = 0$ means $h' = 0$ and $Ah' = Bh''$, so that $h'' = 0$ because B is one-to-one. Proposition 1.21 now implies that X is a lattice-isomorphism, from which we deduce that $X(X^{-1}(\mathcal{H}' \oplus \{0\}))$ is dense in $\mathcal{H}' \oplus \{0\}$. Since

$$\begin{aligned} X(X^{-1}(\mathcal{H}' \oplus \{0\})) &= \{h' \oplus 0 \colon h' \in \mathcal{H}' \quad \text{and} \quad Ah' \in B\mathcal{H}''\} \\ &= A^{-1}(B\mathcal{H}') \oplus \{0\}, \end{aligned} \tag{6.11}$$

we see that (i) is true. The proposition is proved.

6.12. COROLLARY. *Let* T, T', T'', A, *and* B *be as in Proposition 6.9. If* T' *is multiplicity-free then* $A^{-1}(B\mathcal{H}'')$ *contains a cyclic vector of* T'.

PROOF. Define the operator X by (6.10), and denote by P the projection of $\mathcal{H}' \oplus \mathcal{H}$ onto $\mathcal{H}'$. Then (6.11) shows that $A^{-1}(B\mathcal{H}'') = PX(X^{-1}(\mathcal{H}' \oplus \{0\}))$ so that the dense set $A^{-1}(B\mathcal{H}'')$ is the range of the Hilbert space $X^{-1}(\mathcal{H}' \oplus \{0\})$ under a continuous operator. The corollary now follows from Theorem 2.3.7 and Corollary 3.2.7.

We are now ready to prove the additivity of the index for C_0-semi-Fredholm operators.

6.13. THEOREM. *Assume that T, T', and T'' are operators of class C_0, $A \in s\mathcal{F}(T,T')$, and $B \in s\mathcal{F}(T',T'')$. If the sum* $\operatorname{ind}(A) + \operatorname{ind}(B)$ *is defined then $BA \in s\mathcal{F}(T,T'')$ and* $\operatorname{ind}(BA) \sim \operatorname{ind}(A) + \operatorname{ind}(B)$.

PROOF. Assume that T, T', and T'' act on $\mathcal{H}$, $\mathcal{H}'$, and $\mathcal{H}''$, respectively. We note first that

$$(B((A\mathcal{H})^- + \ker(B)))^- = (B(A\mathcal{H})^-)^- = (BA\mathcal{H})^- = (B(B^{-1}(BA\mathcal{H})^-))^-.$$

Since the closed subspaces $((A\mathcal{H})^- + \ker(B))^-$ and $B^{-1}((BA\mathcal{H})^-)$ both contain $\ker(B)$, the fact that B is C_0-semi-Fredholm implies the equality

$$B^{-1}((BA\mathcal{H})^-) = ((A\mathcal{H})^- + \ker(B))^-. \tag{6.14}$$

Denote

$$\mathcal{K} = ((A\mathcal{H})^- + \ker(B))^- \ominus (A\mathcal{H})^- \in \operatorname{Lat}_{1/2}(T'), \tag{6.15}$$

and remark that $T'_{\mathcal{K}}$ necessarily has property (P). Indeed, the assumption that $\operatorname{ind}(A) + \operatorname{ind}(B)$ is defined implies that at least one of the operators $T' \mid \ker(B)$ and $T'_{\ker(A^*)}$ must have property (P). If $T'_{\ker(A^*)}$ has property (P) then $T'_{\mathcal{K}} = T'_{\ker(A^*)} \mid \mathcal{K}$ has property (P) by Corollary 1.17. If $T' \mid \ker(B)$ has property (P) then the operator $Y\colon \ker(B) \to \mathcal{K}$ defined by $Y = P_{\mathcal{K}} \mid \ker(B)$ has dense range and satisfies the relation $T'_{\mathcal{K}}Y = Y(T' \mid \ker(B))$. Hence T' has property (P) because $(T'_{\mathcal{K}})^* \prec^i (T' \mid \ker(B))^*$.

After this preparation, we begin by proving that $BA \mid (\mathcal{H} \ominus \ker(BA))$ is a lattice-isomorphism. To do this it will suffice to show that

$$(BA)_*(\operatorname{Lat}(T)) \supseteq \operatorname{Lat}(T'' \mid (BA\mathcal{H})^-). \tag{6.16}$$

Indeed, this will prove that the map

$$(BA \mid (\mathcal{H} \ominus \ker(BA))_*\colon \operatorname{Lat}(T_{\mathcal{H} \ominus \ker(BA)}) \to \operatorname{Lat}(T'' \mid (BA\mathcal{H})^-) \tag{6.17}$$

is surjective. The same argument applied to $(BA)^* = A^*B^*$ will show that

$$((BA)^* \mid (\mathcal{H}'' \ominus \ker((BA)^*)))_*\colon \operatorname{Lat}((T'' \mid (BA\mathcal{H})^-)^*) \to \operatorname{Lat}(T_{\mathcal{H} \ominus \ker(BA)})$$

is onto. Then an application of Lemma 1.20 implies that the mapping in (6.17) is a bijection.

We now proceed to the proof of (6.16). Assume that $\mathcal{L} \in (BA\mathcal{H})^-$ is an arbitrary invariant subspace for $T'' \mid (BA\mathcal{H})^-$. Since B is C_0-semi-Fredholm, we deduce the existence of a subspace $\mathcal{M} \in \operatorname{Lat}(T)$ such that

$$(B\mathcal{M})^- = \mathcal{L}. \tag{6.18}$$

Of course, (6.14) implies that $\mathcal{M} \subseteq ((A\mathcal{H})^- + \ker(B))^-$. We now apply Proposition 6.9, with T, T', T'', A, and B of that proposition replaced by $T'_{\mathcal{K}}$, $T' \mid \mathcal{M}$, $T' \mid (A\mathcal{H})^-$, $P_{\mathcal{K}} \mid \mathcal{M}$, and $P_{\mathcal{K}} \mid (A\mathcal{M})^-$, respectively. The hypotheses of Proposition 6.9 are easily verified because we showed that $T'_{\mathcal{K}}$ has property (P), and clearly $(P_{\mathcal{K}}(A\mathcal{H})^-)^- = \mathcal{M}$. Thus the subspace $P_{\mathcal{K}}^{-1}(P_{\mathcal{K}}(A\mathcal{H})^-) \cap \mathcal{M}$ is dense in $\mathcal{M}$ and, since $\ker(P_{\mathcal{K}}) = \ker(B)$, this means that

$$\mathcal{M} = (\mathcal{M} \cap ((A\mathcal{H})^- + \ker(B)))^-. \tag{6.19}$$

Let us set $\mathcal{N} = \{h \in (A\mathcal{H})^- : h + h' \in \mathcal{M} \text{ for some } h' \in \ker(B)\}$. Then $\mathcal{N}$ is clearly an invariant linear manifold, $\mathcal{N}^- \subseteq (A\mathcal{H})^-$,

$$B\mathcal{N} \subseteq B\mathcal{M} \subseteq \mathcal{L}, \tag{6.20}$$

and (6.19) can be rewritten as

$$\mathcal{M} \subseteq (\mathcal{N} + \ker(B))^-. \tag{6.21}$$

We now use the fact that A is C_0-semi-Fredholm to find a subspace $\mathcal{R} \in \operatorname{Lat}(T)$ such that $(A\mathcal{R})^- = \mathcal{N}^-$. We then have

$$(BA\mathcal{R})^- = (B(A\mathcal{R})^-)^- = (B\mathcal{N}^-)^- = (B\mathcal{N})^-,$$

and (6.18), (6.20), and (6.21) show that

$$\mathcal{L} = (B\mathcal{M})^- \supseteq (B\mathcal{N})^- = (B(\mathcal{N} + \ker(B))^-)^- \supseteq (B\mathcal{M})^- = \mathcal{L}.$$

We conclude that $(BA\mathcal{R})^- = \mathcal{L}$, and this concludes the proof of (6.16).

To conclude the proof of the theorem we must study the structure of $\ker(BA)$ and $\ker((BA)^*)$. Let us consider the space $\mathcal{K}_1 = (A\mathcal{H})^- \cap \ker(B) \in \operatorname{Lat}(T' \mid (A\mathcal{H})^-)$. We clearly have $\ker(BA) = A^{-1}(\ker(B)) = A^{-1}(\mathcal{K}_1)$ and, because A is a C_0-semi-Fredholm operator, we have $(A(\ker(BA)))^- = \mathcal{K}_1$. It follows that $A \mid \ker(BA) \ominus \ker(A)$ is a quasiaffinity from $\ker(BA) \ominus \ker(A)$ to $\mathcal{K}_1$, and the obvious relation

$$(T' \mid \mathcal{K}_1)(A \mid \ker(BA) \ominus \ker(A)) = (A \mid \ker(BA) \ominus \ker(A))T_{\ker(BA)\ominus\ker(A)}$$

implies that $T' \mid \mathcal{K}_1$ and $T_{\ker(BA)\ominus\ker(A)}$ are quasisimilar. Thus, if

$$T \mid \ker(BA) = \begin{bmatrix} T \mid \ker(A) & X \\ 0 & T_1 \end{bmatrix}$$

is the triangularization of $T \mid \ker(BA)$ with respect to the decomposition $\ker(BA) = \ker(A) \oplus (\ker(BA) \ominus \ker(A))$, we have $T_1 \sim T' \mid \mathcal{K}_1$.

Now assume that $T \mid \ker(A)$ and $T' \mid \ker(B)$ have property (P). Then $T' \mid \mathcal{K}_1 = (T' \mid \ker(B)) \mid \mathcal{K}_1$ also has property (P) by Corollary 1.17. Hence T_1 has property (P) by Corollary 1.16 and finally $T \mid \ker(BA)$ has property (P) by Corollary 1.17. We also note that in this case we have

$$\begin{aligned} \gamma_T(\ker(BA)) &= \gamma_T(\ker(A)) + \gamma_T(\ker(BA) \ominus \ker(A)) \\ &= \gamma_T(\ker(A)) + \gamma_{T'}(\mathcal{K}_1) \end{aligned} \tag{6.22}$$

by Corollary 4.21. Thus we have proved that $BA \in \mathcal{SF}(T, T'')$ if $T \mid \ker(A)$ and $T' \mid \ker(B)$ have property (P).

In an analogous manner, if $T'_{\ker(A^*)}$ and $T''_{\ker(B^*)}$ have property (P), then $T''_{\ker((BA)^*)}$ has property (P), and

$$\gamma_{T''}(\ker((BA)^*)) = \gamma_{T''}(\ker(B^*)) + \gamma_{T'}(\mathcal{K}_1'), \tag{6.23}$$

where $\mathcal{K}_1' = (B^*\mathcal{H}'')^- \cap \ker(A^*) \in \operatorname{Lat}(T'^*)$. Thus we also have $BA \in \mathcal{SF}(T, T'')$ in this case.

The hypothesis that $\text{ind}(A) + \text{ind}(B)$ is defined implies that either $T \mid \ker(A)$ and $T' \mid \ker(B)$ have property (P), or $T'_{\ker(A^*)}$ and $T''_{\ker(B^*)}$ have property (P). Indeed, the only difficulties may arise when both $\text{ind}(A)$ and $\text{ind}(B)$ are infinite. But, if $\text{ind}(A) = \text{ind}(B) = -\infty$ then $T \mid \ker(A)$ and $T' \mid \ker(B)$ have property (P), while $\text{ind}(A) = \text{ind}(B) = +\infty$ implies that $T'_{\ker(A^*)}$ and $T''_{\ker(B^*)}$ have property (P).

Now assume that $\text{ind}(A) = +\infty$, so that $T \mid \ker(A)$ does not have property (P). Then

$$T \mid \ker(BA) = \begin{bmatrix} T \mid \ker(A) & X \\ 0 & T_1 \end{bmatrix}$$

does not have property (P) by Corollary 1.17, and hence $\text{ind}(BA) = +\infty = \text{ind}(A) + \text{ind}(B)$. Analogously, if $\text{ind}(B) = -\infty$, $T''^*_{\ker((BA)^*)}$ does not have property (P), and in that case $\text{ind}(BA) = -\infty = \text{ind}(A) + \text{ind}(B)$. Assume next that $\text{ind}(B) = +\infty$, thus $T' \mid \ker(B)$ does not have property (P) and, necessarily, $T'_{\ker(A^*)}$ has property (P). Then the decomposition

$$T' \mid \ker(B) = \begin{bmatrix} T' \mid \mathscr{K}_1 & Z \\ 0 & T'_{\ker(B) \ominus \mathscr{K}_1} \end{bmatrix}$$

shows that either $T' \mid \mathscr{K}_1$ or $T'_{\ker(B)\ominus\mathscr{K}_1}$ fails to have property (P). Note the relation $T'_{\ker(A^*)}(P_{\ker(A^*)} \mid \ker(B) \ominus \mathscr{K}_1) = (P_{\ker(A^*)} \mid \ker(B) \ominus \mathscr{K}_1)T'_{\ker(B)\ominus\mathscr{K}_1}$, which follows from the fact that $P_{\ker(A^*)}$ is zero on $\ker(B) \ominus \mathscr{K}_1$. Moreover, $P_{\ker(A^*)} \mid \ker(B) \ominus \mathscr{K}_1$ is one-to-one, thus yielding the relations $T'_{\ker(B)\ominus\mathscr{K}_1} \prec^i T'_{\ker(A^*)}$. This implies, as usual, that $T'_{\ker(B)\ominus\mathscr{K}_1}$ has property (P) and henceforth that $T' \mid \mathscr{K}_1$ does not have property (P). We conclude that $T_1 \sim T' \mid \mathscr{K}_1$ fails to have property (P), thus $T' \mid \ker(BA)$ fails to have property (P), and $\text{ind}(BA) = +\infty = \text{ind}(A) + \text{ind}(B)$. The relation $\text{ind}(BA) = -\infty = \text{ind}(A) + \text{ind}(B)$ is proved analogously in case $\text{ind}(A) = -\infty$. Thus we have $\text{ind}(BA) = \text{ind}(A) + \text{ind}(B)$ whenever one of the indices $\text{ind}(A)$ and $\text{ind}(B)$ is infinite.

We conclude by treating the case in which both $\text{ind}(A)$ and $\text{ind}(B)$ belong to $\tilde{\Gamma} \times \tilde{\Gamma}$. Introduce the spaces

$$\mathscr{K}_2 = \ker(B) \ominus \mathscr{K}_1 = \ker(B) \ominus ((A\mathscr{H})^- \cap \ker(B)) \in \text{Lat}_{1/2}(T')$$

and

$$\mathscr{K}_2' = \ker(A^*) \ominus \mathscr{K}_1' = \ker(A^*) \ominus ((B^*\mathscr{H}'')^- \cap \ker(A^*)) \in \text{Lat}_{1/2}(T').$$

Note the obvious equalities

$$T_{\ker(A^*)}(P_{\ker(A^*)} \mid \ker(B)) = (P_{\ker(A^*)} \mid \ker(B))T \mid \ker(B),$$

$$\mathscr{K}_1 = \ker(P_{\ker(A^*)} \mid \ker(B)),$$

and

$$\mathscr{K}_1' = \ker(P_{\ker(B)} \mid \ker(A^*)) = \ker(P_{\ker(A^*)} \mid \ker(B))^*.$$

These equalities clearly imply that $P_{\ker(A^*)} \mid \mathscr{K}_2$ is one-to-one, has dense range in $\mathscr{K}_2'$, and $T'_{\mathscr{K}_2'}(P_{\mathscr{K}_2'} \mid \mathscr{K}_2) = (P_{\mathscr{K}_2'} \mid \mathscr{K}_2)T'_{\mathscr{K}_2}$. Thus the operators $T'_{\mathscr{K}_2'}$ and

$T'_{\mathscr{K}_2}$ are quasisimilar and, in particular, $\gamma_{T'}(\mathscr{K}_2') = \gamma_{T'}(\mathscr{K}_2)$. We denote by $\gamma = \gamma_{T'}(\mathscr{K}_2') = \gamma_{T'}(\mathscr{K}_2)$ the common value. The equalities

$$\gamma_{T'}(\ker(B)) = \gamma_{T'}(\mathscr{K}_1) + \gamma_{T'}(\mathscr{K}_2) = \gamma_{T'}(\mathscr{K}_1) + \gamma$$

and

$$\gamma_{T'}(\ker(A^*)) = \gamma_{T'}(\mathscr{K}_1') + \gamma_{T'}(\mathscr{K}_2') = \gamma_{T'}(\mathscr{K}_1') + \gamma$$

can now be combined with (6.22) and (6.23) to yield

$$\gamma_T(\ker(BA)) + \gamma = \gamma_T(\ker(A)) + \gamma_{T'}(\ker(B)),$$
$$\gamma + \gamma_{T''}(\ker((BA)^*)) = \gamma_{T'}(\ker(A^*)) + \gamma_{T''}(\ker(B^*)),$$

and hence

$$\gamma_T(\ker(BA)) + \gamma_{T'}(\ker(A^*)) + \gamma_{T''}(\ker(B^*)) + \gamma$$
$$= \gamma_{T''}(\ker((BA)^*)) + \gamma_T(\ker(A)) + \gamma_{T'}(\ker(B)) + \gamma.$$

Since clearly $\gamma \leq \gamma_{T'}(\ker(B)) \wedge \gamma_{T'}(\ker(A^*))$, Lemma 6.2(ii) shows that we can cancel γ and obtain

$$\gamma_T(\ker(BA)) + \gamma_{T'}(\ker(A^*)) + \gamma_{T''}(\ker(B^*))$$
$$= \gamma_{T''}(\ker((BA)^*)) + \gamma_T(\ker(A)) + \gamma_{T'}(\ker(B$$

or, equivalently, $\operatorname{ind}(BA) \sim \operatorname{ind}(A) + \operatorname{ind}(B)$. The proof is now complete.

We conclude this section with some partial converses to Theorem 6.13. The basic step is the following lemma.

6.24. LEMMA. *Assume that T, T', and T'' are operators of class C_0 acting on $\mathscr{H}$, $\mathscr{H}'$, and $\mathscr{H}''$, respectively. If $A \in \mathscr{I}(T, T')$ and $B \in \mathscr{I}(T', T'')$ are such that $(BA)_*$ is onto* $\operatorname{Lat}(T'' \mid (BA\mathscr{H})^-)$, *and if $T''_{(B\mathscr{H}')^- \ominus (BA\mathscr{H})^-}$ has property (P), then B_* is onto* $\operatorname{Lat}(T'' \mid (B\mathscr{H}')^-)$.

PROOF. It clearly suffices to show that every cyclic space for T'', contained in $(B\mathscr{H}')^-$, belongs to the range of B_*. Assume therefore that $\mathscr{L} \subseteq (B\mathscr{H}')^-$ is a cyclic space for T'' (i.e., $T'' \mid \mathscr{L}$ has a cyclic vector) and denote $Q = P_{(B\mathscr{H}')^- \ominus (BA\mathscr{H})^-} \mid (B\mathscr{H}')^-$; of course, $Q \in \mathscr{I}(T'' \mid (B\mathscr{H}')^-, T''_{(B\mathscr{H}')^- \ominus (BA\mathscr{H})^-})$. The operator QB has dense range, in particular $(QB\mathscr{H}')^- \supseteq Q\mathscr{L}$. An application of Corollary 6.12 (with A and B of that corollary replaced by Q and QB, respectively) shows the existence of a cyclic vector g for $T'' \mid \mathscr{L}$ such that $Qg \in QB\mathscr{H}'$. That is, there exists $h \in \mathscr{H}'$ such that $Q(g - Bh) = 0$ or, equivalently, $g - Bh \in (BA\mathscr{H})^-$. Now denote $\mathscr{M} = \bigvee_{n \geq 0} T''^n (g - Bh)$. Since $\mathscr{M} \subseteq (BA\mathscr{H})^-$, the assumption about $(BA)_*$ implies the existence of a subspace $\mathscr{N} \in \operatorname{Lat}(T)$ such that $(BA\mathscr{N})^- = \mathscr{M}$. If we now set $\mathscr{R} = (A\mathscr{N})^- \vee (\bigvee_{n \geq 0} T'^n h)$ then $(B\mathscr{R})^- = \mathscr{M} \vee (\bigvee_{n \geq 0} T''^n Bh)$ coincides with the invariant subspace for T'' generated by $g - Bh$ and Bh. Consequently we have $\mathscr{L} \subseteq (B\mathscr{R})^-$, and $T'' \mid (B\mathscr{R})^-$ has property (P). We can apply Proposition 6.9, with A and B of that proposition replaced by $I_{\mathscr{L}}$ and $B \mid \mathscr{R} : \mathscr{R} \to (B\mathscr{R})^-$, respectively. We infer the existence of a subspace $\mathscr{S} \subseteq \mathscr{R}$, $\mathscr{S} \in \operatorname{Lat}(T')$, such that $(B\mathscr{S})^- = \mathscr{L}$, and thus $\mathscr{L} = B_*(\mathscr{S})$. The lemma is proved.

6.25. COROLLARY. *Assume that T, T', and T'' are operators of class C_0 acting on $\mathscr{H}$, $\mathscr{H}'$, and $\mathscr{H}''$, respectively. If $A \in \mathscr{I}(T,T')$ and $B \in \mathscr{I}(T',T'')$ are such that $(BA \mid (\mathscr{H} \ominus \ker(BA)))_*$ is one-to-one, and if $T_{(A^*\mathscr{H})^- \ominus (A^*B^*\mathscr{H}'')^-}$ has property (P), then $(A \mid (\mathscr{H} \ominus \ker(A)))_*$ is one-to-one.*

PROOF. Easily follows from Lemmas 1.20 and 6.24.

6.26. PROPOSITION. *Assume that T, T', T'', and T''' are operators of class C_0, and $A \in \mathscr{I}(T,T')$, $B \in \mathscr{I}(T',T'')$, $C \in \mathscr{I}(T'',T''')$.*

(i) *If CB and BA are C_0-Fredholm, then B is C_0-Fredholm.*

(ii) *If CB and BA are C_0-semi-Fredholm, $\operatorname{ind}(CB) = \operatorname{ind}(BA) = -\infty$, and $T'_{\ker(A^*)}$ has property (P), then B is C_0-semi-Fredholm.*

(iii) *If CB and BA are C_0-semi-Fredholm, $\operatorname{ind}(CB) = \operatorname{ind}(BA) = +\infty$, and $T'' \mid \ker(C)$ has property (P), then B is C_0-semi-Fredholm.*

PROOF. Assume that T, T', T'', and T''' act on $\mathscr{H}$, $\mathscr{H}'$, $\mathscr{H}''$, and $\mathscr{H}'''$, respectively. The obvious inclusions $\ker(B) \subseteq \ker(BA)$ and $\ker(B^*) \subseteq \ker((CB)^*)$ show that we need only to prove that $B \mid (\mathscr{H}' \ominus \ker(B))$ is a lattice-isomorphism. By virtue of Lemma 6.25, it suffices to show that $T''_{(B\mathscr{H}')^- \ominus (BA\mathscr{H})^-}$ and $T'_{(B^*\mathscr{H}'')^- \ominus (B^*C^*\mathscr{H}''')^-}$ have property (P). We deal only with the first of these two operators because the treatment of the second is very similar. In cases (i) and (iii) the operator $T''_{\mathscr{H}'' \ominus (BA\mathscr{H})^-}$ has property (P), and hence $T''_{(B\mathscr{H}')^- \ominus (BA\mathscr{H})^-}$ has property (P) because $(B\mathscr{H}')^- \ominus (BA\mathscr{H})^- \in \operatorname{Lat}(T''_{\mathscr{H}'' \ominus (BA\mathscr{H})^-})$. In case (ii) we note that the operator $S \in \mathscr{I}(T'_{\mathscr{H}' \ominus (A\mathscr{H})^-}, T''_{(B\mathscr{H}')^- \ominus (BA\mathscr{H})^-})$ defined by $S = P_{(B\mathscr{H}')^- \ominus (BA\mathscr{H})^-} B \mid (\mathscr{H}' \ominus (A\mathscr{H})^-)$ has dense range. We conclude that $T''_{(B\mathscr{H}')^- \ominus (BA\mathscr{H})^-}$ is quasisimilar to some compression of $T'_{\mathscr{H}' \ominus (A\mathscr{H})^-} = T'_{\ker(A^*)}$, and hence it has property (P) by Corollary 1.17. This concludes the proof of our proposition.

Exercises

1. Show that "$\sim$" is an equivalence relation when restricted to $\Gamma_0 \times \Gamma_0$.

2. Assume that T and T' are operators of class C_0, both having property (P), and $X \in \mathscr{I}(T',T)$. If $\gamma_T = \gamma_{T'}$, is it always true that $\operatorname{ind}(X) \sim (0,0)$?

3. Let T be an operator of class C_0 and $X \in \mathscr{F}(T)$. Assume that $\ker(X) = \{0\}$ and X has closed range. Does X always have a left inverse in $\{T\}'$?

4. Answer the question in Exercise 3 under the assumption that T is an algebraic operator.

5. Let T be an algebraic operator and $X \in s\mathscr{F}(T)$. Show that X is a classical semi-Fredholm operator, i.e., X has closed range and either $\ker(X)$ or $\ker(X^*)$ is finite-dimensional.

6. Let T be an operator of class C_0 and $u \in K_T^\infty$, i.e., $u \in H^\infty$ and $u \wedge m_T \equiv 1$. Show that $u(T) \in \mathscr{F}(T)$ and $\operatorname{ind}(u(T)) = (0,0)$.

7. Construct Blaschke products b, θ, and $\{b_n : n \geq 1\}$ with the following properties: $b \wedge \theta \equiv 1$, $b_n \wedge \theta \not\equiv 1$, $n \geq 1$, and $\lim_{n\to\infty} \|b - b_n\|_\infty = 0$.

8. Let b, θ, and $\{b_n : n \geq 1\}$ be as in Exercise 7, and set $T = S(\theta)^{(\omega)}$. Use the operators $b(T)$ and $b_n(T)$ to show that $\mathscr{F}(T)$ and $s\mathscr{F}(T)$ are not open subsets of $\{T\}'$.

9. Verify that Corollary 6.25 really follows from Lemmas 1.20 and 6.24.

10. Verify that, under the conditions of Proposition 6.26, the operator $T'_{(B^*\mathscr{H}'')^- \ominus (B^*C^*\mathscr{H}''')^-}$ has property (P).

7. Perturbation theorems. The bicommutant. We start with a perturbation theorem which corresponds, in the classical case, with the invariance of Fredholm operators under finite-rank perturbations.

7.1. THEOREM. *Let $T \in \mathscr{L}(\mathscr{H})$ and $T' \in \mathscr{L}(\mathscr{H}')$ be operators of class C_0, and X, $Y \in \mathscr{I}(T,T')$. If $T \in s\mathscr{F}(T,T')$, and if $T' \mid (Y\mathscr{H})^-$ has property (P), then $X + Y \in s\mathscr{F}(T,T')$ and*

$$\operatorname{ind}(X+Y) \sim \operatorname{ind}(X) + (\gamma_{T'}((Y\mathscr{H})^-), \gamma_{T'}((Y\mathscr{H})^-)).$$

PROOF. The restriction $Y \mid (\mathscr{H} \ominus \ker(Y))$ is a quasiaffinity from $\mathscr{H} \ominus \ker(Y)$ to $(Y\mathscr{H})^-$ and, in addition,

$$(T' \mid (Y\mathscr{H})^-)(Y \mid (\mathscr{H} \ominus \ker(Y))) = (Y \mid (\mathscr{H} \ominus \ker(Y)))T_{\mathscr{H}\ominus\ker(Y)}.$$

We conclude that $T_{\mathscr{H}\ominus\ker(Y)}$ is quasisimilar to $T' \mid (Y\mathscr{H})^-$, and hence it has property (P). Let us set $\mathscr{H}_0 = \ker(Y)$, $\mathscr{H}_0' = \mathscr{H}' \ominus (Y\mathscr{H})^-$, and denote by $J \in \mathscr{L}(\mathscr{H}_0, \mathscr{H})$ and $Q \in \mathscr{L}(\mathscr{H}', \mathscr{H}_0')$ the inclusion and projection operators, respectively. The obvious relations $TJ = J(T \mid \mathscr{H}_0)$ and $T'_{\mathscr{H}_0'}Q = QT'$, combined with the fact that $T' \mid (Y\mathscr{H})^-$ and $T_{\mathscr{H}\ominus\ker(Y)}$ have property (P), show that J and Q are C_0-Fredholm operators. We note that

$$\operatorname{ind}(J) = (0,\delta), \qquad \operatorname{ind}(Q) = (\delta, 0), \tag{7.2}$$

where $\delta = \gamma_{T'}((Y\mathscr{H})^-) = \gamma_T(\mathscr{H} \ominus \ker(Y))$. Observe next that $YJ = 0$, $QY = 0$, and hence the operators

$$(X+Y)J = XJ \quad \text{and} \quad Q(X+Y) = QX \tag{7.3}$$

are C_0-semi-Fredholm by Theorem 6.13. An application of Proposition 6.26, with A, B, C of that proposition replaced by J, $X+Y$, Q, respectively, shows that $X+Y$ is a C_0-semi-Fredholm operator. It remains to prove that the index formula holds. If $\operatorname{ind}(X) = \pm\infty$ then $\operatorname{ind}((X+Y)J) = \pm\infty$ and hence $\operatorname{ind}(X+Y) = \pm\infty$ because J is C_0-Fredholm. Finally, let us consider the case in which X is C_0-Fredholm. Then Proposition 6.26 shows that $X+Y$ is also C_0-Fredholm, and Theorem 6.13 gives

$$\operatorname{ind}(XJ) \sim \operatorname{ind}(X+Y) + (0,\delta), \qquad \operatorname{ind}(XJ) \sim \operatorname{ind}(X) + (0,\delta), \tag{7.4}$$

by virtue of (7.2) and (7.3). Denoting $\operatorname{ind}(XJ) = (\gamma_1, \gamma_2)$, relations (7.4) can be rewritten as

$$\gamma_1 + \delta + \gamma_{T'}(\ker((X+Y)^*)) = \gamma_2 + \gamma_T(\ker(X+Y))$$

and

$$\gamma_2 + \gamma_T(\ker(X)) = \gamma_1 + \delta + \gamma_{T'}(\ker(X^*)),$$

respectively. Adding these two relations we get

$$\begin{aligned}\gamma_1 + \gamma_2 + \delta + \gamma_T(\ker(X)) + \gamma_{T'}(\ker((X+Y)^*)) \\ = \gamma_1 + \gamma_2 + \delta + \gamma_{T'}(\ker(X^*)) + \gamma_T(\ker(X+Y)),\end{aligned} \tag{7.5}$$

and to conclude the proof it suffices to show that γ_1 and γ_2 can be cancelled in (7.5). By Lemma 6.2, it suffices to show that

$$\begin{aligned}\gamma_1 + \gamma_2 \le (\delta + \gamma_T(\ker(X)) + \gamma_{T'}(\ker((X+Y)^*))) \\ \wedge (\delta + \gamma_{T'}(\ker(X^*)) + \gamma_T(\ker(X+Y))).\end{aligned} \tag{7.6}$$

It was seen in the proof of Theorem 6.13 that $\gamma(\ker(BA)) \le \gamma(\ker(B)) + \gamma(\ker(A))$. By virtue of (7.3) we get $\gamma_1 = \gamma_T(\ker(XJ)) \le \gamma_T(\ker(X))$, $\gamma_1 = \gamma_T(\ker((X+Y)J)) \le \gamma_T(\ker(X+Y))$, $\gamma_2 = \gamma_T(\ker(J^*X^*)) \le \delta + \gamma_{T'}(\ker(X^*))$, and $\gamma_2 = \gamma_T(\ker((J^*(X+Y)^*))) \le \delta + \gamma_{T'}(\ker((X+Y)^*))$. Inequality (7.6) clearly follows from these relations. The theorem is proved.

If the operator X is invertible, then the index relation in Theorem 7.1 can be replaced by $\operatorname{ind}(X+Y) \sim \operatorname{ind}(X) = (0,0)$. This easily follows from the following particular case.

7.7. PROPOSITION. *Assume that $T \in \mathscr{L}(\mathscr{H})$ is an operator of class C_0, $Y \in \{T\}'$, and $T \mid (Y\mathscr{H})^-$ has property (P). Then $I+Y \in \mathscr{F}(T)$ and* $\operatorname{ind}(Y) \sim (0,0)$.

PROOF. We already know from Theorem 7.1 that $I+Y$ is C_0-Fredholm. We will show that $(T \mid \ker(I+Y))\rho T_{\ker((I+Y)^*)}$, and this clearly implies the index relation. Denote $\mathscr{H}' = (Y\mathscr{H})^-$, $T' = T \mid \mathscr{H}'$, and $Y' = Y \mid \mathscr{H}'$. The assumption that T' has property (P) implies that $(T' \mid \ker(I+Y'))\rho T'_{\ker((I+Y')^*)}$, so that we can conclude the proof by showing that $T' \mid \ker(I+Y') = T \mid \ker(I+Y)$, while $T_{\ker((I+Y)^*)}$ is similar to $T'_{\ker((I+Y')^*)}$. To do this we first note that for $h \in \ker(I+Y)$ we have $h = -Yh \in \mathscr{H}'$, and hence $\ker(I+Y) = \ker(I+Y')$. Consequently $T \mid \ker(I+Y) = T \mid \ker(I+Y') = T' \mid \ker(I+Y')$, as desired. Finally, note that the relations

$$(I+Y')^* P_{\mathscr{H}'} = P_{\mathscr{H}'}(I+Y)^* \quad \text{and} \quad T'^* P_{\mathscr{H}'} = P_{\mathscr{H}'} T^*$$

show that $P_{\mathscr{H}'} \ker((I+Y)^*) \subseteq \ker((I+Y')^*)$ and, moreover,

$$\begin{aligned}(T'^* \mid \ker((I+Y')^*))(P_{\mathscr{H}'} \mid \ker((I+Y)^*)) \\ = (P_{\mathscr{H}'} \mid \ker((I+Y)^*))(T^* \mid \ker((I+Y)^*)).\end{aligned} \tag{7.8}$$

In addition, the reader will have no difficulty verifying that the operator $R\colon \ker((I+Y')^*) \to \ker((I+Y)^*)$ defined by $Ru = u - Y^*u$, $u \in \ker((I+Y')^*)$, is an inverse to $P_{\mathscr{H}'} \mid \ker((I+Y)^*) \colon \ker((I+Y)^*) \to \ker((I+Y')^*)$. Thus (7.8) shows that $T'_{\ker((I+Y')^*)}$ and $T_{\ker((I+Y)^*)}$ are similar. The proposition follows.

We proceed now to the study of C_0-Fredholm operators in the bicommutant $\{T\}''$ of a given operator T of class C_0. We recall the fact, from Theorem 4.1.2, that every $X \in \{T\}''$ can be written under the form $X = (u/v)(T)$ with u, $v \in H^\infty$ and $v \wedge m_T \equiv 1$.

7.9. LEMMA. *Let T be an operator of class C_0 and $X \in \{T\}''$. If u and v are in H^∞, $v \wedge m_T \equiv 1$, and $X = (u/v)(T)$, then $X_* = (u(T))_*$, i.e., $(X\mathscr{M})^- = (u(T)\mathscr{M})^-$ for every $\mathscr{M} \in \operatorname{Lat}(T)$. In particular, $\ker(X) = \ker(u(T))$ and $\ker(X^*) = \ker(u(T)^*)$.*

PROOF. Assume that T is acting on $\mathscr{H}$. If we know that $X_* = u(T)_*$ then, in particular, $(X\mathscr{H})^- = (u(T)\mathscr{H})^-$ and hence $\ker(X^*) = \ker(u(T)^*)$. We must also have $u(T)_*(\ker(X)) = \{0\}$, whence $\ker(X) \subseteq \ker(u(T))$, and therefore $\ker(X) = \ker(u(T))$ by a symmetrical argument.

To prove the equality $X_* = u(T)_*$, take an arbitrary $\mathscr{M}$ in $\operatorname{Lat}(T)$. Then $(X\mathscr{M})^- \in \operatorname{Lat}(T)$, and $v(T) \mid (X\mathscr{M})^-$ is a quasiaffinity. Indeed, $m_{T|(X\mathscr{M})^-}$ divides m_T and hence $v \wedge m_{T|(X\mathscr{M})^-} \equiv 1$. Therefore we have

$$
\begin{aligned}
u(T)_*(\mathscr{M}) = (u(T)\mathscr{M})^- &= (v(T)X\mathscr{M})^- \\
&= (v(T)(X\mathscr{M})^-)^- = (X\mathscr{M})^- = X_*(\mathscr{M}),
\end{aligned}
$$

as required. The lemma follows.

7.10. COROLLARY. *Under the conditions of Lemma 7.9, X is C_0-semi-Fredholm if and only if $u(T)$ is C_0-semi-Fredholm.*

PROOF. We have to show that $X \mid \mathscr{H} \ominus \ker(X)$ is a lattice-isomorphism if and only if $u(T) \mid \mathscr{H} \ominus \ker(u(T))$ is a lattice-isomorphism. But, for $\mathscr{M} \in \operatorname{Lat}(T_{\mathscr{H}\ominus\ker(X)})$ we have $\mathscr{M} + \ker(X) \in \operatorname{Lat}(T)$, and therefore

$$
\begin{aligned}
(X\mathscr{M})^- &= (X(\mathscr{M} + \ker(X)))^- \\
&= (u(T)(\mathscr{M} + \ker(X)))^- \\
&= (u(T)(\mathscr{M} + \ker(u(T))))^- \\
&= (u(T)\mathscr{M})^-,
\end{aligned}
$$

and this implies the desired conclusion.

7.11. LEMMA. *Assume that T and T' are two quasisimilar operators of class C_0. Then $T \mid \ker(u(T))$ and $T' \mid \ker(u(T'))$ are quasisimilar for every $u \in H^\infty$.*

PROOF. Let X and Y be quasiaffinities satisfying $T'X = XT$ and $TY = YT'$. Then we must also have $u(T')X = Xu(T)$ and $u(T)Y = Yu(T')$ for $u \in H^\infty$, from which we deduce the inclusions $X\ker(u(T)) \subseteq \ker(u(T'))$ and $Y\ker(u(T')) \subseteq \ker(u(T))$. But these inclusions clearly imply that $T \mid \ker(u(T)) \prec^i T' \mid \ker(u(T'))$ and $T' \mid \ker(u(T')) \prec^i T \mid \ker(u(T))$, and thus $T \mid \ker(u(T)) \sim T' \mid \ker(u(T'))$ by Proposition 3.5.32. The lemma is proved.

7.12. COROLLARY. *For every operator T of class C_0 and every X in $\{T\}''$ the operators $T \mid \ker(X)$ and $T_{\ker(X^*)}$ are quasisimilar.*

PROOF. Lemma 7.9 shows that we may assume that $X = u(T)$ for some $u \in H^\infty$. Furthermore, by virtue of Lemma 7.11, we may replace T by its Jordan model $\bigoplus_\alpha S(\theta_\alpha)$. The corollary follows now, because both $S(\theta_\alpha) \mid \ker(u(S(\theta_\alpha)))$ and $S(\theta_\alpha)_{\ker(u(S(\theta_\alpha))^*)}$ are unitarily equivalent to $S(\theta_\alpha \wedge u)$ by Corollary 3.1.12.

7.13. COROLLARY. *For every operator T of class C_0 we have $s\mathscr{F}(T) \cap \{T\}'' = \mathscr{F}(T) \cap \{T\}''$ and $\operatorname{ind}(X) \sim (0,0)$ for $X \in \mathscr{F}(T) \cap \{T\}''$.*

PROOF. Follows immediately from the preceding corollary.

In order to study the C_0-Fredholm operators of the form $u(T)$ we need an additional arithmetical property of H^∞.

7.14. DEFINITION. Let ϕ and ψ be two inner functions in H^∞. We say that ϕ is *absolutely continuous with respect to* ψ, and we write $\phi \ll \psi$, if $\psi \wedge \theta \equiv 1$ implies $\phi \wedge \theta \equiv 1$ for every inner function θ.

7.15. PROPOSITION. *Assume that ϕ and ψ are two inner functions in H^∞. There exist inner functions ϕ_{ac} and ϕ_s such that $\phi = \phi_{ac}\phi_s$, $\phi_s \wedge \psi \equiv 1$, and $\phi_{ac} \ll \psi$. Moreover, ϕ_{ac} and ϕ_s are uniquely determined up to a scalar factor of absolute value one.*

PROOF. Write $\gamma(\phi) = (\mu_0, \nu_0)$ and $\gamma(\psi) = (\mu_1, \nu_1)$ (cf. §4). The Lebesgue decomposition theorem allows us to write $\nu_0 = \nu_{ac} + \nu_s$ with $\nu_{ac} \ll \nu_1$ and $\nu_s \wedge \nu_1 = 0$. Furthermore, we can write $\mu_0 = \mu_{ac} + \mu_s$, where $\mu_{ac}(\lambda) = \mu_0(\lambda)$ if $\mu_1(\lambda) \neq 0$, and $\mu_{ac}(\lambda) = 0$ otherwise. We clearly can find inner functions ϕ_{ac} and ϕ_s such that $\gamma(\phi_{ac}) = (\mu_{ac}, \nu_{ac})$, $\gamma(\phi_s) = (\mu_s, \nu_s)$, and $\phi = \phi_{ac}\phi_s$. The relation $\phi_s \wedge \psi \equiv 1$ is obvious, so let us prove that $\phi_{ac} \ll \psi$. If θ is an arbitrary inner function and $\gamma(\theta) = (\mu, \nu)$, the relation $\theta \wedge \psi \equiv 1$ is equivalent to $\nu \wedge \nu_1 = 0$ and $\mu \wedge \mu_1 = 0$. This means that μ and μ_1 have disjoint supports, and ν and ν_1 are also supported by disjoint Borel sets. Since the support of μ_{ac} is contained in the support of μ_1, and any support set for ν_1 also serves as a support set for ν_{ac}, the relations $\nu \wedge \nu_{ac} = 0$ and $\mu \wedge \mu_{ac} = 0$ follow at once. Therefore $\theta \wedge \phi_{ac} \equiv 1$, as desired.

Assume that $\phi = \phi'_{ac}\phi'_s$ is a second decomposition such that $\phi'_{ac} \ll \psi$ and $\phi'_s \wedge \psi \equiv 1$. The condition $\phi'_s \wedge \psi \equiv 1$ implies that $\phi'_s \wedge \phi_{ac} \equiv 1$ because $\phi_{ac} \ll \psi$. We deduce that $\phi'_s \vee \phi_{ac} \equiv \phi'_s\phi_{ac}$ and, since $\phi'_s \vee \phi_{ac}$ must divide ϕ, we have $\phi'_s\phi_{ac} \mid \phi = \phi_s\phi_{ac}$. Therefore $\phi'_s \mid \phi_s$, and a symmetrical argument shows that in fact $\phi'_s \equiv \phi_s$. The uniqueness assertion, and hence the proposition, is proved.

We are now ready for the main result on $\mathscr{F}(T) \cap \{T\}''$.

7.16. THEOREM. *Assume that T is an operator of class C_0 and $X \in \{T\}''$. Then X is C_0-Fredholm if and only if $T \mid \ker(X)$ has property (P).*

PROOF. The necessity of the condition that $T \mid \ker(X)$ have property (P) is obvious, so let us assume that $T \mid \ker(X)$ has property (P). Let $T \in \mathscr{L}(\mathscr{H})$, and let $T' = \bigoplus_\alpha S(\theta_\alpha)$ be the Jordan model of T. Let $u,\ v \in H^\infty$ be such that $v \wedge m_T \equiv 1$ and $X = (u/v)(T)$. We know from Lemma 7.9 and Corollary 7.10 that $\ker(X) = \ker(u(T))$ and $X \in \mathscr{F}(T)$ if and only if $u(T) \in \mathscr{F}(T)$. It suffices therefore to prove that $u(T)$ is in $\mathscr{F}(T)$. By Lemma 7.11, the assumption that $T \mid \ker(u(T))$ has property (P) is equivalent to saying that $T' \mid \ker(u(T'))$ has property (P). As noted in the proof of Corollary 7.12, the restriction $T' \mid \ker(u(T'))$ is unitarily equivalent to $\bigoplus_\alpha S(\theta_\alpha \wedge u)$ and since this last operator is clearly (unitarily equivalent to) a Jordan operator, we deduce that $T \mid \ker(u(T))$ has property (P) if and only if

$$\bigwedge_{\alpha<\omega} (\theta_\alpha \wedge u) \equiv \left(\bigwedge_{\alpha<\omega} \theta_\alpha \right) \wedge u \equiv 1. \tag{7.17}$$

Let us decompose $m_T = m_{ac}m_s$ such that $m_{ac} \ll u$ and $m_s \wedge u \equiv 1$. Then (7.17) implies that

$$1 \equiv \left(\bigwedge_{\alpha<\omega} \theta_\alpha \right) \wedge m_{ac} \equiv \bigwedge_{\alpha<\omega} (\theta_\alpha \wedge m_{ac}),$$

and by the above argument (with m_{ac} in place of u) we deduce that $T \mid \ker(m_{ac}(T))$ has property (P). By Corollary 7.12, $T_{\ker(m_{ac}(T)^*)}$ must also have property (P). Furthermore, the relation $m_{ac}(T)m_s(T) = 0$ implies that $(m_s(T)\mathscr{H})^- \subseteq \ker(m_{ac}(T))$, and hence $T \mid (m_s(T)\mathscr{H})^-$ must have property (P). In an analogous manner, $T_{\mathscr{H} \ominus \ker(m_s(T))}$ is also seen to have property (P).

After this preparation, we are able to apply an argument similar to that in the proof of Theorem 7.1. We set $\mathscr{H}_0 = \ker(m_s(T))$, $\mathscr{H}_1 = \mathscr{H} \ominus \ker(m_{ac}(T))$, $T_0 = T \mid \mathscr{H}_0$, $T_1 = T_{\mathscr{H}_1}$, and denote by $J \in \mathscr{L}(\mathscr{H}_0, \mathscr{H})$ and $Q \in \mathscr{L}(\mathscr{H}, \mathscr{H}_1)$ the inclusion and projection operators, respectively. The relations $TJ = JT_0$ and $T_1Q = QT$ are obvious, and J and Q are clearly C_0-Fredholm operators. Indeed, they have closed ranges, $\ker(J) = \{0\}$, $\ker(J^*) = \mathscr{H} \ominus \ker(m_s(T))$, $\ker(Q) = \ker(m_{ac}(T))$, $\ker(Q^*) = \{0\}$, and it was noted before that $T \mid \ker(m_{ac}(T))$ and $T_{\mathscr{H} \ominus \ker(m_s(T))}$ have property (P). By Proposition 6.26, with A, B, C of that proposition replaced by J, $u(T)$, Q, respectively, we can conclude the proof by showing that $u(T)J$ and $Qu(T)$ are C_0-Fredholm. Now, we clearly have $u(T)J = Ju(T_0)$, $Qu(T) = u(T_1)Q$, so that by Theorem 6.13 it suffices to show that $u(T_0)$ and $u(T_1)$ are C_0-Fredholm; indeed, we have just shown that J and Q are C_0-Fredholm. Now, clearly m_{T_0} and m_{T_1} divide m_s; in fact $m_{T_0} \equiv m_{T_1} \equiv m_s$ by Proposition 2.4.4. Thus $u \wedge m_{T_0} \equiv u \wedge m_{T_1} \equiv 1$, hence it suffices to show that $u(S)$ is C_0-Fredholm whenever S is of class C_0 and $u \wedge m_s \equiv 1$. This last statement is easy to verify. If $\mathscr{M} \in \operatorname{Lat}(S)$ then $u \wedge m_{S|\mathscr{M}} \equiv 1$, hence $u(S \mid \mathscr{M})$ is a quasiaffinity, and therefore

$$u(S)_*(\mathscr{M}) = (u(S)\mathscr{M})^- = (u(S \mid \mathscr{M})\mathscr{M})^- = \mathscr{M}.$$

Thus $u(S)_*$ is the identity on $\operatorname{Lat}(S)$ and clearly $\ker(u(S)) = \ker(u(S)^*) = \{0\}$. The proof is now complete.

We conclude by showing that condition (i) in Definition 6.1 cannot generally be replaced by the condition that X have closed range, even for operators in $\{T\}''$.

7.18. PROPOSITION. *For every nonalgebraic operator T of class C_0 there exist operators $X \in \mathscr{F}(T) \cap \{T\}''$ that do not have closed range.*

PROOF. By Corollary 4.1.6, there exists a function $v \in H^\infty$ such that $v \wedge m_T \equiv 1$, and every X in $\{T\}''$ can be written as $X = (u/v)(T)$ for some $u \in H^\infty$. Now choose an inner function $p \in H^\infty$ such that $p \wedge m_T \equiv 1$ but

$$\inf\{|p(\lambda)| + |m_T(\lambda)| : |\lambda| < 1\} = 0. \tag{7.19}$$

This can be done because m_T is not a finite Blaschke product. Indeed, it suffices to define p as a Blaschke product whose zeros $\{\alpha_n : n \geq 0\}$ satisfy the conditions $m_T(\alpha_n) \neq 0$, $n \geq 0$, and $\lim_{n\to\infty} m_T(\alpha_n) = 0$.

Now define $X = (pv)(T)$ and note that X is a quasiaffinity since $(pv) \wedge m_T \equiv 1$. Thus $X \in \mathscr{F}(T)$ by Theorem 7.16. Assume that X has closed range, so that $X^{-1} \in \{T\}''$, and write $X^{-1} = (u/v)(T)$ for some $u \in H^\infty$. Equivalently, $(pv)(T)u(T) = v(T)$ or $v(T)[(pu)(T) - I] = 0$, which implies $(pu - 1)(T) = 0$ because $v(T)$ is a quasiaffinity. Thus we see that m_T divides $pu - 1$, say $pu - 1 = m_T g$ for some $g \in H^\infty$. The relation $pu + m_T g = 1$, is, however, in contradiction with (7.19), and this contradiction shows that X cannot have closed range. The proposition follows.

Exercises

1. Assume that T is an operator of class C_0 and $X \in \{T\}''$ is a quasiaffinity. Show that X is a lattice-isomorphism.

2. Let U be a bilateral shift of multiplicity one. Show that there exist quasiaffinities $X \in \{U\}''$ which are not lattice-isomorphisms.

3. Let $A \in \mathscr{L}(\mathscr{H})$ be a noninvertible quasiaffinity such that $\|A\| < 1$. Define $X = T \in \mathscr{L}(\mathscr{H} \oplus \mathscr{H})$ by $T(h \oplus k) = Ak \oplus 0$. Show that T is of class C_0, $X \in \{T\}''$, and $T_{(\mathscr{H}\oplus\mathscr{H})\ominus\ker(X)}$ is not a quasiaffinity. Relate this with Theorem 7.16.

4. Assume that T is an operator of class C_0 acting on $\mathscr{H}$. Show that there exists a largest subspace $\mathscr{M} \in \mathrm{Lat}(\{T\}')$ such that $T \mid \mathscr{M}$ has property (P). Show that there exists another hyperinvariant subspace $\mathscr{N}$ for T such that $\mathscr{M} \vee \mathscr{N} = \mathscr{H}$ and $\mathscr{M} \cap \mathscr{N} = \{0\}$.

5. Is a result similar to that in Exercise 4 true for invariant subspaces of T?

CHAPTER 8

Miscellaneous Applications

Sections 1 and 2 are concerned with a characterization of compact operators in the commutant of a contraction, or intertwining two contractions. Section 1 contains a study of the compact Hankel operators and of operators intertwining contractions of class C_{00}. Some preliminaries about algebras of the form $H^\infty + C$ are also covered. In Section 2 we consider the case of operators of class C_0 and show that in this case there are many compact operators in the commutant, especially if the defect operators are compact. In Section 3 we study a generalization of isometric dilations. It turns out that operators of class C_0 (or, more precisely, uniform Jordan operators) appear in the solution of two natural problems in this generalized dilation theory. Section 4 is concerned with an application of the class C_0 to stochastic realization theory. We do not give the background from realization theory, but formulate the problem in terms of state spaces. We show how operators of class C_0 with property (P) can provide an elegant description of state spaces.

1. Compact intertwining operators. We had occasion to use the spaces $H^p(\mathscr{F})$ and $H^\infty(\mathscr{L}(\mathscr{F},\mathscr{G}))$ associated with two separable Hilbert spaces $\mathscr{F}$ and $\mathscr{G}$. The space $H^p(\mathscr{F})$ was identified with a subspace of $L^p(\mathscr{F})$, but no such identification was made for the space $H^\infty(\mathscr{L}(\mathscr{F},\mathscr{G}))$. We begin by introducing the space $L^\infty(\mathscr{L}(\mathscr{F},\mathscr{G}))$ of all (classes of) functions $\Phi : \mathbf{T} \to \mathscr{L}(\mathscr{F},\mathscr{G})$ that are weakly measurable and essentially bounded, with the norm

$$\|\Phi\|_\infty = \operatorname{ess\,sup}_{\varsigma\in\mathbf{T}} \|\Phi(\varsigma)\|.$$

Here weak measurability means that for every $f \in \mathscr{F}$ and $g \in \mathscr{G}$ the function $\varsigma \to (\Phi(\varsigma)f, g)$ is Lebesgue measurable on $\mathbf{T}$. Note that if Φ is measurable, so is the function $\varsigma \to \|\Phi(\varsigma)\|$, $\varsigma \in \mathbf{T}$. Indeed, if $\{f_n : n \geq 0\}$ and $\{g_n : n \geq 0\}$ are dense sequences in the unit balls of $\mathscr{F}$ and $\mathscr{G}$, respectively, then

$$\|\Phi(\varsigma)\| = \sup_n \sup_m |(\Phi(\varsigma)f_n, g_m)|, \quad \varsigma \in \mathbf{T}.$$

It is convenient sometimes to use the fact that the functions $\varsigma \to \Phi(\varsigma)f$, $\varsigma \in \mathbf{T}$, are measurable for $f \in \mathscr{F}$; this follows immediately from the Pettis theorem since $\mathscr{G}$ is separable.

For a function $\Phi \in L^\infty(\mathscr{L}(\mathscr{F},\mathscr{G}))$ we define the Fourier coefficients $c_n(\Phi) \in \mathscr{L}(\mathscr{F},\mathscr{G})$ by

$$c_n(\Phi) = \frac{1}{2\pi}\int_0^{2\pi} e^{-int}\Phi(e^{it})\,dt, \quad n \in \mathbf{Z}.$$

The integral must be understood in the strong sense, i.e.,

$$c_n(\Phi)f = \frac{1}{2\pi}\int_0^{2\pi} e^{-int}\Phi(e^{it})f\,dt, \quad f \in \mathscr{F}.$$

We have clearly

$$\|c_n(\Phi)\| \leq \|\Phi\|_\infty$$

for all integers n. We would like to identify $H^\infty(\mathscr{L}(\mathscr{F},\mathscr{G}))$ with the subspace of $L^\infty(\mathscr{L}(\mathscr{F},\mathscr{G}))$ consisting of those functions Φ whose Fourier coefficients $c_n(\Phi)$ vanish if $n < 0$. Assume indeed that $\Phi \in L^\infty(\mathscr{L}(\mathscr{F},\mathscr{G}))$ and $c_n(\Phi) = 0$ for $n < 0$. We can then extend Φ inside the unit disc by setting

$$\Phi(\lambda) = \sum_{n=0}^{\infty} \lambda^n c_n(\Phi), \quad \lambda \in \mathbf{D},$$

where the series clearly defines an analytic function on $\mathbf{D}$. The easily verified Poisson formula

$$\Phi(\lambda) = \frac{1}{2\pi}\int_0^{2\pi} \frac{1-|\lambda|^2}{|e^{it}-\lambda|^2}\Phi(e^{it})\,dt, \quad \lambda \in \mathbf{D},$$

shows immediately that

$$\|\Phi(\lambda)\| \leq \|\Phi\|_\infty, \quad \lambda \in \mathbf{D}.$$

Thus $\lambda \mapsto \Phi(\lambda)$ is a bounded analytic function on $\mathbf{D}$. Conversely, it was shown in §5.1 that, given a bounded analytic function $\Phi\colon \mathbf{D} \to \mathscr{L}(\mathscr{F},\mathscr{G})$, one can define the boundary values $\Phi(\varsigma)$ for almost every $\varsigma \in \mathbf{T}$ such that

$$\operatorname{ess\,sup}_{\varsigma\in\mathbf{T}} \|\Phi(\varsigma)\| \leq \sup_{\lambda\in\mathbf{D}} \|\Phi(\lambda)\|.$$

Thus we can view the elements in $H^\infty(\mathscr{L}(\mathscr{F},\mathscr{G}))$ as bounded analytic functions or, alternatively, as elements of the space $L^\infty(\mathscr{L}(\mathscr{F},\mathscr{G}))$. We will use this identification without further discussion.

For a function $\Phi \in L^\infty(\mathscr{L}(\mathscr{F},\mathscr{G}))$ we can introduce the partial sums $s_n\Phi \in L^\infty(\mathscr{L}(\mathscr{F},\mathscr{G}))$ defined by

$$(s_n\Phi)(\varsigma) = \sum_{k=-n}^{n} \varsigma^k c_k(\Phi), \quad n = 0,1,2,\ldots,$$

and the averages $\sigma_n\Phi \in L^\infty(\mathscr{L}(\mathscr{F},\mathscr{G}))$ of $s_0\Phi, s_1\Phi, \ldots, s_{n-1}\Phi$; thus

$$(\sigma_n\Phi)(\varsigma) = \sum_{k=-n}^{n}\left(1-\frac{|k|}{n}\right)\varsigma^k c_k(\Phi), \quad n = 1,2,\ldots.$$

The following result is a classical result of Fejér.

1.1. THEOREM. *For every* $\Phi \in L^\infty(\mathscr{L}(\mathscr{F},\mathscr{G}))$ *we have*

$$\|\sigma_n \Phi\|_\infty \le \|\Phi\|_\infty, \quad n = 1, 2, \ldots.$$

If, in addition, Φ *is continuous, then*

$$\lim_{n\to\infty} \|\Phi - \sigma_n \Phi\|_\infty = 0.$$

We will denote by $C(\mathscr{L}(\mathscr{F},\mathscr{G}))$ the subspace of $L^\infty(\mathscr{L}(\mathscr{F},\mathscr{G}))$ consisting of all continuous functions, and by $C(\mathscr{K}(\mathscr{F},\mathscr{G}))$ the subspace of those functions $\Phi \in C(\mathscr{L}(\mathscr{F},\mathscr{G}))$ such that $\Phi(\varsigma)$ belongs to the space $\mathscr{K}(\mathscr{F},\mathscr{G})$ of compact operators for every $\varsigma \in \mathbf{T}$. Both $C(\mathscr{L}(\mathscr{F},\mathscr{G}))$ and $C(\mathscr{K}(\mathscr{F},\mathscr{G}))$ are closed subspaces of $L^\infty(\mathscr{L}(\mathscr{F},\mathscr{G}))$. The following is an immediate consequence of Fejér's theorem.

1.2. COROLLARY.

(i) $C(\mathscr{L}(\mathscr{F},\mathscr{G}))$ *coincides with the closure in* $L^\infty(\mathscr{L}(\mathscr{F},\mathscr{G}))$ *of all trigonometric polynomials, i.e., of all functions of the form*

$$\Phi(\varsigma) = \sum_{k=-n}^{n} \varsigma^k C_k, \quad \varsigma \in \mathbf{T},\ C_k \in \mathscr{L}(\mathscr{F},\mathscr{G}).$$

(ii) *A function* $\Phi \in C(\mathscr{L}(\mathscr{F},\mathscr{G}))$ *belongs to* $C(\mathscr{K}(\mathscr{F},\mathscr{G}))$ *if and only if the Fourier coefficients* $c_n(\Phi)$ *are compact for all* n.

(iii) $C(\mathscr{K}(\mathscr{F},\mathscr{G}))$ *coincides with the closure in* $L^\infty(\mathscr{L}(\mathscr{F},\mathscr{G}))$ *of all trigonometric polynomials with compact operator coefficients.*

We will need to prove that certain subspaces of $L^\infty(\mathscr{L}(\mathscr{F},\mathscr{G}))$ are closed, and the following result is the abstract tool that is required.

1.3. LEMMA. *Let* Y *and* Z *be two closed subspaces of a Banach space* X, *and let* $\{T_\alpha : \alpha \in A\}$ *be a family of operators on* X. *Suppose that*

(i) $\sup\{\|T_\alpha\| : \alpha \in A\} < \infty$;

(ii) $T_\alpha X \subset Z$ *and* $T_\alpha Y \subset Y$, $\alpha \in A$;

(iii) *for every* $z \in Z$ *and* $\varepsilon > 0$ *there exists* $\alpha \in A$ *such that* $\|z - T_\alpha z\| < \varepsilon$.

Then the subspace $Y + Z$ *is closed in* X.

PROOF. Suppose that $y_n \in Y$, $z_n \in Z$, and $\|y_n + z_n\| < 2^{-n}$, $n \ge 1$. It suffices to show that $\sum_{n=1}^\infty (y_n + z_n) \in Y + Z$. Choose for each n an index $\alpha_n \in A$ such that

$$\|z_n - T_{\alpha_n} z_n\| < 2^{-n},$$

and write

$$y_n + z_n = (y_n - T_{\alpha_n} y_n) + (z_n + T_{\alpha_n} y_n).$$

Then

$$\|z_n + T_{\alpha_n} y_n\| = \|z_n - T_{\alpha_n} z_n + T_{\alpha_n}(y_n + z_n)\| \le 2^{-n}(1 + M),$$

where $M = \sup\{\|T_\alpha\| : \alpha \in A\}$, and hence

$$\|y_n - T_{\alpha_n} y_n\| \le \|y_n + z_n\| + 2^{-n}(1 + M) < 2^{-n}(2 + M).$$

It follows that $\sum_{n=1}^\infty (y_n + z_n) = y + z$, where $y = \sum_{n=1}^\infty (y_n - T_{\alpha_n} y_n) \in Y$ and $z = \sum_{n=1}^\infty (z_n + T_{\alpha_n} y_n) \in Z$. The lemma is proved.

1.4. THEOREM. *The subspaces* $H^\infty(\mathscr{L}(\mathscr{F},\mathscr{G})) + C(\mathscr{L}(\mathscr{F},\mathscr{G}))$ *and* $H^\infty(\mathscr{L}(\mathscr{F},\mathscr{G})) + C(\mathscr{K}(\mathscr{F},\mathscr{G}))$ *are closed in* $L^\infty(\mathscr{L}(\mathscr{F},\mathscr{G}))$. *If* $\mathscr{F} = \mathscr{G}$ *then these subspaces are subalgebras. A function* Φ *in* $H^\infty(\mathscr{L}(\mathscr{F},\mathscr{G}))+C(\mathscr{L}(\mathscr{F},\mathscr{G}))$ *belongs to* $H^\infty(\mathscr{L}(\mathscr{F},\mathscr{G})) + C(\mathscr{K}(\mathscr{F},\mathscr{G}))$ *if and only if* $c_{-k}(\Phi) \in \mathscr{K}(\mathscr{F},\mathscr{G})$ *for* $k \geq 1$.

PROOF. Once we know that $H^\infty(\mathscr{L}(\mathscr{F})) + C(\mathscr{L}(\mathscr{F}))$ is closed, it is immediate that it is an algebra. Indeed, by Corollary 1.2, $H^\infty(\mathscr{L}(\mathscr{F})) + C(\mathscr{L}(\mathscr{F}))$ is the closure of $H^\infty(\mathscr{L}(\mathscr{F}))+C_0$, where C_0 is the set of all trigonometric polynomials in $L^\infty(\mathscr{L}(\mathscr{F}))$. Since $H^\infty(\mathscr{L}(\mathscr{F})) + C_0$ is an algebra, so is its closure. The same observation applies, word for word, to $H^\infty(\mathscr{L}(\mathscr{F})) + C(\mathscr{K}(\mathscr{F}))$.

To prove that the subspaces are closed we apply Lemma 1.3 with $X = L^\infty(\mathscr{L}(\mathscr{F},\mathscr{G}))$, $Y = H^\infty(\mathscr{L}(\mathscr{F},\mathscr{G}))$, and $Z = C(\mathscr{L}(\mathscr{F},\mathscr{G}))$ or $Z = C(\mathscr{K}(\mathscr{F},\mathscr{G}))$. If $Z = C(\mathscr{L}(\mathscr{F},\mathscr{G}))$ we can take as family $\{T_\alpha : \alpha \in A\}$ the family $\{\sigma_n : n \geq 1\}$. The hypotheses of Lemma 1.3 are readily verified in this case. If $Z = C(\mathscr{K}(\mathscr{F},\mathscr{G}))$ the situation is a little bit more complicated. We may of course assume that $\mathscr{F}$ and $\mathscr{G}$ are infinite-dimensional, since otherwise $\mathscr{K}(\mathscr{F},\mathscr{G}) = \mathscr{L}(\mathscr{F},\mathscr{G})$. Let $\{f_n : n \geq 1\}$ and $\{g_n : n \geq 1\}$ be orthonormal bases in $\mathscr{F}$ and $\mathscr{G}$, respectively, and define operators R_n on $L^\infty(\mathscr{L}(\mathscr{F},\mathscr{G}))$ by

$$(R_n\Phi)(\varsigma) = P_n\Phi(\varsigma)Q_n, \quad \varsigma \in \mathbf{T},\ n = 1, 2, \ldots,$$

where P_n and Q_n denote the orthogonal projections onto the spaces generated by $\{g_1, g_2, \ldots, g_n\}$ and $\{f_1, f_2, \ldots, f_n\}$, respectively. We claim that the spaces $X, Y, Z = C(\mathscr{K}(\mathscr{F},\mathscr{G}))$ and the family $\{R_m\sigma_n : m, n \geq 1\}$ satisfy the conditions of Lemma 1.3. Conditions (i) and (ii) are obvious (in fact $\|R_m\sigma_n\| \leq 1$). To verify (iii) take $\Phi \in C(\mathscr{K}(\mathscr{F},\mathscr{G}))$ and $\varepsilon > 0$. By Fejér's theorem we can find n such that

$$\|\Phi - \sigma_n\Phi\|_\infty < \varepsilon/2.$$

Furthermore, since $\sigma_n\Phi$ is a trigonometric polynomial with compact coefficients, we have

$$\|\sigma_n\Phi - R_m\sigma_n\Phi\|_\infty < \varepsilon/2$$

for sufficiently large m. Clearly then $\|\Phi - R_m\sigma_n\Phi\| < \varepsilon$, as desired.

To verify the final assertion suppose that $\Phi \in C(\mathscr{L}(\mathscr{F},\mathscr{G}))$ and $c_{-k}(\Phi) \in \mathscr{K}(\mathscr{F},\mathscr{G})$ for $k = 1, 2, \ldots$. Then clearly $\sigma_n\Phi \in H^\infty(\mathscr{L}(\mathscr{F},\mathscr{G}))+C(\mathscr{K}(\mathscr{F},\mathscr{G}))$, and hence $\Phi \in H^\infty(\mathscr{L}(\mathscr{F},\mathscr{G}))+C(\mathscr{K}(\mathscr{F},\mathscr{G}))$ because this space is closed. The general case of $\Phi \in H^\infty(\mathscr{L}(\mathscr{F},\mathscr{G})) + C(\mathscr{L}(\mathscr{F},\mathscr{G}))$ follows immediately. The theorem is proved.

1.5. COROLLARY. *Assume that* $\Phi \in H^\infty(\mathscr{L}(\mathscr{F},\mathscr{G}))$ *and* $f : \mathbf{T} \to \mathbf{C}$ *is a continuous function. If* $c_n(\Phi)$ *is a compact operator for every* $n \geq 0$ *then the function* $f\Phi$ *defined by* $(f\Phi)(\varsigma) = f(\varsigma)\Phi(\varsigma)$, $\varsigma \in \mathbf{T}$, *belongs to* $H^\infty(\mathscr{L}(\mathscr{F},\mathscr{G}))+ C(\mathscr{K}(\mathscr{F},\mathscr{G}))$.

PROOF. Since $H^\infty(\mathscr{L}(\mathscr{F},\mathscr{G})) + C(\mathscr{K}(\mathscr{F},\mathscr{G}))$ is closed and $\|f\Phi\|_\infty \leq \|f\|_\infty\|\Phi\|_\infty$, it suffices to prove the corollary for trigonometric polynomials f. If

f is a trigonometric polynomial then a simple calculation shows that $c_n(f\Phi)$ is compact for all n. In addition, $c_n(f\Phi) = 0$ for all negative n, except for a finite number of values of n. Thus clearly $f\Phi = \psi_1 + \psi_2$, where $\psi_1 \in H^\infty(\mathcal{L}(\mathcal{F}, \mathcal{G}))$ and ψ_2 is a trigonometric polynomial with compact coefficients. The corollary follows.

Given a function $\Phi \in L^\infty(\mathcal{L}(\mathcal{F}, \mathcal{G}))$ one can define a multiplication operator $M_\Phi \colon L^2(\mathcal{F}) \to L^2(\mathcal{G})$ by

$$(M_\Phi f)(\varsigma) = \Phi(\varsigma)f(\varsigma), \qquad f \in L^2(\mathcal{F}),\ \varsigma \in \mathbf{T}.$$

This definition extends the one given in §5.1 for the case in which $\Phi \in H^\infty(\mathcal{L}(\mathcal{F}, \mathcal{G}))$. It is quite easy to verify that $M_\Phi f$ is indeed measurable for $f \in L^2(\mathcal{F})$, and M_Φ is a bounded operator with norm $\|M_\Phi\| \le \|\Phi\|_\infty$.

1.6. DEFINITION. Given $\Phi \in L^\infty(\mathcal{L}(\mathcal{F}, \mathcal{G}))$ we define the *Toeplitz operator* $T_\Phi \colon H^2(\mathcal{F}) \to H^2(\mathcal{G})$ and the *Hankel operator* $H_\Phi \colon H^2(\mathcal{F}) \to H^2_\perp(\mathcal{G}) = L^2(\mathcal{G}) \ominus H^2(\mathcal{G})$ by

$$T_\Phi = P_{H^2(\mathcal{G})} M_\Phi \mid H^2(\mathcal{F})$$

and

$$H_\Phi = P_{H^2_\perp(\mathcal{G})} M_\Phi \mid H^2(\mathcal{F}).$$

We collect in the following result some basic properties relevant in the study of Hankel operators. We will denote by $\tilde{S}_\mathcal{G}$ the compression of the bilateral shift $U_\mathcal{G}$ to $H^2_\perp(\mathcal{G})$, i.e.,

$$\tilde{S}_\mathcal{G} = P_{H^2_\perp(\mathcal{G})} U_\mathcal{G} \mid H^2_\perp(\mathcal{G}).$$

Note that $\tilde{S}^*_\mathcal{G}$ is unitarily equivalent to a forward shift, and $U_\mathcal{G}$ is the minimal unitary dilation of $\tilde{S}_\mathcal{G}$.

1.7. THEOREM. (i) *An operator $X \colon L^2(\mathcal{F}) \to L^2(\mathcal{G})$ satisfies the relation*

$$U_\mathcal{G} X = X U_\mathcal{F}$$

if and only if $X = M_\Phi$ for some $\Phi \in L^\infty(\mathcal{L}(\mathcal{F}, \mathcal{G}))$. The function Φ is uniquely determined by X and $\|X\| = \|\Phi\|_\infty$.

(ii) *An operator $X \colon H^2(\mathcal{F}) \to H^2_\perp(\mathcal{G})$ satisfies the relation*

$$\tilde{S}_\mathcal{G} X = X S_\mathcal{F}$$

if and only if $X = H_\Phi$ for some $\Phi \in L^\infty(\mathcal{L}(\mathcal{F}, \mathcal{G}))$. The function Φ can be chosen such that $\|\Phi\|_\infty = \|X\|$.

(iii) *For $\Phi \in L^\infty(\mathcal{L}(\mathcal{F}, \mathcal{G}))$ we have $H_\Phi = 0$ if and only if $\Phi \in H^\infty(\mathcal{L}(\mathcal{F}, \mathcal{G}))$. In general,*

$$\begin{aligned}\|H_\Phi\| &= \operatorname{dist}(\Phi, H^\infty(\mathcal{L}(\mathcal{F}, \mathcal{G})) \\ &= \inf\{\|\Phi - \Psi\|_\infty \colon \Psi \in H^\infty(\mathcal{L}(\mathcal{F}, \mathcal{G}))\}.\end{aligned}$$

PROOF. (i) We noted already that $\|M_\Phi\| \le \|\Phi\|_\infty$, and the equality $U_\mathcal{G} M_\Phi = M_\Phi U_\mathcal{F}$ is obvious for $\Phi \in L^\infty(\mathcal{L}(\mathcal{F}, \mathcal{G}))$. Conversely, assume that X satisfies the relation $U_\mathcal{G} X = X U_\mathcal{F}$. Clearly $U^*_\mathcal{G} X = X U^*_\mathcal{F}$, and hence we have

$X(pf) = pXf$ for every $f \in L^2(\mathscr{F})$ and every trigonometric polynomial p. Since trigonometric polynomials are boundedly weak* dense in L^∞, it follows from the Lebesgue dominated convergence theorem that

$$X(uf) = uXf, \quad f \in L^2(\mathscr{F}),\ u \in L^\infty.$$

Let $\{f_n\}$ be an orthonormal basis in $\mathscr{F}$, and for each n choose a function (rather than a class) $g_n(\varsigma)$ in $L^2(\mathscr{G})$ such that

$$(Xf_n)(\varsigma) = g_n(\varsigma)$$

for almost every $\varsigma \in \mathbf{T}$ (here f_n is regarded as a constant function in $L^2(\mathscr{F})$). For every $\varsigma \in \mathbf{T}$ define a linear map $\Phi_0(\varsigma)\colon \mathscr{F}_0 \to \mathscr{G}$, where $\mathscr{F}_0$ denotes the linear span of $\{f_n\}$, such that

$$\Phi_0(\varsigma)f_n = g_n(\varsigma).$$

We claim that $\|\Phi_0(\varsigma)\| \le \|X\|$ for almost every ς. If this were not true, it would follow that there exist $\varepsilon > 0$, a vector $f = \sum a_n f_n \in \mathscr{F}_0$ with rational a_n, and a set $\sigma \subset \mathbf{T}$ of positive measure, such that

$$\|\Phi_0(\varsigma)f\| \ge (\|X\| + \varepsilon)\|f\|, \qquad \varsigma \in \sigma.$$

But then we have

$$\begin{aligned}
\|X(\chi_\sigma f)\|^2 &= \|\chi_\sigma X f\|^2 \\
&= \frac{1}{2\pi}\int_0^{2\pi} \|\chi_\sigma(e^{it})\Phi_0(e^{it})f\|^2\, dt \\
&\ge (\|X\| + \varepsilon)^2 \frac{1}{2\pi}\int_0^{2\pi} \|\chi_\sigma(e^{it})f\|^2\, dt \\
&= (\|X\| + \varepsilon)^2 \|\chi_\sigma f\|^2,
\end{aligned}$$

a contradiction. This shows that there is a function $\Phi \in L^\infty(\mathscr{L}(\mathscr{F},\mathscr{G}))$ such that $\|\Phi\|_\infty \le \|X\|$, and $\Phi_0(\varsigma) = \Phi(\varsigma)\,|\,\mathscr{F}_0$ almost everywhere. It is now easy to verify that $X = M_\Phi$. Note that this proof also shows that $\|\Phi\|_\infty \le \|M_\Phi\|$ for this particular Φ. The uniqueness of Φ and the equality $\|\Phi\|_\infty = \|M_\Phi\|$ for all $\Phi \in L^\infty(\mathscr{L}(\mathscr{F},\mathscr{G}))$ are left to the reader.

(ii) Let X satisfy the relation $\tilde{S}_{\mathscr{G}}X = XS_{\mathscr{F}}$. It follows from the commutant lifting theorem (Theorem 1.1.10) that there exists $Y\colon L^2(\mathscr{F}) \to L^2(\mathscr{G})$ such that $U_{\mathscr{G}}Y = YU_{\mathscr{F}}$,

$$X = P_{H^2_\perp(\mathscr{G})}Y\,|\,H^2(\mathscr{F}),$$

and $\|Y\| = \|X\|$. By (i) above, we have $Y = M_\Phi$ for some $\Phi \in L^\infty(\mathscr{L}(\mathscr{F},\mathscr{G}))$, and $\|\Phi\|_\infty = \|Y\| = \|X\|$. Clearly $X = H_\Phi$.

(iii) If $\Phi \in H^\infty(\mathscr{L}(\mathscr{F},\mathscr{G}))$ then $M_\Phi H^2(\mathscr{F}) \subset H^2(\mathscr{G})$, and hence $H_\Phi = 0$. Conversely, if $H_\Phi = 0$ we have $M_\Phi H^2(\mathscr{F}) \subset H^2(\mathscr{G})$. In particular, if $f \in \mathscr{F}$, $M_\Phi f \in H^2(\mathscr{G})$, and hence

$$\frac{1}{2\pi}\int_0^{2\pi} e^{-int}\Phi(e^{it})f\, dt = 0, \quad n < 0;$$

in other words $\Phi \in H^\infty(\mathscr{L}(\mathscr{F},\mathscr{G}))$. For arbitrary $\Phi \in L^\infty(\mathscr{L}(\mathscr{F},\mathscr{G}))$ and $\Psi \in H^\infty(\mathscr{L}(\mathscr{F},\mathscr{G}))$ we have

$$\|H_\Phi\| = \|H_{\Phi+\Psi}\| \leq \|\Phi + \Psi\|_\infty,$$

so that

$$\|H_\Phi\| \leq \operatorname{dist}(\Phi, H^\infty(\mathscr{L}(\mathscr{F},\mathscr{G}))).$$

On the other hand, by (ii) there exists $\Phi' \in L^\infty(\mathscr{L}(\mathscr{F},\mathscr{G}))$ such that $H_{\Phi'} = H_\Phi$, and $\|\Phi'\|_\infty = \|H_\Phi\|$. We then have $\Phi' - \Phi \in H^\infty(\mathscr{L}(\mathscr{F},\mathscr{G}))$ so that

$$\operatorname{dist}(\Phi, H^\infty(\mathscr{L}(\mathscr{F},\mathscr{G}))) \leq \|\Phi'\|_\infty = \|H_\Phi\|.$$

The theorem is proved.

Note that the above proof shows that the infimum

$$\inf\{\|\Phi + \Psi\|_\infty \colon \Psi \in H^\infty(\mathscr{L}(\mathscr{F},\mathscr{G}))\}$$

is actually attained for every $\Phi \in L^\infty(\mathscr{L}(\mathscr{F},\mathscr{G}))$.

1.8. THEOREM. *For a function $\Phi \in L^\infty(\mathscr{L}(\mathscr{F},\mathscr{G}))$ the operator H_Φ is compact if and only if $\Phi \in H^\infty(\mathscr{L}(\mathscr{F},\mathscr{G})) + C(\mathscr{K}(\mathscr{F},\mathscr{G}))$.*

PROOF. We have $H^2(\mathscr{F}) = \bigoplus_{n=0}^\infty U_{\mathscr{F}}^n \mathscr{F}$ and $H_\perp^2(\mathscr{G}) = \bigoplus_{n=1}^\infty U_{\mathscr{G}}^{-n}\mathscr{G}$, where $\mathscr{F}$ is identified with the set of constant functions in $H^2(\mathscr{F})$. According to these decompositions, the matrix entries of H_Φ are calculated as follows:

$$\begin{aligned}(H_\Phi U_{\mathscr{F}}^n f, U_{\mathscr{G}}^{-m} g) &= (H_\Phi S_{\mathscr{F}}^n f, U_{\mathscr{G}}^{-m} g)\\ &= (\tilde{S}_{\mathscr{G}}^n H_\Phi f, U_{\mathscr{G}}^{-m} g)\\ &= (H_\Phi f, U_{\mathscr{G}}^{-m-n} g)\\ &= (M_\Phi f, U_{\mathscr{G}}^{-m-n} g)\\ &= (c_{-m-n}(\Phi) f, g),\end{aligned}$$

where $f \in \mathscr{F}$, $g \in \mathscr{G}$, $n \geq 0$, and $m \geq 1$. A similar calculation shows that

$$\tilde{S}_{\mathscr{G}}^n H_\Phi = H_{\chi^n \Phi},$$

where $\chi(\varsigma) = \varsigma$, $\varsigma \in \mathbf{T}$. If Φ is a trigonometric polynomial with compact coefficients, then H_Φ has finitely many nonzero matrix entries, and all of these are compact; thus H_Φ is compact. Since the map $\Phi \to H_\Phi$ is continuous, it follows that H_Φ is compact for $\Phi \in C(\mathscr{K}(\mathscr{F},\mathscr{G}))$, and hence H_Φ is compact for all $\Phi \in H^\infty(\mathscr{L}(\mathscr{F},\mathscr{G})) + C(\mathscr{K}(\mathscr{F},\mathscr{G}))$.

Conversely, assume that H_Φ is compact. Then $c_{-k}(\Phi)$ are matrix entries of H_Φ, and hence they are compact for $k \geq 1$. To conclude the proof it suffices, by Theorem 1.4, to show that $\Phi \in H^\infty(\mathscr{L}(\mathscr{F},\mathscr{G})) + C(\mathscr{L}(\mathscr{F},\mathscr{G}))$. Observe that

$$\begin{aligned}\operatorname{dist}(\chi^n\Phi, H^\infty(\mathscr{L}(\mathscr{F},\mathscr{G}))) &= \|H_{\chi^n\Phi}\|\\ &= \|\tilde{S}_{\mathscr{G}}^n H_\Phi\|\end{aligned}$$

and, since $\tilde{S}_{\mathscr{G}}^n \to 0$ strongly and H_Φ is compact, we conclude that

$$\operatorname{dist}(\chi^n\Phi, H^\infty(\mathscr{L}(\mathscr{F},\mathscr{G}))) \to 0$$

as $n \to \infty$. Thus there are $\Psi_n \in H^\infty(\mathcal{L}(\mathcal{F},\mathcal{G}))$ such that

$$\|\chi^n\Phi - \Psi_n\|_\infty = \|\Phi - \chi^{-n}\Psi_n\|_\infty \to 0$$

as $n \to \infty$. Since $\chi^{-n}\Psi_n \in H^\infty(\mathcal{L}(\mathcal{F},\mathcal{G})) + C(\mathcal{L}(\mathcal{F},\mathcal{G}))$, we conclude that $\Phi \in H^\infty(\mathcal{L}(\mathcal{F},\mathcal{G})) + C(\mathcal{L}(\mathcal{F},\mathcal{G}))$, as desired. The proof of the theorem is complete.

We conclude this section with a characterization of the compact operators intertwining two contractions of class C_{00}. Assume therefore that $\Theta \in H^\infty(\mathcal{L}(\mathcal{F}))$ and $\Theta' \in H^\infty(\mathcal{L}(\mathcal{G}))$ are two-sided inner functions. We recall that every operator X such that

$$S(\Theta')X = XS(\Theta)$$

is given by a function $\Phi \in H^\infty_{\Theta,\Theta'}$ by the formula

$$X = P_{\mathcal{H}(\Theta')}M_\Phi \,|\, \mathcal{H}(\Theta), \tag{1.9}$$

where $H^\infty_{\Theta,\Theta'}$ consists of all functions $\Phi \in H^\infty(\mathcal{L}(\mathcal{F},\mathcal{G}))$ with the property that

$$M_\Phi M_\Theta H^2(\mathcal{F}) \subset M_{\Theta'}H^2(\mathcal{G})$$

(cf. Corollary 5.1.27). If $\Phi \in H^\infty(\mathcal{L}(\mathcal{F},\mathcal{G}))$, we denote by $\Theta'^*\Phi$ the function in $L^\infty(\mathcal{L}(\mathcal{F},\mathcal{G}))$ defined by

$$(\Theta'^*\Phi)(\varsigma) = \Theta'(\varsigma)^*\Phi(\varsigma)$$

for almost every $\varsigma \in \mathbf{T}$. The operator X given by (1.9) is intimately related with the Hankel operator $H_{\Theta'^*\Phi}$.

1.10. LEMMA. *For every $\Phi \in H^\infty_{\Theta,\Theta'}$ we have*

$$\ker(H_{\Theta'^*\Phi}) \supset M_\Theta H^2(\mathcal{F}) = H^2(\mathcal{F}) \ominus \mathcal{H}(\Theta),$$
$$\operatorname{ran}(H_{\Theta'^*\Phi}) \subset M^*_{\Theta'}H^2(\mathcal{G}) \ominus H^2(\mathcal{G}) = M^*_{\Theta'}\mathcal{H}(\Theta),$$

and

$$H_{\Theta'^*\Phi} \,|\, \mathcal{H}(\Theta) = M^*_{\Theta'}P_{\mathcal{H}(\Theta')}M_\Phi \,|\, \mathcal{H}(\Theta).$$

PROOF. The first inclusion follows because

$$\begin{aligned} M_{\Theta'^*\Phi}M_\Theta H^2(\mathcal{F}) &= M^*_{\Theta'}M_\Phi M_\Theta H^2(\mathcal{F}) \\ &\subset M^*_{\Theta'}M_{\Theta'}H^2(\mathcal{G}) \\ &= H^2(\mathcal{G}), \end{aligned}$$

and the second inclusion follows from

$$M_{\Theta'^*\Phi}H^2(\mathcal{F}) = M^*_{\Theta'}M_\Phi H^2(\mathcal{F}) \subset M^*_{\Theta'}H^2(\mathcal{G}). \tag{1.11}$$

Finally, we have

$$\begin{aligned} H_{\Theta'^*\Phi} \,|\, \mathcal{H}(\Theta) &= P_{H^2_\perp(\mathcal{G})}M_{\Theta'^*}M_\Phi \,|\, \mathcal{H}(\Theta) \\ &= P_{M^*_{\Theta'}\mathcal{H}(\Theta')}M_{\Theta'^*}M_\Phi \,|\, \mathcal{H}(\Theta) \\ &= P_{M^*_{\Theta'}\mathcal{H}(\Theta)}M^*_{\Theta'}P_{\mathcal{H}(\Theta')}M_\Phi \,|\, \mathcal{H}(\Theta) \\ &= M^*_{\Theta'}P_{\mathcal{H}(\Theta')}M_\Phi \,|\, \mathcal{H}(\Theta), \end{aligned}$$

where we used (1.11) in the second equality, and the inclusion

$$\operatorname{ran} M_{\Theta'^*} P_{M_{\Theta'} H^2(\mathcal{F})} \subset H^2(\mathcal{G})$$

in the third equality. The lemma follows.

1.12. COROLLARY. *For $\Phi \in H^\infty_{\Theta,\Theta'}$, the operator $P_{\mathcal{H}(\Theta')} M_\Phi \,|\, \mathcal{H}(\Theta)$ is compact if and only if*

$$\Theta'^*\Phi \in H^\infty(\mathcal{L}(\mathcal{F},\mathcal{G})) + C(\mathcal{K}(\mathcal{F},\mathcal{G})).$$

PROOF. Since $M^*_{\Theta'}$ is a unitary operator, $P_{\mathcal{H}(\Theta')} M_\Phi \,|\, \mathcal{H}(\Theta)$ is compact if and only if $H_{\Theta'^*\Phi}$ is compact. The corollary now follows immediately from Theorem 1.8.

Exercises

1. Let $\Phi \in L^\infty(\mathcal{L}(\mathcal{F},\mathcal{G}))$, and assume that $\Phi(\varsigma) \in \mathcal{K}(\mathcal{F},\mathcal{G})$ for almost every $\varsigma \in \mathbf{T}$. Show that $c_n(\Phi) \in \mathcal{K}(\mathcal{F},\mathcal{G})$ for all n. Is the converse true?

2. Assume that $\mathcal{F}$ and $\mathcal{G}$ are infinite-dimensional, $\{f_i : i \geq 1\}$ and $\{g_i : i \geq 1\}$ are orthonormal bases in $\mathcal{F}$ and $\mathcal{G}$, and P_n, Q_n denote the orthogonal projections onto the spaces generated by $\{f_1, f_2, \ldots, f_n\}$ and $\{g_1, g_2, \ldots, g_n\}$, respectively. For $\Phi \in L^\infty(\mathcal{L}(\mathcal{F},\mathcal{G}))$ set $\Phi_n(\varsigma) = P_n\Phi(\varsigma)Q_n$, $\varsigma \in \mathbf{T}$. Show that $\lim_{n\to\infty} \|\Phi_n - \Phi\|_\infty = 0$ if $\Phi \in C(\mathcal{K}(\mathcal{F},\mathcal{G}))$.

3. Calculate the essential norm of a Hankel operator in function-theoretic terms.

4. Let $\theta(\lambda) = \exp[(\lambda+1)/(\lambda-1)]$, $\lambda \in \mathbf{D}$. Use Corollary 1.11 to show that $I - S(\theta)$ is compact. (We know from §4.3 that $I - S(\theta)$ is in fact a trace-class operator.)

5. Let $\mathcal{M}$ be a closed subspace of $\mathcal{L}(\mathcal{F},\mathcal{G})$, and denote by $H^\infty(\mathcal{M})$ the space of all functions $\Phi \in H^\infty(\mathcal{L}(\mathcal{F},\mathcal{G}))$ such that $\Phi(\lambda) \in \mathcal{M}$ for all $\lambda \in \mathbf{D}$. Show that $\Phi \in H^\infty(\mathcal{M})$ if and only if $c_n(\Phi) \in \mathcal{M}$ for all $n \geq 0$.

6. Show that $H^\infty(\mathcal{M}) + C(\mathcal{M})$ is closed in $L^\infty(\mathcal{L}(\mathcal{F},\mathcal{G}))$, and it is a module over the space C of all continuous functions on $\mathbf{T}$.

7. Deduce a proof of Corollary 1.5 from the preceding exercise.

2. Compact intertwining operators; the C_0 case. We begin by producing compact operators in the weakly closed algebra generated by certain operators of class C_0.

2.1. PROPOSITION. *Let $\Theta \in H^\infty(\mathcal{L}(\mathcal{F}))$ be an inner function such that $I - \Theta(\lambda)$ is compact for every $\lambda \in \mathbf{D}$, and $T = S(\Theta)$ is an operator of class C_0. Let ψ be an inner divisor of m_T, and let $\varphi \in H^\infty \cap C(\mathbf{T})$ be such that $\varphi \,|\, K \equiv 0$, where $\overline{K} = \mathbf{T} \cap \operatorname{supp}(m_T/\psi)$. Then the operator $\varphi\psi(T)$ is compact.*

PROOF. Let $\Omega \in H^\infty(\mathcal{L}(\mathcal{F}))$ satisfy

$$\Theta(\lambda)\Omega(\lambda) = \Omega(\lambda)\Theta(\lambda) = m_T(\lambda)I, \quad \lambda \in \mathbf{D},$$

from which we deduce that

$$\Theta(\varsigma)^* = \overline{m_T(\varsigma)}\Omega(\varsigma)$$

for almost every $\varsigma \in \mathbf{T}$ (recall that $\Theta(\varsigma)$ is unitary almost everywhere). We have therefore

$$\begin{aligned}\varphi\psi\Theta^* &= \varphi\psi I + \varphi\psi(\Theta^* - I)\\ &= \varphi\psi I + \varphi\psi\Theta^*(I - \Theta)\\ &= \varphi\psi I + \varphi\psi\overline{m_T}\Omega(I - \Theta)\end{aligned}$$

almost everywhere. Now, we have

$$\varphi\psi\overline{m_T} = \varphi\overline{(m_T/\psi)},$$

and the assumption about φ implies that $\varphi\psi\overline{m_T}$ is in $C = C(\mathbf{T})$. We have $\Omega(I-\Theta) \in H^\infty(\mathscr{L}(\mathscr{F}))$, and the coefficients of Θ, and hence those of $\Omega(I-\Theta)$, are all compact operators. Corollary 1.5 now implies

$$\varphi\psi\overline{m_T}\Omega(I - \Theta) \in H^\infty(\mathscr{L}(\mathscr{F})) + C(\mathscr{K}(\mathscr{F})).$$

Therefore $\varphi\psi\Theta^* \in H^\infty(\mathscr{L}(\mathscr{F}))+C(\mathscr{K}(\mathscr{F}))$, and $(\varphi\psi)(T)$ is compact by Corollary 1.12.

In order for the hypothesis concerning φ to be verified by some $\varphi \neq 0$ it is necessary, of course, that the Lebesgue measure of K be zero. Conversely, if the Lebesgue measure of K is zero, then there exists an outer function φ in $H^\infty \cap C$ such that $\varphi \neq 0$ but $\varphi \mid K \equiv 0$ (this fact was proved by Fatou). Replacing φ by $(\varepsilon\varphi)^{1/n}$ with appropriate $\varepsilon > 0$ and natural n, we may assume that $\|\varphi\|_\infty \leq 1$ and $|\varphi(0)|$ is as close to one as we want.

2.2. LEMMA. *Given an inner function θ and $\varepsilon > 0$, there exists an inner divisor ψ of θ such that $\psi(0) > 1 - \varepsilon$, and $\mathbf{T} \cap \operatorname{supp}(\theta/\psi)$ has Lebesgue measure zero.*

PROOF. Let $\theta = \gamma b_\mu s_\nu$ be the canonical decomposition of θ, where μ is some Blaschke function and ν is a singular measure on $\mathbf{T}$. For a fixed compact set $K \subset \mathbf{T}$, of Lebesgue measure zero, and a fixed $r < 1$, consider the function $\psi = b_{\mu'} s_{\nu'}$, where

$$\begin{aligned}\mu'(\lambda) &= 0 && \text{if } |\lambda| \leq r,\\ &= \mu(\lambda) && \text{if } |\lambda| > r,\ \lambda \in \mathbf{D},\end{aligned}$$

and $\nu' = \nu \mid \mathbf{T}\backslash K$. We have

$$\mathbf{T} \cap \operatorname{supp}(\theta/\psi) \subset K,$$

and

$$\psi(0) = \left(\prod_{\substack{|\lambda|>r\\ \mu(\lambda)\neq 0}} |\lambda|^{\mu(\lambda)}\right) \exp(-\nu(\mathbf{T}\backslash K)).$$

Since ν is regular and singular, K can be chosen such that $\nu(\mathbf{T}\backslash K)$ is as small as we want. Thus, for r sufficiently large and conveniently chosen K, ψ satisfies the conditions of the lemma.

2.3. THEOREM. *Let T be an operator of class C_0 such that D_T is compact. There exists a sequence $\{u_n : n \geq 1\} \subset H^\infty$ such that $\|u_n\|_\infty \leq 1$, $u_n(T)$ is compact for $n \geq 1$, and $u_n(T)$ converges ultraweakly to I.*

PROOF. It follows from Corollary 6.3.9 that we may assume without loss of generality that $T = S(\Theta)$, where $\Theta \in H^\infty(\mathscr{L}(\mathscr{F}))$ is inner, and $I - \Theta(\lambda) \in \mathscr{K}(\mathscr{F})$ for all $\lambda \in \mathbf{D}$. For each integer $n \geq 1$ choose an inner divisor ψ_n of m_T such that $\overline{K_n} = \mathbf{T} \cap \operatorname{supp}(m_T/\psi_n)$ has Lebesgue measure zero, and $\psi_n(0) > 1 - 1/n$. By the remark preceding Lemma 2.2, there exist functions $\varphi_n \in H^\infty \cap C$ such that $\|\varphi_n\|_\infty \leq 1$, $\varphi_n \,|\, K_n \equiv 0$, and $\varphi_n(0) > 1 - 1/n$, $n \geq 1$. We set $u_n = \varphi_n \psi_n$, and note that $u_n(T)$ is compact by Proposition 2.1. In view of the continuity of the functional calculus, to conclude the proof it suffices to show that $u_n \to 1$ in the weak* topology of H^∞. Since $\|u_n\|_\infty \leq 1$, it suffices to show that every weak* limit point of $\{u_n : n \geq 1\}$ is the function 1. Now, $1 > u_n(0) > (1 - 1/n)^2$, so $\lim_{n\to\infty} u_n(0) = 1$. Thus every weak* limit point v of $\{u_n : n \geq 1\}$ satisfies $\|v\|_\infty \leq 1$ and $v(0) = 1$. By the maximum modulus principle $v \equiv 1$, and this concludes the proof.

Let us note that an operator T that has property (P) has compact defect operators, and hence Theorem 2.3 applies to T. This follows from Exercise 7.1.9 and Theorem 6.4.5.

2.4. COROLLARY. *Let T and T' be operators of class C_0 such that either D_T is compact or $D_{T'}$ is compact. Then every operator in $\mathscr{I}(T', T)$ is the ultraweak limit of a sequence of compact operators in $\mathscr{I}(T', T)$.*

PROOF. Assume that D_T is compact, and let u_n be chosen as in Theorem 2.33. If $X \in \mathscr{I}(T', T)$ then $u_n(T)X$ converge ultraweakly to X. The case in which $D_{T'}$ is compact is treated analogously.

Exercises

1. Let T be an operator of class C_0, $\{\varphi_n : n \geq 1\}$ a sequence of inner divisors of m_T, and set $T_n = T \,|\, (\operatorname{ran} \varphi_n(T))^-$. Assume that $\varphi_n \to 1$ in the weak* topology of H^∞ and D_{T_n} is compact for every n. Show that there exists a sequence $\{u_n : n \geq 1\}$ of functions in H^∞ such that $\|u_n\|_\infty \leq 1$, $u_n(T)$ is compact, and $u_n(T) \to I$ ultraweakly.

2. Let T be an operator of class C_0. Prove that $\{T\}'$ contains compact operators.

3. Let T be an operator of class C_0. Show that the ultraweak closure of compact operators in $\{T\}'$ contains a quasiaffinity.

4. Show that the property "$\mathscr{A}_T$ contains a nonzero compact operator" is not invariant under quasisimilarity.

5. Denote by $\mathscr{K}(T_1, T_2)$ the set of compact operators in $\mathscr{I}(T_1, T_2)$. Assume that every operator in $\mathscr{I}(T_1, T_2)$ is the weak limit of a sequence from $\mathscr{K}(T_1, T_2)$. Show that $\mathscr{I}(T_1, T_2)$ can be identified with the second dual of $\mathscr{K}(T_1, T_2)$ such that the inclusion map $\mathscr{K}(T_1, T_2) \to \mathscr{I}(T_1, T_2)$ coincides with the natural embedding of a space into its second dual.

6. Let X be a linear subspace of $\mathscr{K}(\mathscr{F}, \mathscr{G})$. Show that the second dual of X can be identified naturally with the ultraweak closure Y of X, such that the inclusion map $X \to Y$ coincides with the natural embedding of a space into its second dual.

3. Subisometric dilations. We saw already that isometric dilations play an important role in the study of contraction operators. Here we will study another class of dilations, the subisometric ones, in which the class C_0 appears in a natural manner.

3.1. DEFINITION. Let T be a contraction on the space $\mathscr{H}$, and let $U_+ \in \mathscr{L}(\mathscr{K}_+)$ be the minimal isometric dilation of T. Suppose that $\mathscr{M} \supset \mathscr{H}$ is an invariant subspace for U_+^*. The operator A on $\mathscr{M}$ given by $A^* = U_+^* | \mathscr{M}$ is called a *subisometric dilation* of T. The subisometric dilation A of T is said to be *minimal* if $U_+ | \mathscr{K}_+ \ominus \mathscr{H}$ is a minimal isometric dilation of $A | \mathscr{M} \ominus \mathscr{H}$. Let $A \in \mathscr{L}(\mathscr{M})$ and $A_1 \in \mathscr{L}(\mathscr{M}_1)$ be two subisometric dilations of T. These dilations are said to be *isomorphic* if there exists a surjective isometry $V: \mathscr{M} \to \mathscr{M}_1$ such that $A_1 V = VA$ and $Vh = h$, $h \in \mathscr{H}$.

It is easy to see that, if $\mathscr{M}$ and $\mathscr{M}_1$ are embedded in the same minimal isometric dilation space, then A and A_1 are equivalent if and only if $\mathscr{M} = \mathscr{M}_1$.

We will be interested in the uniqueness and commutant lifting properties for subisometric dilations. In order to formulate these properties, let $A \in \mathscr{L}(\mathscr{M})$ be a subisometric dilation of $T \in \mathscr{L}(\mathscr{H})$. With respect to the decomposition $\mathscr{M} = \mathscr{H} \oplus (\mathscr{M} \ominus \mathscr{H})$ we can write

$$A = \begin{bmatrix} T & 0 \\ X & T' \end{bmatrix},$$

where $T' = A | (\mathscr{H} \ominus \mathscr{M})$. Note that the minimal isometric dilation of T' is a unilateral shift, and hence T' is of class $C_{\cdot 0}$. We will concentrate first on those operators T' that have the following uniqueness property.

3.2. DEFINITION. An operator B of class $C_{\cdot 0}$ has the *dilation uniqueness property* if for every contraction T, and every pair (A, A_1) of minimal subisometric dilations of T such that

$$A = \begin{bmatrix} T & 0 \\ X & B \end{bmatrix}, \qquad A_1 = \begin{bmatrix} T_1 & 0 \\ X_1 & B \end{bmatrix},$$

it follows that A and A_1 are isomorphic.

To make the above definition clearer, assume that A and A_1 act on $\mathscr{M}$ and $\mathscr{M}_1$, respectively. The occurrence of B in both matrices means that $A\,|\,\mathscr{M}\ominus\mathscr{H}$ and $A_1\,|\,\mathscr{M}_1\ominus\mathscr{H}$ are both unitarily equivalent to B.

In the sequel we will call a *uniform Jordan operator* the direct sum of an arbitrary number of copies of a Jordan block $S(\theta)$, where θ is an inner function (possibly a constant inner function).

3.3. THEOREM. *An operator B of class $C_{\cdot 0}$ has the dilation uniqueness property if and only if it is either a uniform Jordan operator or a unilateral shift (of arbitrary multiplicity).*

PROOF. Let $A\in\mathscr{L}(\mathscr{M})$ and $A_1\in\mathscr{L}(\mathscr{M}_1)$ be two minimal subisometric dilations of $T\in\mathscr{L}(\mathscr{H})$. We may assume that $\mathscr{M},\mathscr{M}_1\subset\mathscr{K}_+$, where $\mathscr{K}_+$ is the space of a minimal isometric dilation U_+ of T. We set $V=U_+\,|\,\mathscr{K}_+\ominus\mathscr{H}$, $T'=A\,|\,\mathscr{M}\ominus\mathscr{H}$, and $T_1'=A_1\,|\,\mathscr{M}_1\ominus\mathscr{H}$. Then V is a minimal isometric dilation of both T' and T_1'. By the uniqueness property of minimal isometric dilations, it follows that T' and T_1' are unitarily equivalent if and only if there exists a unitary operator W on $\mathscr{K}_+\ominus\mathscr{H}$, commuting with V, such that $W(\mathscr{M}\ominus\mathscr{H})=\mathscr{M}_1\ominus\mathscr{H}$ (and consequently $W(\mathscr{K}_+\ominus\mathscr{M})=\mathscr{K}_+\ominus\mathscr{M}_1$). Conversely, given a minimal subisometric dilation A and a unitary operator W commuting with V, we can construct a new subisometric dilation A_1 on $\mathscr{M}_1$ such that $\mathscr{M}_1\ominus\mathscr{H}=W(\mathscr{M}\ominus\mathscr{H})$. Note, however, that A and A_1 are not isomorphic unless $\mathscr{M}=\mathscr{M}_1$.

Now, V is a unilateral shift, and any unilateral shift can occur as $U_+\,|\,\mathscr{K}_+\ominus\mathscr{H}$ for some minimal isometric dilation U_+. Since $\mathscr{K}_+\ominus\mathscr{M}$ is invariant for V, our theorem is reduced to the following statement about unilateral shifts.

Let $V\in\mathscr{L}(\mathscr{K})$ be a unilateral shift, and $\mathscr{N}\subset\mathscr{K}$ an invariant subspace for V. If $\mathscr{N}$ is invariant under every unitary operator commuting with V then either $\mathscr{N}=\theta(V)\mathscr{K}$ for some inner function θ, or $\mathscr{N}=\{0\}$. We conclude the proof by verifying this last statement. Assume therefore that $\mathscr{N}$ is invariant under V and under all unitaries commuting with V. Then $\mathscr{N}$ is also invariant under the selfadjoints commuting with V; indeed, if P is selfadjoint then $\mathscr{N}$ is invariant under the unitary semigroup generated by iP. We now write $\mathscr{K}=\bigoplus_i\mathscr{K}_i$, such that $V\,|\,\mathscr{K}_i$ is a unilateral shift of multiplicity one. Since $\mathscr{N}$ is invariant under $P_{\mathscr{K}_i}$, it follows that $\mathscr{N}=\bigoplus_i\mathscr{N}_i$, with $\mathscr{N}_i$ invariant for $V\,|\,\mathscr{K}_i$. By Beurling's theorem we have $\mathscr{N}_i=\theta_i(V\,|\,\mathscr{K}_i)\mathscr{K}_i$, where θ_i is either inner or the zero function. Since there are unitaries shuffling the components $\mathscr{N}_i$, we conclude that $\theta_i=\theta$ does not depend on i, and this implies the desired conclusion. The theorem is proved.

We note for further use that the subspaces $\theta(V)\mathscr{K}$ are also hyperinvariant for V. This is obvious since $\theta(V)$ belongs to the algebra $\mathscr{A}_V$.

3.4. DEFINITION. Let $A\in\mathscr{L}(\mathscr{M})$ be a minimal subisometric dilation of $T\in\mathscr{L}(\mathscr{H})$. We say that A has the *commutant lifting property* if for every $Z\in\{T\}'$ there exists $Y\in\{A\}'$ such that $\|Y\|=\|Z\|$, and $Y^*\,|\,\mathscr{H}=Z^*$.

3.5. THEOREM. *An operator B of class $C_{\cdot 0}$ is such that every minimal subisometric dilation $A = \begin{bmatrix} T & 0 \\ X & B \end{bmatrix}$ has the commutant lifting property if and only if B is either a uniform Jordan operator or a unilateral shift (of arbitrary multiplicity).*

PROOF. Suppose that B is a uniform Jordan operator, or a unilateral shift, and $A = \begin{bmatrix} T & 0 \\ X & B \end{bmatrix}$ is a minimal subisometric dilation. Then (using the notation in the proof of Theorem 3.3) V is a minimal isometric dilation of B, and $\mathscr{K}_+ \ominus \mathscr{M}$ must have the form $\theta(V)(\mathscr{K}_t \ominus \mathscr{H})$ for some θ. Given $Z \in \{T\}'$ we can find by Theorem 1.1.10 an operator $Q \in \{U_+\}'$ such that $Q(\mathscr{K}_+ \ominus \mathscr{H}) \subset \mathscr{K}_+ \ominus \mathscr{H}$, $\|Q\| = \|Z\|$, and $Q^* \,|\, \mathscr{H} = Z$. As we noted before, $\mathscr{K}_+ \ominus \mathscr{M}$ is hyperinvariant for V, hence it is invariant under $Q \,|\, \mathscr{K}_+ \ominus \mathscr{H}$. Consequently the operator Y defined by $Y^* = Q^* \,|\, \mathscr{M}$ is such that $Y^* \,|\, \mathscr{H} = Z^*$, and

$$\|Z\| \leq \|Y\| \leq \|Q\| = \|Z\|.$$

Conversely, let $B \in \mathscr{L}(\mathscr{N})$ be a given operator of class $C_{\cdot 0}$, and suppose that any minimal subisometric dilation $\begin{bmatrix} T & 0 \\ X & B \end{bmatrix}$ has the commutant lifting property. We construct a particular minimal subisometric dilation as follows. Let $U \in \mathscr{L}(\mathscr{K})$ denote the minimal unitary dilation of B, set

$$\mathscr{M} = \bigvee_{n=0}^{\infty} U^{*^n} \mathscr{N}, \qquad \mathscr{H} = \mathscr{M} \ominus \mathscr{N},$$

and define $T \in \mathscr{L}(\mathscr{H})$ and $A \in \mathscr{L}(\mathscr{M})$ by $T^* = U^* \,|\, \mathscr{H}$ and $A^* = U^* \,|\, \mathscr{M}$. It is obvious that U is the minimal isometric dilation of A, and $U \,|\, \mathscr{K} \ominus \mathscr{H}$ is the minimal isometric dilation of B. To verify that A is a minimal subisometric dilation of T, we must check that U is also the minimal isometric dilation of T. Now, since B is of class $C_{\cdot 0}$ and $U \,|\, \mathscr{K} \ominus \mathscr{H}$ is the minimal isometric dilation of B, it follows that $U \,|\, \mathscr{K} \ominus \mathscr{H}$ is a unilateral shift whose minimal unitary extension is U. Thus T is a backward shift, and U is the minimal isometric dilation of T, as desired. It also follows easily at this point that for every operator $Z \in \{T\}'$ there is a unique $W \in \{U\}'$ such that $W^* \,|\, \mathscr{H} = Z^*$. Indeed, W^* is uniquely determined on the set $\bigcup_{n=0}^{\infty} U^n \mathscr{H} \subset \mathscr{K}$ because

$$W^* U^n h = U^n W^* h = U^n Z^* h, \quad h \in \mathscr{H}.$$

Analogously, each operator $X \in \{U \,|\, \mathscr{K} \ominus \mathscr{H}'\}'$ has a unique extension to an operator $W \in \{U\}'$.

We will conclude the proof of our theorem by showing that $\mathscr{K} \ominus \mathscr{M}$ is hyperinvariant for $U \,|\, \mathscr{K} \ominus \mathscr{H}$, and hence must either be equal to $U \,|\, \mathscr{K} \ominus \mathscr{H}$ or be a uniform Jordan operator. Let X be in $\{U \,|\, \mathscr{K} \ominus \mathscr{H}\}'$, and let $W \in \{U\}'$ be its unique extension. Define $Z \in \{T\}'$ by $Z^* = W^* \,|\, \mathscr{H}$. By assumption, there exists $Y \in \{A\}'$ such that $Y^* \,|\, \mathscr{H} = Z^*$. Since U is a unitary dilation of A, there is $W' \in \{U\}'$ such that $W'^* \,|\, \mathscr{M} = Y^*$. Now, $W'^* \,|\, \mathscr{H} = Z^* = W^* \,|\, \mathscr{H}$, and the uniqueness property shown above implies that $W' = W$. Consequently W^* leaves $\mathscr{M}$ invariant, and hence W leaves $\mathscr{K} \ominus \mathscr{M}$ invariant. Thus any $X \in \{U \,|\, \mathscr{K} \ominus \mathscr{H}\}'$ leaves $\mathscr{K} \ominus \mathscr{M}$ invariant, i.e., $\mathscr{K} \ominus \mathscr{M}$ is hyperinvariant for $U \,|\, \mathscr{K} \ominus \mathscr{H}$. The proof is now complete.

Exercises

1. When is T a minimal subisometric dilation of itself?

2. Let $U_+ \in \mathscr{L}(\mathscr{K}_+)$ be the minimal isometric dilation of T, and let $\mathscr{M}$, $\mathscr{M}_1 \in \operatorname{Lat} U_+^*$ be such that $A = P_{\mathscr{M}} U_+ | \mathscr{M}$ and $A_1 = P_{\mathscr{M}_1} A | \mathscr{M}_1$ are isomorphic subisometric dilations of T. Prove that $\mathscr{M} = \mathscr{M}_1$.

3. Let $A = \left[\begin{smallmatrix} T & 0 \\ X & T' \end{smallmatrix}\right]$ be a contraction on $\mathscr{H} \oplus \mathscr{H}'$. Show that there exists an operator $C\colon \mathscr{H} \to \mathscr{H}'$ such that $C\mathscr{H} \subset \mathscr{D}_{T'^*}$, $C^*\mathscr{H}' \subset \mathscr{D}_{T^*}$, $\|C\| \leq 1$, and $X = D_{T'^*} C D_T$.

4. Let A, T, T', X, C be as above. Show that A is an isometry if and only if T' and $C | \mathscr{D}_T$ are isometries. Show that A is the minimal isometric dilation of T if and only if T' is a unilateral shift and C is unitary from $\mathscr{D}_T$ to $\mathscr{D}_{T'^*}$.

5. With the notation above, assume that $C | \mathscr{D}_T$ is isometric. Prove that $I - A^*A = D^*D$, where $D = [-T'^* C D_T, D_{T'}]$.

6. With the notation above, assume that C is unitary from $\mathscr{D}_T$ to $\mathscr{D}_{T'^*}$. Prove that A is a minimal subisometric dilation of T.

7. Is the converse to Exercise 6 true?

8. Let θ_1 and θ_2 be two inner functions such that $\theta_1 H^\infty + \theta_2 H^\infty = H^\infty$, and set $T = S(\theta_1) \oplus S(\theta_2)$. Prove that all minimal subisometric dilations of T have the commutant lifting property. Show that not all of these dilations have the form $A = \left[\begin{smallmatrix} T & 0 \\ X & T' \end{smallmatrix}\right]$ with a uniform Jordan operator T'.

9. Is the conclusion of Exercise 8 true without the condition $\theta_1 H^\infty + \theta_2 H^\infty = H^\infty$?

10. Let T be an operator such that all minimal subisometric dilations $A = \left[\begin{smallmatrix} T & 0 \\ X & T' \end{smallmatrix}\right]$ have the property that T' is either a shift or a uniform Jordan operator. What can be said about T?

11. Let $A = \left[\begin{smallmatrix} T & 0 \\ X & T' \end{smallmatrix}\right]$ and $A_1 = \left[\begin{smallmatrix} T_1 & 0 \\ X_1 & T_1' \end{smallmatrix}\right]$ be two minimal subisometric dilations such that T_1 and T_1' are uniform Jordan operators with the same minimal function. Show that every operator in $\mathscr{I}(T, T_1)$ can be lifted to an operator in $\mathscr{I}(A, A_1)$.

4. State spaces. We consider a triple $(\mathscr{K}, \mathscr{X}, U)$, where $\mathscr{K}$ is a Hilbert space, $\mathscr{X}$ a subspace of $\mathscr{K}$, and U a unitary operator on $\mathscr{K}$. It is assumed that

$$\mathscr{K} = \bigvee_{n=-\infty}^{\infty} U^n \mathscr{X}.$$

4.1. Definition. A *state space* for $(\mathscr{K}, \mathscr{X}, U)$ is a subspace $\mathscr{H} \subset \mathscr{K}$ such that $\mathscr{X} \subset \mathscr{H}$ and $\mathscr{H}$ is semi-invariant for U. The compression $U_{\mathscr{H}}$ of U to a state space $\mathcal{H}$ is called a *state space operator*. A state space is *minimal* if it contains no proper subspace which is a state space.

We will be interested in understanding the minimal state spaces. The study of minimal state spaces is equivalent to certain problems from stochastic realization theory, which we do not discuss here.

For an arbitrary subspace $\mathcal{Y} \subset \mathcal{K}$ we will set

$$\mathcal{Y}_+ = \bigvee_{n=0}^{\infty} U^n \mathcal{Y}, \qquad \mathcal{Y}_- = \bigvee_{n=-\infty}^{0} U^n \mathcal{Y}.$$

4.2. DEFINITION. A state space $\mathcal{H}$ for $(\mathcal{K}, \mathcal{X}, U)$ is *observable* [resp., *constructible*] if $\mathcal{H} = \bigvee_{n=0}^{\infty} U_{\mathcal{H}}^n \mathcal{X}$ [resp., $\mathcal{H} = \bigvee_{n=0}^{\infty} U_{\mathcal{H}}^{*^n} \mathcal{X}$] or, equivalently, if $\mathcal{H} = (P_{\mathcal{H}} \mathcal{X}_+)^-$ [resp., $\mathcal{H} = (P_{\mathcal{H}} \mathcal{X}_-)^-$].

Minimality is related to observability and constructibility in the following way.

4.3. PROPOSITION. *A state space $\mathcal{H}$ for $(\mathcal{K}, \mathcal{X}, U)$ is minimal if and only if it is constructible and observable.*

PROOF. Assume that $\mathcal{H}$ is minimal. Then

$$\mathcal{H}_o = \bigvee_{n=0}^{\infty} U_{\mathcal{H}}^n \mathcal{X}$$

is invariant for $U_{\mathcal{H}}$, hence semi-invariant for U, and $\mathcal{X} \subset \mathcal{H}_o$. By minimality we deduce that $\mathcal{H} = \mathcal{H}_o$, and hence $\mathcal{H}$ is observable. Analogously,

$$\mathcal{H}_c = \bigvee_{n=0}^{\infty} U_{\mathcal{H}}^{*^n} \mathcal{X}$$

is invariant for $U_{\mathcal{H}}^*$, hence semi-invariant for U, and $\mathcal{H}_c \supset \mathcal{X}$. As before, $\mathcal{H} = \mathcal{H}_c$, and hence $\mathcal{H}$ is also constructible.

Conversely, let us assume that $\mathcal{H}$ is observable and constructible, and let $\mathcal{H}' \subset \mathcal{H}$ be another state space. Then $\mathcal{H}'$ is semi-invariant for $U_{\mathcal{H}}$, and hence $\mathcal{H}' = \mathcal{U} \ominus \mathcal{V}$, where $\mathcal{U}$ and $\mathcal{V}$ are invariant for $U_{\mathcal{H}}$ and $\mathcal{U} \supset \mathcal{V}$. Now, $\mathcal{X} \subset \mathcal{H}' \subset \mathcal{U}$, and $\mathcal{U}$ is invariant, so that $\mathcal{H} = \bigvee_{n=0}^{\infty} U_{\mathcal{H}}^n \mathcal{X} \subset \mathcal{U}$. Thus $\mathcal{H}' = \mathcal{H} \ominus \mathcal{V}$ is invariant for $U_{\mathcal{H}}^*$. But then, by the constructibility of $\mathcal{H}$, $\mathcal{H} = \bigvee_{n=0}^{\infty} U_{\mathcal{H}}^{*^n} \mathcal{X} \subset \mathcal{H}'$, so that $\mathcal{H}' = \mathcal{H}$ and therefore $\mathcal{H}$ is minimal. The proposition follows.

The preceding result allows us to construct explicitly minimal state spaces. Of course, the existence of such spaces can also be proved by Zorn's lemma.

4.4. COROLLARY. *Let $\mathcal{H}$ be a state space, and set*

$$\mathcal{H}_o = \bigvee_{n=0}^{\infty} U_{\mathcal{H}}^n \mathcal{X}, \qquad \mathcal{H}_{oc} = \bigvee_{n=0}^{\infty} U_{\mathcal{H}_o}^{*^n} \mathcal{X},$$

$$\mathcal{H}_c = \bigvee_{n=0}^{\infty} U_{\mathcal{H}}^{*^n} \mathcal{X}, \qquad \mathcal{H}_{co} = \bigvee_{n=0}^{\infty} U_{\mathcal{H}_c}^n \mathcal{X}.$$

Then $\mathscr{H}_{\alpha}$ and $\mathscr{H}_{co}$ are minimal state spaces. In particular,

$$\mathscr{P} = \mathscr{K}_{co} = (P_{\mathscr{X}_-}\mathscr{X}_+)^- \quad \text{and} \quad \mathscr{F} = \mathscr{K}_{\alpha} = (P_{\mathscr{X}_+}\mathscr{X}_-)^-$$

are minimal state spaces.

PROOF. The space $\mathscr{H}_{\alpha}$ is clearly constructible, so we need only to show it is observable to prove it is minimal. Indeed, since $\mathscr{H}_{\alpha}$ is invariant for $U^*_{\mathscr{H}_o}$, we have

$$\begin{aligned}\mathscr{H}_{\alpha} = P_{\mathscr{H}_{\alpha}}\mathscr{H}_o = P_{\mathscr{H}_{\alpha}}\left(\bigvee_{n=0}^{\infty} U^n_{\mathscr{H}_o}\mathscr{X}\right) \\ = \bigvee_{n=0}^{\infty} P_{\mathscr{H}_{\alpha}} U^n_{\mathscr{H}_o}\mathscr{X} = \bigvee_{n=0}^{\infty} U^n_{\mathscr{H}_{\alpha}} P_{\mathscr{H}_{\alpha}}\mathscr{X} \\ = \bigvee_{n=0}^{\infty} U^n_{\mathscr{H}_{\alpha}}\mathscr{X}.\end{aligned}$$

The case of $\mathscr{H}_{co}$ is treated analogously. The formulas for $\mathscr{P} = \mathscr{K}_{co}$ and $\mathscr{F} = \mathscr{K}_{\alpha}$ are immediate since $\mathscr{K}_c = \mathscr{X}_-$ and $\mathscr{K}_o = \mathscr{X}_+$.

We collect below some useful facts about constructible and observable state spaces.

4.5. PROPOSITION. *Let $\mathscr{H}$ be a state space for $(\mathscr{K}, \mathscr{X}, U)$.*

(i) *If $\mathscr{H}$ is observable then $\mathscr{H}_+ \subset \mathscr{P}_+$.*
(ii) *If $\mathscr{H}$ is constructible then $\mathscr{H}_- \subset \mathscr{F}_-$.*
(iii) *If $\mathscr{H}$ is minimal then $\mathscr{H}_+ \subset \mathscr{P}_+$ and $\mathscr{H}_- \subset \mathscr{F}_-$.*
(iv) *$(\mathscr{H}_c)_+ = \mathscr{H}_+$ and $(\mathscr{H}_o)_- = \mathscr{H}_-$.*
(v) *If $\mathscr{H}$ is constructible and $\mathscr{H}_+ \subset \mathscr{P}_+$ then $U_{\mathscr{H}} \prec U_{\mathscr{P}}$.*
(vi) *If $\mathscr{H}$ is observable and $\mathscr{H}_- \subset \mathscr{F}_-$ then $U_{\mathscr{F}} \prec U_{\mathscr{H}}$.*
(vii) *If $\mathscr{H}$ is minimal then $U_{\mathscr{F}} \prec U_{\mathscr{H}} \prec U_{\mathscr{P}}$.*
(viii) *All minimal state spaces have the same dimension.*
(ix) *If there is a minimal state space operator of class C_0, then all minimal state space operators are pairwise quasisimilar operators of class C_0.*

PROOF. (i) Because U is a minimal unitary dilation of $U_{\mathscr{P}}$, we have

$$(\mathscr{P}_+)^{\perp} = \mathscr{P}_- \ominus \mathscr{P} = \mathscr{X}_- \ominus \mathscr{P} = \mathscr{X}_- \cap (\mathscr{X}_+)^{\perp}.$$

Analogously,

$$(\mathscr{H}_+)^{\perp} = \mathscr{H}_- \ominus \mathscr{H} = \mathscr{H}_- \ominus (P_{\mathscr{H}_-}\mathscr{X}_+) = \mathscr{H}_- \cap (\mathscr{X}_+)^{\perp},$$

where we used the observability of $\mathscr{H}$. Furthermore, $\mathscr{X}_- \subset \mathscr{H}_-$ because $\mathscr{X} \subset \mathscr{H}$, hence

$$(\mathscr{P}_+)^{\perp} = \mathscr{X}_- \cap (\mathscr{X}_+)^{\perp} \subset \mathscr{H}_- \cap (\mathscr{X}_+)^{\perp} = (\mathscr{H}_+)^{\perp},$$

as desired.

(ii) Is proved analogously.

(iii) Follows at once from (i), (ii), and Proposition 4.3.

(iv) Note that $\mathscr{H}_o$ is invariant for $U_{\mathscr{H}}$, and therefore

$$\mathscr{P}_{\mathscr{H}}(\mathscr{H}_o)_+ = \bigvee_{n=0}^{\infty} \mathscr{P}_{\mathscr{H}} U^n \mathscr{H}_o = \bigvee_{n=0}^{\infty} U^n_{\mathscr{H}} \mathscr{H}_o = \mathscr{H}_o.$$

We deduce that $(\mathscr{H}_o)_+ \ominus \mathscr{H}_o \subset \mathscr{H}_+ \ominus \mathscr{H}$. On the other hand, from the fact that U is a minimal unitary dilation of $U_{\mathscr{H}}$ and $U_{\mathscr{H}_o}$, we have

$$(\mathscr{H}_o)_+ \ominus \mathscr{H}_o = ((\mathscr{H}_o)_-)^\perp, \qquad \mathscr{H}_+ \ominus \mathscr{H} = (\mathscr{H}_-)^\perp,$$

so that $(\mathscr{H}_o)_- \supset \mathscr{H}_-$. The inclusion $(\mathscr{H}_o)_- \subset \mathscr{H}_-$ is obvious since $\mathscr{H}_o \subset \mathscr{H}$. We proved that $(\mathscr{H}_o)_- = \mathscr{H}_-$, and the equality $(\mathscr{H}_o)_+ = \mathscr{H}_+$ is proved analogously.

(v) We will prove that $Y = P_{\mathscr{H}} \,|\, \mathscr{P}$ is a quasiaffinity intertwining $U^*_{\mathscr{H}}$ and $U^*_{\mathscr{P}}$. To do this we note that, by observability, $\mathscr{H} = (P_{\mathscr{H}} \mathscr{X}_-)^- = (P_{\mathscr{H}_+} \mathscr{X}_-)^-$, where we again use the fact that $\mathscr{X}_- \subset \mathscr{H}_-$, and hence $\mathscr{X}_-$ is orthogonal to $(\mathscr{H}_-)^\perp = \mathscr{H}_+ \ominus \mathscr{H}$. Thus the operator $X = P_{\mathscr{H}_+} \,|\, \mathscr{X}_-$, which clearly satisfies $U^*_{\mathscr{H}_+} X = X U^*_{\mathscr{X}_-}$, has dense range in $\mathscr{H}$. Moreover, $\ker X = \mathscr{X}_- \cap (\mathscr{H}_+)^\perp$, and since $\mathscr{X}_+ \subset \mathscr{H}_+ \subset \mathscr{P}_+$, we have

$$\mathscr{H}_- \cap (\mathscr{X}_+)^\perp \subset \ker X \subset \mathscr{X}_- \cap (\mathscr{P}_+)^\perp.$$

But we saw before that

$$\mathscr{X}_- \cap (\mathscr{P}_+)^\perp = \mathscr{P}_- \cap (\mathscr{P}_+)^\perp = \mathscr{P}_- \ominus \mathscr{P} = \mathscr{X}_- \ominus \mathscr{P} = \mathscr{X}_- \cap (\mathscr{X}_+)^\perp,$$

and hence $\ker X = \mathscr{X}_- \cap (\mathscr{P}_+)^\perp = \mathscr{X}_- \ominus \mathscr{P}$. It follows that $Y = X \,|\, \mathscr{P} \colon \mathscr{P} \to \mathscr{H}$ is a quasiaffinity and $U^*_{\mathscr{H}} Y = Y U^*_{\mathscr{P}}$. Thus $U_{\mathscr{H}} \prec U_{\mathscr{P}}$, as claimed.

(vi) Is proved analogously.

(vii) Follows immediately from (i), (ii), (v), and (vi).

(viii) Follows from (vii) since a quasiaffinity must act between spaces with the same dimension.

(ix) This follows from (vii) and Proposition 3.5.32. The proposition is proved.

A more precise description of the minimal state spaces is possible when the minimal state space operators are of class C_0 and have property (P). We leave the following result as an exercise to the reader.

4.6. LEMMA. *Let $T \in \mathscr{L}(\mathscr{H})$ be an operator of class C_0 with property (P), and let $\mathscr{M} \subset \mathscr{H}$ be a semi-invariant subspace for T. If $T_{\mathscr{M}} \sim T$ then $\mathscr{M} = \mathscr{H}$.*

4.7. THEOREM. *Assume that $(\mathscr{K}, \mathscr{X}, U)$ has a minimal state space operator of class C_0 with property (P).*

(i) *If $\mathscr{H}$ is a state space for $(\mathscr{K}, \mathscr{X}, U)$ and $U_{\mathscr{H}} \sim U_{\mathscr{P}} (\sim U_{\mathscr{F}})$ then $\mathscr{H}$ is minimal.*

(ii) *Let $\mathscr{M}$ be an invariant subspace for U such that*

$$\mathscr{F}_+ \subset \mathscr{M} \subset \mathscr{P}_+. \tag{4.8}$$

Then $(P_{\mathscr{M}} \mathscr{X}_-)^-$ is a minimal state space for $(\mathscr{K}, \mathscr{X}, U)$, and the map $\mathscr{M} \to (P_{\mathscr{M}} \mathscr{X}_-)^-$ is a bijection between invariant subspaces satisfying (4.8) and minimal state spaces.

(iii) *Let $\mathscr{N}$ be an invariant subspace for U^* such that*

$$\mathscr{P}_- \subset \mathscr{N} \subset \mathscr{F}_-. \tag{4.9}$$

Then $(P_{\mathscr{N}}\mathscr{X}_+)^-$ is a minimal state space for $(\mathscr{K},\mathscr{X},U)$, and the map $\mathscr{N} \to (P_{\mathscr{N}}\mathscr{X}_+)^-$ is a bijection between invariant subspaces (for U^*) satisfying (4.9) *and minimal state spaces.*

PROOF. (i) Let $\mathscr{H}' \subset \mathscr{H}$ be a minimal state space. Then $U_{\mathscr{H}'} \sim U_{\mathscr{P}}$ by Proposition 4.5(ix). Moreover, $\mathscr{H}'$ is semi-invariant for $U_{\mathscr{H}}$, and $U_{\mathscr{H}'} \sim U_{\mathscr{H}}$. Since $U_{\mathscr{H}}$ has property (P) (by the hypothesis and by Proposition 4.5(ix)), Lemma 4.6 implies that $\mathscr{H}' = \mathscr{H}$. Thus $\mathscr{H}$ is indeed minimal.

(ii) Observe that $\mathscr{F}_+ \subset \mathscr{X}_+$ since $\mathscr{F} \subset \mathscr{X}_+$, and $\mathscr{F}_+ \supset \mathscr{X}_+$ since $\mathscr{F} \supset \mathscr{X}$. Thus the first condition in (4.8) is equivalent to the inclusion $\mathscr{X} \subset \mathscr{M}$. Suppose that $\mathscr{M}$ satisfies (4.8). Then $(P_{\mathscr{M}}\mathscr{X}_-)^- = \mathscr{M}_c$ is a constructible state space, and $(\mathscr{M}_c)_+ = \mathscr{M}_+ = \mathscr{M}$ by Proposition 4.5(iv). Moreover, Proposition 4.5(v) shows that $U_{\mathscr{M}_c} \prec U_{\mathscr{P}}$. It follows that $U_{\mathscr{M}_c}$ is of class C_0 and quasisimilar to $U_{\mathscr{P}}$, hence $\mathscr{M}_c$ is minimal by part (i) of this theorem. Incidentally, we have also shown that $\mathscr{M}$ can be recovered from $\mathscr{M}_c$ by $\mathscr{M} = (\mathscr{M}_c)_+$, so that the map $\mathscr{M} \to \mathscr{M}_c$ is one-to-one. To see that it is onto, let $\mathscr{H}$ be an arbitrary minimal state space and set $\mathscr{M} = \mathscr{H}_+$. We have $\mathscr{M} \subset \mathscr{P}_+$ by Proposition 4.5(i), and clearly $\mathscr{X} \subset \mathscr{M}$, so $\mathscr{M}$ satisfies (4.8). We also have

$$\mathscr{H} = (P_{\mathscr{H}}\mathscr{X}_-)^- = (P_{\mathscr{M}}\mathscr{X}_-)^-$$

because $\mathscr{X}_-$ is orthogonal to $\mathscr{M} \ominus \mathscr{H}$. This shows that the map $\mathscr{M} \to (P_{\mathscr{M}}\mathscr{X}_-)^-$ is onto.

(iii) Is proved analogously.

Exercises

1. Suppose that U is the minimal unitary dilation of $U_{\mathscr{H}}$. Prove that $\mathscr{H}_+ \ominus \mathscr{H} = (\mathscr{H}_-)^\perp$. (This easy fact was used repeatedly.)

2. Prove Lemma 4.6.

3. Let U be a unitary operator on $\mathscr{K}$, and $\mathscr{X}$ a subspace of $\mathscr{K}$. Show that the following properties are equivalent:

   (i) U is a dilation of $U_{\mathscr{X}}$;
   (ii) $\mathscr{X}$ is semi-invariant for U;
   (iii) $P_{\mathscr{X}_-}\mathscr{X}_+ = P_{\mathscr{X}}\mathscr{X}_+ = \mathscr{X}$;
   (iv) $P_{\mathscr{X}_+}\mathscr{X}_- = P_{\mathscr{X}}\mathscr{X}_- = \mathscr{X}$.

4. Let $\mathscr{H}$ be a state space for $(\mathscr{K},\mathscr{X},U)$. Show that $U_{\mathscr{H}}$ is of class $C_{\cdot 0}$ if and only if $\bigcap_{n=0}^\infty U^{*^n}\mathscr{H}_- = \{0\}$. Formulate and prove a similar statement about the class $C_{0\cdot}$.

5. Assume that $\mathscr{X}$ is a finite-dimensional space, and $(\mathscr{K}, \mathscr{X}, U)$ admits a minimal state space operator of class C_{00}. Show that Theorem 4.7 applies in this situation.

6. Define $x \in L^2$ by $x(\varsigma) = e^{\varsigma}$, $\varsigma \in \mathbf{T}$, and let $\mathscr{X}$ be the one-dimensional space generated by x. Denote by U the bilateral shift on L^2.

(i) Show that $\bigvee_{n=-\infty}^{\infty} U^n \mathscr{X} = L^2$.
(ii) Prove that $\mathscr{X}_+ = \mathscr{F}_+ = H^2$.
(iii) Show that $\mathscr{F} = H^2$.
(iv) Show that $\mathscr{H}_- = x(L^2 \ominus UH^2)$.
(v) Prove that $\mathscr{P} = \mathscr{X}_-$ and $\mathscr{P}_+ = L^2$.
(vi) Show that the only minimal state spaces for $(\mathscr{K}, \mathscr{X}, U)$ are $\mathscr{P}$ and $\mathscr{F}$.
(vii) Show that there are invariant subspaces $\mathscr{M}$ for U, $\mathscr{F}_+ \neq \mathscr{M} \neq \mathscr{P}_+$, which satisfy (4.8).

APPENDIX

Notes and Comments

Operators of class C_0 first appeared in the study of functional models. The subject originated in the paper [**1**] of Sz.-Nagy and Foiaş. It soon became clear that operators in the class C_0 exhibit remarkable properties, which make them easier to study than general functional model operators. This book does not follow exactly the historical development of the subject because it tries to present results in their definitive form. In the notes to follow we try to indicate the precursors and sources of the more important theorems. It should be noted that all the results about the class C_0 apply to algebraic operators as well. Indeed, if T is algebraic then αT is of class C_0 if α is a sufficiently small positive scalar. However, certain results were first proved for algebraic operators. See, for instance, Turner [**1**], Kaplansky [**1**], and Apostol, Douglas, and Foiaş [**1**]. A good reference for the study of finite matrices from an operator theoretical point of view is Gohberg, Lancaster, and Rodman [**1**].

Chapter 1. All the material in Sections 1 and 2 is covered in the book [**6**] by Sz.-Nagy and Foiaş. We refer to this book for historical comments. We will mention only a few papers that played a fundamental role in the development of dilation theory. The existence of unitary dilations was proved by Sz.-Nagy [**1**]. Sarason [**2**] proved an interpolation theorem which can be viewed as the commutant lifting theorem in a particular case. The general commutant lifting theorem was proved by Sz.-Nagy and Foiaş [**4**]. The fact that the unitary dilation of a completely nonunitary contraction is absolutely continuous was proved by Sz.-Nagy [**2**].

Problem 1.1 is from Sarason [**3**].

Theorem 3.7 is from Sz.-Nagy and Foiaş [**3**]. The result was subsequently generalized by Wu [**6**], but we do not require this extension.

Problem 3.4 is related to Jordan models, but it can be proved directly without difficulty. See Apostol, Douglas, and Foiaş [**1**] for instance.

Chapter 2. Most properties of H^∞ used here can be found in either Duren [**1**] or Hoffman [**1**]. For the structure of the weak*-closed ideals in H^∞ see Gamelin [**1**]. A more extensive discussion of the functional calculus, and much of the contents of Sections 1 and 4, can be found in Sz.-Nagy and Foiaş [**6**].

The class C_0 was first defined by Sz.-Nagy and Foiaş [**1**]; see Sz.-Nagy and Foiaş [**10**] for the local definition. The minimal function was defined by Sz.-Nagy and Foiaş [**1**], where they also proved its existence.

Theorem 1.11 is from Beurling [**1**].

Lemma 1.12 and Proposition 1.18 are from Sz.-Nagy and Foiaş [**1**].

The properties of inner functions in Section 2 are well known. We record them for the reader's convenience.

Proposition 2.14 and its application, Lemma 3.3, are due to Sherman [**1**, **2**].

Lemmas 3.4 and 3.5 and Theorem 3.6 are due to Sz.-Nagy and Foiaş; see [**8**, **10**]. The existence of maximal vectors was proved independently by Herrero [**2**].

Theorem 3.7 first appeared in Bercovici, Foiaş, and Sz.-Nagy [**1**].

Exercises 3.5, 3.6, and 3.7 are from Sz.-Nagy and Foiaş [**10**, **7**].

Most of the material in Section 4 is due to Sz.-Nagy and Foiaş [**1**]; see also [**6**]. The spectra of Jordan blocks were determined by Moeller [**1**]. Proposition 4.13 is from Foiaş [**2**]. Jafarian [**1**] extended this result to show that weak contractions are decomposable.

Chapter 3. Most of Section 1 can be viewed as the theory of a special class of functional models (corresponding with scalar-valued inner functions). Thus, much of this material is contained implicitly in Chapter VI of the book [**6**] by Sz.-Nagy and Foiaş.

Proposition 1.8 is certainly known, but it does not seem to have been recorded before.

Theorem 1.14 is due to Sz.-Nagy [**6**] (see the Main Lemma).

Corollary 1.20 is Sarason's interpolation theorem from [**2**]. Proposition 1.21 and Exercise 1.6 are also due to Sarason [**2**]. See also Rosenoer [**1**].

Exercise 1.18 is due to Fuhrman [**2**].

The results in Section 2 are due to Sz.-Nagy and Foiaş. See [**5**] for the case of operators of class $C_0(N)$ (i.e., with finite defect indices) and [**8**] for the general case.

Theorem 3.1 was stated in Bercovici, Foiaş, and Sz.-Nagy [**1**] and proved in Bercovici, Foiaş, Kérchy, and Sz.-Nagy [**1**]; see also Bercovici [**6**]. For the case of operators with finite defect indices see Sz.-Nagy and Foiaş [**7**].

Theorem 3.6 is from Bercovici [**9**].

Theorem 3.8 is due to Sz.-Nagy and Foiaş [**9**].

Jordan operators were defined by Sz.-Nagy and Foiaş (see [**7**, **9**]) for the case of finite multiplicity, by Bercovici, Foiaş, and Sz.-Nagy [**1**] for the general separable case, and by Bercovici [**2**, **9**] in the nonseparable case. In Sections 4 and 5 we follow the approach from Bercovici [**9**].

Lemmas 4.7 and 4.11, and the finite-multiplicity cases of Theorems 4.12 and 4.13, are from Sz.-Nagy and Foiaş [**7**]. The separable case of Theorems 4.12 and 4.13 is from Bercovici, Foiaş, and Sz.-Nagy [**1**] and the nonseparable case from Bercovici [**2**, **9**].

Theorem 5.1 is in Sz.-Nagy and Foiaş [**7**, **9**] for operators with finite multiplicity and in Bercovici, Foiaş, and Sz.-Nagy [**1**] for the general separable case.

The remaining part of Section 5 (from Corollary 5.9 on) is in Bercovici [**2**, **9**]. Exercise 5.15 is due to Apostol [**1**].

The question of Jordan models for algebraic operators was considered by Apostol, Douglas, and Foiaş [**1**] (for the nilpotent case) and by Williams [**1**, **2**] in general.

The existence of approximate decompositions was proved by Sz.-Nagy and Foiaş [**7**] for operators with finite defect indices. The results in Section 6 are from Bercovici [**9**]. For the notion of a quasidirect sum see Brodskiĭ [**1**], where it is applied to the study of Volterra operators.

Chapter 4. Theorem 1.2 was proved by Wu [**1**] for the case of finite defect indices and by Sz.-Nagy and Foiaş [**12**] in the general case. However, the equality $\{T\}'' = \mathscr{F}_T$ was proved by Sz.-Nagy and Foiaş [**7**, **9**] for operators with finite multiplicity, by Bercovici, Foiaş, and Sz.-Nagy [**1**] for the separable case, and by Bercovici [**2**] for the nonseparable case. See also Turner [**1**] for the case of algebraic operators.

Theorems 1.23 and 1.25 are from Bercovici, Foiaş, and Sz.-Nagy [**1**]. See also Wu [**4**] for the case of finite defect indices and Bercovici [**6**].

Reflexive finite matrices were characterized by Deddens and Fillmore [**1**]. Brickman and Fillmore [**1**] proved Theorem 1.2 for the finite-dimensional case. For the reflexivity of nilpotent operators see Apostol, Douglas, and Foiaş [**1**].

Problem 1.18 is from Conway and Wu [**1**].

The results in Section 2 are in Bercovici [**8**]. For the case of operators with finite defect indices, Proposition 2.1, 2.6, and Corollaries 2.5 and 2.15 were proved earlier by Uchiyama [**1**]. See also Ando and Sekiguchi [**1**] and Fillmore, Herrero, and Longstaff [**1**] for the case of operators on a finite-dimensional space. Herrero [**5**] gave an example of nonisomorphic hyperinvariant subspace lattices for quasisimilar nilpotent operators.

The general properties of semigroups and their relations to accretive operators are well known. See also the book [**6**] of Sz.-Nagy and Foiaş for an exposition.

The unitary equivalence relating V_τ and $S(e_\tau)$ was first noted (for $\tau = 1$) by Sarason [**1**]. Sarason also gave a proof of Theorem 3.26 using this unitary equivalence.

The renorming technique at the beginning of the proof of Theorem 3.22 is from Foiaş [**1**]. The classification of semigroups with $T(\tau) = 0$ is from Bercovici [**2**].

The material in Section 4 is from Foiaş [**1**] and Bercovici [**2**]. Foiaş showed that a nondegenerate representation of $L^1(0,1)$ is given by integration against a strongly continuous semigroup of operators with $T(1) = 0$. He also proved that, modulo a quasiaffine transform, the semigroup can be assumed to be contractive. The classification of representations of $L^1(0,1)$ was then completed by Bercovici, who also replaced quasiaffine transforms by quasisimilarities.

Chapter 5. The material on functional models in Section 7 is contained in Sz.-Nagy and Foiaş [**6**], Chapter VI. However, we consider only the case of contractions of class $C_{\cdot 0}$, and this simplifies the presentation considerably.

The representations of $\mathscr{L}(\mathscr{H})$ used in Section 2 were first considered by Weyl [**1**], and in the infinite-dimensional case by Segal [**1**] and Kirillov [**1**].

Theorems 2.4 and 2.7 are due to Weyl; see [**1**] for the finite-dimensional case. See also Segal [**1**] for the extension to infinite dimensions.

The C^*-algebra results in Section 2 are quite standard. See for instance Dixmier [**1**] for the necessary background.

Lemmas 2.20 and 2.21 are from Bercovici and Voiculescu [**1**].

Propositions 3.2 and 3.4 and Corollary 3.3 are due to Sz.-Nagy and Foiaş [**1**]; see also [**6**].

The remaining part of Section 3 is from Bercovici and Voiculescu [**1**]. This work was motivated in part by Nordgren [**1**] and Moore and Nordgren [**1**].

Chapter 6. Exterior powers and algebraic adjoints are mentioned in Gohberg and Krein [**1**]. Our treatment here follows Bercovici and Voiculescu [**1**].

The results in Section 2 are from Bercovici and Voiculescu [**1**]. However, for finite-dimensional $\mathscr{F}$, Proposition 2.9 was proved by Sz.-Nagy and Foiaş (see [**6**], Chapter VI) for $k = 0$ and by Moore and Nordgren [**1**] for $k \geq 1$.

The results in Sections 3 and 4 are also mostly from Bercovici and Voiculescu [**1**]. However, the fact that a contraction T of class C_{00}, such that $I - T^*T$ is trace-class, is necessarily of class C_0 was proved by Takahashi and Uchiyama [**1**].

Lemma 3.3 is well known (see Sz.-Nagy and Foiaş [**6**], Chapters I and IX).

Corollary 3.18(ii) was proved by Sz.-Nagy and Foiaş [**7**] for operators with finite defect indices.

Lemma 3.20 is from Sz.-Nagy and Foiaş [**2**].

Proposition 3.23 may be new.

The particular case of Theorem 3.7 when T is multiplicity-free and $m_T(\lambda) = \exp[(\lambda+1)/(\lambda-1)]$ was proved (in an equivalent form) by Isaev [**1**].

The invariant factors for finite matrices over H^∞ were introduced by Nordgren [**1**], who also proved Theorem 5.28 for finite matrices. The general form of Theorem 5.28 was proved by Sz.-Nagy [**6**]. See also Szücs [**1**] for an abstract diagonalization theorem extending Nordgren's result.

Moore and Nordgren [**1**] gave a proof of the existence of Jordan models (in the case of finite defect indices) based on the diagonalization theorem. Consequently they deduced Proposition 5.11.

Proposition 5.8 was first mentioned by Müller [**1**].

Müller [**2**] gave a proof of the existence of Jordan models in the separable case, based on Sz.-Nagy's diagonalization theorem.

Theorem 6.10 was proved by Frazho [**1**] when $\mathscr{F}$ is finite-dimensional. The general case was proved by Bercovici [**12**].

Chapter 7. The fact that operators of class C_0 with finite multiplicity have property (P) was proved by Sz.-Nagy and Foiaş [**13**]. They conjectured that in fact such contractions have property (Q), but this conjecture was shown to be false by Bercovici [**3**], Uchiyama [**2**], and Wu [**5**]. Uchiyama proved that $T \mid \ker X$ and $(T^* \mid \ker X^*)^*$ are quasisimilar if $X \in \{T\}''$ and T is of class C_0 with finite defect indices. This result was extended to arbitrary contractions T of class C_0 by Bercovici [**8**].

The material in Sections 1, 2, 3, 4, 6, and 7 is contained in Bercovici [**3**, **8**, **10**].

Proposition 5.2 is from Bercovici [**8**], where it is also shown that the equality $\gamma_T = \gamma_{T'}$ implies $T \rho T''$ and $T'' \rho T'$ for some T''.

Theorem 5.18 (the converse of Proposition 5.2) was proved by Kérchy [**1**].

Chapter 8. The fact that $H^\infty + C$ is closed was proved by Sarason [**2**]. This result was extended by Zalcman [**1**], and Zalcman's method was put in an abstract framework by Rudin [**1**]. Thus, Lemma 1.3 is from Rudin [**1**].

The boundedness of Hankel operators in the scalar case was studied by Nehari [**1**] and their compactness by Hartman [**1**]. For the case of operator-valued functions see Page [**1**]. See also Power [**1**] for more information about Hankel operators.

Corollary 1.12 was proved by Muhly [**1**]; he considered only the case $\Theta = \Theta'$, but this is a minor restriction.

The compact operators commuting with a Jordan block were studied by Sarason [**2**], who proved Theorem 2.3 in this particular case.

Theorem 2.3 was proved by Nordgren [**3**]. Moore and Nordgren [**3**] had shown that $\mathscr{A}_T$ contains compact operators if D_T is compact. See also Clancey and Moore [**1**] for the case of finite defect indices.

Kriete, Moore, and Page [**1**] proved Corollary 2.4 when T and T' are Jordan blocks.

Exercise 2.2 is from Sz.-Nagy [**4**].

Exercises 2.5 and 2.6 are essentially from Gellar and Page [**1**].

The material about subisometric dilations (Section 3) is from Douglas and Foiaş [**1**].

The material about state spaces (Section 4) is from Foiaş and Frazho [**1**], to which we refer for motivation and further references.

Additional notes. There are results about operators of class C_0 that were not mentioned in the text or exercises.

Apostol, Bercovici, Foiaş, and Pearcy [**1**] showed the role of algebraic operators in the theory of quasiaffine transforms.

Arveson [**1**] gives some results about the unitary equivalence of Jordan blocks.

Barria, Kim, and Pearcy [**1**] discuss the reflexivity of algebraic operators which generate "good" algebras.

Bercovici [**7**] presents a version of the class C_0 for dissipative operators.

Bercovici [**11**] shows that quasisimilarity is a "good" equivalence relation in the class C_0.

Bercovici and Foiaş [1] study the relationship between the structure of an operator of class C_0 and properties of the algebras it generates.

Cassier [1] provides an example of a multiplicity-free operator T of class C_0, whose minimal function is a Blaschke product with simple zeros, such that the weak and weak* topologies do not coincide on the algebra generated by T.

Foiaş [3] uses a result of Stampfli [1] to prove the reflexivity of certain Jordan blocks with singular minimal functions given by "smooth" singular measures.

Foiaş, Tannenbaum, and Zames [1] give norm calculations for functions of a Jordan block.

Kérchy [2] has results concerning contractions with defect operators in a given norm-ideal.

Kérchy [3, 4] studies property (P) for a different class of contractions (the class C_{11}).

Kissilevski [1] studies a Jordan model theory for dissipative Volterra operators. See also Brodskiĭ [1]. This theory can be viewed now as a particular case of the theory of operators of class C_0.

Moore and Nordgren [3] apply operators of class C_0 to the transitive algebra problem.

Sz.-Nagy and Foiaş [11], Moore and Nordgren [2], and Sz.-Nagy [6] give an application of diagonalization theory to the classification of operators of class $C_{\cdot 0}$ with at least one finite defect index.

Wu [2] extended the Jordan model to weak contractions, not necessarily of class C_{00}.

Wu [3] studies the spectrality of contractions of class $C_0(N)$.

References

T. Ando and T. Sekiguchi

1. *Hyperinvariant subspaces of a nilpotent operator* (unpublished preprint).

C. Apostol

1. *Inner derivations with closed range*, Rev. Roumaine Math. Pures Appl. **21** (1976), 249–265.

C. Apostol, H. Bercovici, C. Foiaş, and C. Pearcy

1. *Quasiaffine transforms of operators*, Michigan Math. J. **29** (1982), 243–255.

C. Apostol, R. G. Douglas, and C. Foiaş

1. *Quasisimilar models for nilpotent operators*, Trans. Amer. Math. Soc. **224** (1976), 407–415.

W. B. Arveson

1. *Subalgebras of C^*-algebras*, Acta Math. **123** (1969), 141–224.
2. *Subalgebras of C^*-algebras.* II, Acta Math. **128** (1972), 271–308.

J. Barria, H. Kim, and C. Pearcy

1. *Algebraic operators, reflexivity, and the properties* $(B_{m,n})$ (preprint).

H. Bercovici

1. *Jordan model for some operators*, Acta. Sci. Math. (Szeged) **38** (1976), 275–279.
2. *On the Jordan model of C_0 operators*, Studia Math. **60** (1977), 267–284.
3. *C_0-Fredholm operators.* I, Acta Sci. Math. (Szeged) **41** (1979), 15–27.
4. *Modèle de Jordan pour des contractions sur l'espace de Hilbert*, Exposé No. 6, Seminaire d'Analyse Fonctionnelle, Palaiseau (1979/1980), 12pp.
5. *Théorie de l'indice pour des opérateurs non-Fredholm*, Exposé No. 7, Seminaire d'Analyse Fonctionnelle, Palaiseau (1979/1980), 12pp.
6. *Teoria operatorilor de clasa C_0*, Stud. Cerc. Mat. **31** (1979), 657–704.
7. *Modele ale operatorilor maximal disipativi*, Stud. Cerc. Mat. **32** (1980), 235–280.
8. *C_0-Fredholm operators.* II, Acta Sci. Math. (Szeged) **42** (1980), 3–42.

9. *On the Jordan model of C_0 operators*. II, Acta Sci. Math. (Szeged) **42** (1980), 43–56.
10. *C_0-Fredholm operators*. III, Michigan Math. J. **30** (1983), 355–360.
11. *Three test problems for quasisimilarity*, Canad. J. Math., **39** (1987), 880–892.
12. *Jordan models and *-cyclic sets*, J. Math. Anal. Appl., to appear.

H. Bercovici and C. Foiaş
1. *Representations of dual algebras and operators of class C_0*, Libertas Math. **6** (1986), 107–116.

H. Bercovici, C. Foiaş, L. Kérchy, and B. Sz.-Nagy
1. *Compléments à l'étude des opérateurs de classe C_0*. IV, Acta Sci. Math. (Szeged) **41** (1979), 29–31.

H. Bercovici, C. Foiaş, and B. Sz.-Nagy
1. *Compléments à l'étude des opérateurs de classe C_0*. III, Acta Sci. Math. (Szeged) **37** (1975), 313–322.
2. *Reflexive and hyper-reflexive operators of class C_0*, Acta Sci. Math. (Szeged) **43** (1981), 5–13.

H. Bercovici and D. Voiculescu
1. *Tensor operations on characteristic functions of C_0 contractions*, Acta Sci. Math. (Szeged) **39** (1977), 205–231.

A. Beurling
1. *On two problems concerning linear transformations on Hilbert space*, Acta. Math. **81** (1949), 239–255.

L. Brickman and P. A. Fillmore
1. *The invariant subspace lattice of a linear transformation*, Canad. J. Math. **19** (1967), 89–93.

M. S. Brodskiĭ
1. *Triangular and Jordan representations of linear operators* (in Russian), Nauka, Moscow, 1969, 287pp.

G. Cassier
1. *Un example d'opérateur pour lequel les topologies faible et ultrafaible ne coincident pas sur l'algèbre duale*, J. Operator Theory **16** (1986), 325–333.
2. *Sur la classification des algèbres duales de H. Bercovici, C. Foiaş, et C. Pearcy*, C. R. Acad. Sci. Paris Sér. I Math. **303** (1986), 69–71.

K. Clancey and B. Moore, III
1. *Operators of class $C_0(N)$ and transitive algebras*, Acta Sci. Math. (Szeged) **36** (1974), 215–218.

J. B. Conway and P. Y. Wu
1. *The splitting of $\mathcal{A}(T_1 \oplus T_2)$ and related questions*, Indiana Univ. Math. J. **26** (1977), 41–56.

J. Daughtry

1. *The inaccessible subspaces of certain* C_0 *operators* (preprint).

J. Dazord

1. *The growth of the resolvent of a contraction of class* C_0, Rev. Roumaine Math. Pures Appl. **24** (1979), 213–219.

J. A. Deddens and P. A. Fillmore

1. *Reflexive linear transformations*, Linear Algebra and Appl. **10** (1975), 89–93.

J. Dixmier

1. *Les* C^**-algèbres et leurs représentations*, Gauthier-Villars, Paris, 1969.

R. G. Douglas

1. *Canonical models*, Topics in Operator Theory, Math. Surveys, No. 13, Amer. Math. Soc., Providence, R.I., 1974, pp. 161–218.

R. G. Douglas and C. Foiaş

1. *Subisometric dilations and the commutant lifting theorem*, Operator Theory: Advances and Appl. **12** (1984), 129–139.

R. G. Douglas and C. Pearcy

1. *On a topology for invariant subspaces*, J. Funct. Anal. **2** (1968), 323–341.

R. G. Douglas, H. S. Shapiro, and A. L. Shields

1. *Cyclic vectors and invariant subspaces for the backward shift*, Ann. Inst. Fourier (Grenoble) **20** (1971), 37–76.

P. L. Duren

1. *Theory of* H^p *spaces*, Academic Press, New York, 1970.

P. Fatou

1. *Séries trigonometriques et séries de Taylor*, Acta Math. **30** (1906), 335–400.

A. Feintuch

1. *On diagonal coefficients of* C_0 *operators*, J. London Math. Soc. (2) **17** (1978), 507–510.
2. *Causal* C_0*-operators and feedback stability*, Math. Systems Theory **12** (1978), 283–288.

P. A. Fillmore, D. A. Herrero, and W. E. Longstaff

1. *The hyperinvariant subspace lattice of a linear transformation*, Linear Algebra Appl. **17** (1977), 125–132.

C. Foiaş

1. *On certain contraction semigroups, related with the representations of convolution algebras* (in Russian), Rev. Roumaine Math. Pures Appl. **7** (1962), 319–325.

2. *The class C_0 in the theory of decomposable operators*, Rev. Roumaine Math. Pures Appl. **14** (1969), 1433–1440.
3. *On the scalar parts of decomposable operators*, Rev. Roumaine Math. Pures Appl. **17** (1972), 1181–1198.
4. *La maximalité de H^∞ dans le calcul fonctionnel*, An. Univ. Timişoara, Ser. Mat.-Fiz. **2** (1964), 77–81.

C. Foiaş and A. E. Frazho
1. *A note on unitary dilation theory and state spaces*, Acta Sci. Math. (Szeged) **45** (1983), 165–175.

C. Foiaş, A. Tannenbaum, and G. Zames
1. *On decoupling the H^∞-optimal sensitivity problem for products of plants*, Systems and Control Letters **7** (1986), 239–245.

A. E. Frazho
1. *On stochastic realization theory*, Stochastics **7** (1982), 1–27.
2. *Infinite-dimensional Jordan models and Smith McMillan forms*, Integral Equations Operator Theory **5** (1982), 184–192.
3. *Infinite-dimensional Jordan models and Smith McMillan forms.* II, Acta Sci. Math. (Szeged) **46** (1983), 317–321.

P. A. Fuhrman
1. *On the Corona theorem and its application to spectral problems in Hilbert spaces*, Trans. Amer. Math. Soc. **132** (1968), 55–66.
2. *On sums of Hankel operators*, Proc. Amer. Math Soc. **46** (1974), 65–68.
3. *Linear systems and operators in Hilbert space*, McGraw-Hill, New York, 1981.

T. Gamelin
1. *Uniform algebras*, Prentice-Hall, Englewood Cliffs, N.J., 1969.

R. Gellar and L. Page
1. *A new look at some familiar spaces of intertwining operators*, Pacific J. Math. **47** (1973), 435–441.

I. C. Gohberg and M. G. Krein
1. *Introduction to the theory of nonselfadjoint operators on Hilbert space*, Nauka, Moscow, 1965.

I. C. Gohberg, P. Lancaster, and L. Rodman
1. *Invariant subspaces of matrices with applications*, Wiley-Interscience, New York, 1986.

P. Hartman
1. *On completely continuous Hankel matrices*, Proc. Amer. Math. Soc. **9** (1958), 362–366.

D. A. Herrero
1. *Inner function-operators*, Ph.D. Thesis, Univ. of Chicago, 1970.

2. *The exceptional set of a C_0 contraction*, Trans. Amer. Math. Soc. **173** (1972), 93–115.
3. *Inner functions under uniform topology*, Pacific J. Math. **51** (1974), 167–175.
4. *Inner functions under uniform topology.* II, Rev. Un. Mat. Argentina **28** (1976), 23–35.
5. *Quasisimilarity does not preserve the hyperlattice*, Proc. Amer. Math. Soc. **65** (1977), 80–84.

K. Hoffman

1. *Banach spaces of analytic functions*, Prentice-Hall, Englewood Cliffs, N.J., 1962.

L. E. Isaev

1. *On a class of operators with spectrum reduced to the origin*, Dokl. Akad. Nauk SSSR **178** (1968), 783–785.

A. A. Jafarian

1. *Weak contractions of Sz.-Nagy and Foiaş are decomposable*, Rev. Roumaine Math. Pures Appl. **22** (1977), 489–494.

R. V. Kadison and I. M. Singer

1. *Three test problems in operator theory*, Pacific J. Math. **7** (1957), 1101–1106.

I. Kaplansky

1. *Infinite abelian groups*, Univ. of Michigan Press, Ann Arbor, 1969.

J. L. Kelley and J. H. Spanier

1. *Euler characteristics*, Pacific J. Math. **26** (1968), 317–339.

L. Kérchy

1. *On C_0 operators with property (P)*, Acta Sci. Math. (Szeged) **42** (1980), 109–116.
2. *On p-weak contractions*, Acta Sci. Math. (Szeged) **42** (1980), 281–297.
3. *On the commutant of C_{11}-contractions*, Acta Sci. Math. (Szeged) **43** (1981), 15–26.
4. *On invariant subspace lattices of C_{11}-contractions*, Acta Sci. Math. (Szeged) **43** (1981), 281–293.

A. A. Kirillov

1. *Representations of the infinite-dimensional unitary group*, Dokl. Akad. Nauk. SSSR **212** (1973), 288–290.

G. E. Kissilevski

1. *On a generalization of the Jordan theory to a class of linear operators on Hilbert space*, Dokl. Akad. Nauk SSSR **176** (1967), 768–770.

T. L. Kriete, III, B. Moore, III, and B. L. Page

1. *Compact intertwining operators*, Michigan Math. J.**18** (1971), 115–119.

J. W. Moeller

1. *On the spectra of some translation invariant subspaces*, J. Math. Anal. Appl. **4** (1962), 276–296.

B. Moore, III and E. A. Nordgren

1. *On quasiequivalence and quasisimilarity*, Acta Sci. Math. (Szeged) **34** (1973), 311–316.
2. *Remark on the Jordan model for contractions of class* $C_{\cdot 0}$, Acta Sci. Math. (Szeged) **37** (1975), 307–312.
3. *On transitive algebras containing* C_0 *operators*, Indiana Univ. Math. J. **24** (1974/75), 777–784.

P. S. Muhly

1. *Compact operators in the commutant of a contraction*, J. Funct. Anal. **8** (1971), 197-224.

V. Müller

1. *On Jordan models of* C_0*-contractions*, Acta Sci. Math. (Szeged) **40** (1978), 309–313.
2. *Jordan models and diagonalization of the characteristic function*, Acta Sci. Math. (Szeged) **43** (1981), 321–322.

Z. Nehari

1. *On bounded bilinear forms*, Ann. of Math. **65** (1957), 153–162.

E. A. Nordgren

1. *On quasiequivalence of matrices over* H^∞, Acta. Sci. Math (Szeged) **34** (1973), 301–310.
2. *The ring* N^+ *is not adequate*, Acta Sci. Math. (Szeged) **36** (1974), 203–204.
3. *Compact operators in the algebra generated by essentially unitary* C_0 *operators*, Proc. Amer. Math. Soc. **51** (1975), 159–162.

L. B. Page

1. *Bounded and compact vectorial Hankel operators*, Trans. Amer. Math. Soc. **150** (1970), 529–539.

S. C. Power

1. *Hankel operators on Hilbert space*, Pitman, Boston, 1982.

S. Rosenoer

1. *Note on operators of class* $C_0(1)$, Acta Sci. Math. (Szeged) **46** (1983), 287–293.

W. Rudin

1. *Spaces of type* $H^\infty + C$, Ann. Inst. Fourier (Grenoble) **25** (1975), 99–125.
2. *Functional analysis*, McGraw-Hill, New York, 1973.

D. Sarason

1. *A remark on the Volterra operator*, J. Math. Anal. Appl. **12** (1965), 244–246.
2. *Generalized interpolation in H^∞*, Trans. Amer. Math. Soc. **127** (1967), 179–203.
3. *On spectral sets having connected complement*, Acta. Sci. Math. (Szeged) **26** (1965), 289–299.

I. E. Segal

1. *The structure of a class of representations of the unitary group on a Hilbert space*, Proc. Amer. Math. Soc. **8** (1957), 197–203.

M. J. Sherman

1. *Invariant subspaces containing all analytic directions*, J. Funct. Anal. **3** (1969), 164–172.
2. *Invariant subspaces containing all constant directions*, J. Funct. Anal. **8** (1971), 82–85.

J. G. Stampfli

1. *A local spectral theory for operators.* III: *Resolvents, spectral sets and similarity*, Trans. Amer. Math. Soc. **168** (1972), 133–151.

B. Sz.-Nagy

1. *Sur les contractions de l'espace de Hilbert*, Acta. Sci. Math. (Szeged) **15** (1953), 87–92.
2. *Sur les contractions de l'espace de Hilbert.* II, Acta Sci. Math. (Szeged) **18** (1957), 1–15.
3. *Quasisimilarity of operators of class C_0*, in Hilbert space operator algebras, Proc. Internat. Conf. Tihany, 1970, Colloq. Math. Soc. Jànos Bolyai, No. 5, North-Holland, Amsterdam, 1972, pp. 513–517.
4. *On a property of operators of class C_0*, Acta Sci. Math. (Szeged) **36** (1974), 219–220.
5. *Models of Hilbert space operators*, Proc. Roy. Irish Acad. Sect. A **74** (1974), 263–270.
6. *Diagonalization of matrices over H^∞*, Acta Sci. Math. (Szeged) **38** (1976), 233–258.

B. Sz.-Nagy and C. Foiaş

1. *Sur les contractions de l'espace de Hilbert.* VII. *Triangulations canoniques, fonctions minimum*, Acta Sci. Math. (Szeged) **25** (1964), 12–37.
2. *Forme triangulaire d'une contraction et factorization de la fonction caractéristique*, Acta Sci. Math. (Szeged) **28** (1967), 201–212.
3. *Vecteurs cycliques et quasiaffinités*, Studia Math. **31** (1968), 35–42.
4. *Dilatations des commutants d'opérateurs*, C. R. Acad. Sci. Paris Ser. A **266** (1968), 493–495.
5. *Opérateurs sans multiplicité*, Acta Sci. Math. (Szeged) **30** (1969), 1–18.

6. *Harmonic analysis of operators on Hilbert space*, North-Holland, Amsterdam, 1970.
7. *Modèle de Jordan pour une classe d'opérateurs de l'espace de Hilbert*, Acta Sci. Math. (Szeged) **31** (1970), 91–115.
8. *Compléments à l'étude des opérateurs de classe C_0*, Acta Sci. Math. (Szeged) **31** (1970), 281–296.
9. *Compléments à l'étude des opérateurs de classe C_0. II*, Acta Sci. Math. (Szeged) **33** (1971), 113–116.
10. *Local characterization of operators of class C_0*, J. Funct. Anal. **8** (1971), 76–81.
11. *Jordan model for contractions of class $C_{\cdot 0}$*, Acta Sci. Math. (Szeged) **36** (1974), 305–322.
12. *Commutants and bicommutants of operators of class C_0*, Acta Sci. Math. (Szeged) **38** (1976), 311–315.
13. *On injections intertwining operators of class C_0*, Acta Sci. Math. (Szeged) **40** (1978), 163–167.

J. Szücs

1. *Diagonalization theorems over certain domains*, Acta Sci. Math. (Szeged) **36** (1974), 193–201.

K. Takahashi and M. Uchiyama

1. *Every C_{00} contraction with Hilbert-Schmidt defect operators is of class C_0*, J. Operator Theory **10** (1983), 331–335.

T. R. Turner

1. *Double commutants of algebraic operators*, Proc. Amer. Math. Soc. **33** (1972), 415–419.

M. Uchiyama

1. *Hyperinvariant subspaces of operators of class $C_0(N)$*, Acta Sci. Math. (Szeged) **39** (1977), 179–184.
2. *Quasisimilarity of restricted C_0 contractions*, Acta Sci. Math. (Szeged) **41** (1979), 429–433.
3. *Contractions and unilateral shifts*, Acta Sci. Math. (Szeged) **46** (1983), 345–356.
4. *Contractions with (σ, c) defect operators*, J. Operator Theory **12** (1984), 221–233.

H. Weyl

1. *The classical groups*, Princeton Univ. Press, Princeton, N.J., 1939.

L. R. Williams

1. *Similarity invariants for a class of nilpotent operators*, Acta Sci. Math. (Szeged) **38** (1976), 423–438.
2. *A quasisimilarity model for algebraic operators*, Acta Sci. Math. (Szeged) **40** (1978), 185–188.

P. Y. Wu

1. *Commutants of $C_0(N)$ contractions*, Acta Sci. Math. (Szeged) **38** (1976), 193–202.
2. *Jordan model for weak contractions*, Acta Sci. Math. (Szeged) **40** (1978), 189–196.
3. *Conditions for completely nonunitary contractions to be spectral*, J. Funct. Anal. **31** (1979), 1–12.
4. *On the reflexivity of $C_0(N)$ contractions*, Proc. Amer. Math. Soc. **79** (1980), 405–409.
5. *On a conjecture of Sz.-Nagy and Foiaş*, Acta Sci. Math. (Szeged) **42** (1980), 331–338.
6. *Hyponormal operators quasisimilar to an isometry*, Trans. Amer. Math. Soc. **291** (1985), 229–239.

M. Zajac

1. *Hyperinvariant subspace lattice of some C_0-contractions*, Math. Slovaca **31** (1981), 397–404.

L. Zalcman

1. *Bounded analytic functions on domains of infinite connectivity*, Trans. Amer. Math. Soc. **144** (1969), 241–269.

List of Notation

$\mathscr{L}(\mathscr{H})$	algebra of bounded linear operators on the Hilbert space $\mathscr{H}$
$\mathscr{L}(\mathscr{H},\mathscr{K})$	space of bounded linear operators from $\mathscr{H}$ to $\mathscr{K}$
$P_{\mathscr{M}}$	orthogonal projection onto the space $\mathscr{M}$
$A \mid \mathscr{M}$	restriction of the operator A to the space $\mathscr{M}$
$\bigvee_i \mathscr{F}_i$	closed linear span of $\bigcup_i \mathscr{F}_i$
$\bigvee M$	closed linear span of the set M
$\mathscr{F} \vee \mathscr{G}$	closed linear span of $\mathscr{F} \cup \mathscr{G}$
$\oplus$	orthogonal sum
$\ominus$	orthogonal complement
$(,)$	scalar product
$\ker A$	kernel of A
$\operatorname{ran} A$	range of A
$A^{1/2}$	positive square root of the positive operator A
$\{T\}'$	commutant of $T = \{X: TX = XT\}$
$\{T\}''$	double commutant or bicommutant of $T = \{X: XY = YX \text{ for all } Y \in \{T\}'\}$
μ_T	cyclic multiplicity of T
$A \prec B$	A is a quasiaffine transform of B
$A \sim B$	A and B are quasisimilar
H^2	Hardy space on the unit disc
$\mathbf{C}$	the set of complex numbers
L^∞	$L^\infty(\mathbf{T}, dt)$
H^∞	$L^\infty \cap H^2$
$\mathbf{D}$	open unit disc
$\mathbf{T}$	unit circle
$\tilde{u}$	$\tilde{u}(\lambda) = \overline{u(\bar{\lambda})}$
m_T	minimal function of T
m_h	minimal function of the restriction of T to a cyclic subspace
$\setminus$	set theoretical difference
$\mathscr{H}(\theta)$	$H^2 \ominus \theta H^2$, the space on which a Jordan block acts

$S(\theta)$	Jordan block
$u \mid v$	u divides v
$u \vee v$	least common inner multiple of u and v
$u \wedge v$	greatest common inner divisor of u and v
$\mathbf{N}$	set of natural numbers
s_ν	singular inner function with singular measure ν
b_μ	Blaschke product with multiplicity function μ
$u \equiv v$	u and v differ by a scalar factor of absolute value one
$\operatorname{supp} \nu$	closed support of ν
$\operatorname{int}(M)$	interior of the set M
$\overline{M}$ or M^-	closure of M
$m_{\mathscr{K}}$	least common inner multiple of $\{m_h : h \in \mathscr{K}\}$
$\mathbf{C}[X]$	polynomial ring
$\sigma(T)$	spectrum of T
$\sigma_p(T)$	point spectrum of T
$\mathscr{I}(T, T')$	intertwining operators between T' and $T = \{X : T'X = XT\}$
$\mathscr{A}_T$	weak*-closed algebra generated by T and I
$\mathscr{W}_T$	weakly closed algebra generated by T and I
$\mathscr{F}_T$	set of all bounded operators of the form $(u/v)(T)$
$\operatorname{card}(\alpha)$, $\operatorname{card}(M)$	cardinality
$S(\Theta)$	Jordan operator with model function Θ or functional model
$A \prec^i B$	there exists an injection in $\mathscr{I}(B, A)$
$\operatorname{Alg}(\mathscr{L})$	operators leaving invariant the subspaces of a lattice $\mathscr{L}$
$\operatorname{Lat}(\mathscr{A})$	invariant subspace lattice of the set $\mathscr{A}$
e_t	$\exp(t(\lambda + 1)/(\lambda - 1))$
V_τ	Volterra operator on $L^2(0, \tau)$
$f * g$	convolution of f and g
$H^2(\mathscr{F})$	vector-valued Hardy space
$L^2(\mathscr{F})$	vector-valued Lebesgue space
$S_{\mathscr{F}}$	unilateral shift on $H^2(\mathscr{F})$
$U_{\mathscr{F}}$	bilateral shift on $L^2(\mathscr{F})$
$H^\infty(\mathscr{L}(\mathscr{F}, \mathscr{F}'))$	set of bounded $\mathscr{L}(\mathscr{F}, \mathscr{F}')$-valued analytic functions on $\mathbf{D}$
T_Φ	Toeplitz operator with symbol Φ
M_Φ	multiplication (Laurent) operator with symbol Φ
Θ_T	characteristic function of T
$\otimes$	tensor product
Γ_n	a representation of $\mathscr{L}(\mathscr{F})$ on $\bigotimes^n \mathscr{F}$
$\mathscr{S}_n$	symmetric group on n objects
$\bigwedge^k \mathscr{F}$	kth exterior power of the space
$\bigwedge^k A$	kth exterior power of the operator A

$A^{\mathrm{Ad}\,k}$	kth algebraic adjoint of A
$\det(A)$	determinant of A
$C_1(\mathscr{F})$	trace-class operators on $\mathscr{F}$
$\lVert A\rVert_1$	trace norm of A
$\mathrm{tr}(X)$	trace of X
$\mathscr{D}_k(\Theta)$	elementary divisor of Θ
$\mathscr{E}_k(\Theta)$	invariant factor of Θ
$\mathrm{Lat}_{1/2}(T)$	semi-invariant subspaces of T
$T_{\mathscr{M}}$	compression of T to $\mathscr{M}$
$d_T(\mathscr{M})$	determinant function of $T_{\mathscr{M}}$
$\widetilde{\Gamma}$	the set of generalized inner functions
sort	sorting operation on sequences
$\mathscr{F}(T',T)$, $s\mathscr{F}(T',T)$	the set of C_0-Fredholm and C_0-semi-Fredholm operators, respectively

Subject Index